Species

The Systematics Association Special Volume Series

Series Editor

David M. John

Department of Botany, The Natural History Museum, Cromwell Road, London SW7 5BD, UK.

The Systematics Association provides a forum for discussing systematic problems and integrating new information from cytogenics, ecology and other specific fields into taxonomic concepts and activities. It has achieved great success since the Association was founded in 1937 by promoting major meetings covering all areas of biology and palaeontology, supporting systematic research and training courses through the award of grants, production of a membership newsletter and publication of review volumes by its publisher Chapman & Hall. Its membership is open to both amateurs and professional scientists in all branches of biology who are entitled to purchase its volumes at a discounted price.

The first of the Systematics Association's publications, *The New Systematics*, edited by its then president Sir Julian Huxley, was a classic work. Over 50 volumes have now been published in the Association's 'Special Volume' series often in rapidly expanding areas of science where a modern synthesis is required. Its *modus operandi* is to encourage leading exponents to organise a symposium with a view to publishing a multi-authored volume in its series based upon the meeting. The Association also publishes volumes that are not linked to meetings in its 'Volume' series.

Anyone wishing to know more about the Systematics Association and its volume series are invited to contact the series editor.

Forthcoming titles in the series:

Relationships of the Major Arthropod Groups
Edited by R. A. Fortey and R. H. Thomas

Other Systematics Association publications are listed after the index for this volume.

The Systematics Association Special Volume Series 54

Species

The units of biodiversity

Edited by

M. F. Claridge
School of Pure and Applied Biology,
University of Wales Cardiff, UK

H. A. Dawah
School of Pure and Applied Biology,
University of Wales Cardiff, UK

and

M. R. Wilson
Department of Zoology,
National Museums and Galleries of Wales,
Cardiff, UK

CHAPMAN & HALL
London · Weinheim · New York · Tokyo · Melbourne · Madras

Published by Chapman & Hall,
2–6 Boundary Row, London SE1 8HN, UK

Chapman & Hall, 2–6 Boundary Row, London SE1 8HN, UK

Chapman & Hall GmbH, Pappelallee 3, 69469 Weinheim, Germany

Chapman & Hall USA, 115 Fifth Avenue, New York, NY 10003, USA

Chapman & Hall Japan, ITP-Japan, Kyowa Building, 3F, 2-2-1 Hirakawacho, Chiyoda-ku, Tokyo 102, Japan

Chapman & Hall Australia, 102 Dodds Street, South Melbourne, Victoria 3205, Australia

Chapman & Hall India, R. Seshadri, 32 Second Main Road, CIT East, Madras 600 035, India

First edition 1997

Typeset by Saxon Graphics Ltd, Derby

Printed by TJ International, Padstow, Cornwall

ISBN 0 412 63120 2

A catalogue record for this book is available from the British Library

Library of Congress Catalog Number: 96-70872

Printed on acid-free text paper, manufactured in accordance with ANSI/NISO Z39.48-1992 (Permanence of Paper).

Contents

Contributors

C. M. Brasier
Forest Research Station, Alice Holt Lodge, Farnham, Surrey GU10 4LH, UK.

J. Chun
Department of Microbiology, The Medical School, Framlington Place, Newcastle upon Tyne NE2 4HH, UK.

M. F. Claridge
School of Pure and Applied Biology, University of Wales Cardiff, P.O. Box 915, Cardiff CF1 3TL, Wales, UK.

G. B. Corbet
Little Dumbarnie, Upper Largo, Leven, Fife KY8 6JQ, Scotland, UK.

J. Cracraft
Department of Ornithology, American Museum of Natural History, Central Park West at 79th Street, New York, New York 10024, USA.

H. A. Dawah
School of Pure and Applied Biology, University of Wales Cardiff, P.O. Box 915, Cardiff CF1 3TL, Wales, UK.

T. Martin Embley
Department of Zoology, The Natural History Museum, Cromwell Road, London SW7 5BD, UK.

D. Foddai
Università di Padova, Dipartimento di Biologia, Via Trieste 75, I35121 Padova, Italy.

R. G. Foottit
Eastern Cereal and Oilseed Research Centre, Research Branch, Agriculture and Agri-Food Canada, K.W. Neatby Bldg., Central Experimental Farm, Ottawa, Ontario, KIA OC6, Canada.

M. Goodfellow
Department of Microbiology, The Medical School, Framlington Place, Newcastle upon Tyne NE2 4HH, UK.

R. J. Gornall
School of Biological Sciences, Department of Botany, University of Leicester, University Road, Leicester LE1 7RH, UK.

J. G. Hawkes
School of Continuing Studies, The University of Birmingham, Edgbaston, Birmingham B15 2TT, UK.

D. L. Hull
Department of Philosophy, Northwestern University, Evanston, IL 60208, USA.

D. J. Hunt
International Institute of Parasitology, 395A, Hatfield Road, St Albans AL4 OXU, UK.

D. M. John
Department of Botany, The Natural History Museum, Cromwell Road, London SW7 5BD, UK.

N. Knowlton
Smithsonian Tropical Research Institute, Apartado 2072, Balboa, Republic of Panama.

R. Lane
Department of Entomology, The Natural History Museum, Cromwell Road, London SW7 5BD, UK.

C. A. Maggs
School of Biology and Biochemistry, The Queen's University of Belfast, Belfast BT9 7BL, Northern Ireland, UK.

G. Paulo Manfio
Department of Microbiology, The Medical School, Framlington Place, Newcastle upon Tyne NE2 4HH, UK.

R. L. Mayden
Department of Biological Sciences, P.O. Box 0344, University of Alabama, Tuscaloosa, AL 35487, USA.

A. Minelli
Università di Padova, Dipartimento di Biologia, Via Trieste 75, I35121 Padova, Italy.

O. W. Purvis
Department of Botany, The Natural History Museum, Cromwell Road, London SW7 5BD, UK.

E. Stackebrandt
DSM-German Collection of Micro-organisms and Cell Cultures Gmbh, Mascheroder Weg 1 b, 38124 Braunschweig, Germany.

M. H. V. Van Regenmortel
Institut de Biologie Moleculaire et Cellulaire, CNRS, 15 rue René Descartes, 67084 Strasbourg, France.

L. A. Weigt
Smithsonian Tropical Research Institute, Apartado 2072, Balboa, Republic of Panama; now at The Field Museum, Roosevelt Road at Lake Shore Drive, Chicago, IL 60605, USA.

M. R. Wilson
Department of Zoology, National Museums and Galleries of Wales, Cathays Park, Cardiff CF1 3NP, Wales, UK.

Preface

The idea of bringing together a wide diversity of specialists on different groups of living organisms to discuss the practical nature of species had been a longstanding ambition of one of us (M.F.C.). A recommendation 'that an interdisciplinary clarification of species and population biology concepts be accorded high priority, and that a special meeting devoted to this topic be convened at an early date' was unanimously agreed at the 2nd Workshop on the Identification and Characterization of Pest Organisms, held at the International Mycological Institute in 1993 [D.L. Hawksworth (ed.) (1994), *The Identification and Characterization of Pest Organisms*, CAB International, p.475]. This recommendation provided the catalyst and final stimulus for us to organize a Systematics Association Symposium on 'The Units of Biodiversity: Species in Practice'. This was eventually held at University Hall, a conference centre of the University of Wales Cardiff, from 19–21 April, 1995. A total of 66 participants heard 20 different papers on a diversity of topics concerning species concepts and their application in a very wide selection of living organisms. Most importantly there was the opportunity for extensive informal discussion. The present volume represents the considered and revised proceedings of that meeting. Unfortunately, two contributors were unable to provide us with manuscripts and so are not included in this book. We are particularly grateful to Richard Mayden who kindly offered to present his overview on species concepts for publication, though he did not present it formally at the meeting.

As organizers of the Cardiff meeting we are deeply indebted to the enormous number of colleagues, students and friends who gave freely of their time to make the meeting a success. In particular we thank Gareth Holmes and Stella Shackel who played vital roles in welcoming and seeing to the needs of participants, David and Diana Edwards who designed and produced the programme cover, John Morgan who ensured the functioning of the audiovisual systems, Dr M. Al-Yaseen for assistance with converting word processing systems, and Rosemary Jones who provided efficient secretarial assistance. We are also indebted to the chairmen of the formal sessions for keeping control, but not limiting too much the very lively discussions – David Hawksworth (CAB International Mycological Institute), Joel Cracraft (American Museum of Natural History) and Bob Foottit (Agriculture and Agri-Food Canada). Of critical significance to the success of the meeting and of the subsequent book were the generous financial contributions from the Wellcome Trust, the UK Federation of Culture Collections, and the Systematics Association. We are deeply indebted to those organizations for their generosity and support.

We are very grateful to all of the contributors to our book for their forebearance and patience. We hope that the resulting volume justifies their efforts. The quality of the book has certainly been enhanced by the work of our independent reviewers, to whom we offer our sincere thanks. They were: Dr Kathryn Benson-Evans, Prof. Arthur J. Cain, Prof. John C. Fry, Prof. David L. Hawksworth, Prof. Peter Morgan, Dr Graham Oliver, Prof. Sir Ghillean T. Prance, Dr Jeremy A. Roberts and Prof. Harford H. Williams.

M.F.Claridge,
H.A. Dawah,
M.R.Wilson

1

Practical approaches to species concepts for living organisms

M. F. Claridge, H. A. Dawah and M. R. Wilson

Contacting address: School of Pure and Applied Biology, University of Wales Cardiff, P.O. Box 915, Cardiff CF1 3TL, Wales, UK.

ABSTRACT

From a practical viewpoint species are generally the units of biodiversity. Traditionally since before Linnaeus species have been defined in terms of clear morphological differentiation – the morphospecies. In practice most species are still described on a basis of dead preserved material and are therefore morphospecies.

The increasing recognition by naturalists, geneticists and evolutionists over the past 200 years that species occur as reproductively isolated natural entities in the field led to the various biological species concepts. Reproductively isolated species are separate evolutionary entities characterized by unique specific mate recognition systems. An important consequence of the biological species is the recognition of reproductively isolated sibling species that show no clear morphological differentiation but which are reproductively isolated. In practice biological species are diagnosed by markers that may be morphological, cytological, behavioural, molecular, etc., but which indicate the presence of high levels of reproductive isolation.

The biological species can only be applied to biparental sexually reproducing organisms, or at least organisms that regularly exchange genetic material. Thus, only some form of morphospecies is available for asexual and obligately parthenogenetic forms (agamospecies). Also application of biological species to populations isolated in space – allopatry – is difficult and usually subjective.

These difficulties and the desire to apply cladistic techniques at the species level have led to widespread rejection of the biospecies by systematists in favour of a broadly phylogenetic species. Here, species are essentially equated with diagnosably distinct clades. Advantages are that allopatric and asexual populations can be treated in the same way

Species: The Units of Biodiversity. Edited by M.F. Claridge, H.A. Dawah and M.R. Wilson. Published in 1997 by Chapman & Hall. ISBN 0 412 63120 2

as sympatric sexually reproducing ones. Disadvantages include the difficulties of deciding objectively on what is a diagnosably distinct clade and the possibility of ignoring sibling species. There is clearly common ground between these two general concepts for describing biological diversity and together they form a unitary taxonomic or evolutionary species.

1.1 INTRODUCTION

> 'There is probably no other concept in biology that has remained so consistently controversial as the species concept'. (Mayr, 1982)
> 'What are species? Perhaps no other issue in comparative or evolutionary biology has provoked quite so much disparate opinion as this simple question'. (Eldredge, 1995)

These views of two of the most influential of recent evolutionary biologists are borne out by the plethora of publications on species concepts in recent years. After a lapse in the 1950s and 1960s when a consensus seemed to have been achieved, the basic philosophy and biology of species concepts has once again been opened up widely for discussion. This renewed interest is exemplified by the publication of review volumes (Otte and Endler, 1989; Lambert and Spencer, 1995; Wheeler and Meier, 1977) and reprinted collections of classic papers (Ereshefsky, 1992). Many authors now apparently feel the need to come up with yet other and apparently new personal species concepts. Mayden (1997: Chapter 19) has identified 22 concepts to date that he recognizes as distinct, though some of us may regard many of them as essentially synonymous. These concepts include a variety of approaches, some purely theoretical and some entirely empirical. Hull (1997: Chapter 18) has attempted to bring some of these approaches together.

The prolonged wrangle among scientists and philosophers over the nature of species has recently taken on added and wider significance. The belated recognition of the importance of biological diversity to the survival of mankind and the sustainable use of our natural resources makes it a matter of very general and urgent concern. Species are normally the units of biodiversity and conservation (Wilson, 1992) so it is important that we should know what we mean by them. One major concern has been with estimating the total number of species of living organisms that currently inhabit the earth (May, 1990). In addition, many authors have attempted to determine the relative contributions of different groups of living organisms to the totality of living biodiversity in which usually some sort of morphological species concept is used (Figure 1.1). Unless we have some agreed criteria for species such discussions are of only limited value. Above all we need to know whether species units are comparable between different major groups of organisms. This volume is an attempt to find common ground in the practical use of species in documenting biodiversity by bringing together specialists on as wide a range of organisms as possible.

1.2 LINNAEUS AND EARLY SPECIES CONCEPTS

The term species derives from classical Greek logic. As Cain (1958) has emphasized, it would have been natural for all scholars of the 17th and 18th centuries to adopt the Aristotelian system of logic with its precise set of terms, including Definition, Genus, Differentia, and Species, in attempting to classify living organisms. In this system the Genus is that part of the Definition which refers to the general kind while Species refers to the particular, as qualified by the Differentia.

Linnaeus was not only the founder of the modern binomial system of nomenclature but he was also one of the few 18th century systematists who wrote down precisely what he was doing and what he thought he should do (Cain, 1958). To Linnaeus species were the lowest particular kinds of organisms in his classifications, though varieties were also sometimes noted. From a careful analysis of Linnaeus's many writings Ramsbottom (1938) showed that his notion of species was characterized by three different attributes. To Linnaeus, species were:

1. Distinct and monotypic.
2. Immutable and created as such.
3. Breeding true.

Later, however, he developed a complex theory of speciation by hybridization (see Cain, 1993).

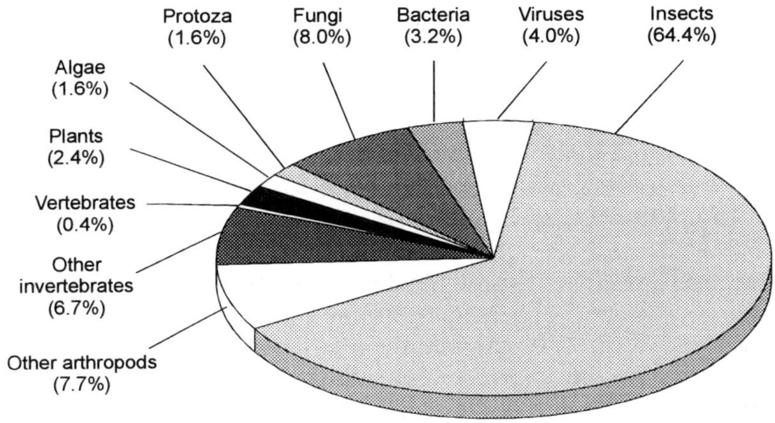

Figure 1.1 Estimates of proportions of species of major groups of organisms contributing to the total of living biological diversity. (Adapted from data in Hammond, 1992.)

These features are interesting because in his practical work Linnaeus was more and more confronted with large numbers of specimens from all over the world which he had to describe. Thus, usually all that was available to him were characteristics of dead museum specimens so that in practice he relied almost exclusively on the obvious morphological features of his material. However, even in the mid-18th century the notion that these characters were markers representing some sort of breeding units was obviously present in however vague a form.

Until the times of Linnaeus most early taxonomists were familiar with the organisms with which they worked as living entitities in the field. During the later 18th and the 19th centuries the natural history and museum traditions were to become more and more separate as increasingly large amounts of material needing description arrived in museums from all parts of the known world. Inevitably taxonomists were forced back more and more to describing dead specimens, usually with little knowledge of the habits and habitats of their organisms. Thus, the tradition was reinforced that species, and indeed higher taxa, must be based on morphological characters recognizable in preserved specimens. This is the morphological species concept or morphospecies which clearly arose as an empirical approach to a practical problem. It is not truly a concept but a technique of description. Nevertheless, most species in the most species-rich groups of living organisms are still today effectively morphospecies, often known from little more than preserved specimens. The amount of difference required to allow the recognition of different species is wholly determined by the subjective judgement of the individual taxonomist. This is epitomized by the well-known statement of Regan (1926) that 'A species is a community, or a number of related communities, whose distinctive morphological characters are, in the opinion of a competent systematist, sufficiently definite to entitle it, or them to a specific name'.

Naturalists immediately following Linnaeus were well aware that species had some biological reality in the field irrespective of degrees of morphological distinctiveness. For example, Gilbert White (1789) recognized the morphologically very similar species of breeding song birds in Britain, the Willow Warbler (*Phylloscopus trochilus*) and Chiffchaff (*P. collybita*), as different species on a basis of their songs, now recognized as elements of their specific mate recognition systems. Unfortunately the two traditions diverged during the late 18th and 19th centuries with naturalists continuing to emphasize breeding criteria and species as reproductive communities, but with systematists tending to emphasize morphological differences.

Darwin and Wallace both represented the natural history tradition from which emerged the theory of natural selection and the overwhelming evidence for descent with modification. Modern species then became the end terms of lines of descent. The controversies surrounding evolution itself meant that the nature of species was not widely regarded as an important problem in the mid-1800s. Indeed, Darwin himself seemed to regard species as rather arbitrary stages in the process of evolutionary divergence.

In the early 20th century Poulton (1908) made what was probably the most important advance towards what has since become known as the biological species, originally in his 1904 Presidential Address to the Entomological Society of London. In this he emphasized the importance of interbreeding in nature as the species criterion. He also very clearly differentiated this from simple hybridization. This was the modern biological species in all but name.

1.3 BIOLOGICAL SPECIES CONCEPTS

The most significant development in the establishment of a broadly biological species concept was the unification of genetics, systematics and evolutionary biology in the 1930s and 1940s, as exemplified by the publication of *Genetics and the Origin of Species* by Theodosius Dobzhansky (1937) and *Systematics and the Origin of Species* by Ernst Mayr (1942). Of the various definitions provided during this period probably the most useful is – 'species are groups of actually or potentially interbreeding natural populations which are reproductively isolated from other such groups' (Mayr, 1942: 120). Reproductive isolation is seen to be maintained by what Dobzhansky originally termed species isolating mechanisms which are any properties of species populations that reduce the likelihood of interbreeding with other species and thus the breakdown of co-adapted genetic systems. Of course it was recognized early on that reproductive isolation may not always be complete, but it must always be adequate to maintain the essential integrities of the interacting species populations. Thus, species were seen as beginning at the stage when different lineages become separate evolutionary entities. Speciation, the origin of new biological species, was, according to Dobzhansky, the origin of reproductive isolation and thus of distinct isolating mechanisms. The biological species has been developed and refined over the past 50 years or so, particularly by Cain (1954) and Mayr (1963, 1982), and has been accepted and widely used, mostly by field biologists, geneticists and evolutionists.

A major set of criticisms of the biological species has been developed in a series of papers by Hugh Paterson (summarized in 1985, 1993). In these he developed what he regards as a new concept which he termed the recognition concept in contrast to the biological species *sensu* Mayr which he prefers to call the isolation concept.

Paterson's main concern was with the conception of species being reproductively isolated by isolating mechanisms. The clear implication of this is that these mechanisms have been evolved under natural selection in order to achieve and maintain such isolation. They would thus obviously be primary adaptations. In this, Paterson clearly identified an area of imprecise thinking in the original formulation of the biological species concept. While Dobzhansky was a major advocate of the evolution of isolating mechanisms by reinforcement, Mayr has always regarded them as the result of incidental divergence between evolving nascent species (Mayr, 1988). To avoid this controversy Avise (1994) has suggested the more neutral term reproductive isolating barriers.

Paterson views species as groups of organisms with common fertilization systems – 'We can, therefore, regard as a species that most inclusive population of individual biparental organisms which share a common fertilization system'. He further recognizes an important adaptive subset of the fertilization system, the specific mate recognition system (SMRS) which is 'involved in signalling between mating partners or their cells' (Paterson, 1985). In most groups of animals the SMRS is well exemplified by exchanges of signals during courtship sequences, as described by ethologists. A classic example is the courtship sequence of the three-spined stickleback (*Gasterosteus aculeatus*), a small freshwater fish, first described and analysed by Tinbergen (1951). In such courtship sequences successive signals release successive responses via tuned receptors in the opposite sex, usually initiated first by the male. Unless the appropriate responses are made at each stage and the signals are recognized as appropriate, then the exchange will be terminated and fertilization will not be achieved (Figure 1.2). Thus, to Paterson the defining properties of particular species are their unique specific mate recognition systems and the process of speciation is the evolution of new and different SMRSs. He has on several occasions strongly differentiated what he terms his recognition concept from the earlier biological species of Mayr, Dobzhansky and Cain, which he prefers to term the isolation concept (Paterson, 1985). Vrba (1995) has recently cogently argued the same case, but others (Claridges 1988, 1995a; Coyne *et al.*, 1988; Mayr, 1988) have doubted the validity of the extreme contrast between the two concepts. Indeed, Mayr had earlier referred to what Paterson and his followers now term specific mate recognition systems as species recognition, 'the exchange of appropriate stimuli between male and female to ensure mating of conspecific individuals' (Mayr, 1963: 95).

The differentiation between the isolation and recognition concepts has generated much controversy and is discussed at length in a recent multiauthor volume (Lambert and Spencer, 1995) and by Mayr (1988) and Coyne *et al.* (1988). We take the view that, in practice, the two concepts are similar and will usually allow the recognition of the same entities as species (Claridge, 1988, 1995a). Thus, a broadened or composite biological species concept recognizes that different species are characterized by

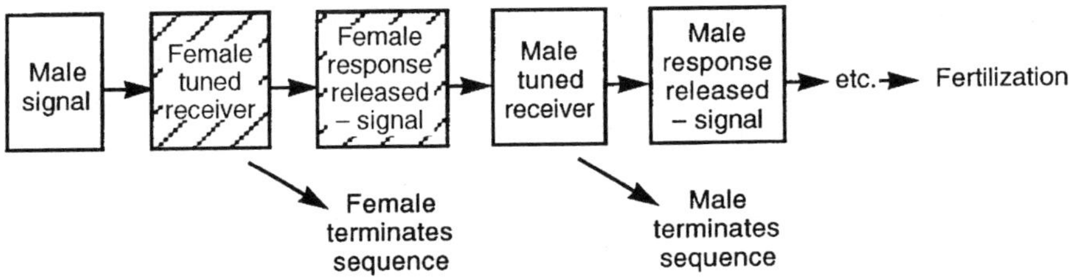

Figure 1.2 Diagrammatic representation of the exchange of signals and responses that constitute a specific mate recognition system (SMRS) (after Paterson, 1985).

distinct SMRSs which result in the reproductive isolation observed between different sympatric species in the field. This species concept is widely used in various groups of biparentally reproducing organisms.

In practice, species are rarely recognized by direct studies of the SMRS though that must remain the ultimate arbiter for biological species (Claridge, 1988, 1994; Claridge *et al.*, 1997: Chapter 12). Usually, biological species are recognized by distinctive markers not necessarily directly associated with the SMRS. These most often in the past have been diagnostic morphological differences between species, but in more recent years have increasingly included characteristics of cytology, behaviour, biochemistry, etc. Most excitingly, molecular markers involving characteristics of DNA, the hereditary material itself, are more and more being used (Avise, 1994). However, all of the diversity of characters now available to taxonomists are used as indicators of levels of reproductive isolation under the biological species concept. Molecular techniques can often be used directly to measure levels of gene flow between populations, and therefore of reproductive isolation.

One very important consequence of the broad concept of biological species advocated here is that reproductive isolation may occur without associated morphological or other obvious differentiation. Thus, real biological species may exist that show little or no obvious differentiation to the human observer – so-called sibling or cryptic species. Such species are well known and abundant in many groups of organisms (Claridge, 1988; Claridge *et al.*, 1997: Chapter 12; Knowlton, 1997: Chapter 10; Lane, 1997: Chapter 13). These are just as real biological entities as are morphologically recognizable species, but they pose practical and theoretical difficulties if non-biological species concepts are used.

A large body of opinion among systematists has always been unhappy with the biological species (Sokal and Crovello, 1970) and has preferred either an overtly morphospecies approach or some sort of phenetic system. In particular, botanists have been widely critical because of the frequent occurrence in the field of interspecific hybrids between plant species (Gornall, 1997: Chapter 8). However Mayr (1992) showed that in at least one North American local flora the biological species was easily applicable to all but between 6% and 7% of the plants present.

There is certainly no dispute even among the most ardent advocates of biological species that there are two important areas where application is either very difficult or impossible:

1. **Asexual or parthenogenetic forms**. The biological species can only apply to biparental sexually reproducing organisms in which a SMRS is present leading to reproductive isolation. Neither asexual nor parthenogenetic organisms can have SMRSs, since functional mating and fusion of gametes does not take place. Thus, the biological species concept cannot be used. Such organisms exist as clones which may differ in features of morphology, biochemistry, behaviour, etc. (Foottit, 1997: Chapter 14). Distinctive and diagnosable clones may be recognized as species but they can not be biological species. Cain (1954) coined the term agamospecies for such entities and De Bach (1969), in a useful contribution, termed them uniparental species. Clearly the very many groups of microorganisms which have no exchange of gametes or other genetic material in reproduction, can only be agamospecies.

2. **Allopatric forms**. A further difficulty with biological species concepts is that reproductive isolation in the field can only be observed between sympatric populations; in these alone opportunities exist for testing the effectiveness of presumed SMRSs. The problem of determining the status of allopatric populations is a longstanding one. Varying degrees of observed differentiation from virtually nothing to large differences at least comparable with those between sympatric species, may be observed. The criterion of gene flow and degree of reproductive isolation in the field cannot be tested. The polytypic nature of biological species has long been recognized and a series of categories from superspecies to subspecies has been advocated in attempts to document such essentially continuous variation (Mayr, 1942; Cain, 1954). However, the allocation of allopatric forms within this series has always been largely subjective. This problem is undoubtedly a weakness in the applicability of biological species concepts.

These difficulties, together with a desire to eliminate the priority of one set of characteristics of organisms – the SMRS and reproductive isolation – over others have led many modern systematists to reject completely biological species and to favour some variant of what may be termed a general phylogenetic species concept.

1.4 PHYLOGENETIC SPECIES AND RELATED CONCEPTS

With the development and widespread acceptance of cladistic methods by English-speaking systematists over the past 30 years or so, dissatisfaction with the biological species has developed. Curiously Hennig (1966), the founder of cladistic methodologies, regarded species as reproductive communities and his species concept was close to the biological species of Ernst Mayr (Nixon and Wheeler, 1990). Of course Hennig was also concerned with extending species back in time as diagnosable clades for which the biological species is not adequate. In this he was continuing what Simpson (1951) had begun by developing a broader evolutionary concept which has since been taken up by others, including Cain (1954), Wiley (1978) and Mayden (1997: Chapter 19).

Opinion has been widely divided among cladists on the nature and status of species. Some even take the view that species have no special significance and are no different from higher taxa ('A species is only a taxon'; Nelson, 1989), but this certainly does not seem to be the general view. Hennig (1966) saw species as the very important level in the systematic hierarchy above which cladistic methods were applicable and below which they were not. At and above the species level phylogenetic relationships were important. Below the species level interbreeding relationships dominate and these he differentiated from phylogenetic, as tokogenetic relationships. As Nixon and Wheeler (1990) state, 'Species are uniquely different from higher level taxa, in that species do not have resolvable internal phylogenetic structure among the individual organisms included'.

Various authors have given definitions of phylogenetic species. Perhaps the most widely cited and influential is that of Cracraft (1983) in which species were defined as the 'smallest diagnosable cluster of individual organisms within which there is a parental pattern of ancestry and descent'. Contrary to the views of some critics (e.g. Avise, 1994; Mallet, 1995), it is clear that this concept can and does apply to populations (Cracraft, 1997: Chapter 16), in which it therefore resembles the biological species. It is, though, applicable also to non-sexually reproducing organisms. This is emphasized in a more recent refinement of the concept in which Nixon and Wheeler (1990) define species as the 'smallest aggregation of populations (sexual) or lineages (asexual) diagnosable by a unique combination of character states in comparable individuals'. Thus, the essence of the phylogentic species concept is concerned with the recognition of diagnosably distinct clades. The major question that has to be asked in practice is – What precisely is diagnostic? How different do two populations or lineages have to be to be regarded as diagnosably distinct? What is distinct to one worker may not be so to another.

Any of the markers discussed above as indicative of biological species may be used to characterize phylogenetic species, though in most groups of larger organisms the characters used are generally mainly morphological ones. In microorganisms they are more usually biochemical and molecular.

As presently applied (Cracraft, 1997: Chapter 17) the differences in practice between a phylogenetic concept and a broadly biological species do not seem to us to be very great. The advantages of the former are clear:

- Phylogenetic species can be applied to non-sexual and parthenogenetic forms and will depend on the diagnosability of the different clones or lineages. They will effectively be the same as agamospecies (Cain, 1954).
- Phylogenetic species may also be applied to allopatric forms.

Diagnosably distinct allopatric populations will therefore be regarded as separate species, presumably however slight the difference. The result is almost always that more allopatric populations are recognized as different species than has traditionally been the case for the polytypic biological species. For example Cracraft (1992), in reviewing the Birds-of-Paradise (Aves, Paradisaeidae), recognized more than twice as many phylogenetic species (90) as had previous applications of the biological species to the same data (40+). However, this decision still remains a matter of personal subjective judgement. In our view there is no fundamental difference of principle between the two concepts on this matter, but it would seem best to us, when in doubt, to err on the side of recognizing more rather than less species.

A major disadvantage of phylogenetic species concepts that is not often recognized is the improbability that they will reveal the existence in many groups of organisms of complexes of sibling species. The emphasis on the criterion of reproductive isolation and specific mate recognition in the biological species concept means that sibling species will be revealed there. The philosophy of the phylogenetic species gives no reason or incentive to search for further divisions of existing diagnosably distinct forms. In view of the significance of sibling species in many groups this a major weakness. The phylogenetic species should have little difficulty in accommodating sibling species once discovered, but only application of biological species concepts will positively identify them.

1.5 SPECIES CONCEPTS AND SPECIATION: PATTERN AND PROCESS

Most authors will presumably agree that in recognizing species we are attempting to provide a framework for describing and understanding the diversity of living organisms and their evolutionary relationships. However, there have been longstanding controversies about the interaction between particular species concepts and the theories of speciation upheld by different workers. In principle, it is surely necessary to establish a system for describing the diversity of living organisms that is independent of the various modes by which that diversity may have evolved. Many evolutionary biologists do not agree with this. For example, Guy Bush has for long been a proponent of theories of sympatric speciation, in which, contrary to widely supported theories of allopatric speciation (Mayr, 1942, 1963; Cain, 1954), no period of spatial isolation is required before initial genetic divergence can take place (Bush, 1975, 1993, 1994). These interesting ideas are controversial, but are clearly supported by some studies, particularly on specialist feeders and parasites. Our understanding of the pattern of diversity in the field should not vary according to which theory of speciation we support. The recognition of biological species and the inevitable difficult cases, where some level of gene flow occurs between species, should be completely compatible with either allopatric or sympatric models of speciation (Claridge, 1995b, but see also Bush, 1995).

A similar argument in which species concepts and models of speciation have become completely intertwined is that of the recognition concept (here regarded as part of a wider biological species). Paterson (1985, 1993) is a passionate advocate of what we might term pure allopatric speciation which, despite his differences of opinion with Mayr, follows closely the original proposals of Mayr (1942, 1963). In these it is envisaged that speciation is completed in allopatry, often after the isolation of a relatively small subgroup of populations from the original ancestral group. Species differences thus evolve completely in allopatry. Paterson's rejection of the concept of species isolating mechanism follows since, if such characteristics evolve entirely in isolation, then obviously they cannot be adaptations to ensure reproductive isolation.

Contrary to theories of speciation in complete allopatry, Wallace (1889), Dobzhansky (1937) and others, have developed the theory of reinforcement of species isolating mechanisms after diverging incipient species populations have once again become sympatric. Any differences so evolved could truly be isolating mechanisms as a result of natural selection favouring homogametic matings. The weight of evidence currently is generally thought to be against the importance of reinforcement (Butlin, 1989), but recent theoretical studies once again suggest that it may be sometimes more significant than has been thought (Liou and Price, 1994; Butlin, 1995).

These important and interesting controversies should not affect the species units that we recognize in nature and that form the basis for speciation studies. This matter has been well discussed at length by Chandler and Gromko (1989). We believe that the biological species concept must be formulated to allow any of the theories of speciation to be fairly tested without prior assumptions concerning speciation. Thus, in principle, we agree with many cladists on the particular point that it is necessary to describe patterns in nature, so far as possible, independent of theories concerning the origins of those patterns (Nixon and Wheeler, 1990).

However, we would not go so far as Wheeler and Nixon (1990) who state that 'the responsibility for species concepts lies solely with systematists'. If we take an evolutionary view of species then we cannot separate species also from genetics and evolutionary biology.

Though we emphasize the need to separate so far as possible the recognition of pattern, in the form of species, from process in the form of modes of speciation, we cannot – and indeed should not – divorce species concepts from evolution. If we accept the generality of evolution and species as the result of evolutionary divergence then the species itself must be an evolutionary concept.

1.6 SPECIES IN PRACTICE – AN EVOLUTIONARY SYNTHESIS

We are concerned here entirely with species as used for extant living organisms and not with the difficult problems of using species in a time dimension when fossil forms are studied. However, the different species concepts we have briefly reviewed all attempt to describe the extant terminal branches of different evolutionary lineages. Simpson (1951) was one of the first to develop an explicitly evolutionary species concept, by integrating species as used for living organisms with species as segments of evolutionary lineages so as to include fossils. Cain (1954) took up this integration. Thus in Cain's system the taxonomic species includes palaeospecies which are named lineages over time, which might be akin to the internodal species of Kornet (1993) and other similar recent concepts (see Mayden, 1997: Chapter 19). The biological species are modern sexually reproducing forms evolving separately from their closest relatives and from which they are reproductively isolated. Other separate lineages of organisms lacking biparental sexual reproduction are clones which may be recognized as separate agamospecies on a basis of particular characteristics. Thus, these forms are all part of the general taxonomic species concept and what Simpson (1951) first developed as an evolutionary species concept. This useful umbrella term was taken up more recently by Wiley (1978) and Mayden (1997: Chapter 19), who resuscitates it in this volume as a general species concept that we can all use. The development of phylogenetic species concepts in recent years (Cracraft, 1997: Chapter 16) has established a framework which many taxonomists find useful. However, we do not regard it as so different to the biological species and agamospecies in the above system (after Cain, 1954).

The detailed considerations by specialists on a wide diversity of living organisms in this volume give reason for some hope that species may be used consistently as the basic units of biodiversity. We agree with Cracraft that this is even more important now that high priority is rightly being given to the conservation of biodiversity. It is vital that some general agreement on species for all groups of organisms should be achieved. We believe that there are no insuperable difficulties to using biological species, agamospecies and phylogenetic species in different groups and particular situations as appropriate. They all represent part of the taxonomic or evolutionary species. The particular problem of the status of allopatric populations is common to all these approaches and should have similar resolutions.

Acknowledgements

We thank all colleagues who have devoted many hours to discussing ideas on species concepts with us. In particular we are grateful to all of the contributors and participants at the 1995 meeting in Cardiff who clarified much of our thinking.

We are particularly indebted to Arthur Cain for his critical comments on a draft of this chapter, John Morgan for helping in the preparation of the figures and Rosemary Jones for preparing the manuscript.

1.7 REFERENCES

Avise, J.C. (1994) *Molecular Markers, Natural History and Evolution*. Chapman & Hall, New York.

Bush, G.L. (1975) Modes of animal speciation. *Annual Review of Ecology and Systematics*, **6**, 339–64.

Bush, G.L. (1993) A reaffirmation of Santa Rosalia, or why are there so many kinds of small animals, in *Evolutionary Patterns and Processes* (eds D.R. Lees and D. Edwards), Academic Press, London, pp. 229–49.

Bush, G.L. (1994) Sympatric speciation. *Trends in Ecology and Evolution*, **9**, 285–8.

Bush, G.L. (1995) Species and speciation, reply from G. Bush. *Trends in Ecology and Evolution*, **10**, 38.

Butlin, R. (1989) Reinforcement of premating isolation, in *Speciation and its Consequences*, (eds D. Otte and J.A. Endler), Sinauer Associates, Sunderland, Massachusetts, pp. 158–79.

Butlin, R. (1995) Reinforcement: an idea evolving. *Trends in Ecology and Evolution*, **10**, 432–4.

Cain, A. J. (1954) *Animal Species and their Evolution*. Hutchinson, London.

Cain, A.J. (1958) Logic and memory in Linnaeus's system of taxonomy. *Proceedings of the Linnean Society of London*, **169**, 144–63.

Cain, A.J. (1993) Linnaeus's *Ordines naturales*. *Archives of Natural History*, **20**, 405–15.

Chandler, C.R. and Gromko, M.H. (1989) On the relationship between species concepts and speciation processes. *Systematic Zoology*, **38**, 116–25.

Claridge, M.F. (1988) Species concepts and speciation in parasites, in *Prospects in Systematics* (ed. D.L. Hawksworth), Clarendon Press, Oxford, pp 92–111.

Claridge, M.F. (1995a) Species concepts and speciation in insect herbivores: planthopper case studies. *Bolletino di Zoologia*, **62**, 53–8.

Claridge, M.F. (1995b) Species and speciation. *Trends in Ecology and Evolution*, **10**, 38.

Claridge, M.F., Dawah, H.A. and Wilson, M.R. (eds) (1997) *Species: the Units of Biodiversity*, Chapman & Hall, London.

Coyne, J.A., Orr, H.A. and Futuyma, D.J. (1988) Do we need a new species concept? *Systematic Zoology*, **37**, 190–200.

Cracraft, J. (1983) Species concepts and speciation analysis, in *Current Ornithology* (ed. R. Johnston), Plenum, New York, pp. 159–87.

Cracraft, J. (1992) The species of the birds-of-paradise (Paradisaeidae): applying the phylogenetic species concept to complex patterns of diversification. *Cladistics*, **8**, 1–43.

Cracraft, J. (1997) Species concepts in systematics and conservation – an ornithological viewpoint, in *Species: the Units of Biodiversity* (eds M.F. Claridge, H.A. Dawah and M.R. Wilson), Chapman & Hall, London, pp. 325–39.

De Bach, P. (1969) Uniparental, sibling and semi-species in relation to taxonomy and biological control. *Israel Journal of Entomology*, **4**, 11–27.

Dobzhansky, T. (1937) *Genetics and the Origin of Species*. Columbia University Press, New York.

Eldredge, N. (1995) Species, selection, and Paterson's concept of the specific-mate recognition system, in *Speciation and the Recognition Concept* (eds D.M. Lambert and H.G. Spencer), The Johns Hopkins University Press, Baltimore, London, pp. 464–77.

Ereshefsky, M. (ed.) (1992) *The Units of Evolution: Essays on the Nature of Species*. MIT Press, New York.

Foottit, R. (1997) Recognition of parthenogenetic insect species, in *Species: the Units of Biodiversity* (eds M.F. Claridge, H.A. Dawah and M.R. Wilson), Chapman & Hall, London, pp. 291–307.

Gornall, R. J. (1997) Practical aspects of the species concept in plants, in *Species: the Units of Biodiversity* (eds M.F. Claridge, H.A. Dawah and M.R. Wilson), Chapman & Hall, London, pp. 171–90.

Hammond, P.M. (1992) Species inventory, in *Global Biodiversity: Status of the earth's Living Resources* (ed. B. Groombridge), Chapman & Hall, London, pp. 17–39.

Hennig, W. (1966) *Phylogenetic Systematics*. University of Illinois Press, Urbana.

Hull, D.L. (1997) The ideal species concept – and why we can't get it, in *Species: the Units of Biodiversity* (eds M.F. Claridge, H.A. Dawah and M.R. Wilson), Chapman & Hall, London, pp. 357–80.

Knowlton, N. (1997) Species of marine invertebrates: current practices and options for the future, in *Species: the Units of Biodiversity* (eds M.F. Claridge, H.A. Dawah and M.R. Wilson), Chapman & Hall, London, pp. 199–219.

Kornet, D. (1993) Permanent splits as speciation events: a forward reconstruction of the internodal species concept. *Journal of Theoretical Biology*, **164**, 407–35.

Lambert, D.M. and Spencer, H.G. (eds) (1995) *Speciation and the Recognition Concept*. The Johns Hopkins University Press, Baltimore, London.

Lane, R. (1997) The species concept in bloodsucking vectors of human diseases, in *Species: the Units of Biodiversity* (eds M.F. Claridge, H.A. Dawah and M.R. Wilson), Chapman & Hall, London, pp. 273–89.

Liou, L.W. and Price, T.D. (1994) Speciation by reinforcement of premating isolation. *Evolution*, **48**, 1451–9.

Mallet, J. (1995). A species definition for the modern synthesis. *Trends in Ecology and Evolution*, **10**, 294–9.

May, R. (1990) How many species?. *Philosophical Transactions of the Royal Society* B, **330**, 293–304.

Mayden, R. L. (1997) A hierarchy of species concepts: the denouement in the saga of the species problem, in *Species: the Units of Biodiversity* (eds M.F. Claridge, H.A. Dawah and M.R. Wilson), Chapman & Hall, London, pp. 381–424.

Mayr, E. (1942) *Systematics and the Origin of Species from the Viewpoint of a Zoologist.* Columbia University Press, New York.

Mayr, E. (1982) *The Growth of Biological Thought.* Harvard University Press, Cambridge, Massachusetts.

Mayr, E. (1988) The why and how of species. *Biology and Philosophy*, **3**, 431–41.

Mayr, E. (1992) A local flora and the biological species concept. *American Journal of Botany*, **79**, 222–38.

Mayr, E.(1963) *Animal Species and Evolution.* Harvard University Press, Cambridge, Massachusetts.

Nelson, G. (1989) Cladistics and evolutionary models. *Cladistics*, **5**, 275–89.

Nixon, K. and Wheeler, Q.D. (1990) An amplification of the phylogenetic species concept. *Cladistics*, **6**, 211–23.

Otte, D. and Endler, J.A. (eds) (1989) *Speciation and its Consequences.* Sinauer Associates, Sunderland, Massachusetts.

Paterson, H.E.H. (1985) The recognition concept of species, in *Species and Speciation* (ed. E.S. Vrba), Transvaal Museum, Pretoria, pp. 21–9.

Paterson, H.E.H. (1993) *Evolution and the Recognition concept of Species, Collected Writings of H.E.H. Paterson* (ed. S.F. McEvey), The Johns Hopkins University Press, Baltimore, London.

Poulton, E.B. (1908) What is a species? in *Essays on Evolution 1889–1907*, Clarendon Press, Oxford, pp. 46–94.

Ramsbottom, J. (1938) Linnaeus and the species concept. *Proceedings of the Linnean Society of London*, **150**, 192–219.

Regan, C.T. (1926) Organic Evolution. *Report of the British Association for the Advancement of Science*, 1925, 75–86.

Simpson, G.G. (1951) The species concept. *Evolution*, **5**, 285–98.

Sokal, R.R. and Crovello T.J. (1970) The biological species concept: a critical evaluation. *The American Naturalist*, **104**, 127 –53.

Tinbergen, N. (1951) *The Study of Instinct*, Oxford University Press, Oxford.

Vrba, E. (1995) Species as habitat-specific, complex systems, in *Speciation and the Recognition Concept* (eds D.M. Lambert and H.G. Spencer), The Johns Hopkins University Press, Baltimore, London, pp. 3–44.

Walker, F. (1832) Monographia Chalcidum. *Entomological Magazine*, **1**, 12–29.

Wallace, A.R. (1889) *Darwinism: An exposition of the Theory of Natural Selection.* Macmillan, London.

Wheeler, Q.D. and Meier, R. (eds) (1997) *Species Concepts and Phylogenetic Theory: A Debate.* Columbia University Press, New York (in press).

Wheeler, Q.D. and Nixon, K.C. (1990) Another way of looking at the species problem: a reply to de Queiroz and Donoghue. *Cladistics*, **6**, 77–81.

White, G. (1789) *The Natural History of Selborne.* White, Cochran, London.

Wiley, E. (1978) The evolutionary species concept reconsidered. *Systematic Zoology*, **27**, 17–26.

Wilson, E.O. (1992) *The Diversity of Life.* Harvard University Press, Cambridge, Massachusetts.

2

Viral species

M. H. V. Van Regenmortel

Contacting address: Institut de Biologie Moleculaire et Cellulaire, CNRS, 15 rue René Descartes, 67084 Strasbourg, France

ABSTRACT

Species is the universally accepted term for the lowest taxonomic cluster of living organisms. It has been argued that species taxa should be regarded as individuals and not as classes or categories because species change during evolution while classes are immutable and timeless. This viewpoint is based on the notion that species correspond to so-called Aristotelian classes or universal classes that can be defined by one or more properties that are both necessary and sufficient for class membership. However, because of the inherent variability of the organisms constituting a species taxon, the species category does not fit the classical notion of class but is more like a fuzzy set with no clear-cut boundaries.

Although viruses are not living organisms, it is possible to use the species concept in virology because viruses are biological entities, not simply chemicals. Viruses have genomes, replicate, evolve and occupy particular ecological niches. In 1991, the International Committee on Taxonomy of Viruses (ICTV) accepted the following definition of virus species: A virus species is a polythetic class of viruses that constitutes a replicating lineage and occupies a particular ecological niche. The definition incorporates the notions of genome, biological replication and natural selection, since the term 'replicating lineage' indicates an inherited genealogy extending over many generations and unified by a common descent. The reference to 'ecological niche occupancy' in the definition brings in the role played by environmental determinants such as host, tissue and vector tropisms in maintaining species identity. This definition does not provide a list of diagnostic properties for recognizing members of particular virus species. The characters most commonly used for recognizing members of individual species are certain features of genome, the presence of antigenic cross-reac-

Species: The Units of Biodiversity. Edited by M.F. Claridge, H.A. Dawah and M.R. Wilson. Published in 1997 by Chapman & Hall. ISBN 0 412 63120 2

tions and various biological properties such as host range and reactions, tissue tropism, type of vector and transmission route.

It should be stressed that the definition of species as a polythetic class rules out the possibility of finding a single character that could be used as a differential diagnostic property. The members of a virus species do not have a single defining property in common that is necessary and sufficient for class membership (i.e. a property that is common and peculiar only to members of the species). Certain common properties such as morphological features or genome composition are shared also by members of other virus species and such properties define higher categories such as genera and families and are not species-defining properties. A single diagnostic property such as a particular level of genome homology, the extent of antigenic similarity or a certain host reaction will always fail as a criterion for membership of a particular virus species.

The classification of viruses should not be confused with the classification of viral genome sequences. Viruses are biological entities and the notion of ecological niche is a crucial component for demarcating individual viral species.

2.1 INTRODUCTION

For many years the world community of virologists could not agree on the status and nomenclature of the taxa to be used in virus classification (Matthews, 1983, 1985). The virologists who study the viruses that infect plants were particularly reluctant to apply the species concept in virology, arguing that entities that reproduce by clonal means could not be accommodated within the classical definition of biological species (Harrison, 1985; Milne, 1985). Those plant virologists who were opposed to the use of the species concept in virology took the view that the only legitimate definition of species was that of biological species characterized by gene pools and reproductive isolation, and applicable only to sexually reproducing organisms. After several years of vigorous debate concerning the validity of various alternative species concepts (Matthews, 1983; Bishop, 1985; Kingsbury, 1988; Milne, 1988; Van Regenmortel, 1989, 1990), the International Committee on Taxonomy of Viruses (ICTV) agreed in 1991 that the usual categories of species, genus and family should also be used in virus classification (Pringle, 1991; Van Regenmortel *et al.*, 1991). Eventually, it was accepted that the species concept is applicable in virology because viruses have genomes, replicate, evolve and occupy particular ecological niches. The following definition of virus species was endorsed by the ICTV: 'A virus species is a polythetic class of viruses that constitutes a replicating lineage and occupies a particular ecological niche' (Van Regenmortel, 1990). The earlier reluctance of some virologists to accept viral species was due in part to the well-known difficulties posed by the classification of asexual organisms (Mayr, 1982; Holman, 1987). The difficulties experienced by virologists in arriving at an acceptable definition of virus species are thus an illustration of the more general issues that arise in any classification when attempts are made to deal with organisms that reproduce in a clonal or parthenogenic manner.

2.2 SEMANTICS

Part of the confusion surrounding the debates about species is of a semantic nature. The term 'species' can be used to refer to a taxonomic category in which case it corresponds to an abstract concept devoid of any spatiotemporal location. This use of the word species either as a class of categories used in taxonomy or as a class of organisms is often confused with another meaning of the word, namely that of a concrete collective entity made up of real organisms localized in space and time, i.e. a taxon. When the word species is used to refer to a practical entity of real organisms, it may enter into what logicians call part–whole relations applicable only to spatiotemporally localized entities (Hull, 1976). The part–whole relation applies for instance to a particular dog which is part of the taxon dog comprising all animals with dog features. On the other hand, the abstract concept of species cannot enter into part–whole relations but instead can take part in relations known as class-inclusion or class-membership. Although an organism may thus be considered as a member of a species (viewed as a class), it is logically impossible for it to be part of an entity of different logical type such as the abstract concept of class.

Confusion between the abstract and practical usages of the term species is responsible for innumerable idle debates about the reality of species. Many biologists readily accept that genera and families are artificial, abstract constructions of the mind, but insist that species are real, i.e. endowed with an objective reality and individuality. The inability to conceive of species as a conceptual construction led Milne (1984) for instance to assert 'Linnaeus did not create species, he found them'. The same reluctance to view species as classes led to the proposal that particular species should be regarded as individuals. According to this viewpoint, species are constituted of organisms in the same way that an individual organism is constituted of cells and organs (Ghiselin, 1974; Hull, 1976). The proposal that species should be regarded as individuals, i.e. as practical entities and not as abstract classes, stems from the belief that all classes are necessarily Aristotelian classes, immutable and timeless. Since species taxa change during evolution, they cannot correspond to universal Artistotelian classes and this is advanced as an argument against viewing species as a class. However, as discussed below, the concepts of polythetic class and fuzzy set make it possible to reconcile phylogenetic change with class membership and this removes the rationale for considering species only as real individuals. The suggestion that biological classification is concerned with individuals and not with classes was not extended to higher taxa although there is, of course, no reason why genera could not be considered as individuals constituted of species. Presumably the protagonists of the species-as-individual thesis recognized that if genera, families and kingdoms are not allowed to be classes, any taxonomy becomes impossible.

As pointed out by Quine (1987), universal classes and properties are related abstract entities. Ascribing a property to a thing, for instance spherical shape or possession of an RNA genome, amounts to assigning the thing to a universal class, i.e. the class of spheres and of entities containing a RNA genome. In this sense, viruses or organisms can be members of various universal classes corresponding to higher taxonomic categories such as genera or families. These taxonomic classes correspond to Aristotelian classes defined by a single property or by a set of properties necessary and sufficient for membership in the class. In contrast, species are not universal classes and their members do not have a single defining property in common (Beckner, 1959). A final note of caution about the distinction between practical species taxa and abstract species classes should be made. When species are viewed as taxa it is impossible to define them. They can only be given proper names in an arbitrary manner analogous to baptism (Kitts, 1984). Only when they are viewed as abstract classes can species be defined. However, such a definition of the concept is of little help for identifying the members of a particular species. For example, the definition of biological species in terms of gene pools and reproductive isolation is of little use for identifying members of the species. The diagnostic properties of real objects should not be confused with the theoretical, defining properties of abstract classes (Ghiselin, 1984).

2.3 CONTINUITY VERSUS DISCONTINUITY AND THE PROBLEM OF SPECIES DEMARCATION

Ever since the demise of the theory of spontaneous generation of living organisms, it has been recognized that life is a continuum linking individuals to their parents, unicellular organisms to multicellular organisms and animals to man. Although the study of biology is thus concerned with a continuous historical development embodied in the uninterrupted chain of replicating DNA molecules, there is also clearly a need to recognize discontinuities among different life forms. Species undergo continuous variation that is regulated by discontinuous single nucleotide changes. Transition from one species to another during evolution occurs within the continuity of gene pools. In spite of this continuity, however, so-called evolutionary species are considered separate at one particular point in time (Lovtrup, 1979). Difficulties in reconciling continuous and discontinuous frames of references are not unique to biology. There is, for example, vigorous debate concerning the possible continuous or discontinuous nature of space and time. On a less abstract note the Swiss sometimes find it difficult to say exactly where the Jungfrau and neighbouring Monch mountains start and stop, but this does not lead them to doubt the reality of these two mountains because their limits are unclear. Although one cannot make absolutely clear distinctions where none exists, it is useful to identify and give names to parts of a continuum, whether it be a geological rock formation, a colour in the visible spectrum of electromagnetic waves or a particular influenza virus responsible for a worldwide pandemic.

The classical concept of biological species demands that we identify clear-cut breeding discontinuities that are mostly absent. Similarly, the concept of evolutionary species requires that we demarcate boundaries in time that coincide with the appearance of new species, a task equally impossible to do in practice. The solution lies in viewing species as polythetic classes since this does away with the need to use any single discontinuity as criterion for demarcation.

2.4 SPECIES AS POLYTHETIC CLASSES

The polythetic species concept was introduced by Beckner (1959) to replace the classical notion of universal class. He gave the name polytypic (later changed to polythetic) to classes that are defined by a combination of characters, each of which may occur also outside the given class and may be absent in any member of the class. The nature of polythetic classes can be illustrated by the following example (Sattler, 1986). Suppose a species is defined by a set of five properties F1, F2, F3, F4 and F5. If these properties are distributed in the way shown in Table 2.1, the class will be polythetic. This example represents a polythetic class because each individual possesses a large number of the properties (i.e. four out of five), each property is possessed by a large number of individuals and no property is possessed by all individuals. Contrary to the situation with universal classes, no single property is either necessary or sufficient for membership in a polythetic class. The concept of polythetic class is extremely useful for dealing with biological entities endowed with intrinsic variability, since it can accommodate individual members that lack one or other character considered typical of the class. In this kind of class, certain elements may evolve and there is no difficulty in reconciling class membership with phylogenetic change. This makes a polythetic species similar to a fuzzy set (Beatty, 1982; Kosko, 1994) with boundaries that are modifiable and not uniquely defined. The view that species are sets has been elaborated by Kitcher (1984).

Table 2.1 Distribution of five properties, F1 to F5, among five members of a polythetic class

Individual	*Properties*				
1	F1	F2	F3	F4	
2	F1	F2	F3		F5
3	F1	F2		F4	F5
4	F1		F3	F4	F5
5		F2	F3	F4	F5

2.5 SPECIES FUZZINESS

It is generally accepted that species sets consist of organisms that are phylogenetically connected, although in the case of structures as simple as some viruses it cannot be excluded that the same pathogenic entity might have arisen or evolved from a different parent more than once during biological evolution. The fuzziness inherent in species sets arises from the rejection of so-called binary or bivalent logic, according to which every statement or sentence is either true or false (i.e. A or not A). This type of logical dichotomy is absent in fuzzy logic, which on the contrary assumes that everything is a matter of degree, including truth and set membership. Fuzzy logic and fuzzy sets were introduced by Lofti Zadeh (1965) and have been popularized by Bart Kosko (1994). When fuzzy logic is applied to the diagnostic characters that are used for identifying members of a species, these properties are taken as multivalent and capable of assuming several intermediate states. Life itself is a fuzzy concept, as shown for instance by the ambiguities that result when its beginning in man is defined at conception and its end is made to coincide with brain damage. Viruses are an interesting example of the difficulty of defining life since they represent a type of 'borrowed' life. Viruses are entirely dependent on living cells for their replication and are devoid of any metabolic activity, although they are endowed with life-like properties such as genetic continuity and adaptability. It is also a sobering thought that as the amount of information increases together with greater precision of collected data, the fuzziness actually increases rather than decreases (Kosko, 1994). It used to be believed that when the entire genome sequences of many different viruses would become available, a simple examination of these sequences expressing the complete viral blueprint would enable one to assess if individual viruses belonged to the same or to different species. This in fact has not happened and it is now evident that the boundary between two viral species in terms of percentage of sequence identity cannot be drawn in a non-fuzzy manner. In the family Potyviridae, for example, it is accepted that different potyvirus species exhibit coat protein sequence identities of

40–70% while strains of the same potyvirus have sequence identities of 90–99%. However, two potyviruses showing 85% coat protein sequence identity could be considered either as two different species or as two strains of the same species (Van Regenmortel, 1992). There is no single sequence identity figure that can be stated as the absolute cut-off between species and strains but there is a range of values (80–90%) within which the alternative categories can be considered sensible options, depending on the relative significance of other properties (Shukla *et al.*, 1994: 207).

2.6 SPECIES OR QUASISPECIES

It should be stressed that it is no longer accepted that a virus species can be defined by a single genome sequence. Because RNA viruses have genomes that replicate in the absence of repair mechanisms, they evolve very rapidly with a mutation frequency per nucleotide site in the viral genome of 10^{-3} to 10^{-5}. Since RNA viral genomes usually contain about 10^4 nucleotides, a clone of an RNA virus consisting of say 10^{12} particles will always consist of a complex mixture of millions of different genomes all of which compete during replication of the clone (Holland *et al.*, 1992). In view of their genome plasticity, RNA viruses are usually considered to be quasispecies populations. The term quasispecies was introduced by Eigen (for review, see Eigen, 1993) to describe the distribution of self-replicating RNAs believed to be the first genes on earth. A quasispecies population consists of a master sequence corresponding to the most fit genome sequence with respect to a given environment together with innumerable competing virus mutants. It should be noted that the use of the term quasispecies for a virus population does not imply that there exists an entity called virus species with a single, invariant genome sequence. The current status of virus classification, including a discussion of the various taxonomic categories used in virology, can be found in Shukla *et al.* (1994) and Murphy *et al.* (1995).

2.7 REFERENCES

Beatty, J. (1982) Classes and cladists. *Systematic Zoology*, **31**, 25–34.

Beckner, M. (1959) *The Biological Way of Thought*. Columbia University Press, New York.

Bishop, D.H.L. (1985) The genetic basis for describing viruses as species. *Intervirology*, **24**, 79–93.

Eigen, M. (1993) Viral quasispecies. *Scientific American*, **269**, 32–9.

Ghiselin, M.T. (1974) A radical solution to the species problem. *Systematic Zoology*, 23, 536–44.

Ghiselin, M.T. (1984) Definition, character and other equivocal terms. *Systematic Zoology*, **33**, 104–10.

Harrison, B.D. (1985) Usefulness and limitations of the species concept for plant viruses. *Intervirology*, **24**, 71–8.

Holland, J.J., de la Torre, J.C. and Steinhauer, D.A. (1992) RNA virus populations as quasispecies, in *Genetic Diversity of RNA Viruses* (ed. J.J. Holland), Springer-Verlag, Vienna, pp. 1–20.

Holman, E.W. (1987) Recognizability of sexual and asexual species of rotifers. *Systematic Zoology*, **36**, 381–6.

Hull D.L. (1976) Are species really individuals? *Systematic Zoology*, **25**, 174–91.

Kingsbury, D.W. (1988) Biological concepts in virus classification. *Intervirology*, **29**, 242–53.

Kitcher, P. (1984) Species. *Philosophy of Science*, **51**, 308–33.

Kitts, D.B. (1984) The names of species: a reply to Hull. *Systematic Zoology*, **33**, 112–15.

Kosko, B. (1994) *Fuzzy Thinking*. HarperCollins, London.

Lovtrup, S. (1979) The evolutionary species: fact or fiction? *Systematic Zoology*, **28**, 386–92.

Matthews, R.E.F. (1983) *A Critical Appraisal of Viral Taxonomy*. CRC Press, Boca Raton, Florida.

Matthews, R.E.F. (1985) Viral taxonomy for the nonvirologist. *Annual Review of Microbiology*, **39**, 451–74.

Mayr, E. (1982) *The Growth of Biological Thought. Diversity, Evolution, and Inheritance*. Harvard University Press, Cambridge, Massachusetts.

Milne, R.G. (1984) The species problem in plant virology. *Microbiological Science*, **1**, 113–22.

Milne, R.G. (1985) Alternatives to the species concept for virus taxonomy. *Intervirology*, **24**, 94–8.

Milne, R.G. (1988) Species concept should not be universally applied to virus taxonomy – but what to do instead? *Intervirology*, **29**, 254–9.

Murphy, F.A., Fauquet, C.M., Bishop, D.H.L., Ghabrial, S.A., Jarvis, A.W., Martelli, G.P., Mayo, M.A. and Summers, M.D. (eds) (1995) *Virus Taxonomy, Sixth Report of the International Committee on Taxonomy of Viruses*. Springer-Verlag, Vienna.

Pringle, C.R. (1991) The 20th meeting of the executive committee of ICTV. Virus species, higher taxa, a universal database and other matters. *Archives of Virology*, **119**, 303–4.

Quine, W.V. (1987) Classes versus properties, in *Quiddities*, Belknap Press of Harvard University Press, Cambridge, pp. 22–24.

Sattler, R. (1986) Biophilosophy. *Analytic and Holistic Perspectives*. Springer, Berlin.

Shukla, D.D., Ward, C.W. and Brunt, A.A. (eds) (1994) *The Potyviridae*, CAB International, Cambridge, UK.

Van Regenmortel, M.H.V. (1989) Applying the species concept to plant viruses. *Archives of Virology*, **104**, 1–17.

Van Regenmortel, M.H.V. (1990) Virus species, a much overlooked but essential concept in virus classification. *Intervirology*, **31**, 241–54.

Van Regenmortel, M.H.V. (1992) What is a virus? in *Potyvirus Taxonomy* (ed. O.W. Barnett), *Archives of Virology*, Suppl. **5**, Springer-Verlag, Vienna, pp. 47–53.

Van Regenmortel, M.H.V., Maniloff, J. and Calisher, C. (1991) The concept of virus species. *Archives of Virology*, **120**, 313–14.

Zadeh, L.A. (1965) Fuzzy sets. *Information and Control*, **8**, 338–53.

3

Towards a practical species concept for cultivable bacteria

M. Goodfellow, G. P. Manfio and J. Chun

Contacting address: Department of Microbiology, The Medical School, Framlington Place, Newcastle upon Tyne NE2 4HH, UK

ABSTRACT

The basic unit in bacteria systematics has long been recognized as the species. However, despite this, there is still no universally accepted definition of species in bacteriology. The traditional view is that bacterial species can be distinguished by correlated phenotypic characters and, as such, members of a given species have a combination of characters peculiar to it. In practice, the number of such phenotypic species in a genus is influenced by the aims of the taxonomist, the extent to which the taxon has been studied, the criteria adopted to define the species and the ease by which strains can be brought into pure culture. The phenotypic species concept has been useful in practice but has severe limitations.

Bacterial species can now be defined in molecular terms. Indeed, DNA : DNA relatedness is often seen as the gold standard for the circumscription of bacterial species. This method is attractive as it can be applied to all prokaryotes, irrespective of their growth requirements. Although the exact level below which organisms are considered to belong to different species varies, extensive studies with the family Enterobacteriaceae and related taxa have led to the recommendation that genomic species should encompass strains with approximately 70% or more DNA : DNA relatedness with a difference of 5°C or less in thermal stability. 16S ribosomal RNA is now routinely used to highlight novel – that is, previously undescribed – species, but it is not always possible to detect diverged species in this way.

It is now becoming increasingly accepted that the integrated use of genotypic and phenotypic characteristics – that is, polyphasic taxonomy – is necessary for the delineation of bacterial taxa, including species. This polyphasic species concept will be considered with reference to suitable examples of organisms of medical and industrial importance.

Species: The Units of Biodiversity. Edited by M.F. Claridge, H.A. Dawah and M.R. Wilson. Published in 1997 by Chapman & Hall. ISBN 0 412 63120 2

3.1 INTRODUCTION

It is widely recognized that the species is the basic unit in biological classification. This emphasis on the species raises the intriguing problem of the nature and comparability of species across the whole range of biological diversity. The most widely accepted species concept is the biological species which can be considered as an interbreeding or potentially interbreeding community of populations. A major problem with the biological species concept is that it is not universally applicable, even in sexual organisms! It is, for example, not possible to test members of all possible pairs of species for their ability to interbreed. The small size of bacteria (including Archaea), their mainly asexual reproductive behaviour and the dearth of knowledge on the genetics of bacterial populations necessitate a pragmatic approach to the bacterial species concept. More generally, the term species implies distinctness between organisms, an approach which encompasses all organisms irrespective of whether they are classified as Archaea, Bacteria or Eukarya.

Little is known about the taxonomic structure of bacteria (Sneath, 1985). It is generally accepted that the vast majority of strains fall into distinct phenetic clusters separated by definite gaps. However, some (Cowan, 1955, 1962) consider that bacteria might instead form a continuous spectrum. This may prove to be the case with some groups though it seems highly unlikely that there are no gaps at all as this would mean that all possible combinations of properties would occur among bacteria.

Recent attempts to define bacterial species have tended to reflect the methods used to classify individual strains. This is highlighted by the dramatic impact which modern taxonomic methods have had on the ways in which bacteria are classified (Stackebrandt and Goodfellow, 1991; Goodfellow and O'Donnell, 1993, 1994). Technique-driven approaches to the circumscription of bacterial species are reasonably sound in an operational sense, but they overlook the fact that species are the product of biological processes. It is the pattern of distinctive properties shown by bacteria not the process which gave rise to them which is currently seen to be paramount in bacterial systematics.

The development of a universally accepted species concept for bacteria is proving to be a formidable task. In practice, bacterial species are usually taken to be groups of strains which individually show high levels of biochemical, genetical, morphological, nutritional and structural similarity. This somewhat ill-defined operational species concept is widely applied in bacteriology. It is, for example, used by diagnostic bacteriologists to tease out the complex taxonomy of new and emerging pathogens (McNeil and Brown, 1994), by industrial microbiologists searching for novel, commercially significant microbial products (Bull *et al.*, 1992), by molecular ecologists monitoring the impact of genetically manipulated bacteria released into the environment (Edwards, 1993), by soil microbiologists trying to establish relationships between microbial diversity and sustainable land management (Hawksworth, 1991), by molecular biologists engaged in reconstructing bacterial evolution (Woese, 1987, 1992), and by bacterial systematists intent on unravelling the extent of bacterial diversity in natural habitats (O'Donnell *et al.*, 1994).

This contribution is designed to demonstrate how recent developments in bacterial systematics are being used to provide an improved operational species concept for cultivable bacteria. However, it must be remembered that the number of bacterial species known and described represents only a tiny fraction of the estimated species diversity (Bull *et al.*, 1992; Embley and Stackebrandt, 1997: Chapter 4).

3.2 EARLY SPECIES CONCEPTS

The species concept is still a difficult and controversial issue in bacterial systematics (Cowan, 1955, 1959, 1962; Ravin 1960; Gordon, 1978; Goodfellow and O'Donnell, 1993). Early definitions of bacterial species were often based on monothetic groups described by subjectively selected sets of phenotypic properties. This species concept had severe limitations as strains which varied in key characters could not be accommodated in existing taxonomies. Moreover, such classifications often lacked uniformity as different criteria were frequently used by different investigators for the same group of organisms. Another consequence of this approach was that the rank of species became very unevenly defined over the whole range of bacterial variation. The legacy of this approach is still apparent. Typical examples are the current treatments of the families Bacillaceae and Enterobacteriaceae with organisms in the first group underdifferentiated (Rainey *et al.*, 1993; White *et al.*, 1993), whereas in the latter taxon different generic descriptions are preserved for bacteria related at the species level (Palleroni, 1993).

It is evident from the early literature that the number of species in a genus was influenced by the aims of the taxonomist, the extent to which the taxon had been studied, the criteria adopted to define the species and the ease by which the strains could be brought into pure culture (Williams *et al.*, 1984). Some idea of the variations in the number of species within bacterial genera can be gleaned from Table 3.1. In general, members of ecologically and medically important genera have been underclassified and those in industrially significant taxa overclassified. This latter point is well illustrated by reference to the genus *Streptomyces*. Over 3000 species have been named, mainly in the patent literature, on the basis of their capacity to produce novel secondary metabolites (Trejo, 1970). However, Pridham and Tresner (1974) only recognized 463 *Streptomyces* species using a limited number of standardized phenotypic tests.

Table 3.1 Number of species in selected bacterial genera

Genus	*Bergey's Manual*	*Approved lists*	
	1974	*1980*	*1995*
A. Medically important bacteria			
Actinomyces	5	6	20
Aeromonas	3	3	15
Campylobacter	3	4	15(9)*
Clostridium	61	75	123(12)
Helicobacter	–	–	12
Legionella	–	1	34(6)
Mycobacterium	29	41	70
Mycoplasma	36	57	98(6)
Nocardia	23	20	15(11)
Staphylococcus	3	13	33
Streptococcus	21	27	42(16)
Vibrio	5	9	35(3)
B. Ecologically important bacteria			
Azospirillum	–	2	5
Azotobacter	6	6	6
Beijerinckia	4	4	4
Bradyrhizobium	–	–	3
Curtobacterium	6	6	6
Nitrobacter	1	1	1
Phyllobacterium	–	–	2
Rhizobium	6	6	11(3)
Rhodospirillium	5	5	6
C. Industrially important bacteria			
Actinoplanes	6	9	21(1)
Amycolatopsis	–	–	9
Bacillus	48	31	82(14)
Lactobacillus	27	35	60(11)
Micromonospora	16	12	15
Rhodococcus	–	10	11(11)
Saccharomonospora	–	1	4
Streptomyces	463	337	463(3)
Streptosporangium	11	13	13
Thermoactinomyces	4	5	8

* The numbers in parentheses refer to the number of species transferred to other genera.

Cowan (1968) recognized the subjective nature of the traditional species concept when he whimsically described a species as 'a group of organisms defined more or less subjectively by the criteria chosen by the taxonomist to show to best advantage and as far as possible put into practice his individual concept of what a species is'. Other attempts to define bacterial species were also descriptive. Stanier *et al.* (1986), for example, stated that, 'a species consists of an assemblage of clonal populations that share a high degree of phenotypic similarity coupled with an appreciable dissimilarity from other assemblages of the same general kind'. The subjective nature of the species concept led not just to the circumscription of phenotypic species but also to the classification of taxonomists into lumpers and splitters! Lumpers tend to emphasize the similarities between strains, thereby recognizing relatively few taxa, whereas splitters highlight differences in the belief that the clarity inherent in small groups is paramount.

The limitations of the classical approach to circumscribing phenotypic species are readily exemplified. Thus, the sugar tests used to separate *Bacillus circulans* from other facultative bacilli are of no value in distinguishing between other members of the genus, such as *Bacillus sphaericus* which have an oxidative metabolism and as such cannot metabolize sugars (Priest, 1993). This particular approach to the delineation of *Bacillus* species was intrinsically flawed as most sugar-utilizing strains formed homogeneous species whereas their oxidative counterparts belonged to genetically diverse groups.

Buchanan (1955), a firm believer in the nomenclatural type concept, considered that 'A bacterial species may (then) be defined as the type culture together with such other cultures or strains of bacteria as are accepted by bacteriologists as sufficiently closely related'. This definition acknowledges the fact that the type strain of a species need not be the most characteristic strain of that species. Gordon (1967, 1978) took a very different view; she considered that species descriptions should be based on type strains, other authentic reference strains, fresh isolates, and old stock cultures and their variants.

Three very practical kinds of species were recognized by Ravin (1963): namely genospecies, which encompass mutually interfertile forms and correspond most closely to the biological species concept; nomenspecies, made up of individuals resembling the nomenclatural type strain; and taxospecies, groups of strains which share a high proportion of common properties. Genospecies should not be confused with genomic species, that is, with organisms which share high DNA relatedness values.

The genetic concept of species based on fertility has contributed little towards a more precise definition of bacterial species (Ravin, 1960; Jones, 1989). Genetic material can pass between bacteria in two ways: (i) vertically, from mother to daughter cell during asexual reproduction by binary fission; and (ii) horizontally between cells through the main genetic exchange mechanisms, transduction, transformation and conjugation. Populations in which vertical transmission predominates are clonal. Thus, as bacteria divide and spread, new alleles derived from mutations are restricted to the direct descendants of the cell in which they arose and cannot reassort with alleles of other genes into novel genetic combinations. In contrast, horizontal genetic exchange provides a mechanism for sexual transmission of genetical material allowing the reassortment of mutant alleles among strains and thereby promoting the genetic diversity of the species (Maynard Smith *et al.*, 1991; Maynard Smith, 1995). This process works against the natural tendency of an asexual population to be clonal. Consequently, there is a spectrum of population structures that bacterial populations can exhibit ranging from highly clonal to panmictic (Maynard Smith *et al.*, 1993). Evidence of genetic isolation between members of closely related taxa has been observed (Maynard Smith *et al.*, 1991; Roberts *et al.*, 1994) and could be due to physical barriers to gene exchange, codon usage patterns and host restriction/modification mechanisms.

In general, genetic exchange plays a limited role as a force of cohesion among prokaryotes (Cohan, 1994). Horizontal gene transfer has mainly been recorded for members of genera which have a special capacity for the uptake and incorporation of chromosomal DNA (Maynard Smith *et al.*, 1991; Maynard Smith 1995). Such naturally transformable bacteria are found among both Gram-negative (e.g. *Haemophilus* and *Neisseria*) and Gram-positive bacteria (e.g. *Bacillus* and *Streptococcus*). Bacteria differing in DNA sequence by up to 25% can and do exchange chromosomal DNA. The exchange is usually local, often involving only a few hundred base pairs, and hence does not destroy clonal structures. It is possible that most of the genes which are exchanged frequently confer some immediate adaptive advantage, such as those encoding cell-surface proteins or restriction–modification systems, in contrast to 'housekeeping' genes, such as those encoding enzymes studied by multilocus enzyme electrophoresis (Selander *et al.*, 1994).

Some phages and plasmids mediate transfer between closely related strains (Jones, 1989; Harwood, 1993). This might appear to provide a basis on which to develop a universal species concept for bacteria but other phages and plasmids have a broad host range (Holloway, 1993). Promiscuous plasmids have the potential to transfer gene operons across widely divergent taxa, even across phyla. The biological basis of such extensive host ranges is not known, though some plasmids appear to have developed complex and highly flexible replicative systems. However, wide host-range plasmids do not necessarily display conjugal competence in members of diverse species.

It is perhaps not surprising given the problems outlined above that little interest has been shown recently in the bacterial species concept. In practice, species level taxonomy has been based implicitly or explicitly on the detection of phenotypic discontinuities. The species concept was never formally abandoned but it would appear to have become a concept that dare not speak its name!

3.3 THE NEW BACTERIAL SYSTEMATICS

The current renaissance in bacterial systematics can be traced to the introduction and application of new concepts and methods, notably to the emergence of numerical taxonomy (Sokal and Sneath, 1963), chemosystematics (Goodfellow and Minnikin, 1985) and molecular systematics (Stackebrandt and Goodfellow, 1991). Views on how bacterial species should be circumscribed have radically changed in light of these developments.

3.3.1 Numerical taxonomy

In the late 1950s, the call for a more objective taxonomy led to the development of numerical taxonomy (Sokal and Sneath, 1963; Sneath and Sokal, 1973). As the name implies, numerical taxonomy consists of applying various mathematical procedures to numerically encoded character state data; organisms are assigned to groups on the basis of overall similarity. In the 1960s and early 1970s, bacterial classifications, considered almost entirely in phenetic terms, were generated from numerical analyses of biochemical, nutritional and physiological data.

The primary objective of early numerical taxonomic studies was to assign individual bacterial strains to homogeneous groups or clusters, which could be equated with taxospecies, using large sets of phenotypic data. The resultant quantitative data on numerically defined taxospecies were used to design improved identification schemes. When introduced, the numerical taxonomic procedure was in sharp contrast to the prevailing orthodoxy as species were recognized using many equally weighted features, not by a few subjectively chosen morphological, staining and behavioural properties. The theoretical basis of numerical taxonomy is well documented (Sneath and Sokal, 1973; Goodfellow *et al.*, 1985; Sackin and Jones, 1993); the main steps involved in numerical classification and identification are outlined in Figure 3.1. Numerical taxonomic data are usually stored and managed using computer systems given the widespread availability of specialized software (Canhos *et al.*, 1993; Sackin and Jones, 1993) and the need to study large numbers of strains and properties.

Numerical taxonomic procedures have been applied to most groups of cultivable bacteria (Goodfellow and Dickinson, 1985; Sackin and Jones, 1993) both to revise existing classifications and to classify unknown strains isolated from diverse habitats. It is evident from these studies that numerical taxonomic procedures are effective in delineating taxospecies. The method has been less successful in the construction of higher taxonomic

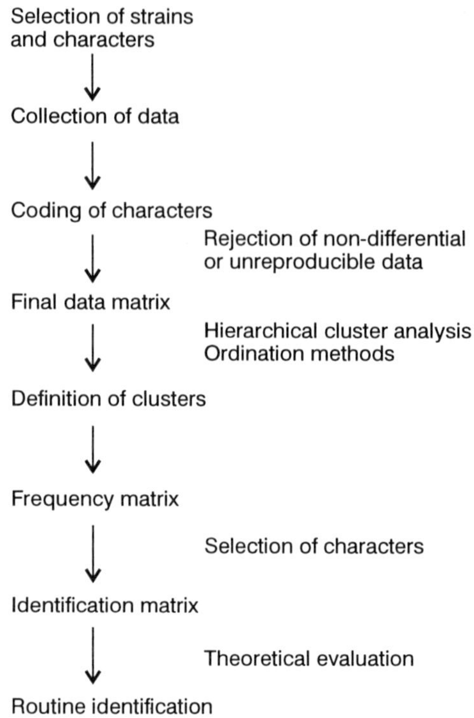

Figure 3.1 Major steps in numerical classification and identification.

ranks, but this is almost certainly due to the types of data used rather than to fundamental flaws in numerical methods. Thus, representative strains from diverse genera may have different metabolisms and growth requirements which can make studies across generic boundaries difficult. Numerical taxonomic surveys have been used to circumscribe many taxospecies, including those encompassed in taxonomically complex taxa such as *Bacillus* (White *et al.*, 1993; Nielson *et al.*, 1995), *Mycobacterium* (Wayne, 1985), *Pseudomonas* (Strenstöm *et al.*, 1990) and *Streptomyces* (Kämpfer *et al.*, 1991).

Taxonomic clusters or taxospecies are 'operator unbiased' representations of natural relationships between strains though group composition may be influenced by the choice of strains and tests, experimental procedures, test error and statistics (Sackin and Jones, 1993). It is, therefore, essential to evaluate the taxonomic integrity of taxospecies by examining representative strains using independent taxonomic criteria derived from the application of chemotaxonomic and molecular systematic methods.

There is evidence that Curie-point pyrolysis mass spectrometry provides a quick and effective way of evaluating the taxonomic status of taxospecies circumscribed in numerical phenetic surveys (Goodfellow *et al.*, 1994a).

Comprehensive databases – one of the end-products of numerical taxonomy – contain extensive information on the biochemical, nutritional, physiological and tolerance properties of test strains. These data can readily be arranged into tables which list the percentage of strains in each taxospecies that are positive for each unit character (percentage positive tables). These data can be used for several purposes, notably:

- To construct frequency matrices for the identification of unknown bacteria.
- To design media for the selective isolation of target organisms from natural habitats.
- To choose representative strains for additional taxonomic studies.

A frequency matrix is a reduced version of a percentage positive matrix in which a combination of unit characters is selected for the identification of unknown isolates to taxospecies. The selection of an optimal combination of features is achieved using intuitive mathematical and statistical routines (Sneath, 1979a,b, 1980a,b,c). The practical and theoretical developments in computer-assisted numerical identification have been described in detail elsewhere (Pankhurst, 1991; Priest and Williams, 1993).

The polythetic nature of probabilistic identification matrices gives them several advantages over conventional identification keys and diagnostic tables as no single property is either sufficient or necessary for the identification of a strain to a previously defined group. Frequency matrices are also theoretically robust as they can be used to accommodate natural variation in test results presented by bacteria isolated from diverse sources. Theoretically sound and practically useful frequency matrices are available for the identification of industrially and medically important taxospecies (Bryant, 1993; Canhos *et al.*, 1993), including campylobacters (On *et al.*, 1996), slowly growing mycobacteria (Wayne *et al.*, 1984), acidophilic actinomycetes (Seong *et al.*, 1995) and neutrophilic streptomycetes (Kämpfer and Kroppenstedt, 1991). Some probabilisitic identification matrices can be accessed through the internet (Canhos *et al.*, 1993; http://www.bdt.org.br/cgl-bin/msdn/matrices). Many commercial diagnostic kits and automated instruments used for the identification of unknown pathogenic bacteria are based on numerical taxonomic methods.

Commercially available automated bacterial identification systems are now available for the routine identification of isolates in the laboratory and for the construction of databases (Mauchline and Keevil, 1991). Identification tests are assembled in microtitre plates or disposable kits and results read and collected automatically by a plate reader connected to a microcomputer or visually by the operator (Bochner, 1989). Rapid automated identification systems are used for rapid and reliable identification of clinical isolates, but few systems are available for the identification of environmentally important strains (Klinger *et al.*, 1992).

The wealth of information held in numerical taxonomic databases can also be used to determine the nutritional and tolerance limits of members of individual taxospecies in order to devise media formulations selective for one or more taxospecies. This taxonomic approach to selective isolation has been used to isolate target and novel streptomycete species of potential industrial importance from soils (Williams *et al.*, 1984; Williams and Vickers, 1988; Goodfellow *et al.*, 1994a). These studies clearly show that there is no such thing as a 'general' isolation medium for streptomycetes; several selective media must be used to gain a more accurate picture of the qualitative nature of bacterial populations in environmental samples. The taxonomic approach to selective isolation has a crucial part to play in establishing the extent of bacterial variation in natural habitats (Bull *et al.*, 1992).

It is evident that the ability to delineate taxospecies has had a profound influence in bacteriology, notably for the identification of pathogenic bacteria. Indeed, the strengths of the numerical taxonomic procedure far outweigh its limitations (Table 3.2). Any tendency to see numerical taxonomy as a method with a long past and an uncertain future should be resisted. Improved methods and automated data acquisition systems will facilitate the generation of high-quality phenotypic databases for a variety of purposes. It can, for example, be anticipated that with the developing interest in bacterial species diversity, these databases, which often reflect the functional diversity of a habitat, will be put to other more fundamental uses.

3.3.2 Molecular systematics

The motive force in contemporary bacterial systematics owes much to developments in molecular biology. These advances underpin an impressive array of methods which can be used to derive taxonomically useful information from analyses of nucleic acids. The estimation of the mean overall base composition of DNA, indirect comparisons of nucleotide sequences by DNA : DNA pairing and sequence analyses of conserved homologous genes, notably 16S rRNA, all provide grist for the circumscription of bacterial species.

(a) DNA base composition

In bacteria, DNA base composition values range from ~ 25 to 80 mol% GC (Tamaoka, 1994). The DNA base composition of members of well circum-

Table 3.2 Advantages and disadvantages of numerical classification and identification

Advantages	
	Circumscription of taxospecies
	Provision of simple phenotypic properties for identification
	Formulation of taxon-specific selective media
	Choice of representative strains for additional studies
Disadvantages	
	Collection of data time-consuming and laborious
	Some diagnostic tests are difficult to read
	Little value for classification above the genus level
Needed	
	Automated data acquisition systems

scribed taxospecies does not usually differ by more than 3 mol% GC, whereas members of species within a genus should not usually differ from one another by more than about 10 mol% GC. Firm guidelines have yet to be set for the range of DNA base compositions that can be encompassed at these taxonomic ranks but a range of > 15 mol% GC can be taken to indicate heterogeneity within a genus. *Bacillus* (33–64 mol% GC), *Clostridium* (22–55 mol% GC) and *Flavobacterium* (31–68 mol% GC), as defined in the current edition of *Bergey's Manual of Systematic Bacteriology* (Krieg and Holt, 1984; Sneath *et al.*, 1986), are all examples of heterogeneous taxa.

DNA base composition data need to be interpreted with care as the choice of analytical method and experimental conditions influence results (Tamaoka, 1994). The sensitivity and reproducibility of the thermal melting point method provide sufficiently good data for taxonomic purposes though discrepancies occur between different laboratories. The high-pressure liquid chromatography (HPLC) method, which is more accurate, should be adopted when DNA base composition data are used to determine hybridization conditions for DNA : DNA relatedness studies (Kusunoki and Ezaki, 1992).

(b)DNA : DNA pairing

A unique property of DNA and RNA is the ability for reassociation or hybridization . The complementary strands of DNA, once denatured, can, under appropriate experimental conditions, reassociate to reform native duplex structures. The specific pairings are between the base pairs, adenine with thymine and guanine with cytosine, and the overall pairing of the nucleic acid fragments is dependent upon similar linear arrangements of these bases along the DNA. When comparing nucleic acids from different organisms, the amount of molecular hybrid formed and its thermal stability provide an average measurement of nucleotide sequence similarity.

Since the 1970s information from DNA : DNA pairing studies has been used extensively to delineate bacterial species (Krieg, 1988; Stackebrandt and Goebel, 1994). An important practical advantage of DNA : DNA pairing was that the continua occasionally found between phenotypically defined groups were usually resolved as organisms tended to be either closely related or not. The rationale of using DNA : DNA pairing as the gold standard for the delineation of bacterial species is based on the results of numerous investigations where good agreement was found between DNA relatedness values and corresponding results based on numerical phenetic and chemotaxonomic data (Nielsen *et al.*, 1995).

The term genomic species is applied to strains which show DNA : DNA relatedness values greater than some specified figure and thermal denaturation values less than some specific rating. Extensive studies with members of the family Enterobacteriaceae and related taxa led to the recommendation that genomic species should encompass strains with approximately 70% or more DNA : DNA relatedness with a difference of 5°C or less in thermal stability (ΔT_m; Wayne *et al.*, 1987). A definition based on these values has been used to clarify species relationships in many bacterial genera. Genetically closely related bacteria that diverge in phenotype can be designated subspecies. Members of related subspecies show approximately 70–85% DNA : DNA relatedness and ΔT_m values below 3%. Values from 30% to 70% reflect a moderate degree of relationship, but values become increasingly unreliable once they fall below the 30% level as they can be attributed to experimental artefacts.

Caution is sometimes needed in reconciling classifications at the species level based on phenotypic and DNA : DNA relatedness data. The ideal situation is where an equivalence is found between genomic species and taxospecies. DNA : DNA hybridization data have usually confirmed the taxonomic status of phenotypically defined species, thereby showing that the phenotypic characters used to define them were predictive. In addition, many newly described species have been proposed using the genomic group as a criterion. However, there are several examples of taxospecies which show marked genetic diversity. For example, in *Xanthomonas*, hybridization values range from 0 to 100% between pathovars (Hildebrandt *et al.*, 1990) known to be indistinguishable using biochemical tests (Van den Mooter and Swings, 1990). Similar incongruities have been reported with *Bacillus* (Priest, 1993; Nielsen *et al.*, 1995), *Flavobacterium* (Ursing and Bruun, 1991) and *Pseudomonas* species (Rosselló *et al.*, 1991). In the first instance, representatives of such taxospecies need to be examined to find new phenotypic features for distinguishing between the constituent genomic species. This strategy has been applied successfully to unravelling the classification of organisms classified as *Bacillus circulans* (Nakamura, 1984) and *Bacillus sphaericus* (Alexander and Priest, 1990). Thus, strains previously known as *Bacillus circulans* were assigned to five species, *B. amylolyticus, B. circulans sensu stricto, B. lautus, B. pabuli* and *B. validus,* when conclusions drawn from DNA relatedness experiments were reinforced by phenotypic features.

The genomic species concept may be difficult to apply where there is extensive overlap between DNA relatedness values. This seems to be the case with members of the genus *Xanthomonas* (Hildebrandt *et al.*, 1990) and the family Enterobacteriaceae (Gavini *et al.*, 1989). It is possible that chromosomal rearrangements may result in changes in phenotype without influencing corresponding DNA : DNA pairing values (Egel *et al.*, 1991). Similarly, chromosomal rearrangement might change relatedness values without affecting phenotypic expression (Krawiec, 1985). Such possibilities underline the need for prudence in setting arbitrary relatedness values to define species.

DNA : DNA pairing studies are hampered by technical difficulties and results impaired by experimental error (Sneath, 1983; Johnson, 1991a). Studies employing complete matrices of DNA : DNA relatedness values are the exception rather than the rule. The usual practice is to take a few organisms as reference strains, and to compare all of the other strains against this restricted set. The challenge is to recover a taxonomic structure that is in accordance with the position which would have been reached from the analysis of a complete matrix. This is not straightforward for it has been shown (Sneath, 1983; Hartford and Sneath, 1988) that information on the underlying taxonomic structure is necessarily lost when a small number of reference strains are studied. It is important that reference strains are widely spaced and representative of the constituent species. Such problems together with the diversity of available experimental procedures (Figure 3.2) emphasize the need for a critical appraisal of all DNA : DNA pairing data.

It is not always appreciated that DNA : DNA relatedness values do not reflect the actual degree of sequence similarity at the level of primary structure. Thermal stabilities of heteroduplexes have been shown to fall by 1 to 1.5°C for every 1% of unpaired bases (Britten and Kohne, 1968). This means that in bacteria DNA heteroduplexes will not be formed, even under non-stringent conditions, unless the strains show DNA relatedness values of more than 50 to 70%. It has been estimated that organisms which have 70% or more DNA similarity will also have at least 96% DNA sequence identity (Stackebrandt and Goebel, 1994). These authors also noted that significant differences in the phenotype of *Escherichia coli* strains could be anticipated given a 2% difference in the nucleotide base sequence of the genome. It seems unlikely that the primary structure of most of the genes of this

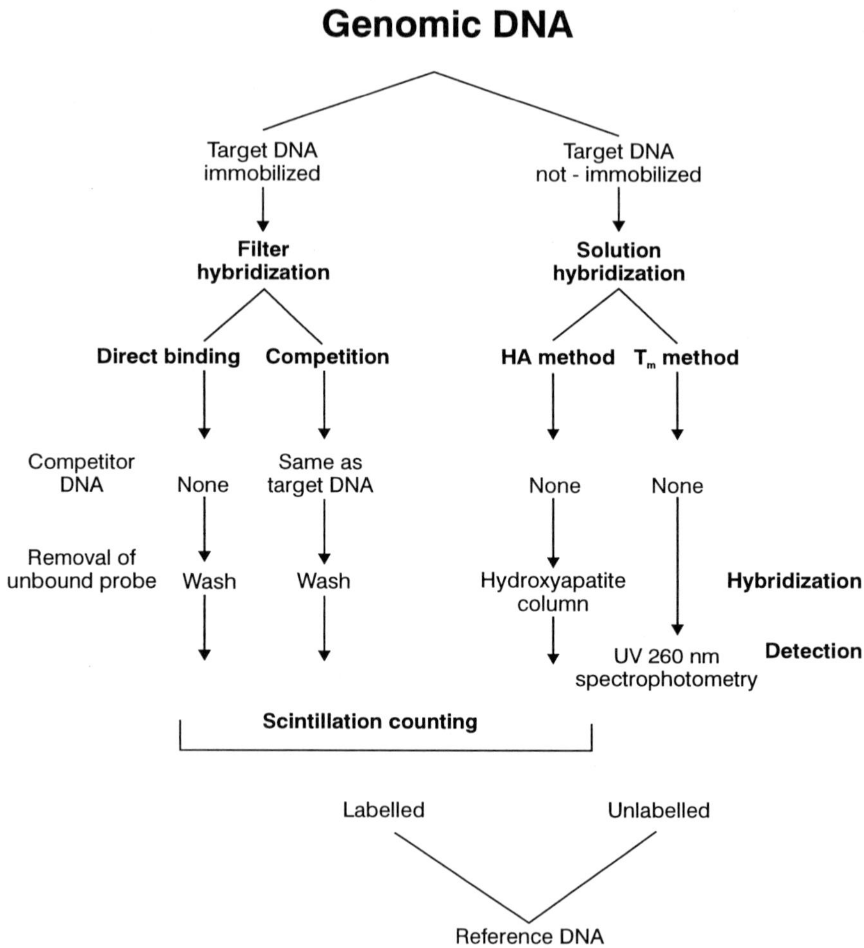

Figure 3.2 Schematic representation of some of the different DNA reassociation assays used to determine DNA sequence similarity. HA, hydroxyapatite; Tm, thermal melting.

organism are affected by genomic rearrangements caused by horizontal gene transfer or by the presence of mobile elements.

Despite the problems, the advantages of DNA : DNA pairing analyses outweigh their limitations. The method is still seen as the final arbiter for delineating species (Wayne *et al.*, 1987; Stackebrandt and Goebel, 1994) and is attractive as it can be applied to all prokaryotes irrespective of their growth requirements. In particular, DNA : DNA pairing studies provide:

- A unified concept for the delineation of bacterial species.
- Confirmation of the taxonomic integrity of phenotypically defined species.
- Detection of taxospecies that accommodate substantial amounts of genetic diversity.
- Recognition that species assigned to the same genus (or even different genera) belong to a single genomic species.

DNA : DNA pairing studies can also be used to detect and identify unknown isolates. However, classical DNA pairing techniques are too finicky and time-consuming to apply to large numbers of strains, since DNA must be extracted and purified from each of the test strains before the pairing experiment (Johnson, 1991b). However, recent studies show that blotting and lysis of whole organisms on membrane filters followed by pairing with labelled DNA from reference strains (chromosomal probes) provides a rapid means of identifying isolates, notably those of medical importance (Schleifer *et al.*, 1993).

3.3.3 Nucleotide sequence analysis

It is well known that the sequences of the genes that code for ribosomal RNA are relatively highly conserved and hence can be used to determine taxonomic relationships between species which show little DNA relatedness. It is becoming increasingly popular to propose new bacterial species using data generated from 16S rRNA sequencing studies (Table 3.3). Two basic assumptions underline this approach, namely that lateral gene transfer has not occurred between 16S rRNA genes, and that the amount of evolution or dissimilarity between 16S rRNA sequences of a given pair of organisms is representative of the variation shown by the corresponding genomes. The good congruence found between phylogenies based on 16S rRNA sequences and those derived from alternative molecules, such as 23S rRNA (Ludwig *et al.*, 1993a), ATPase subunits (Ludwig *et al.*, 1993b), elongation factors (Ludwig *et al.*, 1993b) and RNA polymerases (Zillig *et al.*, 1989), give substance to this latter point. It also seems likely that lateral gene transfer between 16S rRNA genes is rare as this gene is responsible for the maintenance of functional and tertiary structural consistency. However, the possibility of horizontal gene transfer in 16S rDNA should not be overlooked (Sneath, 1993).

Fox *et al.* (1992) pointed out that 16S rRNA molecules from members of closely related species may be so conserved that they cannot be used to differentiate between strains at the species level. This important observation means that strains of related species with identical, or almost identical, 16S rRNA nucleotide sequences may belong to different genomic species. This is the case with species of *Aeromonas* (Martinez-Murcia *et al.*,

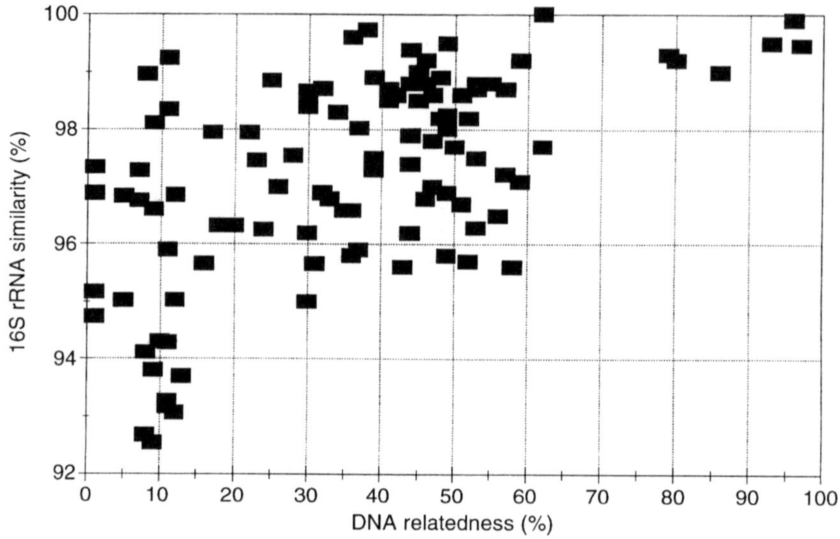

Figure 3.3 Comparison of 16S rRNA and DNA relatedness values of mycolic acid-containing actinomycetes. (Based on data taken from molecular systematic studies on mycolic acid-containing organisms; Chun, 1995.)

Table 3.3 New species described within existing genera or recommended as synonyms in the *International Journal of Systematic Bacteriology*

		Techniques employed		
Year	*DNA:DNA hybridization*	*Serological tests*	*16S rRNA sequencing*	*Others*
1987*	60	10	0	30
1993*	75	8	14	3
1995	68	7	90	3

* Data from Stackebrandt and Goebel (1994).

1992), *Bacillus* (Ash *et al.*, 1991) and *Legionella* (Fry *et al.*, 1991). It is clear from these observations that the resolution of DNA hybridization is higher than that of 16S rRNA sequence analysis and that DNA : DNA pairing remains the method of choice for measuring the degree of relatedness between closely related organisms. Nevertheless, even in this context 16S rRNA sequence data can be used to select appropriate reference strains for the more exacting DNA : DNA pairing studies, thereby reducing the number of reference strains which need to be examined. The terms rRNA species complex and rRNA superspecies have been proposed for organisms which have virtually identical 16S rRNA sequences but can be distinguished using DNA : DNA relatedness data (Fox *et al.*, 1992).

The correlation between 16S rRNA sequence and DNA relatedness data is not linear (Figure 3.3) though rRNA similarity values below 97% invariably correspond to DNA relatedness values below 60%. Similar findings were reported by Stackebrandt and Goebel (1994) who argued that genomic species *sensu* Wayne *et al.* (1987) usually have more than 97% sequence identity. This cut-off point is plausible given the results shown in Figure 3.4.

16S rRNA sequencing analyses are easier and more cost-effective than DNA hybridization studies due to developments in molecular biology, notably the use of the polymerase chain reaction (PCR) and the availability of automatic DNA sequencers (Figure 3.5). It is well known that the 3% or 45 nucleotide sequence differences that can be used to distinguish most species are not evenly scattered along the primary structure of the 16S rRNA macromolecule but tend to be concentrated in hypervariable regions. There is evidence that the hypervariable regions can be taxon-specific (Stackebrandt and Goebel, 1994). It is clear, therefore, that only complete 16S rRNA sequences allow reliable comparisons of novel organisms with available databases containing complete or almost complete nucleotide sequences (Canhos *et al.*, 1993). Information on rRNA sequences can be accessed through the internet (http://www.bdt.org.br/structure/molecular.html).

The taxonomic relationships of both potentially novel and poorly misclassified organisms can readily be determined by comparing their 16S rRNA sequences with corresponding results held in databases. It is, for example, evident from the example (Figure 3.6) that the unknown actinomycete isolated from activated sludge belongs to the genus *Tsukamurella* and that an organism until recently known as *Nocardia amarae* forms a distinct species in the genus *Gordona*. It is also clear that a new taxonomic niche is needed for actinomycetes classified as *Nocardia pinensis* (Chun *et al.*, 1996).

Evolutionary relationships between bacteria need to be interpreted with care as estimates of phylogeny are based on relatively simple assumptions when considered against the complexities of evolutionary processes. All methods of phylogenetic inference are based on certain assumptions that may be violated by the data to a greater or lesser extent (Swoffold and Olsen, 1990; Hillis *et al.*, 1993). Such questions were raised by O'Donnell *et al.* (1993) who also pointed out that potential problems in nucleotide sequence data include alignment artefacts, non-independence of sites, inequalities in base substitution frequencies between sequences, and lineage-dependent inequalities in rates of change.

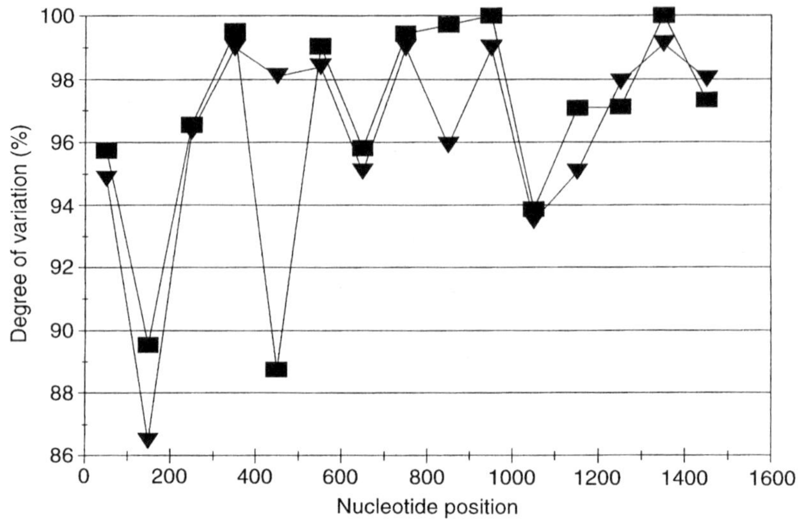

Figure 3.4 Nucleotide sequence variation in different regions of the 16S rRNA molecule. Data points correspond to the average nucleotide sequence variation calculated from aligned 16S rRNA sequences. (From Chun, 1995.)

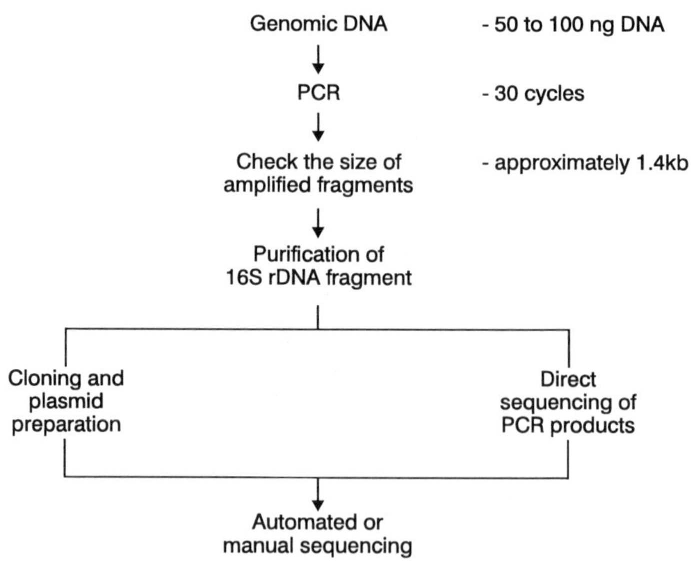

Figure 3.5 Schematic representations of 16S rRNA sequencing methods.

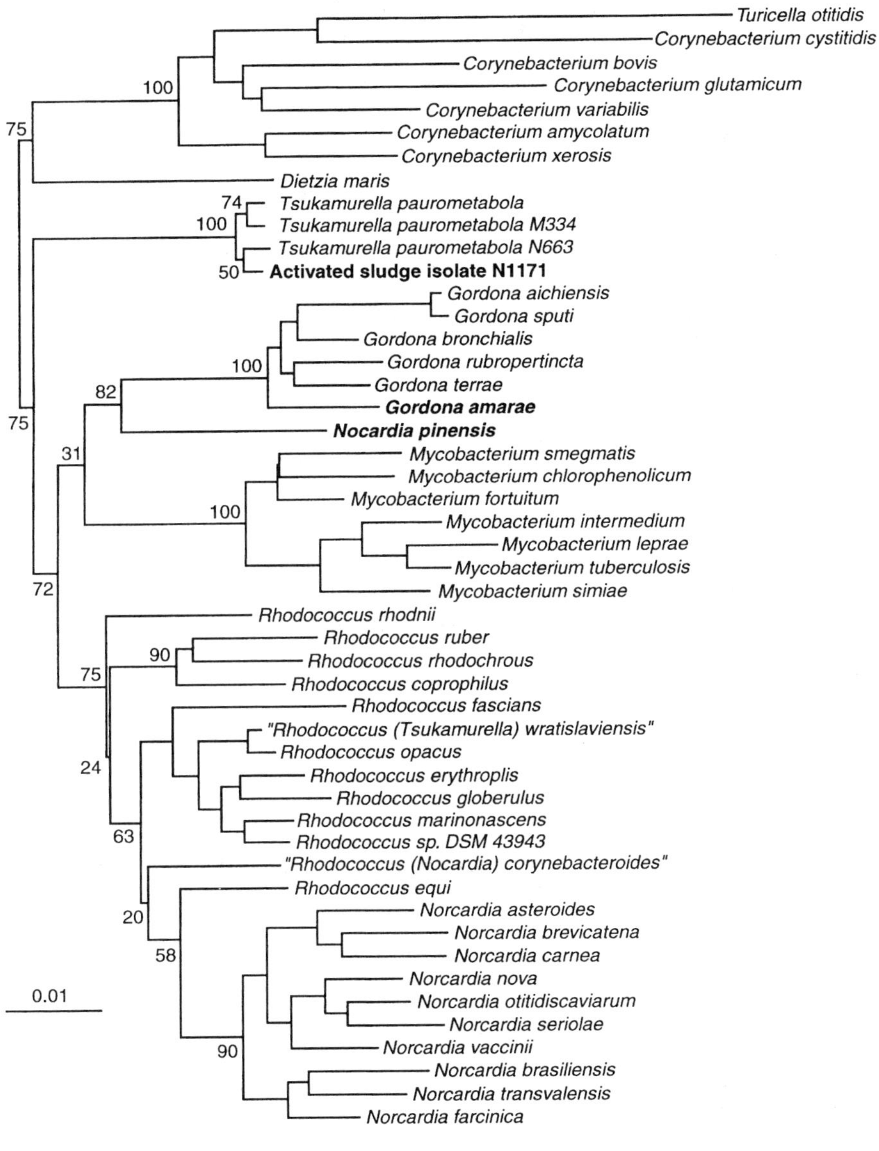

Figure 3.6 An unrooted phylogenetic tree showing relationships between isolate N1171 and other mycolic acid-containing actinomycetes. The tree was constructed by using the Jukes and Cantor distance (1969) and neighbour-joining methods (Saitou and Nei, 1987). The numbers at the nodes indicate the level of bootstrap support on 1000 resamplings. The scale bar indicates 0.01 substitutions per nucleotide position.

Problems such as these are compounded by bacterial taxonomists who regard the various data handling techniques as 'blackboxes' integrated into computer software. The literature is replete with examples of inadequate practices, as witnessed by the following examples taken from a single issue of the *International Journal of Systematic Bacteriology*:

- 'The ODEN program package was used to align the sequences, and phylogenetic distances were calculated by using both the unweighted pair group method and neighbor-joining method' (Ezaki *et al.*, 1994: 130). In this study, the authors did not say how evolutionary distances were calculated.
- 'Evolutionary distances were determined by using the neighbor-joining method' (Briglia *et al.*, 1994: 494). The method used in this study was designed for constructing trees from distance matrices not for generating distances!
- 'The dendrogram was constructed by using the software program PILEUP obtained from the Genetic Computer Group Inc'. (Robertson *et al.*, 1994: 836). The authors did not give any information on how the data were analysed.
- 'An unrooted phylogenetic tree was produced by using the DNADIST and FITCH programs in the PHYLIP package' (Dupuy *et al.*, 1994: 461). It is not clear which distance method was used as the DNADIST program contains four different distance methods.
- A common omission was a failure to mention which distance models were used (e.g. Cai and Collins, 1994; Collins *et al.*, 1994; Jagoueix *et al.*, 1994).

There are, of course, many examples of good practice in the analysis of molecular sequence data (Chun and Goodfellow, 1995; Chun *et al.*, 1996; Kim *et al.*, 1996). Despite the blind applications of computer packages, there is no doubt that the advantages of using 16S rRNA sequencing for helping to delineate new species far outweigh the limitations. However, there can be no bacterial species definition based solely on sequence similarity of rRNAs or their genes as absolute values for delineating species cannot be set because of different rates of sequence divergence. Nevertheless, good agreement is generally found between relationships derived from 16S rRNA sequence data and those based on DNA : DNA pairing data, for example, in species of the genus *Serratia* (Dauga *et al.*, 1990).

The advantages of using 16S rRNA sequencing for the recognition of novel species far outweigh the deficiencies (Table 3.4). Further, the ability to obtain rRNA sequence data from difficult to culture and uncultured bacteria will help in the exploration of the huge diversity of prokaryotic species which await discovery (see Embley and Stackebrandt, 1997; Chapter 4). Nucleic acid sequence data can also be used to design probes for *in situ* hybridizations (Schleifer *et al.*, 1993) and thereby facilitate the development of appropriate selective isolation strategies by showing whether environmental samples contain members of target species.

3.3.4 Chemotaxonomy

Chemical data derived from the analysis of cell components can be used to classify bacteria at different taxonomic ranks according to the pattern of distribution of the different compounds within and between members of different taxa. Chemotaxonomic analyses of chemical macromolecules, particularly amino acids and peptides (e.g. from peptidoglycan and pseudomurein), lipids (lipopolysaccharides), polysaccharides and related polymers (e.g. methanochondroitin, wall sugars), proteins (e.g. bacteriochlorophyll, whole-organism protein patterns), enzymes (e.g. hydrolases, lyases), and other complex polymeric compounds, such as isoprenoid quinones and sterols, all provide valuable data for the chemotaxonomic cornucopia (Goodfellow and O'Donnell, 1994). The base composition of DNA is also a chemical property *sensu stricto* but is usually considered as a molecular feature. Chemical fingerprints of taxonomic value can be obtained using analytical chemical techniques, notably Curie-point pyrolysis mass spectrometry (Goodfellow *et al.*, 1994a). Other promising approaches which provide valuable data for delineating species include analyses of cellular fatty acids (Stead *et al.*, 1992; Vauterin *et al.*, 1996) and whole-organism proteins (Vauterin *et al.*, 1993; Verissimo *et al.*, 1996), and the elucidation of enzyme profiles based on chromogenic and fluorogenic substrates (Manafi *et al.*, 1991).

Developments in molecular systematics should not be seen as a threat to chemosystematics as the two approaches are complementary.

Table 3.4 The advantages and limitations of using 16S rRNA sequence data for the circumscription of bacterial species

Advantages
- Full sequence analysis has become rapid and inexpensive
- Provision of high-quality databases
- More objective definition of species
- Species presented within a supra-generic framework
- Nucleic acid probes for identification

Limitations
- Only near-complete sequences allow reliable comparisons with other near-complete sequences from databases
- Resolution limited when closely related organisms are compared
- Strains belonging to different species may have identical sequences
- Taxonomic relationships are affected by the choice of statistical methods

Phylogenetic data provide a hierarchic framework of relationships among bacterial species but do not give reliable information for the delineation of taxa above the species level. In contrast, chemical markers are unevenly distributed across taxa but rarely give information on the hierarchic rank of taxa. It is very encouraging that good congruence exists between the distribution of chemical markers and the relative positions of species in phylogenetic trees (Goodfellow and O'Donnell, 1994; Chun *et al.*, 1996). Chemical data are not only employed to evaluate existing phylogenies but can also be used to adjudicate between conflicting phylogenetic trees. The phylogenetic positions of activated sludge isolate N1171, *Gordona amarae* and *Nocardia pinensis* (Figure 3.6) are supported by chemotaxonomic evidence (Blackall *et al.*, 1989; Goodfellow *et al.*, 1994b; Chun, 1995).

3.3.5 Towards a new operational species concept

It is evident from the previous section that none of the individual approaches used to circumscribe bacterial species is without problems. Molecular systematic techniques, especially those based on 16S rRNA sequencing and DNA hybridization, will continue to play an important role in the revision of established classifications and in the delineation of new species. Nucleic acid sequence data can be used to generate a framework of the perceived evolutionary relationships between representatives of all bacterial taxa, but this phylogeny needs to be evaluated and refined using independent taxonomic data as phylogenetic relationships between organisms can be distorted by differences in evolutionary rates, as well as by technical and statistical problems. Further, simple phenotypic properties are still at a premium for the identification of unknown strains to validly described species.

A more reliable and comprehensive approach to the delineation of species is emerging based on the integrated use of genotypic and phenotypic data. This approach, known as polyphasic taxonomy, was introduced by Colwell (1970) to signify successive or simultaneous taxonomic studies on groups of organisms using an array of methods designed to provide genotypic and phenotypic data. Polyphasic taxonomic studies can by their nature be expected to yield well-defined species, a stable nomenclature and improved species definitions.

Polyphasic taxonomy remains in its infancy for a variety of reasons. All-embracing approaches to the circumscription of species only recently became practical given the availability of rapid data acquisition systems and improved data handling procedures. A lack of suitably trained taxonomists and a reluctance of funding agencies to support taxonomic research seen to be 'non-molecular' have compounded the situation, despite significant developments in chemo- and numerical taxonomy (Goodfellow and O'Donnell, 1993, 1994). It is, therefore, far from surprising that only limited phenetic data are available to evaluate even the major phylogenetic groups of prokaryotes (Murray *et al.*, 1990). Nevertheless, it is encouraging that most descriptions of new cultivable prokaryotes in the 1995 issues of the *International Journal of Systematic Bacteriology* are based on a judicious selection of genotypic and phenotypic criteria.

The subgeneric classification of several bacterial groups, notably aerobic, endospore-forming bacilli (White *et al.*, 1993), bartonellae (Birtles *et al.*, 1995), flavobacteria (Bernardet *et al.*, 1996), microtetrasporae (Kroppenstedt *et al.*, 1990), mycobacteria (Wayne *et al.*, 1996), oceanospirilla (Pot *et al.*, 1989) and xanthomonads (Vauterin *et al.*, 1990, 1995), have been clarified using the polyphasic taxonomic approach. Similarly, several new species, for instance, *Amycolatopsis methanolica* (De Boer *et al.*, 1990), *Burkholderia vietnamiensis* (Gillis *et al.*, 1995), *Nocardia pseudobrasiliensis* (Ruimy *et al.*, 1996), *Paenibacillus validus* (Heyndrickx *et al.*, 1995), *Rhizobium mediterraneum* (Nour *et al.*, 1995), *Rhodococcus percolatus* (Briglia *et al.*, 1996) and *Tsukamurella inchonensis* (Yassin *et al.*, 1995) have been described using a combination of genotypic and phenotypic data. Such studies, while admirable, have not addressed the question of exactly what type of procedures and analyses constitute a polyphasic taxonomy. At present, polyphasic taxonomic studies tend to reflect the particular expertise of the laboratory or of the investigators.

Ideally, new species should be described using data derived from an analysis of a representative set of strains. Sneath (1977) visualized a species as a cluster of a large number of strains in a character space. As an approximation, this space can be regarded as roughly spherical, and can thereby be described, using sampling theory, by its centre, while its radius defines an envelope that encompasses most of the strains. Sneath considered that about 25 strains were needed to define accurately the centre and radius of a cluster (taxospecies) but went on to suggest that never less than 10 strains be used for this purpose. In practice, most new species are still described on the basis of a few strains, not infrequently on a single strain!

The proposition that nomenclature should reflect genomic relationships (Wayne *et al.*, 1987) and that all pre-conceived notions be re-examined within this context (Murray *et al.*, 1990) has been widely accepted in practice. There are no problems in naming new species given congruence between the various data subsets, notably genotypic and phenotypic data. However, it has already been pointed out that some taxospecies encompass two or more genomic species. Wayne *et al.* (1987) recommended that distinct genomic species which cannot be differentiated from other genomic species using known phenotypic properties should not be formally named until they can be differentiated by suitable phenotypic markers.

This proposition, while laudable, has tended to leave taxospecies which accommodate more than one genomic species in a state of limbo. Such taxospecies can also be considered as nomenspecies.

It has been suggested that genomic species accommodated in nomenspecies should be referred to as genomovars (Rosselló *et al.*, 1991; Ursing *et al.*, 1995). This concept is useful as it indicates that genomic species are an integral part of a nomenspecies and hence should not be overlooked in subsequent taxonomic work. Once phenotypically delineated, genomovars can be given formal names. Rosselló *et al.* (1991) also suggested that genomovars encompassed in nomenspecies should be numbered; the first genomovar being the genomic species that contains the type strain of the nomenspecies. Each of the following genomovars should have a designated reference strain, which should be deposited in a service culture collection.

3.4 MINIMAL AND NOMENCLATURAL STANDARDS FOR THE DESCRIPTION OF NEW SPECIES

Since 1 January 1980, priority of bacterial names has been based on the *Approved Lists of Bacterial Names* (Skerman *et al.*, 1980). Names not included in such lists lost standing in bacterial nomenclature, though old names are available for reuse individually, provided that the provisions for doing so are met. Following the introduction of the new starting date for bacterial nomenclature, valid publication of new names and new combinations can only be achieved in the *International Journal of Systematic Bacteriology*, although they may be effectively published elsewhere and then validated by announcement in the validation lists published periodically in the journal. To date, 56 validation lists have been published. All validated names can be accessed through an electronic database on the internet (http://www.bdt.org.br/cgi-bin/bdtnet/bacterianame). Before publication of the name of a new cultivable species, a culture of the type strain should be deposited in at least one of the permanently established culture collections. It is also a requirement that the accession number designated to the strain by the culture collection is quoted in the published description.

One of the responsibilities of the taxonomic subcommittees of the International Committee on Systematic Bacteriology of the International Union of Microbiological Societies is to recommend minimal standards for the publication of new species (Sneath, 1992). This practice is intended to prevent the literature becoming cluttered with inadequately described species, as was the case before the introduction of the approved lists (Skerman *et al.*, 1980). Such standards need to include tests for the establishment of generic identity and for the delineation of species. In practice, the properties recommended for the description of species in different genera vary as they reflect the biological features of the organisms, but they should follow the tenets of polyphasic taxonomy (Figure 3.7). Guidelines for the minimal requirements necessary for the description of species are available (Sneath, 1977; Trüper and Schleifer, 1992). Minimal standards have been recommended recently for the description of new species of campylobacters (Ursing *et al.*, 1994), methanogenic bacteria (Boone and Whitman, 1988), mollicutes (ICSB Subcommittee on the Taxonomy of Mollicutes, 1995), root and stem-nodulating bacteria (Graham *et al.*, 1991) and slowly growing mycobacteria (Lévy-Frébault and Portaels, 1992).

3.5 CONCLUSIONS

Bacterial systematics as a core discipline is practised by few, but the applications of the subject are important to most – if not all – bacteriologists. It is the implementation of taxonomic concepts and practices which give rise to identification and typing systems, procedures for quality control and risk assessment, protocols for the analysis and characterization of biodiversity, hypotheses about the evolution of prokaryotes, and improved procedures for the selective isolation and use of microorganisms in

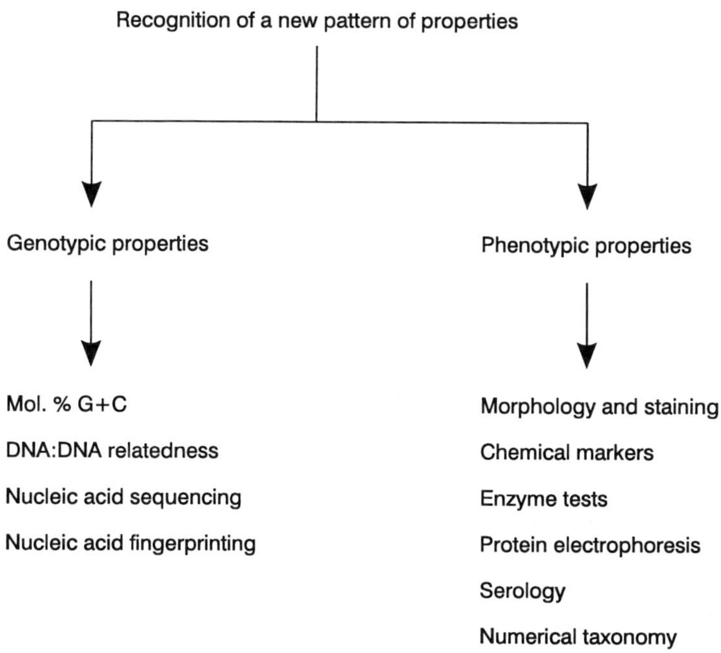

Figure 3.7 Methods relevant to the generation of genotypic and phenotypic properties for setting minimal standards for the description of new species.

biotechnological processes. Consequently, the nature of the bacterial species is not simply a matter for philosophical discourse, but is one of real practical significance.

Recent advances in bacterial systematics have promoted a more unified approach to the delineation of bacterial species. The concept of the 'polyphasic species' has distinct advantages over traditional more descriptive species concepts, especially since it can be expected to yield well-described species, a stable nomenclature and better identification systems. This approach to the circumscription of species assumes a population that is reproducing asexually is adapted to particular microhabitats, and is maintained as a relatively stable entity by natural selection. The ever-increasing availability of new methods for the generation of genotypic and phenotypic data, associated with new software tools, will help expedite polyphasic taxonomic studies. In practice, bacteriologists have a more standard set of comparative taxonomic methods than botanists and zoologists, but cross-checking of findings is still critical to sound work.

The polyphasic species concept is universally applicable and will become ever easier to apply, though the details of the approach need to be tailored to take into account the differing behavioural properties of members of taxonomically diverse genera. The concept does not address the question of the origin of bacterial species, that is, an understanding of the evolutionary processes that generate taxonomic diversity. This is clearly a major omission but is also a fertile area for collaboration between microbial population geneticists and bacterial systematists. However, will a theoretically sound species concept prove to be as practical as the present operational species concept?

Acknowledgements

Thanks are due to Professor P.H.A. Sneath and Professor F.G. Priest for critically reviewing the manuscript.

3.6 REFERENCES

Alexander, B. and Priest, F.G. (1990) Numerical classification and identification of *Bacillus sphaericus* including some strains pathogenic to mosquito larvae. *Journal of General Microbiology*, **136**, 367–76.

Ash, C., Farrow, J.A.E., Wallbanks, S. and Collins, M.D. (1991) Phylogenetic heterogeneity of the genus *Bacillus* revealed by comparative analysis of small subunit-ribosomal RNA sequences. *Letters in Applied Microbiology*, **13**, 202–6.

Bernardet, J.-F., Segers P., Vancanneyt, M., Berthe, F., Kersters, K. and Vandamme, P. (1996) Cutting a Gordonian knot: Emended classification and description of the genus *Flavobacterium*, emended description of the family *Flavobacteriaceae* and proposal of *Flavobacterium hydatis* nom. nov. (Basonym, *Cytophaga aquatilis* Strohl and Tait, 1978). *International Journal of Systematic Bacteriology*, **46**, 128–48.

Birtles, R.J., Harrison, T.G., Saunders, N.A. and Molyneux, D.H. (1995) Proposal to unify the genera *Grahamella* and *Bartonella*, with descriptions of *Bartonella talpae* comb. nov., *Bartonella peromysci* comb. nov. and three new species, *Bartonella grahamii* sp. nov., *Bartonella taylorii* sp. nov., and *Bartonella doshiae* sp. nov. *International Journal of Systematic Bacteriology*, **45**, 1–8.

Blackall, L.L., Parlett, J.H., Hayward, A.C., Minnikin, D.E., Greenfield, P.F. and Habers, A.E. (1989) *Nocardia pinensis* sp. nov., an actinomycete found in activated sludge foams in Australia. *Journal of General Microbiology*, **135**, 1547–58.

Bochner, B.R. (1989) Sleuthing out bacterial identities. *Nature*, **339**, 157–8.

Boone, D.R. and Whitman, W.B. (1988) Proposal of minimal standards for describing new taxa of methanogenic bacteria. *International Journal of Systematic Bacteriology*, **38**, 212–19.

Briglia, M., Eggen, R.I.L., van Elsas, D.J. and De Vos, W.M. (1994) Phylogenetic evidence for transfer of pentachlorophenol-mineralising *Rhodococcus chlorophenolicus* PCP-1 to the genus *Mycobacterium*. *International Journal of Systematic Bacteriology*, **44**, 494–8.

Briglia, M., Rainey, F.A., Stackebrandt, E., Schraa, G. and Salkinoja-Salonen, M.S. (1996) *Rhodococcus percolatus* sp. nov., a bacterium degrading 2,4,6-trichlorophenol. *International Journal of Systematic Bacteriology*, **46**, 23–30.

Britten, R.J. and Kohne, D.E. (1968) Repeated sequences in DNA. *Science*, **161**, 529–40.

Bryant, T.N. (1993) A review of probabilistic identification matrices. A compilation of probabilistic bacterial identification matrices. *Binary*, **5**, 207–10.

Buchanan, R.E. (1955) Taxonomy. *Annual Review of Microbiology*, **9**, 1–20.

Bull, A.T., Goodfellow, M. and Slater, J.H. (1992) Biodiversity as a source of innovation in biotechnology. *Annual Review of Microbiology*, **46**, 219–57.

Cai, J. and Collins, M.D. (1994) Phylogenetic analysis of species of the meso-diaminopimelic acid-containing genera *Brevibacterium* and *Dermabacter*. *International Journal of Systematic Bacteriology*, **44**, 583–5.

Canhos, V.P., Manfio, G.P. and Blaine, L.D. (1993) Software tools and databases for bacterial systematics and their dissemination *via* global networks. *Antoine van Leeuwenhoek*, **64**, 205–29.

Chun, J. (1995) *Computer Assisted Classification and Identification of Actinomycetes.* Unpublished PhD Thesis, University of Newcastle upon Tyne, UK.

Chun, J. and Goodfellow, M. (1995) A phylogenetic analysis of the genus *Nocardia* with 16S rRNA gene sequences. *International Journal of Systematic Bacteriology*, **45**, 240–5.

Chun, J., Kang, S.O., Hah, Y.C. and Goodfellow, M. (1996) Phylogeny of mycolic acid-containing actinomycetes. *Journal of Industrial Microbiology* (in press).

Cohan, F.M. (1994) Genetic exchange and evolutionary divergence in prokaryotes. *Trends in Ecology and Evolution*, **9**, 175–80.

Collins, M.D., Lawson, P.A., Willems, A., Cordoba, J.J., Fernandez-Garayzabal, J., Gercia, P., Cai, J., Hippe, H. and Farrow, J.A.E. (1994) The phylogeny of the genus *Clostridium:* proposal of five new genera and eleven new species combinations. *International Journal of Systematic Bacteriology*, **44**, 812–26.

Colwell, R.R. (1970) Polyphasic taxonomy of bacteria, in *Culture Collections of Microorganisms* (eds H. Iizuka and T. Hasegawa), University of Tokyo Press, Tokyo, pp. 421–36.

Cowan, S.T. (1955) The principles of microbial classification. Introduction: the philosophy of classification. *Journal of General Microbiology*, **12**, 314–19.

Cowan, S.T. (1959) Bacterial classification – problems and developments, in *Microbiology Yesterday and Today* (ed. V. Bryson), Institute of Microbiology, Rutgers, The State University, New Brunswick, pp. 54–79.

Cowan, S.T. (1962) The microbial species – a macromyth? in *Microbial Classification* (eds G.S. Ainsworth and P.H.A. Sneath), Cambridge University Press, Cambridge, pp. 433–55.

Cowan, S.T. (1968) *A Dictionary of Microbial Taxonomic Usage*. Oliver and Boyd, Edinburgh.

Dauga, C., Grimont, F. and Grimont, P.A.D. (1990) Nucleotide sequences of 16S rRNA from 10 *Serratia* species. *Research in Microbiology*, **141**, 1139–49.

De Boer, L., Dijkhuizen, L., Grobben, G., Goodfellow, M., Stackebrandt, E., Parlett, J.H., Whitehead, D. and Witt, D. (1990) *Amycolatopsis methanolica* sp. nov., a facultatively methylotrophic actinomycete. *International Journal of Systematic Bacteriology*, **40**, 194–204.

Dupuy, N., Willems, A., Pot, B., Dewettinck, D., Vandenbruaene, I., Maestrojuan, G., Dreyfus, B., Kersters, K., Collins, M.D. and Gillis, M. (1994) Phenotypic and genotypic characterization of bradyrhizobia nodulating the leguminous tree *Acacia albida*. *International Journal of Systematic Bacteriology*, **44**, 461–73.

Edwards, C. (ed.) (1993) *Monitoring Genetically Manipulated Microorganisms in the Environment*. John Wiley and Sons, Chichester.

Egel, D.S., Graham, J.H. and Stall, R.E. (1991) Genomic relatedness of *Xanthomonas campestris* strains causing diseases of citrus. *Applied and Environmental Microbiology*, **57**, 2724–30.

Embley, T.M. and Stackebrandt, E. (1997) Species in practice: exploring uncultured prokaryote diversity in natural samples, in *Species: the Units of Biodiversity* (eds M.F. Claridge, H.A. Dawah and M.R. Wilson), Chapman & Hall, London, pp. 61–81.

Ezaki, T., Li, N., Hashimoto, Y., Miura, H. and Yamamoto, H. (1994) 16S ribosomal DNA sequences of anaerobic cocci and proposal of *Ruminococcus hansenii* comb. nov. and *Ruminococcus productus* comb. nov. *International Journal of Systematic Bacteriology*, **44**, 130–6.

Felsenstein, J. (1993) PHYLIP (phylogenetic inference package), version 3.5c Department of Genetics, University of Washington, Seattle.

Fry, N.K., Saunders, N.A., Warwick, S. and Embley, T.M. (1991) The use of 16S ribosomal RNA analyses to investigate the phylogeny of the family *Legionellaceae*. *Journal of General Microbiology*, **137**, 1215–22.

Fox, G.E., Wisotzkey, J.D. and Jurtshuk, P., Jr (1992) How close is close: 16S rRNA sequence identity may not be sufficient to guarantee species identity. *International Journal of Systematic Bacteriology*, **42**, 166–70.

Gavini, F., Mergaert, J., Beji, A., Mielcarek, C., Izard, D., Kersters, K. and De Ley, J. (1989) Transfer of *Enterobacter agglomerans* (Beijerinck 1888) Ewing and Fife 1972 to *Pantoea* gen. nov. as *Pantoea agglomerans* comb. nov. and description of *Pantoea dispersa*. *International Journal of Systematic Bacteriology*, **39**, 337–45.

Gillis, M., Van, V.T., Bardin, R., Gvor, M., Hebbar, P., Willemes, A., Segers, P., Kersters, K., Heulin, T. and Fernandez, M.P. (1995) Polyphasic taxonomy in the genus *Burkholderia* leading to an emended description of the genus and proposition of *Burkholderia vietnamiensis* sp. nov. for nitrogen-fixing isolates from rice in Vietnam. *International Journal of Systematic Bacteriology*, **45**, 274–89.

Goodfellow, M. and Dickinson, C.H. (1985) Delineation and description of microbial populations using numerical methods, in *Computer-Assisted Bacterial Systematics* (eds M. Goodfellow, D. Jones and F.G. Priest), Academic Press, Orlando, pp. 165–225.

Goodfellow, M. and Minnikin, D.E. (eds) (1985) *Chemical Methods in Bacterial Systematics.* Academic Press, London.

Goodfellow, M. and O'Donnell, A.G. (eds) (1993) *Handbook of New Bacterial Systematics.* Academic Press, London.

Goodfellow, M. and O'Donnell, A.G. (1994) *Chemical Methods in Prokaryotic Systematics.* John Wiley and Sons, Chichester.

Goodfellow, M., Jones, D. and Priest, F.G. (eds) (1985) *Computer-Assisted Bacterial Systematics.* Academic Press, London.

Goodfellow, M., Chun, J., Atalan, E. and Sanglier, J.-J. (1994a) Curie-point pyrolysis mass spectrometry and its practical application to bacterial systematics, in *Bacterial Diversity and Systematics* (eds F.G. Priest, A. Ramos-Cormenzana and B.J. Tindall), Plenum Press, New York, pp. 87–104.

Goodfellow, M., Chun, J., Stubbs, S. and Toboli, A.S. (1994b) Transfer of *Nocardia amarae* Lechevalier and Lechevalier 1974 to the genus *Gordona* as *Gordona amarae* comb. nov. *Letters in Applied Microbiology*, **19**, 401–5.

Gordon, R.E. (1967) The taxonomy of soil bacteria, in *The Ecology of Soil Bacteria* (eds T.R.G. Gray and D. Parkinson), Liverpool University Press, Liverpool, pp. 293–321.

Gordon, R.E. (1978) A species definition. *International Journal of Systematic Bacteriology*, **28**, 605–7.

Graham, P.H., Sadowsky, M.J., Keyser, H.H., Barnet, Y.M., Bradley, R.S., Cooper, J.E., De Ley, D.J., Jarvis, B.D.W., Roslycky, E.B., Strijdom, B.W. and Young, J.P.W. (1991) Proposed minimal standards for the description of new genera and species of root-and-stem-nodulating bacteria. *International Journal of Systematic Bacteriology*, **41**, 582–7.

Hartford, T. and Sneath, P.H.A. (1988) Distortion of taxonomic structure from DNA relationships due to different choice of reference strains. *Systematic and Applied Microbiology*, **10**, 241–50.

Harwood, C.R. (1993) Plasmids, transposons and gene flux, in *Handbook of New Bacterial Systematics* (eds M. Goodfellow and A.G. O'Donnell), Academic Press, London, pp. 115–50.

Hawksworth, D.L. (ed.) (1991) *The Biodiversity of Microorganisms and Invertebrates: Its Role in Sustainable Agriculture.* Redwood Press, Melksham.

Heyndrickx, M., Vandemeulebroeke, K., Scheldeman, P., Hoste, B., Kersters, K., De Vos, P., Logan, N.A., Aziz, A.M., Ali, M. and Berkeley, R.C.W. (1995) *Paenibacillus* (formerly *Bacillus*) *gordonae* (Pichinoty *et al.* 1986) Ash *et al.* 1994 is a later subjective synonym of *Paenibacillus* (formerly *Bacillus*) *validus* (Nakamura 1984) Ash *et al.* 1994: Amended description of *P. validus*. *International Journal of Systematic Bacteriology*, **45**, 661–9.

Hildebrandt, D.C., Palleroni, N.J. and Schroth, M.N. (1990) Deoxyribonucleic acid relatedness of 24 xanthomonad strains representing 23 *Xanthomonas campestris* pathovars and *Xanthomonas fragariae*. *Journal of Applied Bacteriology*, **68**, 263–9.

Hillis, D.M., Allard, W. and Miyamoto, M.M. (1993) Analysis of DNA sequence data: phylogenetic inference. *Methods in Enzymology*, **224**, 456–87.

Holloway, B.W. (1993) Genetics for all bacteria. *Annual Review of Microbiology*, **47**, 659–84.

ICSB Subcommittee on the Taxonomy of Mollicutes. (1995) Revised minimum standards for description of new species of the class *Mollicutes* (Division *Tenericutes*). *International Journal of Systematic Bacteriology*, **45**, 605–12.

Jagoueix, S., Bové, J.M. and Garnier, M. (1994) The phloem-limited bacteria of greening disease of citrus is a member of the α subdivision of the *Proteobacteria*. *International Journal of Systematic Bacteriology*, **44**, 379–86.

Johnson, J.L. (1991a) DNA reassociation experiments, in *Nucleic Acid Techniques in Bacterial Systematics* (eds M. Goodfellow and A.G. O'Donnell), John Wiley and Sons, Chichester, pp. 21–44.

Johnson, J.L. (1991b) Isolation and purification of nucleic acids, in *Nucleic Acid Techniques in Bacterial Systematics* (eds M. Goodfellow and A.G. O'Donnell), John Wiley and Sons, Chichester, pp. 1–19.

Jones, D. (1989) Bacterial classification. IV: Genetic methods, in *Bergey's Manual of Systematic Bacteriology, Volume 4* (eds S.T. Williams, M.E. Sharpe and J.G. Holt), Williams and Wilkins, Baltimore, pp. 2310–12.

Jukes, T.H. and Cantor, C.R. (1969) Evolution of protein molecules, in *Mammalian Protein Metabolism* (ed. H. N. Munro), Academic Press, New York, pp. 21–132.

Kämpfer, P. and Kroppenstedt, R.M. (1991) Probabilistic identification of streptomycetes using miniaturised physiological tests. *Journal of General Microbiology*, **137**, 1893–902.

Kämpfer, P., Kroppenstedt, R.M. and Dott, W. (1991) A numerical classification of the genera *Streptomyces* and *Streptoverticillium* using miniaturized physiological tests. *Journal of General Microbiology*, **137**, 1831–91.

Kim, D., Chun, J., Sahin, N., Hah, Y.C. and Goodfellow, M. (1996) Analysis of thermophilic clades within the genus *Streptomyces* by 16S rDNA sequence comparisons. *International Journal of Systematic Bacteriology*, **46**, 581–7.

Klinger, J.M., Stowe, R.P. Obenhuber, D.C., Grover, T.O., Mishra, S.K. and Pierson, D.L. (1992) Evaluation of the Biolog automated microbial identification system. *Applied and Environmental Microbiology*, **58**, 2089–92.

Krawiec, S. (1985) Concept of a bacterial species. *International Journal of Systematic Bacteriology*, **35**, 217–20.

Krieg, N.R. (1988) Bacterial classification: an overview. *Canadian Journal of Microbiology*, **34**, 536–40.

Krieg, N.R. and Holt, J.G. (eds) (1984) *Bergey's Manual of Systematic Bacteriology, Volume I*, Williams and Wilkins, Baltimore.

Kroppenstedt, R.M., Stackebrandt, E. and Goodfellow, M. (1990) Taxonomic revision of the actinomycete genera *Actinomadura* and *Microtetraspora*. *Systematic and Applied Microbiology*, **13**, 148–60.

Kusunoki, S. and Ezaki, T. (1992) Proposal of *Mycobacterium peregrinum* sp. nov., and elevation of *Mycobacterium chelonae* subsp. *abscessus* (Kubica *et al.*) to species status: *Mycobacterium abscessus* comb. nov. *International Journal of Systematic Bacteriology*, **42**, 240–5.

Lévy-Frébault, V.V. and Portaels, F. (1992) Proposed minimal standards for the genus *Mycobacterium* and for description of new slowly growing *Mycobacterium* species. *International Journal of Systematic Bacteriology*, **42**, 315–23.

Ludwig, W., Kirchhof, G., Klugbauer, N., Weizenegger, M., Betzyl, D., Ehrmann, M., Hertel, C., Jilg, S., Tatzel, R., Zitzelsberger, H., Liebl, S., Hochberger, M., Shah, J., Lane, D., Wallnöfer, P.R. and Schliefer, K.H. (1993a) Complete 23S ribosomal RNA sequences of Gram-positive bacteria with a low G+C content. *Systematic and Applied Microbiology*, **15**, 487–501.

Ludwig, W., Neumaier, J., Klugbauer, N., Brockmann, E., Roller, C., Jilg, S., Reetz, K., Schachtner, I., Ludvigsen, A., Bachleitner, M., Fischer, U. and Schleifer, J.H. (1993b) Phylogenetic relationships of Bacteria based on comparative sequence analysis of elongation factor Tu and ATP-synthase β-subunit genes. *Antonie van Leeuwenhoek*, **64**, 285–305.

McNeil, M.M. and Brown, J.M. (1994) The medically important actinomycetes: epidemiology and microbiology. *Clinical Microbiology Reviews*, **7**, 357–417.

Manafi, M., Kneifel, W. and Bascomb, S. (1991) Fluorogenic and chromogenic substrates used in bacterial diagnostics. *Microbiological Reviews*, **55**, 335–48.

Martinez-Murcia, A.J., Benllock, S. and Collins, M.D. (1992) Phylogenetic interrelationships of members of the genera *Aeromonas* and *Plesiomonas* as determined by 16S ribosomal DNA sequencing: Lack of congruence with results of DNA: DNA hybridisation. *International Journal of Systematic Bacteriology*, **42**, 412–21.

Mauchline, W.S. and Keevil, C.W. (1991) Development of the BIOLOG substrate utilization system for identification of *Legionella* spp. *Applied and Environmental Microbiology*, **57**, 3345–9.

Maynard Smith, J. (1995) Do bacteria have population genetics? in *Population Genetics of Bacteria* (eds S. Baumberg, J.P.W. Young, E.M.H. Wellington and J.R. Saunders), Cambridge University Press, Cambridge, pp. 1–12.

Maynard Smith, J., Dowson, C.C. and Spratt, B.G. (1991) Localised sex in bacteria. *Nature*, **349**, 29–31.

Maynard Smith, J., Smith, N.H., O'Rourke, M. and Spratt, B.G. (1993) How clonal are bacteria? *Proceedings of the National Academy of Sciences, USA*, **90**, 4384–8.

Murray, R.G.E., Brenner, D.J., Colwell, R.R., De Vos, P., Goodfellow, M., Grimont, P.A.D., Pfennig, N., Stackebrandt, E. and Zavarzin, G.A. (1990) Report of the *ad hoc* committee on approaches to taxonomy within the *Proteobacteria*. *International Journal of Systematic Bacteriology*, **40**, 213–15.

Nakamura, L.K. (1984) *Bacillus amylolyticus* sp. nov., nom. rev., *Bacillus lautus* sp. nov., nom. rev. *Bacillus pabuli* sp. nov., nom. rev., and *Bacillus validus* sp. nov., nom rev. *International Journal of Systematic Bacteriology*, **34**, 224–6.

Nielsen, P., Fritze, D. and Priest, F.G. (1995) Phenetic diversity of alkaliphilic *Bacillus* strains: Proposal for nine new species. *Microbiology*, **141**, 1745–61.

Nour, S.M., Cleyet-Marel, J-C., Normand, P. and Fernandez, M.P. (1995) Genomic heterogeneity of strains nodulating chickpeas (*Cicer arientum* L.) and description of *Rhizobium mediterraneum* sp. nov. *International Journal of Systematic Bacteriology*, **45**, 640–8.

O'Donnell, A.G., Embley, T.M. and Goodfellow, M. (1993) Future of bacterial systematics, in *Handbook of New Bacterial Systematics* (eds M. Goodfellow and A.G. O'Donnell), Academic Press, London, pp. 513–24.

O'Donnell, A.G., Goodfellow, M. and Hawksworth, D.L. (1994) Theoretical and practical aspects of the quantification of biodiversity among microorganisms. *Philosophical Transactions of the Royal Society B*, **345**, 65–73.

On, S.L.W., Holmes, B. and Sackin, M. (1996) A probabilistic identification matrix for campylobacters, helicobacters and allied taxa. *Journal of Applied Bacteriology*, **81**, 425–32.

Palleroni, N.J. (1993) Structure of the bacterial genome, in *Handbook of New Bacterial Systematics* (eds M. Goodfellow and A.G. O'Donnell), Academic Press, London, pp. 57–113.

Pankhurst, R.J. (1991) *Practical Taxonomic Computing.* Cambridge University Press, London.

Pot, B., Gillis, M., Hoste, B., Van de Velde, A., Bekaert, F., Kersters, K. and De Ley, J. (1989) Intra-and intergeneric relationships of the genus *Oceanospirillum. International Journal of Systematic Bacteriology*, **39**, 23–34.

Pridham, T.G. and Tresner, H.D. (1974) Genus I. *Streptomyces* Waksman and Henrici, in *Bergey's Manual of Determinative Bacteriology* (eds R.E. Buchanan and N.E. Gibbons), Williams and Wilkins Co., Baltimore, pp. 748–829.

Priest, F.G. (1993) Systematics and ecology of *Bacillus*, in *Bacillus subtilis and other Gram-Positive Bacteria: Biochemistry, Physiology and Molecular Genetics* (eds A.L. Sonnenheim, J.A. Hoch and R. Losick), American Society for Microbiology, Washington, DC, pp. 3–16.

Priest, F.G. and Williams, S.T. (1993) Computer-assisted identification, in *Handbook of New Bacterial Systematics* (eds M. Goodfellow and A.G. O'Donnell), Academic Press, London, pp. 361–81.

Rainey, F.A., Janssen, P.H., Morgan, H.W. and Stackebrandt, E. (1993) A biphasic approach to the determination of the phenotypic and genotypic diversity of some anaerobic, cellulolytic, thermophilic, rod-shaped bacteria. *Antonie van Leeuwenhoek*, **64**, 341–55.

Ravin, A.W. (1960) The origin of bacterial species: genetic recombination and factors limiting it between bacterial populations. *Bacteriological Reviews*, **24**, 201–20.

Ravin, A.W. (1963) Experimental approaches to the study of bacterial phylogeny. *American Naturalist*, **97**, 307–18.

Roberts, M.S., Nakamura, L.K. and Cohan, F.M. (1994) *Bacillus mojavensis* sp. nov., distinguishable from *Bacillus subtilis* by sexual isolation, divergence of DNA sequence, and differences in fatty acid composition. *International Journal of Systematic Bacteriology*, **44**, 256–64

Robertson, J.A., Howard, L.A., Zinner, C.L. and Stemke, G.W. (1994) Comparison of 16S rRNA genes within the T960 and Parvo biovars of ureaplasmas isolated from humans. *International Journal of Systematic Bacteriology*, **44**, 836–8.

Rosselló, R., Garcia-Valdés, J., Lalucat, J. and Ursing, J. (1991) Genotypic and phenotypic diversity of *Pseudonomas stutzeri. Systematic and Applied Microbiology*, **8**, 124–7.

Ruimy, R., Riegal, P., Carlotti, A., Boiron, P., Bernardin. G., Monteil, H., Wallace, R.J., Jr and Christen, R. (1996) *Nocardia brasiliensis* sp. nov., a new species of *Nocardia* which groups bacterial strains previously identified as *Nocardia brasiliensis* and associated with invasive disease. *International Journal of Systematic Bacteriology*, **46**, 259–64.

Sackin, M.J. and Jones, D. (1993) Computer-assisted classification, in *Handbook of New Bacterial Systematics* (eds M. Goodfellow and A.G. O'Donnell), Academic Press, London, pp. 281–333.

Saitou, N. and Nei, M. (1987) The neighbor-joining method: a new method for constructing phylogenetic trees. *Molecular Biology and Evolution*, **4**, 406–25.

Schleifer, K.H., Ludwig, W. and Amann, R. (1993) Nucleic acid probes, in *Handbook of New Bacterial Systematics* (eds M. Goodfellow and A.G. O'Donnell), Academic Press, London, pp. 463–510.

Selander, R.K., Li, J., Boyd, E.F., Wang, F-S. and Nelson, K. (1994) DNA sequence analysis of the genetic structure of populations of *Salmonella enterica* and *Escherichia coli*, in *Bacterial Diversity and Systematics* (eds F.G. Priest, A. Ramos-Cormenzana and B.J. Tindall), Plenum Press, New York, pp. 17–49.

Seong, C.N., Park, S.K., Goodfellow, M., Kim, S-B. and Hah, Y-C. (1995) Construction of probability identification matrix and selective medium for acidophilic actinomycetes using numerical classification data. *Journal of Microbiology*, **33**, 95–102.

Skerman, V.B.D., McGowan, V. and Sneath, P.H.A. (1980) Approved lists of bacterial names. *International Journal of Systematic Bacteriology*, **30**, 225–420.

Sneath, P.H.A. (1977) The maintenance of large numbers of strains of microorganisms and the implications for culture collections. *FEMS Microbiology Letters*, **1**, 333–4.

Sneath, P.H.A. (1979a) BASIC program for identification of an unknown with presence–absence data against an identification matrix of percent positive characters. *Computers and Geosciences*, 5, 195–213.

Sneath, P.H.A. (1979b) BASIC program for character separation indices from an identification matrix of percent positive characters. *Computers and Geosciences*, **5**, 349–57.

Sneath, P.H.A. (1980a) BASIC program for the most diagnostic properties of groups from an identification matrix of percent positive characters. *Computers and Geosciences*, **6**, 21–6.

Sneath, P.H.A. (1980b) BASIC program for determining the best identification scores possible from the most typical examples when compared with an identification matrix of percentage positive characters. *Computers and Geosciences*, **6**, 27–34.

Sneath, P.H.A. (1980c) BASIC program for determining overlap between groups in an identification matrix of percent positive characters. *Computers and Geosciences*, **6**, 267–78.

Sneath, P.H.A. (1983) Distortions of taxonomic structure from incomplete data on a restricted set of reference strains. *Journal of General Microbiology*, **129**, 1045–73.

Sneath, P.H.A. (1985) Future of numerical taxonomy, in *Computer-Assisted Bacterial Systematics* (eds M. Goodfellow, D. Jones and F.G. Priest), Academic Press, Orlando, pp. 415–431.

Sneath, P.H.A. (1992) *International Code of Nomenclature of Bacteria*. American Society for Microbiology, Washington, DC.

Sneath, P.H.A. (1993) Evidence from *Aeromonas* for genetic crossing-over in ribosomal sequences. *International Journal of Systematic Bacteriology*, **43**, 626–9.

Sneath, P.H.A. and Sokal, R.R. (1973) *Numerical Taxonomy: The Principles and Practice of Numerical Classification*. Freeman, San Francisco.

Sneath, P.H.A., Mair, N.S., Sharpe, M.E. and Holt, J.G. (eds) (1986) *Bergey's Manual of Systematic Bacteriology*, Volume 2. Williams and Wilkins, Baltimore.

Sokal, R.R. and Sneath, P.H.A. (1963) *Principles of Numerical Taxonomy*. Freeman, San Francisco.

Stackebrandt, E. and Goebel, B.M. (1994) Taxonomic note: A place for DNA:DNA reassociation and 16S rRNA sequence analysis in the present species definition in bacteriology. *International Journal of Systematic Bacteriology*, **44**, 846–9.

Stackebrandt, E. and Goodfellow, M. (eds) (1991) *Nucleic Acid Techniques in Bacterial Systematics*. John Wiley and Sons, Chichester.

Stanier, R.Y., Ingraham, J.L., Wheelis, M.L. and Paintes, P.R. (1986) *The Microbial World, Fifth Edition*. Prentice Hall, Englewood Cliffs, New Jersey.

Stead, D.E., Sellwood, J.E., Wilson, J. and Viney, I. (1992) Evaluation of a commercial microbial identification system based on fatty acid profiles for rapid, accurate identification of plant pathogenic bacteria. *Journal of Applied Bacteriology*, **72**, 315–21.

Strenstöm, I.-M., Zakaria, A., Ternström, A. and Molin, G. (1990) Numerical taxonomy of fluorescent *Pseudomonas* associated with tomato roots. *Antonie van Leeuwenhoek*, **57**, 223–36.

Swofford, D.L. and Olsen, G.J. (1990) Phylogenetic reconstruction, in *Molecular Systematics* (eds D. Hillis and C. Moritz), Sinauer Associates, Sunderland, USA, pp. 411–501.

Tamaoka, J. (1994) Determination of DNA base composition, in *Chemical Methods in Prokaryotic Systematics* (eds M. Goodfellow and A.G. O'Donnell), John Wiley and Sons, Chichester, pp. 463–469.

Trejo, W.H. (1970) An evaluation of some concepts and criteria used in the speciation of streptomycetes. *Transactions of the New York Academy of Sciences*, Series II, **32**, 989–97.

Trüper, H.G. and Schleifer, K.H. (1992) Prokaryote characterisation and identification, in *The Prokaryotes* (eds A. Balows, H.G. Trüper, M. Dworkin, W. Harder and K.H. Schleifer), 2nd edn, Springer-Verlag, New York, pp. 126–48.

Ursing, J. and Bruun, B. (1991) Genotypic heterogeneity of *Flavobacterium* group IIb and *Flavobacterium breve* demonstrated by DNA:DNA hybridisation. *Acta Pathologica Micobiologica et Immunologica Scandinavica*, **99**, 780–6.

Ursing, J.B., Lior, H. and Owen, R.J. (1994) Proposal of minimal standards for describing new species of the family Campylobacteriaceae. *International Journal of Systematic Bacteriology*, **44**, 842–5.

Ursing, J.B., Rosselló-Mora, R.A., García-Valdés, E. and Lalucat, J. (1995) Taxonomic note: a pragmatic approach to the nomenclature of phenotypically similar genomic groups. *International Journal of Systematic Bacteriology*, **45**, 604.

Van den Mooter, M. and Swings, J. (1990) Numerical analysis of 295 phenotypic features of 266 *Xanthomonas* strains and related strains and an improved taxonomy of the genus. *International Journal of Systematic Bacteriology*, **40**, 348–69.

Vauterin, L., Vantomme, R., Pot, B., Hoste, B., Swings, J. and Kersters, K. (1990) Taxonomic analysis of *Xanthomonas campestris* pv. *legoniae* and *X. campestris* pv. *pelargonii* by means of phytopathological, phenotypic, protein electrophoretic and DNA hybridisation methods. *Systematic and Applied Microbiology*, **13**, 166–76.

Vauterin, L., Swings, J. and Kersters, K. (1993) Protein electrophoresis and classification, in *Handbook of New Bacterial Systematics* (eds M. Goodfellow and A.G. O'Donnell), Academic Press, London, pp. 251–80.

Vauterin L., Hoste, B., Kersters, K. and Swings, J. (1995) Reclassification of *Xanthomonas*. *International Journal of Systematic Bacteriology*, **45**, 472–89.

Vauterin, L., Yang, P. and Swings, J. (1996) Utilisation of fatty acid methyl esters for the differentiation of new *Xanthomonas* species. *International Journal of Systematic Bacteriology*, **46**, 298–304.

Veríssimo, A., Morais, P.V., Dioga, A., Gomes, C. and da Costa, M.S. (1996) Characterisation of *Legionella* species by numerical analysis of whole-cell protein electrophoresis. *International Journal of Systematic Bacteriology*, **46**, 41–9.

Wayne, L.G. (1985) Computer-assisted analysis of data from co-operative studies on mycobacteria, in *Computer-Assisted Bacterial Systematics* (eds M. Goodfellow, D. Jones and F.G. Priest), Academic Press, Orlando, pp. 91–105.

Wayne, L.G., Krichevsky, M.I., Portyrata, D. and Jackson, C.K. (1984) Diagnostic probability matrix for the identification of slowly growing mycobacteria in clinical laboratories. *Journal of Clinical Microbiology*, **20**, 722–9.

Wayne, L.G., Brenner, D.J, Colwell, R.R., Grimont, P.A.D., Kandler, P., Krichevsky, M.I., Moore, L.H., Moore, W.E.C., Murray, R.G.E., Stackebrandt, E., Starr, M.P. and Trüper, H.G. (1987) Report of the *ad hoc* committee on reconciliation of approaches to bacterial systematics. *International Journal of Systematic Bacteriology*, **37**, 463–4.

Wayne, L.G., Good, R.C., Böttger, E.C., Butler, R., Dorsche, M., Ezaki, T., Gross, W., Jonas, V., Kilburn, J., Kirschner, P., Krichevsky, M.I., Ridell, M., Shinnick, T.M., Springer, B., Stackebrandt, E., Tarnok, I., Tarnok, Z., Tasaka, H., Vincent, V., Warren, N.G., Knott, C.A. and Johnson, R. (1996) Semantide- and chemotaxonomy-based analyses of some problematic phenotypic clusters of slowly-growing mycobacteria, a cooperative study of the International Working Group on Mycobacterial Taxonomy. *International Journal of Systematic Bacteriology*, **46**, 280–97.

White, D., Sharp, R.J. and Priest, F.G. (1993) A polyphasic taxonomic study of thermophilic bacilli from a wide geographical area. *Antonie van Leeuwenhoek*, **64**, 357–86.

Williams, S.T. and Vickers, J.C. (1988) Detection of actinomycetes in the natural environment, in *Biology of Actinomycetes '88* (eds Y. Okami, T. Beppu and H. Ogawara), Japan Scientific Societies Press, Tokyo, pp. 265–270.

Williams, S.T., Vickers, J.C. and Goodfellow, M. (1984) New microbes from old habitats? in *The Microbe 1984: Part II: Prokaryotes and Eukaryotes* (eds D. Kelly and N.G. Carr), Cambridge University Press, Cambridge, pp. 219–56.

Woese, C.R. (1987) Bacterial evolution. *Microbiological Reviews*, **51**, 221–71.

Woese, C.R. (1992) Prokaryotic systematics: the evolution of a science, in *The Prokaryotes* (eds A. Balows, H.G. Trüper, M. Dworkin, W. Harder and K.H. Schleifer), 2nd edn, Springer-Verlag, New York, pp. 1–8.

Yassin, A.F., Rainey, F.A., Brzezinka, H., Burghardt, J., Lee, H.J. and Schaal, K.P. (1995) *Tsukamurella inchonensis* sp. nov. *International Journal of Systematic Bacteriology*, **45**, 522–7.

Zillig, W., Klenk, H.P., Palm, P., Pühler, G., Gropp, F., Garrett, R.A. and Leffers, H. (1989) The phylogenetic relations of DNA-dependent RNA-polymerases of archaebacteria, eukaryotes and eubacteria. *Canadian Journal of Microbiology*, **35**, 73–80.

4

Species in practice: exploring uncultured prokaryote diversity in natural samples

T. Martin Embley and E. Stackebrandt

Contacting address: Department of Zoology, The Natural History Museum, Cromwell Road, London SW7 5BD, UK

ABSTRACT

Traditionally prokaryote species have been recognized using phenetic methods incorporating information from the phenotype and genotype. Molecular methods have now made it possible to use 16S rRNA sequences to analyse the prokaryote community in natural habitats without the need to culture. The sequences recovered can be directly compared with sequences from cultured organisms through the medium of tree diagrams, and relationships can be interpreted in terms of inferred common ancestry. These studies have revealed patterns of prokaryote diversity which were unavailable using traditional microbiological methods. Not only are most of the lineages uncovered new to science, but some are of such clear phylogenetic or ecological importance that their further study is imperative.

4.1 INTRODUCTION

Indeed taxa are all much the same, even if some taxa include others. I hesitate to suggest that if there are taxonomic units of evolution, the units are taxa generally' Nelson (1989: 61)

The main subjects of the present paper are those prokaryotes for which there are no recognized laboratory cultures and hence no phenotypic information. This apparent contradiction occurs because in the past it was necessary first to isolate microorganisms into pure culture, before taxonomic analysis could proceed and a newly discovered prokaryote be recognized as representing a new centre of taxonomic variation. Isolation of a pure culture for study requires that a microorganism will grow under a

Species: The Units of Biodiversity. Edited by M.F. Claridge, H.A. Dawah and M.R. Wilson. Published in 1997 by Chapman & Hall. ISBN 0 412 63120 2

set of more or less artificial conditions, which many, perhaps most, have seemed remarkably reluctant to do.

It is now possible to explore natural prokaryote taxonomic diversity using molecular sequences without the need to isolate and culture microorganisms. This approach has been incredibly successful, but it does raise issues about how this uncultured microbial diversity can be appreciated in familiar terms, and in the context of this volume, whether the term species can be meaningfully applied to prokaryotes for which we only have sequences as evidence. Before we discuss these issues, it is necessary to discuss how species are currently recognized among prokaryotes which can be cultured in the laboratory. It is also useful briefly to consider the treatment of eukaryotes under the biological species concept, since this probably dominates how species are generally viewed and discussed.

4.2 SPECIES AMONG EUKARYOTES

The biological species concept (BSC) was formulated for 'groups of interbreeding natural populations which are reproductively isolated from other such groups' (Mayr, 1970: 12). So in a strict sense it should be used only for those individuals for which these criteria have been satisfied. However, as discussed by Sokal and Crovello (1970) most species, even among Metazoa, are practically recognized using taxonomic information, and reproductive isolation is assumed but seldom tested.

The recognition that taxonomic analysis is the practical means whereby most species are recognized, has frequently raised the question of whether species are somehow different from other taxa (some of this literature has been reviewed by Nelson, 1989; Mallet, 1995). Nelson (1989) has argued persuasively from the perspective of a taxonomist that this is not the case, commenting: 'There seems to be no basic taxonomic unit and no particular taxonomic unit of evolution'. From this perspective the major question transmutes into how best to carry out taxonomic analysis, and this applies equally to all types of data and to all potential taxa – eukaryote or prokaryote. It is of course perfectly legitimate subsequently to try and understand the reasons why a particular group might be recognized in a particular taxonomic analysis.

4.3 SPECIES AMONG CULTURED PROKARYOTES

Gene flow does occur between prokaryotes but it occurs by different mechanisms and has different effects compared with sexual eukaryotes (Maynard-Smith *et al.*, 1993; Cohan, 1994). Put simply, prokaryotes do not produce gametes and do not have meiosis; thus, any concept such as the BSC which requires these characteristics cannot be applied to prokaryotes. Gene transfer among individuals of different bacteria varies in frequency

(Maynard-Smith *et al.*, 1993), and the rate of recombination varies among loci encoding proteins of different types (Selander *et al.*, 1994). At the moment it is difficult to appreciate how bacterial population genetics can provide another 'biological species concept' for prokaryotes (Maynard-Smith, 1995). Bacterial population genetic studies clearly can and do provide characters for fine-resolution taxonomic analysis, and for studying patterns of gene flow between individuals (Selander *et al.*, 1994).

It has been recognized that asexual organisms do not fit the definition of species under the BSC. Indeed, Mayr (1987) has previously suggested using the term paraspecies to describe asexual organisms, but this has not been adopted by prokaryote systematists. In fact, there would have to be very convincing objective reasons for now limiting use of the word species for the description of one particular subset of cellular life. In our opinion such convincing arguments have not been presented. Mayr has suggested that 'trouble' (Mayr, 1987: 165) will be the inevitable consequence of adopting a concept of species that applies to all living organisms and also 'Inevitably in such a broadened concept of species, all those criteria will have to be excluded that are particularly characteristic for the **majority** (our emphasis) of species, that is the characteristics of biological species'. The process of exploration of the natural diversity of prokaryotes has barely begun, so arguments about numbers of species are almost certainly premature. However, it is already apparent (see later) that two of the possibly three fundamental domains of life, the Archaea and Bacteria, are exclusively prokaryote (Woese *et al.*, 1990). Moreover, the sexual habits of many microbial eukaryotes, a not insubstantial fraction of eukaryote diversity and history, are also not well documented (Sogin, 1991).

4.4 PRACTICAL RECOGNITION OF CULTURED PROKARYOTE SPECIES

Since the late 1950s (Sneath, 1957a,b) bacterial taxonomists have mainly used numerical taxonomy to recognize entities which have subsequently formed the basis for new species. The method as generally practised analyses a group of strains for many unweighted characters, codes the responses of each strain, calculates a measure of similarity using one or more coefficients, and organizes the results by way of a phenogram or ordination plot (Sneath and Sokal, 1973).

The characters which have traditionally been used have comprised the ability to degrade or metabolize certain chemicals, the ability to produce, or survive exposure to antibiotics, the ability to grow on particular carbon or nitrogen sources, particular staining reactions and external morphological features (Williams *et al.*, 1983). A serious problem with some of the more traditional tests and their corresponding analysis, are the difficulties in inferring homologous (*sensu* Owen, see Patterson, 1987) features, i.e.

those which it is logically legitimate to compare when attempting to infer historical relationships. For example, different microorganisms might produce a zone of clearing (the feature scored) in a bacterial lawn due to the production of different antimicrobials, or may produce the same indicator reaction in a biochemical test but using different enzymes. The uncertain quality of tests sometimes used has been blamed (Sackin and Jones, 1993) for numerical taxonomy's past short-comings in indicating consistent higher groupings among microorganisms, as opposed to its relative success and convenience for organising large numbers of prokaryotes into homogeneous groupings which can be equated to species. More recently there have been moves towards using protein banding patterns, fatty acid profiles and rapid enzyme tests, which are perhaps more precise in what they measure, and which can be semi-automated to deal with large numbers of strains (Goodfellow and O'Donnell, 1993).

An inferred amount of genomic similarity measured using DNA : DNA reassociation assays was subsequently incorporated with overall phenotypic resemblance into the taxonomic method (the so-called polyphasic approach) for identifying prokaryote species, the stated desire being to provide a common basis for establishing species among different bacterial groups (Johnson, 1973). It has been suggested that strains of a species should share at least 70% DNA : DNA reassociation under optimal conditions with 5°C or less ΔT_m: the difference between the thermal stability of the homologous and heterologous duplex (Wayne *et al.*, 1987). This apparently arbitrary level has been justified by reference to empirical observation that some strains of bacteria which are highly related phenetically share at least this amount of DNA : DNA reassociation (Brenner, 1973), and the observation that heteroduplexes which share less than 70% reassociation are sometimes less thermally stable, implying a greater degree of mispairing than the corresponding homoduplexes (Johnson, 1973).

DNA : DNA pairing assays have advantages in that they theoretically sample a very large number of characters, but different methods may give higher or lower values for the same strains (Grimont *et al.*, 1980), and all methods are subject to significant experimental error, typically in the region of 5–6% (Hartford and Sneath, 1990). The level of DNA : DNA reassociation between strains of established taxonomic species can be greater than 70%. For example, the two species *Neisseria gonorrhoea* and *Neisseria meningitidis* share over 74% DNA : DNA reassociation (Rossau *et al.*, 1989), but are kept as separate species because they can be distinguished phenotypically (Barrett and Sneath, 1994), and because they cause different diseases. So far there have been about 4000 prokaryote taxonomic species validly described using a more or less polyphasic approach.

A potential problem with the phenetic approach to recognize clusters which are then called species, is that these species may not be monophyletic groups; non-monophyletic in that they may contain individuals which resemble each other phenetically, but some of them are more closely related to individuals outside of the species by the criterion of common ancestry. Classification based upon monophyletic groups is considered desirable because only then will a hierarchical taxonomy be isomorphic with a phylogenetic tree (Harvey and Pagel, 1991). Phenetics makes no distinction between synapomorphy (i.e. shared derived characters), homoplasy (false similarity including convergence), or plesiomorphy (ancestral similarity) when it measures resemblance. There is an apparent assumption among some systematists that prokaryote species identified using phenetic methods are necessarily phylogenetic, e.g. 'at present the species is the only unit which can be defined in phylogenetic terms' (Wayne *et al.*, 1987). It has been pointed out previously that in this context the use of the term phylogenetic is potentially confusing (Sneath, 1988).

Sneath and Sokal have argued that phenetic groups defined using 'good' (see previous comments, Sackin and Jones, 1993) characters and rigorous numerical phenetic methods are usually monophyletic (Sneath and Sokal, 1973: 46). However, it has been demonstrated that under some conditions phylogenetic and phenetic analyses give different trees for the same data sets (Farris, 1977; Platnick, 1989). Moreover, in some recently published papers dealing with prokaryotes there are differences between branching diagrams based upon analyses of different character sets or using different coefficients of similarity (e.g. compare composition of some clusters in Figures 1 and 2, Dupuy *et al.*, 1994; Figures 1 and 3, Rainey *et al.*, 1993; Figures 1 and 2, Segers *et al.*, 1994). Unfortunately, there are no estimates of branching pattern reliability (e.g. from bootstrapping) for the figures compared from these cited papers, so it is not possible to gauge the strength of support for the particular differences observed. It is probably unrealistic, because of the effort that would be necessary, to think that monophyly will be practically investigated for strains of the 4000 or so currently recognized prokaryote species, apart from a few model species which have particular medical or ecological importance. It is outside of our expertise to appreciate how widely, if at all, non-monophyly of species is a potential problem for metazoan taxonomy.

4.5 EXPLORING UNCULTURED PROKARYOTE DIVERSITY USING MOLECULAR SEQUENCES

The current polyphasic approach to recognize prokaryote species requires pure cultures. Yet it is possible to extrapolate, from comparing direct and viable counts, and from molecular studies, that much of prokaryote diversity in nature is probably not recognized in pure culture (Ward *et al.*, 1992). Thus, most prokaryotes cannot at present be considered directly in terms of prokaryote species under the polyphasic system. The lack of pure cultures stems from the obvious difficulties in isolating prokaryotes from nature when nothing is known of their growth requirements. There are, however, persuasive arguments based upon consideration of global biodiversity (Hawksworth, 1995), nutrient cycling and ecosystem function (Brock *et al.*, 1994), and understanding the evolution of life itself (Woese, 1994), which suggest that these uncultured prokaryotes are some of the more interesting and important candidates for biological study.

Environmental DNA samples have been analysed using reassociation kinetics to estimate community complexity and to estimate the number of contributing genomes (Torsvik *et al.*, 1990), but this method is too imprecise to identify individual genomes, or to place them within a hierarchical taxonomic framework. In contrast, analyses of molecular sequences, which are now commonly used to infer phylogenetic relationships between established prokaryote species, can be applied to individual uncultured prokaryotes. It is this approach which currently holds greatest promise for the exploration of uncultured natural microbial diversity within a taxonomic framework (Pace *et al.*, 1985).

4.6 16S rRNA ANALYSES TO INFER PHYLOGENETIC RELATIONSHIPS BETWEEN PROKARYOTES

In prokaryote systematics it is comparisons of 16S rRNA sequences (now more likely to be 16S rRNA gene sequences or 16S rDNA) which have been used most to infer phylogenetic relationships between species. The reasons for choosing 16S rRNA, and its useful features for classification and identification of prokaryotes, have been fully discussed in articles by Carl Woese, who pioneered this field of endeavour (Woese, 1987; Woese *et al.*, 1990). They include its functional constancy and universal distribution in cellular life, its high information content, ease of alignment, and its mosaic structure which allows both ancient and recent relationships to be inferred. Before molecular sequence comparisons, prokaryote taxonomists had largely given up on trying to infer phylogeny and prokaryote systematics was an endeavour aimed at most efficiently putting taxa into groups, whereby their identification for medical, ecological or commercial purposes could be best facilitated (Goodfellow and O'Donnell, 1993).

By 1977 sufficient sequence data were available to suggest that cellular life, at least that in culture for study, could be organized into three primary divisions based upon SSU rRNA sequence comparisons, which Woese and Fox (1977) called the Eubacteria, the Archaebacteria and the Urkaryotes (eukaryotes). The discovery of a second great kingdom of prokaryote life – the Archaebacteria – is arguably one of the great achievements of 20th century biology. Sequence comparisons also supported inferences from the fossil record, that the classical higher eukaryote kingdoms Plantae and Animalia are but recent branches in an already highly diversified tree. Most of the branches in the tree of life based upon SSU rRNA sequences, and all of the deepest ones, are represented by microorganisms. To signify the primacy of the three divisions of cellular life over the classical metazoan kingdoms, Woese *et al.* (1990) subsequently suggested that the term domain be used. Thus, Eubacteria became domain Bacteria, Archaebacteria became domain Archaea and eukaryotes became domain Eucarya.

Concerns have been expressed about basing the taxonomic hierarchy for prokaryotes so heavily upon a single molecule (Olsen and Woese, 1993), but it is difficult at present to identify any other molecule likely to usurp it as the most generally useful molecular tool. Anyway, it is important to consider any gene tree as a hypothesis of relationships which should be challenged using other genes. So far there have been few explicit tests of congruence, but in the cases where topologies derived from other genes have been compared with the 16S rRNA tree, the agreement about relationships has sometimes been good, when allowance is made for the different resolving power of different molecules (Ludwig *et al.*, 1993; Springer *et al.*, 1995). At the very deepest levels of relationship, i.e. between domains, it has proved more difficult to reconcile different gene trees and there is a lively debate about different tree topologies and how to interpret them from an evolutionary perspective (Forterre *et al.*, 1993; Gogarten, 1995; Golding and Gupta, 1995).

A second concern is what is the best method for inferring phylogenetic relationships from sequence data, a debate which affects all molecular studies, not just those dealing with prokaryotes. It is outside of the scope of this review to go into details of what are considered the strengths and weaknesses of the different methods, and how they perform for inferring phylogeny (Felsenstein, 1988; Sneath, 1989; Hillis *et al.*, 1993). Computer simulations and experiments have revealed that all methods fail when the assumptions upon which they are based are badly violated, and almost all will detect strong signals in the data which may reflect phylogeny (Jin and Nei, 1991; Hillis *et al.*, 1993; Huelsenbeck and Hillis, 1993).

Most investigations of microbial relationships using 16S rRNA have continued to use relatively simple phenetic methods such as neighbour joining, for making phylogenetic inferences. The apparent greater success of these methods in inferring consistent higher groupings for prokaryotes (see Sackin and Jones, 1993) is probably due in part to data (sequences) which allow more reliable hypotheses of homology to be inferred for the characters (aligned base positions) and sequences compared (but see Patterson, 1987, 1988). There is also increasing comparison, and greater caution in the interpretation, of trees generated using different methods such as parsimony and maximum likelihood (Felsenstein, 1981). Bootstrapping has also served to introduce a necessary degree of caution in interpreting the strength of support for particular groupings (Felsenstein, 1985). Some of these changes in habits are moving towards what Sneath and Sokal (1973) perhaps meant when they argued for rigorous methods in numerical taxonomy. The trend towards increasing sophistication of analysis of molecular data is likely to continue and it can be expected that newer approaches such as spectral analysis (Lento *et al.*, 1995), and transformations akin to paralinear distances (Lake, 1994; Lockhart *et al.*, 1994), will soon make their appearance in the prokaryote literature.

4.7 16S rDNA SEQUENCES TO INVESTIGATE THE RELATIONSHIPS OF UNCULTURED PROKARYOTES IN NATURAL SAMPLES

The methods whereby 16S rDNA sequences may be recovered and analysed from natural populations of prokaryotes, without culturing them, have been extensively reviewed, since Norman Pace and colleagues first suggested the general approach (Pace *et al.*, 1985; Olsen *et al.*, 1986; Ward *et al.*, 1992). The current most common method is based upon PCR amplification of 16S rRNA genes or gene fragments using 16S rRNA gene-specific primers, followed by segregation of individual gene copies by cloning into *Escherichia coli* (Giovannoni *et al.*, 1990). The process produces a library of community 16S rRNA genes, the composition of which can be estimated by sampling clones and comparing their sequences by restriction endonuclease digestion, their reaction to specific probes, or by full or partial sequencing (Ward *et al.*, 1992). This information can be analysed to infer clone abundance and representation in the library. Unique clones can be completely sequenced and their relationships to sequences from cultured taxa in a taxonomic hierarchy based upon rRNA sequences can be discovered.

A concern about using PCR to analyse microbial communities is how accurately does library composition reflect gene composition *in situ*, i.e. how well does PCR sample the pool of natural prokaryote diversity? Are there biases in the system which mean that some sequences will amplify more readily, or will some sequences be under-represented or missed entirely? Simple computer models (Wagner *et al.*, 1994) suggest that individual PCR reactions may produce skewed distributions (PCR drift) of products starting from template mixtures of known composition. PCR drift is due to stochastic processes which occur in early cycles of PCR and it can be reduced but not eliminated, by pooling many PCR reactions. Selection for certain templates in a mixture can also occur if they have different GC base compositions (Reysenbach *et al.*, 1992), and it can be predicted by simulation that some templates in mixtures may have higher replication probabilities for other reasons (discussed in Wagner *et al.*, 1994). The contribution of starting templates from different bacteria will also vary, even if there are the same number of individuals of each species in the original population. Different bacteria will be more or less susceptible to lysis or differential sampling may occur due to shape or size. Different PCR products may be cloned or be retained more or less efficiently in the new host, and different bacteria have different numbers of rRNA genes and different sizes of genome, e.g. cyanobacteria may have between two and six rRNA operons and their genomes may vary between 2.2 and 3.6 $\times$ 10^9 Da (Nichols *et al.*, 1982). It is therefore necessary to treat the distribution of clone sequences in libraries with caution, and to realize that they may not accurately reflect sequence abundance *in situ* (Farelly *et al.*, 1995).

Measures of the relative abundance of a sequence in a sample can be estimated by using probes (to the sequence) to analyse total ribosomal RNA extracts (Giovannoni *et al.*, 1990; DeLong *et al.*, 1994). This approach also has some limitations, including the fact that different prokaryotes may contain different numbers of ribosomes and thus variable amounts of probe target (Ward *et al.*, 1992). A more direct measurement of cell abundance can be obtained by labelling probes with fluorescence, to probe and identify intact microorganisms directly in environmental samples (DeLong *et al.*, 1989). It is this approach that currently offers the greatest potential for accurate counts of organisms in natural samples without the need to culture them. It can also be used to link a sequence to a microbial morpho-type and to identify samples which may contain cells from which a sequence of particular taxonomic interest originates, thus providing a tool for use in isolation strategies (Huber *et al.*, 1995).

4.8 PROKARYOTE 16S rDNA SEQUENCE DIVERSITY IN NATURAL SAMPLES: THE RELATIONSHIP BETWEEN 16S rRNA SEQUENCE SIMILARITY AND PROKARYOTE SPECIES RECOGNIZED USING DNA : DNA PAIRING ASSAYS

Before describing some of the results of community analysis of uncultured prokaryotes using 16S rRNA sequences it is essential, in the context of this symposium, to discuss the relationship between 16S rRNA sequence similarities and prokaryote species as recognized using the polyphasic approach. First of all it is important to realize that there is no precise correlation between 16S rRNA sequence similarity and prokaryote nomenclature (Fox *et al.*, 1992; Martinez *et al.*, 1992). The 16S gene samples only a fraction of the genome, and by itself it tells nothing about the important phenotypic component which in practice often provides a guide to drawing the taxonomic line. However, strains of taxonomic species which share about 70% DNA : DNA reassociation often demonstrate inter-strain 16S rRNA sequence similarities of over 98%, and strains which show less than 98% 16S rRNA sequence similarity seldom demonstrate DNA : DNA pairing values of greater than 60% under optimal conditions (Amann *et al.*,

1992; Stackebrandt and Goebel, 1994). Within this framework, it is a reasonable working hypothesis to infer that any environmental sequence, which shares 98% 16S rRNA sequence or less similarity to sequences from established prokaryote species, potentially originates from a new species of prokaryote.

The 98% value is just a rule of thumb which has been developed using similarities measured between complete or very nearly complete 16S rRNA sequences. Clone sequences from environmental samples are often only fragments. This is because some of the PCR primers designed to be specific for particular groups will amplify only a fragment of the gene, e.g. for beta-proteobacteria ammonia oxidizers (McCaig *et al.*, 1994), or because only short fragments of individual clones are actually sequenced. Ribosomal RNA sequences are mosaics of variable and conserved sequence regions depending on structure and function (Woese *et al.*, 1983), and different partial fragments of sequence will sample this mosaic in different ways. Restriction of the analysis to more conserved stretches or to more variable regions will lead to higher or lower similarity values between sequences, respectively.

In addition to the known mosaic structure of 16S rRNA it is apparent that the position of hypervariable regions may vary in different groups of bacteria. It is therefore difficult to estimate precisely the degree of similarity of complete sequences from partial sequences (Stackebrandt and Rainey, 1995). Using selected sequence stretches from members of the important genus *Vibrio* as an example, i.e. region 50 through 500 and region 550 through 1500, the similarity is about 3% higher and 2% lower, respectively, when compared with the values obtained with complete sequences. In this genus the hypervariable regions are located in the 3′ half of the molecule, while, for example, in actinomycetes these regions are found within the 5′ half of the 16S rDNA molecule. In this group of organisms the similarity values of the 5′ half of the molecule are on the average 1.5% lower than determined for the total sequence. While most studies have analysed the 5′ 500 nucleotides, several studies have derived their conclusions about taxon composition from other parts of the molecule.

A more important concern about analysing only partial fragments of 16S rRNA sequences for the exploration of microbial diversity is the negative effects it has on the ability to infer relationships. One of the great strengths of sequence data is that it accumulates, so that old and new sequences can be directly compared, to produce an increasingly comprehensive taxonomic analysis of the microbial world. However, this can only be done if the stretches of sequence from different studies cover the same parts of the molecule. Trees based upon small fragments of 16S rRNA sequences are often less stable than those based on more characters, and resolving power is lost concerning the relationships of environmental sequences to those from cultured microorganisms (Schmidt *et al.*, 1991).

4.9 RESULTS OF SOME STUDIES USING SSU rRNA SEQUENCES TO EXPLORE PROKARYOTE DIVERSITY IN NATURE

The analysis of uncultured prokaryote communities using 16S rRNA sequences has now become commonplace and it has already been widely reviewed (Ward *et al.*, 1992; Embley *et al.*, 1994; Olsen, 1994), so here we will concentrate on a few generalizations and recent examples which illustrate the power and potential of the approach. Almost all of the sequences recovered so far from natural samples have been new to science. In part this undoubtedly reflects the still small number of cultured prokaryotes which have been analysed for their 16S rRNA sequences, but it also suggests that there is enormous prokaryote diversity in nature which remains to be discovered. Whether the pace of discovery of completely new lineages will continue, is not easy to appreciate from the small number of samples and habitats so far investigated. In samples taken from the oceans it is becoming apparent that sequences which form monophyletic groups with previously discovered sequences (e.g. the SAR 11 and SAR 7 bacterial clusters, and archaeal group I and II clusters) have been found in most samples (DeLong *et al.*, 1994; McInerney *et al.*, 1995; Mullins *et al.*, 1995).

Some of the environmental sequences may be artefacts of the PCR process, which has been shown sometimes to produce chimeric molecules from mixtures of 16S rRNA templates (Liesack *et al.*, 1991). However, as far as can be ascertained by current analytical approaches (Robison-Cox *et al.*, 1995) most of the new sequences appear to be genuine; for example Barns *et al.* (1994) reported that four of 98 clones analysed were chimeras, and Choi *et al.* (1995) found seven chimeras among 81 clones analysed.

Because of their depth of branching in the rRNA tree based upon sequences from cultured organisms, some of the environmental sequences can be predicted to originate from organisms which may represent entirely new major groups of prokaryotes. For example, one of the most interesting recent discoveries is that of 16S rRNA sequences from Archaea in marine samples. Before describing the results of these studies it is worth saying something about the diversity of the Archaea which are already in laboratory culture. Woese and co-workers (Woese *et al.*, 1990) have identified two Archaea kingdoms, the Crenarchaeota and Euryarchaeota, within the SSU rRNA tree for cultured Archaea. The two groups are defined with reference to the root relative to Bacterial outgroup sequences, which split cultured Archaea into these two putative monophyletic groups. The Archaea which make up the Euryarchaeota and Crenarchaeota are physiologically heterogeneous and have traditionally been collected from environments such as acidic hot springs, anaerobic sediments or saturated brines. Because they were seldom isolated from (for example) the open oceans, freshwaters, or agricultural soils, there was a perception that Archaea, with the exception of methanogenic Archaea, might not be important in global ecology (Olsen, 1994).

In 1992 this perception was shown to be fundamentally flawed. Two independent studies recovered two groups (termed group I and group II) of novel Archaea 16S rRNA sequences from small samples of deep and shallow waters taken from the Atlantic and Pacific Oceans (DeLong, 1992; Fuhrman *et al.*, 1992). Analysis of these sequences has revealed that group I marine Archaea sequences are so divergent (Figure 4.1) from those of cultured Archaea that some have suggested that they might be designated a new Archaeal kingdom (Olsen and Woese, 1993). Sequences which form a strongly supported monophyletic group with group I marine Archaea sequences have subsequently been recovered from gut samples from a Holothurian collected from 4870 m in the North Atlantic Abyss (McInerney *et al.*, 1995), and in samples from frigid marine surface waters in Antarctica (DeLong *et al.*, 1994). A recent survey of sequence diversity within a single paddy field in Japan has reported Archaeal sequences which formed a sister group to the marine Archaea group I, suggesting that members of this group may be present in non-marine systems (Ueda *et al.*, 1995). Unfortunately, the sequence fragments for the paddy field clones are rather short (c. 300 bases) and this has made assessment of their detailed position relative to other clones uncertain. Sequences which form a deeply branched monophyletic group with group II marine Archaea (Figure 4.1) have recently been detected in salt marsh sediments on the south-east coast of England, which are sometimes covered during high tides (M. Munson, D. Nedwell and T. M. Embley, unpublished data).

These data, from small samples taken over vast geographic distances, suggest that group I and II marine Archaea are extraordinarily important

Figure 4.1 Tree based upon SSU rRNA sequences from representative cultured Archaea including sequences from clone libraries prepared from different environmental samples. The tree is based upon analysis of 791 sequence positions using the Jukes and Cantor (Jukes and Cantor, 1969) correction and neighbour joining (Saitou and Nei, 1987). Essentially the same topology was recovered in maximum parsimony analysis. Interstingly, analysis using a method (Lockhart *et al.*, 1994) designed to deal with unequal base compositions between sequences, as occurs here, consistently placed environmental clones env. pJP27 and env. pJP78 on the Eucarya branch. Key: env. pJP, clones from mud volcano area of Jim's Black Pool, hot spring, Yellowstone National Park (Barns *et al.*, 1994); env. carna, Carna Bay bacterioplankton, west coast of Ireland, unpublished M. Malarkey and R. Powell, University of Galway; env. pPM7, deep water sample Atlantic, unpublished R. McInerney and R. Powell; env. ANT, Antarctic bacterioplankton (DeLang *et al.*, 1994); env. WHAR, Woods Hole DNA clone (DeLang 1992); env. SBAR, Santa Barbara Channel bacterioplankton (DeLang, 1992); env. C and env. P, coastal salt marsh clones, south-east coast of England, unpublished M. Munson, D. Nedwell and T.M. Embley. Published aligned sequences were obtained from the Ribosomal Data Base Project (Maidak *et al.*, 1994).

in ocean processes, a suggestion confirmed by recent data which suggest that at certain times of the year marine Archaea comprise about 34% of the prokaryote biomass in coastal Antarctic surface waters (DeLong *et al.*, 1994). The published data also suggest that pelagic marine Archaea are broadly represented by these two groups. This apparent limited diversity, in the sense of the number of new monophyletic groups so far discovered, would potentially greatly facilitate analysis of community dynamics using nucleic acid probes (Mullins *et al.*, 1995). None of the marine Archaea has yet been cultured so their role(s) in the ocean systems, and the reasons why they are apparently so extraordinarily successful, are completely unknown.

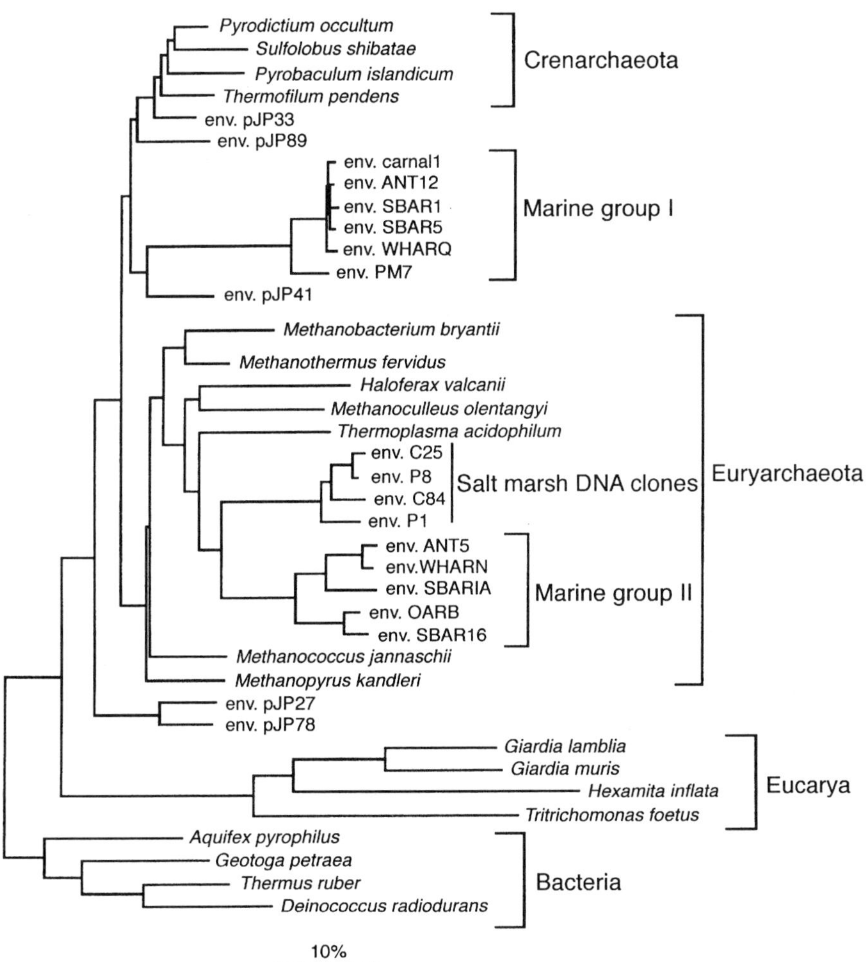

A second major discovery of Archaea relationships came from Barns, Pace and co-workers (Barns *et al.*, 1994) who used primers designed selectively to retrieve a variety of different Archaea SSU rRNA sequences from 5 ml of sediment taken from a hot spring termed 'Jim's Black Pool' in Yellowstone National Park. Several of the recovered sequences showed no close phylogenetic affinity to cultivated species of Archaea (pJP 33, 41, 89, 27, and pJP 78 in Figure 4.1). In different analyses (Barns *et al.*, 1994) these five sequences branched closer to the root of the Crenarchaeota than did sequences from cultured Crenarchaeota. In some analyses JP 27 and JP 78 were the sister group to all other Archaea (as in Figure 4.1 and as in the RDP maximum likelihood tree, Maidak *et al.*, 1994).

Other examples of widely distributed marine monophyletic groups of 16S rRNA sequences (Maidak *et al.*, 1994), are clearly bacterial in character (Mullins *et al.*, 1995). They include a large group of alpha-proteobacterial-like sequences (proteobacteria are a major phylum of cultured bacteria) termed the SAR 11 cluster and a group of cyanobacterial-like sequences called the SAR 7 cluster (Giovannoni *et al.*, 1990), both of which have been recovered from the Atlantic and Pacific Oceans and are probably very important in these systems (Giovannoni *et al.*, 1990; Schmidt *et al.*, 1991; Fuhrman *et al.*, 1993).

An apparently monophyletic group of widely distributed soil clone sequences has also recently been discovered (E. Stackebrandt, unpublished). Related sequences have been isolated from a forest soil in Queensland, Australia, a peat bog in North Germany, and mud from a hot spring in New Zealand. Recent sequence entries in the ribosomal data base project (server@rdp.lefe.uiuc.edu or http://rdp.life.uiuc.edu/RDP/data/ssu.html) suggest that short fragments of similar sequences have now been recovered from paddy field and soybean field samples. In some analyses representatives of these new sequences form a clade at the base of the Actinomycete phylum (a group noted for its abilities to produce antibiotics), along with an iron-oxidizing culture TH3 isolated from a copper bioleaching pond, and a filamentous strain '*Microthrix parvicella*' isolated from activated sludge. Whether or not suitable isolation strategies for uncultured members of this group can be inferred from the limited physiological information on TH3 and '*Microthrix*' remains to be tested.

As well as these highly divergent and geographically widespread monophyletic groups of sequences, most studies have also recovered sequences which bear more resemblance to those from different cultured prokaryote species. A single example will serve to illustrate the potential for discovery in even well-studied habitats. Choi *et al.* (1994) focused on spirochaete bacteria of the medically important genus *Treponema* occupying a gingival crevice in the mouth of a patient with severe destructive periodontitis. After extracting DNA from gingival material and amplifying rRNA genes, 81 clones related to *Treponema* were identified Further analysis revealed that these new sequences fell into 23 clusters defined at 98% or less 16S rRNA sequence similarity, calculated over the 5′ 500 bases of 16S rRNA sequence. Only two of these groups contained representatives of cultured oral *Treponema* species, suggesting that the other 21 clusters represent novel centres of variation, and potentially new species. Goebel and colleagues (unpublished) have subsequently managed to isolate and grow spirochaetes with the same sequences as two of the closely related (about 96.5% similar) sequence groups discovered by Choi *et al.* The isolated strains indeed show significant phenotypic differences from each other, and under current taxonomic practice would be classified as new species.

4.10 CONCLUDING REMARKS

At the beginning of this chapter we posed the question of how uncultured prokaryote diversity can be considered in familiar terms, since current practice in prokaryote systematics requires pure cultures for species description. The most useful approach to the analysis of prokaryote natural diversity is through tree diagrams of relationships based upon taxonomic analysis of sequence characters. Within this context environmental sequences can be directly compared with sequences from cultured taxa, prokaryote or eukaryote, and hypotheses of common ancestry and estimates of sequence divergence can be inferred. In this context, the relationships inferred for uncultured taxa and the information they convey about prokaryote distributions and evolutionary diversity, are no less real because they deal with organisms which have yet to be brought into recognized laboratory culture or classified as prokaryote species. The biological implications of the results so far are enormous, not only in terms of the number of new lineages which have been revealed and the degree of phylogenetic novelty displayed by some of them, but also in the promise of an enormous prokaryote phenotypic diversity still to be discovered. Although taxonomic analyses can proceed using sequences alone, only the isolation of living cultures will allow organism biology to be more fully explored. Some of the new prokaryotes uncovered by molecular analysis of environmental samples, are of such clear biological, phylogenetic or ecological significance that they merit the most determined efforts to do precisely this.

The excitement and potential for a microbiology which succeeds in linking historical analysis through sequence comparisons, with the study of organismal biology have been captured by the naturalist Edward O. Wilson (1994: 364):

> ′ If I could do it all again, and relive my vision for the twenty first century, I would be a microbial ecologist. Ten billion bacteria live in a gram of ordinary soil, a mere pinch held between thumb and forefinger. They represent thousands of species, almost none of which are known to science. Into that world I would go with the aid of

modern microscope and molecular analysis. I would cut my way through clonal forests sprawled across grains of sand, travel in an imagined submarine through drops of water proportionally the size of lakes, and track predators and prey in order to discover new life ways and alien food webs. All this and I need venture no more than ten paces outside my laboratory building. The jaguars, ants, and orchids would still occupy distant forests in all their splendour, but now they would be joined by an even stranger and vastly more complex living world virtually without end.'

4.11 REFERENCES

Amann, R.I., Lin, C., Key, R., Montgomery, L. and Stahl, D.A. (1992) Diversity among *Fibrobacter* isolates: towards a phylogenetic classification. *Systematic and Applied Microbiology*, **15**, 23–31.

Barns, S.M., Fundyga, R.E., Jeffries, M.W. and Pace, N.R. (1994) Remarkable archaeal diversity detected in a Yellowstone National Park hot spring environment. *Proceedings of the National Academy of Sciences, USA*, **91**, 1609–13.

Barrett, S.J. and Sneath, P.H.A. (1994) A numerical phenotypic taxonomic study of the genus *Neisseria*. *Microbiology*, **140**, 2867–91.

Brenner, D.J. (1973) Deoxyribonucleic acid reassociation in the taxonomy of enteric bacteria. *International Journal of Systematic Bacteriology*, **23**, 298–307.

Brock, T.D., Madigan, M.T., Martinko, J.M. and Parker, J. (1994) *Biology of Microorganisms*. Prentice Hall International Inc., New York.

Choi, B.K., Paster, B.J., Dewhirst, F.E. and Gobal, U.B. (1994) Diversity of cultivable and uncultivable oral spirochaeates from a patient with severe destructive periodontitis. *Infection and Immunity*, **62**, 1889–95.

Cohan, F.M. (1994) Genetic exchange and evolutionary divergence in prokaryotes. *Trends in Ecology and Evolution*, **9**, 175–80.

DeLong, E.F. (1992) Archaea in coastal marine environments. *Proceedings of the National Academy of Sciences, USA*, **89**, 5685–9.

DeLong, E.F., Wickham, G.S., and Pace, N.R. (1989) Phylogenetic stains: ribosomal RNA-based probes for the identification of single cells. *Science*, **243**, 1360–3.

DeLong, E.F., Wu, K.Y., Prezelin, B.B. and Jovine, R.V.M. (1994) High abundance of Archaea in Antarctic marine picoplankton. *Nature*, **371**, 695–7.

Dupuy, N., Willems, A., Pot, B., Dewettinck, D., Vandenbruaene, I., Maestrojuan, G., Dreyfus, B., Kersters, K., Collins, M.D. and Gillis, M. (1994) Phenotypic and genotypic characterisation of bradyrhizobia nodulating the leguminous tree *Acacia albida*. *International Journal of Systematic Bacteriology*, **44**, 461–73.

Embley, T.M., Hirt, R.P. and Williams, D.M. (1994) Biodiversity at the molecular level: the domains, kingdoms and phyla of life. *Philosophical Transactions of the Royal Society London B*, **345**, 21–33.

Farelly, V., Rainey, F.A. and Stackebrandt, E. (1995) Effect of genome size and rrn copy number on PCR amplification of 16S rRNA genes from a mixture of bacterial species. *Applied and Environmental Microbiology*, **61**, 2798–801.

Farris, J.S. (1977) On the phenetic approach to vertebrate classification, in *Major Patterns in Vertebrate Evolution* (eds M. K. Hecht, P. Goody and B. M. Hecht), Plenum, New York, pp. 823–50.

Felsenstein, J. (1981) Evolutionary trees from DNA sequences: a maximum likelihood approach. *Journal of Molecular Evolution*, **17**, 368–76.

Felsenstein, J. (1985) Confidence limits on phylogenies: an approach using the bootstrap. *Evolution*, **39**, 783–91.

Felsenstein, J. (1988) Phylogenies from molecular sequences: inference and reliability. *Annual Review of Genetics*, **22**, 521–65.

Forterre, P., Benachenhou-Lahfa, N., Confalonieri, F., Duguet, M., Elie, C. and Labedan, B. (1993) The nature of the last universal ancestor and the root of the tree of life, still open questions. *BioSystems*, **28**, 15–32.

Fox, G.E., Wisotzkey, J.D. and Jurtshuk, P., Jr (1992) How close is close: 16S rRNA sequence identity may not be sufficient to guarantee species identity. *International Journal of Systematic Bacteriology*, **42**, 166–70.

Fuhrman, J.A., McCallum, K. and Davis, A.A. (1992) Novel major archaebacterial group from marine plankton. *Nature*, **356**, 148–9.

Fuhrman, J.A., McCallum, K. and Davis, A.A. (1993) Phylogenetic diversity of subsurface marine microbial communities from the Atlantic and Pacific Oceans. *Applied and Environmental Microbiology*, **59**, 1294–302.

Giovannoni, S.J., Britschgi, T.B., Moyer, C.L. and Field, K. G. (1990) Genetic diversity in Sargasso Sea bacterioplankton. *Nature*, **345**, 60–3.

Gogarten, J.P. (1995) The early evolution of cellular life. *Trends in Ecology and Evolution*, **10**, 147–51.

Golding, G.B. and Gupta, R.S. (1995) Protein-based phylogenies support a chimeric origin for the eukaryotic genome. *Molecular Biology and Evolution*, **12**, 1–6.

Goodfellow, M. and O'Donnell, A.G. (1993) *Handbook of New Bacterial Systematics*. Academic Press, London.

Grimont, P.A.D., Popoff, M.Y., Grimont, F., Coynault, C. and Lemelin, M. (1980) Reproducibility and correlation study of three deoxyribonucleic hybridisation procedures. *Current Microbiology*, **4**, 325–30.

Hartford, T. and Sneath, P.H.A. (1990) Experimental error in DNA–DNA pairing: a survey of the literature. *Journal of Applied Bacteriology*, **68**, 527–42.

Harvey, P.H. and Pagel, M.D. (1991) *The Comparative Method in Evolutionary Biology*. Oxford University Press, Oxford.

Hawksworth, D.L. (1995) *Biodiversity, Measurement and Estimation*. Chapman & Hall, London.

Hillis, D.M., Allard, M.W. and Miyamoto, M.M. (1993) Analysis of DNA sequence data: phylogenetic inference. *Methods in Enzymology*, **224**, 456–87.

Huber, R., Burggraf, S., Mayer, T., Barns, S.M., Rossnagel, P. and Stetter, K.O. (1995) Isolation of a hyperthermophilic archaeum predicted by *in situ* RNA analysis. *Nature*, **376**, 57–8.

Johnson, J.L. (1973) Use of nucleic acid homologies in the taxonomy of anaerobic bacteria. *International Journal of Systematic Bacteriology*, **23**, 308–15.

Jukes, T.H. and Cantor, C.R. (1969) Evolution of protein molecules, in *Mammalian Protein Metabolism* (ed. H.N. Munro), Academic Press, New York, pp. 21–132.

Lake, J.A. (1994) Reconstructing evolutionary trees from DNA and protein sequences: Paralinear distances. *Proceedings of the National Academy of Sciences, USA*, **91**, 1455–9.

Lento, G.M., Hickson, R.E., Chambers, G.K. and Penny, D. (1995) Use of spectral analysis to test hypotheses on the origins of pinnipeds. *Molecular Biology and Evolution*, **12**, 28–52.

Liesack, W., Weyland, H. and Stackebrandt, E. (1991) Potential risks of gene amplification by PCR as determined by 16S ribosomal DNA analysis of a mixed-culture of strict barophilic bacteria. *Microbial Ecology*, **21**, 191–8.

Lockhart, P.J., Steel, M.A., Hendy, M.D. and Penny, D. (1994) Recovering evolutionary trees under a more realistic model of sequence evolution. *Molecular Biology and Evolution*, **11**, 605–12.

Ludwig, W., Neumaier, J., Klugbauer, N., Brockmann, E., Roller, C., Jilg, S., Reetz, K., Schachtner, I., Ludvigsen, A., Bachleitner, M., Fischer, U. and Schleifer, K.H. (1993) Phylogenetic relationships of Bacteria based on comparative sequence analysis of elongation factor Tu and ATP-synthase beta-subunit genes. *Antonie van Leeuwenhoek*, **64**, 285–305.

Maidak, B.L., Larsen, N., McCaughey, J., Overbeek, R., Olsen, G.J., Fogel, K., Blandy, J. and Woese, C.R. (1994) The ribosomal database project. *Nucleic Acids Research*, **22**, 3485–7.

Mallet, J. (1995) A species definition for the modern synthesis. *Trends in Ecology and Evolution*, **10**, 294–9.

Martinez, A.J., Benlloch, S. and Collins, M.D. (1992) Phylogenetic interrelationships of members of the genera *Aeromonas* and *Plesiomonas* as determined by 16S ribosomal DNA sequencing: lack of congruence with results of DNA–DNA hybridisations. *International Journal of Systematic Bacteriology*, **42**, 412–21.

Maynard-Smith, J. (1995) Do bacteria have population genetics? in *Population Genetics of Bacteria* (eds S. Baumberg, J.P.W. Young, E.M.H. Wellington and J.R. Saunders), Cambridge University Press, Cambridge, pp. 1–12.

Maynard-Smith, J., Smith, N.H., O'Rourke, M. and Spratt, B.G. (1993) How clonal are bacteria? *Proceedings of the National Academy of Sciences, USA*, **90**, 4384–8.

Mayr, E. (1970) *Populations, Species and Evolution*. Belknap Press of Harvard, Cambridge, Massachusetts.

Mayr, E. (1987) The ontological status of species: scientific progress and philosophical terminology. *Biology and Philosophy*, **2**, 145–66.

McCaig, A.E., Embley, T.M. and Prosser, J.I. (1994) Molecular analysis of enrichment cultures of marine ammonia oxidisers. *FEMS Microbiology Letters*, **120**, 363–8.

McInerney, J.O., Wilkinson, M., Patching, J.W., Embley, T.M. and Powell, R. (1995) Recovery and phylogenetic analysis of novel archaeal rRNA sequences from a deep-sea deposit feeder. *Applied and Environmental Microbiology*, **61**, 1646–8.

Mullins, T.D., Britschgi, T.B., Krest, R.L. and Giovannoni, S.J. (1995) Genetic comparisons reveal the same unknown bacterial lineages in Atlantic and Pacific bacterioplankton communities. *Limnology and Oceanography*, **40**, 148–58.

Nelson, G. (1989) Species and taxa: systematics and evolution, in *Speciation and its Consequences* (eds D. Otte and J.A. Endler), Sinauer Associates, Sunderland, Massachusetts, pp. 60–81.

Nichols, J.M., Foulds, I.J., Crouch, D.H. and Carr, N.G. (1982) The diversity of cyanobacterial genomes with respect to ribosomal RNA cistrons. *Journal of General Microbiology*, **128**, 2739–46.

Olsen, G.J. (1994) Archaea, Archaea, everywhere. *Nature*, **371**, 657–8.

Olsen, G.J., Lane, D.J., Giovannoni, S.J., Pace, N.R. and Stahl, D.A. (1986) Microbial ecology and evolution: a ribosomal RNA approach. *Annual Review of Microbiology*, **40**, 337–65.

Olsen, G.J. and Woese, C.R. (1993) Ribosomal RNA: a key to phylogeny. *The FASEB Journal*, **7**, 113–23.

Pace, N.R., Stahl, D.A., Lane, D.J. and Olsen, G.J. (1985) Analysing natural microbial populations by rRNA sequences. *American Society for Microbiology News*, **51**, 4–12.

Patterson, C. (1987) Introduction, in *Molecules and Morphology in Evolution: Conflict or Compromise* (ed. C. Patterson), Cambridge University Press, Cambridge, pp. 1–22.

Patterson, C. (1988) Homology in classical and molecular biology. *Molecular Biology and Evolution*, **5**, 603–25.

Platnick, N.I. (1989) Cladistics and phylogenetic analysis today, in *The Hierarchy of Life* (eds B. Fernhom, K. Bremer and X. Jornvall), Excerpta Medica, Amsterdam, pp. 17–24.

Rainey, F.A., Janssen, P.H., Morgan, H.W. and Stackebrandt, E. (1993) A biphasic approach to the determination of the phenotypic and genotypic diversity of some anaerobic, celluloytic, thermophilic, rod-shaped bacteria. *Antonie van Leeuwenhoek*, **64**, 341–55.

Reysenbach, A.-L., Giver, L.J., Wickham, G.S. and Space, N.R. (1992) Differential amplification of rRNA genes by polymerase chain reaction. *Applied and Environmental Microbiology*, **58**, 3417–18.

Robison-Cox, J.F., Bateson, M.M. and Ward, D.M. (1995) Evaluation of nearest neighbor methods for detection of chimeric small-subunit rRNA sequences. *Applied and Environmental Microbiology*, **61**, 1240–5.

Rossau, R., Vandenbussche, G., Thielemans, S., Segers, P., Grosch, H., Gothe, E., Mannheim, W. and De Ley, J. (1989) Ribosomal ribonucleic acid cistron similarities and deoxyribonucleic acid homologies of *Neisseria*, *Kingella*, *Eikenella*, *Simonsiella*, *Alysiella* and centres for disease control groups EF-4 and M-5 in the emended family *Neissseriaceae*. *International Journal of Systematic Bacteriology*, **39**, 185–98.

Sackin, M.J. and Jones, D. (1993) Computer assisted classification, in *Handbook of New Bacterial Systematics* (eds M. Goodfellow and A.G. O'Donnell), Academic Press, London, pp. 281–313.

Saitou, N. and Nei, M. (1987) The neighbor-joining method: new method for reconstructing phylogenetic trees. *Molecular Biology and Evolution*, **4**, 406–25.

Schmidt, T.M., DeLong, E.F. and Pace, N.R. (1991) Analysis of a marine picoplankton community by 16S rRNA gene cloning and sequencing. *Journal of Bacteriology*, **173**, 4371–8.

Segers, P., Vancanneyt, M., Pot, B., Torck, U., Hoste, B., Dewerrinck, D., Falsen, E., Kersters, K. and de Vos, P. (1994) Classification of *Pseudomonas diminuta* Leifson and Hugh 1954 and *Pseudomonas vesicularis* Busing, Doll and Freytag 1953 in *Brevundimonas* gen. nov. as *Brevundimonas diminuta* comb. nov. and *Brevundimonas vesicularis* comb. nov., respectively. *International Journal of Systematic Bacteriology*, **44**, 499–510.

Selander, R.K., Li, J., Boyd, F. and Wang, F. S. (1994) DNA sequence analysis of the genetic structure of populations of *Salmonella enterica* and *Escherichia coli*, in *Bacterial Diversity and Systematics* (eds F.G. Priest, A. Ramos-Cormenzana and B.J. Tindall), Plenum Press, New York, pp. 17–49.

Sneath, P.H.A. (1957a) The application of computers to taxonomy. *The Journal of General Microbiology*, **17**, 201–26.

Sneath, P.H.A. (1957b) Some thoughts on bacterial classification. *Journal of General Microbiology*, **17**, 184–200.

Sneath, P.H.A. (1988) The phenetic and cladistic approaches, in *Prospects in Systematics* (ed. D.L. Hawksworth), Clarendon Press, Oxford, pp. 252–73.

Sneath, P.H.A. (1989) Analysis and interpretation of sequence data for bacterial systematics: the view of a numerical taxonomist. *Systematic and Applied Microbiology*, **12**, 15–31.

Sneath, P.H.A. and Sokal, R.R. (1973) *Numerical Taxonomy*. W.H. Freeman and Company, San Francisco.

Sokal, R.R. and Crovello, T.J. (1970) The biological species concept: a critical evaluation. *American Naturalist*, **104**, 127–53.

Sogin, M.L. (1991) Early evolution and the origin of eukaryotes. *Current Opinion in Genetics and Development*, **1**, 457–63.

Springer, E., Sachs, M.S., Woese, C.R. and Boone, D.R. (1995) Partial gene sequences for the A subunit of methyl-coenzyme M reductase (*mcr*I) as a phylogenetic tool for the family *Methanosarcinaceae*. *International Journal of Systematic Bacteriology*, **45**, 554–9.

Stackebrandt, E. and Goebel, B.M. (1994) Taxonomic note: a place for DNA : DNA reassociation and 16S rRNA sequence analysis in the present species definition in bacteriology. *International Journal of Systematic Bacteriology*, **44**, 846–9.

Stackebrandt, E. and Rainey, F.A. (1995) Partial and complete 16S rDNA sequences, their use in generation of 16S rDNA phylogenetic trees and their implications in molecular ecological studies, in *Molecular Microbial Ecology Manual* (eds A.D.L. Akkermans, J.D. van Elsas and F.J. de Bruin), Kluwer Academic Publishers, Dordrecht, The Netherlands, pp. 1–17.

Torsvik, V., Salte, K., Sorheim, R. and Goksoyr, J. (1990) Comparison of phenotypic diversity and DNA heterogeneity in a population of soil bacteria. *Applied and Environmental Microbiology*, **56**, 776–81.

Ueda, T., Suga, Y. and Matsuguchi, T. (1995) Molecular phylogenetic analysis of a soil microbial community in a soybean field. *European Journal of Soil Science*, **46**, 415–21.

Wagner, A., Blackstone, N., Cartwright, P., Dick, M., Misof, B., Snow, P., Wagner, G.P., Bartels, J., Murtha, M. and Pendleton, J. (1994) Surveys of gene families using polymerase chain reaction: PCR selection and PCR drift. *Systematic Biology*, **43**, 250–61.

Ward, D.M., Bateson, M.M., Weller, R. and Ruff-Roberts, A.L. (1992) Ribosomal RNA analysis of microorganisms as they occur in nature. *Advances in Microbial Ecology*, **12**, 219–86.

Wayne, L.G., Brenner, D.J., Colwell, R.R., Grimont, P.A. D., Kandler, O., Krichevsky, M.I., Moore, L.H., Moore, W.E.C., Murray, R.G.E., Stackebrandt, E., Starr, M.P. and Truper, H.G. (1987) Report of the *ad hoc* committee on reconciliation of approaches to bacterial systematics. *International Journal of Systematic Bacteriology*, **37**, 463–4.

Williams, S.T., Goodfellow, M., Alderson, G., Wellington, E.M.H., Sneath, P.H.A. and Sackin, M.J. (1983) Numerical classification of *Streptomyces* and related genera. *Journal of General Microbiology*, **129**, 1743–813.

Wilson, E.O. (1994) *Naturalist*. Island Press, Washington.

Woese, C.R. (1987) Bacterial evolution. *Microbiological Reviews*, **51**, 221–71.

Woese, C.R. (1994) There must be a prokaryote somewhere: microbiology's search for itself. *Microbiological Reviews*, **58**, 1–9.

Woese, C.R. and Fox, G.E. (1977) Phylogenetic structure of the prokaryotic domain: the primary kingdoms. *Proceedings of the National Academy of Sciences, USA*, **74**, 5088–90.

Woese, C.R., Gutell, R., Gupta, R. and Noller, H.F. (1983) Detailed analysis of the higher-order structure of 16S-like ribosomal ribonucleic acids. *Microbiological Reviews*, **47**, 621–69.

Woese, C.R., Kandler, O. and Wheelis, M.L. (1990) Towards a natural system of organisms: Proposal for the domains Archaea, Bacteria, and Eucarya. *Proceedings of the National Academy of Sciences, USA*, **87**, 4576–9.

5

Species problems in eukaryotic algae: a modern perspective

D. M. John and C. A. Maggs

Contacting address: Department of Botany, The Natural History Museum, Cromwell Road, London SW7 5BD, UK

ABSTRACT

The algae are a very diverse, unnatural assemblage of seven distinct phylogenetic lineages of oxygen-producing photosynthetic organisms. The number of species presently recognized, about 36 000, is estimated to be about 10% of the true worldwide total. Species-level taxonomy is based explicitly or implicitly on the detection of morphological discontinuities in sets of field-collected or cultured algae. New data on phenotypic variation, breeding compatibility, and molecular analyses are clarifying species concepts. Culture studies have demonstrated that the species concept traditionally applied to many morphologically simple algae is too narrow. Polyploidy, for example, can cause spontaneous changes in the morphology of some clonal cultures of green algal groups. The biological species concept has not been widely examined since relatively few algae are available as clonal cultures and sexual reproduction is either unknown or rare and unpredictable in the majority of algal classes. It has been tested most frequently in rhodophytes, chlorophytes and diatoms: congruence has been shown to exist in some genera between morphological data and sexual compatibility. Cryptic variation has been demonstrated by the discovery of mating complexes or sibling species within traditional morphospecies. In diatoms, for example, investigations of sexual compatibility indicate that many 'morphospecies' are masking significant variation. Molecular data are assisting in calibrating or testing the limits of morphospecies and may provide the touchstone for the interpretation of other data. In the future the traditional morphological species concept will increasingly operate alongside less formalized concepts involving data from other disciplines. These data will enable larger suites of concordant characters to be used for calibrating species concepts and defining species boundaries. If cladistic methods are

Species: The Units of Biodiversity. Edited by M.F. Claridge, H.A. Dawah and M.R. Wilson. Published in 1997 by Chapman & Hall. ISBN 0 412 63120 2

applied more frequently in future to study species-level relationships in algae then the 'phylogenetic species concept' will come into wider use.

5.1 INTRODUCTION

It is not an easy task to review species concepts and species delimitation in an unnatural assemblage of oxygen-producing photosynthetic organisms possessing enormous morphological, cytological, molecular and reproductive diversity. Ultrastructural, biochemical and molecular studies have demonstrated the 'algae' to be divisible into seven evolutionary lineages, the chlorophytes, chromophytes, rhodophytes, dinophytes, euglenophytes, cryptophytes and glaucophytes. There is an increasing tendency to adopt a 'protistan' view and to place these lineages of so-called primitive eukaryotes into the Protista, a Kingdom that can no longer be justified (Corliss, 1994). For convenience, all those photosynthetic eukaryotic protists traditionally referred to as algae are treated in this review. Excluded are the 'blue–green algae', a prokaryotic group now more correctly referred to as the Cyanobacteria or 'blue–green bacteria' (see Castenholz, 1992, for a review of species concepts in this group).

Global concern over biodiversity issues, the impact of molecular data on taxonomic decision making, and the increasing adoption of cladistic methods of data analysis are just a few of the reasons why it is timely to debate species concepts in algae. Increasingly, agendas for future environmental research depend upon comparisons of estimates of species diversity. It is tacitly assumed that the units compared are equivalent – an assumption that is clearly untenable when dealing with a diverse and unnatural assemblage like the algae. Despite this non-equivalence, such comparisons continue to be made along with estimates, by taxonomic specialists in particular groups, of the numbers of species still to be described. About 36 000 species have been described and educated 'guesstimates' predict the true global total to be in excess of 200 000 (Andersen, 1992; John, 1994). Recently it has been suggested that the estimate for the diatoms, a chromophyte group, should been increased by a factor of 10 to 200 000 species (Mann and Droop, 1996), so that the number of species currently recognized would be only about 11% of the real total (see Table 5.1).

A vast literature exists on the nature of species and species concepts (see Grant, 1981, for reviews) with no consensus emerging on what constitutes a species, nor agreement on the philosophical approach to be taken towards species as classes or individuals. Towards the end of the 19th century, discontinuities or unevenness in the pattern of morphological variation were the universally accepted criteria for distinguishing species (the 'morphological species concept'). Other species concepts

Table 5.1 Diversity of algae: estimated known number of species and predicted world species totals within the major series or lineages of algae

Taxonomic grouping	*Known species*	*Estimated species*	*Proportion known (%)*
KINGDOM PLANTAE SUBKINGDOM VIRIDIPLANTAE Chlorophyta (four algal classes)	15 345	79 000	19
SUBKINGDOM BILIPHYTA (two algal classes)	4 263	12 800	33
KINGDOM CHROMISTA Chromophyta *sensu lato* (10 algal classes)	13 776	217 200	6
KINGDOM PROTOZOA (three algal classes)	2 853	9 270	31
TOTAL	**36 237**	**318 270**	**11**

developed subsequently include the 'biological species concept' (Mayr, 1992), which has found wide acceptance as the most objective way of defining species limits in sexually reproducing animals. The seven species concepts reviewed by Steussy in 1990 (morphological, biological, genetic, palaeontological, evolutionary, phylogenetic, biosystematic) are all applicable to algae, although the morphological species concept is still the one most frequently applied. It is not our intention to review the long-running debate on the relative merits of adopting one concept as opposed to another, but to reflect on how algal species are dealt with in practice by reference to selected examples.

5.2 THE MORPHOLOGICAL SPECIES CONCEPT

5.2.1 Historical overview

The morphological species concept has dominated algal systematics: the vast majority of species are recognized by discontinuities in morphological characters. It applies implicitly or explicitly even to those genera or groups in which attempts have been made to use other species concepts. The morphological species or 'morphospecies' concept – sometimes referred to as the Linnaean, classical or typological concept – considers a taxon described at one point in time. It is assumed that morphological characters are stable until a quite different form is exhibited under changed physiological conditions or at another growth stage. Of the 14 algal genera recognized by Linnaeus in 1753, only four (*Conferva, Ulva, Fucus, Chara*) are attributable to the 'algae' as defined in a modern sense; these macroalgae were recognized by readily observed morphological criteria. William Hudson was the first English botanist to use the Linnaean system of classification and nomenclature. The species concepts he used when describing a large number of new algae (Hudson, 1762) can still be deduced by reference to his herbarium, literature sources and his own publications (Irvine and Dixon, 1982). In the following century, the period from 1800–1875 has been referred to as the 'Golden Age of plant taxonomy' (Prescott, 1951) because major advances were made in plant systematics, including the description of the majority of the genera recognized today. These advances were in large part due to considerable improvement in optical instruments (e.g. invention of the achromatic lens in 1827) that facilitated critical observations of taxonomically important features of both microscopic and macroscopic algae. Some of the taxonomic problems encountered today when working with microalgae date back to the 19th century. Many are related to the failure of early taxonomists to provide complete species diagnoses, take account of variability in key diagnostic features or preserve material for future examination. The incompleteness of the diagnoses was often due to an examination of limited amounts of field-collected material with the result that some sexual, life-history and ontogenetic stages were not observed. Nevertheless, key characters for identification of macroalgae were recognized by some talented taxonomists working in the early and mid-19th century. For example, the description of *Fucus devoniensis* [now *Ahnfeltiopsis devoniensis* (Greville) Silva et De Cew, see below] by Greville (1821) was based on acute morphological and phenological observations by Mrs Amelia Griffiths; the importance of one of the features she noted was only understood a century-and-a-half later, after a comprehensive study of this group (Maggs *et al.*, 1992).

A new wave of interest in algal systematics in the 1950s followed the introduction of the transmission electron microscope (TEM), improvements in interference light microscopy, and the availability of a vast array of equipment developed for the cell biologist. The introduction in the mid-1960s of the scanning electron microscope (SEM) was of particular significance to phycologists studying algae with interesting surface features. These powerful and versatile research tools led to the discovery of new suites of morphological characters and a re-interpretation of those described earlier. The synthesis and evaluation of ultrastructural data strengthened the morphological species concept and gave rise to a dramatic increase in the number of recognized species. For example, all but one of almost 50 species now attributed to the chrysophyte genus *Paraphysomonas* were described after the advent of the electron microscope, when the fine details of scale morphology became evident (Preisig and Hibberd, 1982). Several other groups (e.g. diatoms, stoneworts) benefited greatly from SEM examination: many new 'morphospecies' were described, while some existing species were reduced to synonymy.

The ability to grow clones of algae (axenically or non-axenically) under rigorously controlled laboratory conditions has permitted great advances in our understanding of algal species by providing new information on morphology and cytology, reproduction, life-history and ontogeny. Laboratory culture of algae had its beginning towards the end of the last century but it was not until the 1920s and 1930s that 'type cultures' were established. During the 1940s, culture-based investigations of algae became fully established through the work of Pringsheim, and they were continued in the following decade by Bold and his co-workers who took full advantage of new culture techniques and media then available (see reviews by Pringsheim, 1967; Bold, 1974). Today, species are often recognized after detailed culture investigations; eventually the synthesis of the results of culture-based studies with other types of data may lead to the development of sounder species concepts and taxonomic systems. Species descriptions should ideally be based on a study of several clones since a single clone will not possess the range of variation found in natural populations (Lewin, 1975). The establishment of formal 'Culture Collections' has been of tremendous significance, particularly as they can hold cultures of the 'type' when new species are described. These are of paramount importance for maintaining populations of microalgae in the same way as herbaria house critical collections of preserved macroalgal material.

In relatively simple algae there is a limited range of morphological characters available for species diagnosis, so other features increase in importance. These include characters associated with nucleic acids, metabolism, descriptive biochemistry, physiology, ontology, phenology, breeding relations, life-history and ecology. As it has only been possible to study some of these aspects relatively recently, microalgal taxonomy is going through a revolutionary period. Attention is often directed towards these investigations in order to resolve problems remaining after the diligent application of alpha taxonomy (naming and describing of organisms). However, to date only a small number of algal taxa have been investigated using a full range of sophisticated molecular, genetic, cytological, breeding and culture techniques. Selected examples that demonstrate the contribution of molecular data and crossability experiments to our understanding of algal species will be considered later.

5.2.2 The type method in algae

The classical taxonomic method involving comparison of material with type specimens ('the type method', see Silva, 1952) is of fundamental importance when studying macroalgae and some microscopic groups like diatoms (Williams, 1993). Diatoms are one of the few classes of microalgae that retain many taxonomically important characters on preservation, because their systematics are based largely on the highly resistant siliceous frustule. For seaweeds in particular, preparation of monographs depends on the macroscopic and microscopic examination of dried herbarium mounts, liquid-preserved material and, when available, living material. Morphologically defined species are considered more reliable and well-founded if the variability of taxonomically important characters has been thoroughly investigated. Assessing the significance of morphological variation is easier in seaweeds and other macroalgae than in microalgae, providing that there are large numbers of herbarium specimens available for examination and that key characters remain unaffected by long-term preservation. However, sometimes staggeringly large numbers of specimens must be examined to assess the full range of variation in potentially important taxonomic characters. For example, Littler and Littler (1992) studied all available herbarium material including types, approximately 1500 specimens, when monographing the green alga *Avrainvillea* in the tropical western Atlantic. They concluded from this analysis and experimental evidence (see below) that although traditionally important in species recognition, some of the more obvious gross morphological traits (e.g. number of stipes per holdfast, blade thickness and texture) exhibit substantial variation in response to substratum features and other environmental parameters. In amending diagnoses, emphasis was placed therefore on more reliable anatomical characters (e.g. form and shape of medullary and cortical filaments) that were well preserved in dried herbarium material.

Recent quantitative morphological studies (Necchi and Zucchi, 1995; Vis *et al.*, 1995) have greatly clarified the systematics of freshwater representatives of the red algal genera *Batrachospermum* and *Audouinella*. Necchi and Zucchi evaluated qualitative and quantitative data on all potentially useful morphological characters for 45 Brazilian populations of *Audouinella* and all relevant type specimens. Although the size ranges of many morphological features overlapped between species, multivariate analysis unequivocally grouped the populations into five clusters, four of which included a type specimen (one type was unavailable). Five species were therefore clearly discriminated by this elegant application of the type method. In contrast, the large majority of microalgal species lose many of their most valuable taxonomic characters on preservation and field-collected material frequently lacks features required for precise determination (e.g. reproductive structures, certain life-history stages). As a consequence, culturing of clones under carefully manipulated laboratory conditions is often indispensable for describing and identifying microalgae. It is common to see references in the literature to 'type cultures'. These are only living elements of the type since, according to the *International Code of Botanical Nomenclature*, Article 9.5, 'Type specimens of names of taxa must be preserved permanently and cannot be living plants or cultures' although it recommends (9A.1) that 'Wherever practicable a living culture should be prepared from the holotype material of the name of a newly described taxon of fungi or algae and deposited in a reputable culture collection'. The maintenance of a cultured element of the type has proven invaluable for comparisons with the original description and for the examination of characters subsequently recognized to be of taxonomic significance. There are many problems associated with using clonal cultures to characterize species, however, particularly if morphological features are thought to be species-specific on the basis of their state in single clonal isolates of each species.

5.2.3 Significance of experimental data

The contribution of experimental studies in the field and laboratory to our understanding of morphologically defined algae cannot be overestimated. Such studies have demonstrated the extraordinary degree of plasticity exhibited by many algae and the extent to which morphological characters are environmentally and genetically controlled. Frequently there has been a failure to recognize just how phenotypically plastic are algae, leading to doubt attaching to the validity of many earlier-described species. Even at the present time insufficient account is taken of the taxonomic, as opposed to the physiological or ecological, significance of transplants and other manipulative field experiments. An exception to this is the study by Littler and Littler (1992) of populations of all western Atlantic species of *Avrainvillea* in which they were able to demonstrate, for example, that *A. levis* Howe develops a bulbous holdfast in sand or soft sediment and a conical holdfast when growing on hard substrata. In several other species of *Avrainvillea* quantitative characters considered of taxonomic importance (e.g. stipe length) were discovered to be unreliable since influenced by light and depth conditions. There are many other genera in which morphological characters show continuous variation, with each character a graded series of expressions that overlap between species. Although continuity of ranges of variation in key characters between putative species is suggestive of conspecificity, this has not always proven to be the case when thoroughly investigated. For example, species discrimination in the notoriously difficult green algal genera *Cladophora* and *Enteromorpha* is often based on overlapping quantitative characters (e.g. cellular dimensions, number of pyrenoids per cell); tables are often more useful than keys for comparing character combinations and identifying species. Several investigations of crossability within these two genera (Bliding, 1963; Hoek, 1963, 1964, 1982) have demonstrated a close correspondence between the results obtained using biological and morphological species criteria.

As already stated, the contribution of culture studies to the evaluation of microalgal morphospecies cannot be overestimated. These have enabled the validity of algal morphospecies originally described from field-collected material to be tested. Several studies have demonstrated that species are mere 'ecomorphs' or 'ecophenes' characteristic of particular environmental conditions. One genus that has received considerable attention is the very polymorphic and widely distributed green alga *Stigeoclonium*. As a result of culture studies, the most reliable and conservative characters for distinguishing its species are those associated with the system of prostrate filaments and germination of the zoospores. Of the 28 species recognized by Islam (1963), only three were recognized by Simons *et al.* (1986). In his review of algal species diversity, Andersen (1992) mentions another well-researched example, *Scenedesmus*, a green algal genus containing 330 species and 1300 varieties according to its most recent revision (Hegewald and Silva, 1988). Culture studies by Trainor and co-workers (Trainor, 1991; Trainor and Egan, 1991) have demonstrated that the principal characters used for species discrimination in *Scenedesmus* (e.g. presence and form of spines, colony size) are temperature-dependent. Many species of *Scenedesmus* are now considered to be 'ecomorphs' that, in response to environmental cues, undergo a seasonal or aseasonal succession termed cyclomorphosis. As a result of these studies, the number of valid species is estimated by Trainor (personal communication, in Andersen, 1992) to lie somewhere between 12 and 30.

Similarly, culture studies have demonstrated considerable phenotypic variation in desmids, a green algal group whose species are typically distinguished on the basis of morphological characteristics observed in field-collected material. In view of the variation exhibited by desmid species, the recent floristic treatment of British desmids by Brook (1997) illustrates populations of cells in order to demonstrate the full form range of each species as encountered in nature. Breeding experiments indicate the possible existence of syngens within desmid morphospecies (Blackburn and Tyler, 1987), as reported in other green algal groups and diatoms (see section 5.3). Desmidologists will continue to define desmids using the traditional typological method since the majority only reproduce asexually. Many culture-based studies on diatoms also reveal that taxonomically important features of frustule morphology are influenced by a wide range of environmental factors that include temperature, salinity, eutrophication and silica supply (see Cox, 1993, for references).

It is important not to fall into the trap of considering all environmentally associated morphological variation to be the result of phenotypic plasticity. Recent studies have borne out Russell and Fielding's (1981) prediction that many variable characters once attributed to phenotypic plasticity will be shown to be at least partly heritable. This has proven the case in *Chondrus crispus* (Stackhouse) Guiry, a very variable red alga to which a large number of infraspecific epithets have been applied. In this alga Guiry (1992) discovered that morphological variation is associated with particular shore habitats. Of the 26 isolates examined by him, all were fully interfertile in culture and yet each maintained its own distinct morphology. It is clear, therefore, that morphological variability in this single biological species is under both environmental and genetical control. Another example is the brown alga *Pilayella littoralis* (L.) Kjellman in which the significance of morphological variation, particularly differences in axis diameter between Baltic and Atlantic populations, has been hotly debated during the past two decades. It has now been demonstrated by Russell (1994) that although, in culture, axis diameter shows a relationship to salinity and other environmental conditions, the limits are genetically determined. Baltic populations have significantly narrower axes, both in the field and in culture, and therefore might be regarded as a geographical subspecies. The species is apomictic, however, so the biological species concept is not applicable and all distinctive variants are analogous to the microspecies of certain higher plants.

An interesting example of taxonomic insights gained from study of a single clone was reported by McCourt and Hoshaw (1990). A clone of the green alga *Spirogyra* underwent spontaneous change after 45 months of culturing to produce by autopolyploidy four distinct sub-clones. These sub-clones were morphotypes distinguished by differences in filament breadth and three corresponded to recognized species. McCourt and Hoshaw (1990) consider each morphotype to be a 'species complex' and argue against describing further species based on minor morphological variations including filament width. Similar complexes are believed to be widespread in the genus and on this basis Hoshaw estimated for Andersen (Andersen, 1992) the number of *Spirogyra* species to be nearer 50 rather than the 386 reported by Kadlubowska in 1984.

A further contribution of culture studies to microalgal taxonomy is to enable species to be defined using suites of characters only evident when clonal cultures are grown under carefully defined laboratory conditions. Many species of aquatic algae and the majority of soil and symbiotic algae have been described only after isolation and study in laboratory culture. For example, the taxonomy of freshwater representatives of

Pseudendoclonium, a filamentous green algal genus, is entirely culture-based. Characters of the prostrate system of filaments as evident in culture have been used to compare and group clonal cultures; groups correspond to species and subgroups to varieties (John and Johnson, 1989). As in the majority of green algae, the species of *Pseudendoclonium* described in culture are 'morphospecies' and the biological species concept is not applicable since sexual reproduction is unknown or, at least, a rare and unpredictable event. The morphologically simple green alga *Trebouxia*, the most common phycobiont of lichens, is often impossible to identify to species level when *in situ*. Only in culture does *Trebouxia* exhibit all the morphological characteristics (e.g. form of the chloroplast, cell size and shape during different growth phases) necessary for recognizing its 25 species (Gärtner, 1985). Phylogenetic analysis of the nuclear-encoded small-subunit ribosomal RNA (18S rRNA) in four genera of green algae, including three *Trebouxia* species (Friedl and Zeltner, 1994), resolved a 'lichen algae group'. One of the *Trebouxia* species was more closely related to members of other genera than to its congeners suggesting the 'genus' to be paraphyletic.

5.2.4 Interpretation and data analysis

The majority of algal taxonomists are still 'Linnaean in their thinking' (Guiry, 1992) since the recognition of species is often based largely on the observer's *a priori* (intuitive) opinion concerning phylogeny. This taxonomic 'method' relies on the experience and competence of the specialist in deciding the importance of each character (whether morphological, biochemical or otherwise) and determining whether differences between populations are sufficient to warrant species-level recognition. This 'evolutionary approach' is highly subjective in terms of the selection and weighting of characters. Taxonomic judgements are particularly difficult to make when putative 'species' are not separated by absolute morphological discontinuities but rather by combinations of overlapping quantitative characters (as in *Cladophora*).

Only in the past three decades has there been significant progress in the development of rigorous and objective methods for understanding and interpreting morphological data, including distinguishing between different kinds of similarity (Hennig, 1966). Of the three analytical approaches (phenetics, evolutionary systematics, cladistics), cladistics is finding widest acceptance. This method, developed from a body of taxonomic theory known as phylogenetic systematics, is organized around three principles: (i) monophyletic taxa are natural; (ii) organisms are related through descent; and (iii) evolutionary modifications uniquely shared by organisms are evidence of their unique phylogenetic history (Hennig, 1966). Use of the method leads to the formation of groups on the basis of nested sets of derived characters. Unlike other approaches, it is scientific in the sense that the results can be repeated and hypotheses tested using the Popperian falsification model. There is no doubt that cladistics is a powerful tool for testing phylogenetic hypotheses, but it is not without conceptual and practical problems (Tehler, 1994).

Several reviews of cladistics (Kociolek *et al.*, 1989; Theriot, 1989, 1992; Williams, 1992) were intended to draw the attention of phycologists to its potential for examining phylogenetic relationships within algae. Cladistics has been widely adopted for addressing taxonomic problems in algae at a level above that of the 'species'. There are a few exceptions amongst macro- and microalgal groups where it has been used at the species level. Several diatom studies (Williams, 1985; Theriot *et al.*, 1987) have used cladistics in order to analyse morphological character sets when revising lower taxonomic groups. In a study by Littler and Littler (1992) of the green alga *Avrainvillea* (see above), phylogenetic relationships were demonstrated by carrying out a cladistic analysis based largely upon anatomical characters. The analysis distinguished only species groups and highlighted the problems of working with morphologically plastic algae.

5.3 THE BIOLOGICAL SPECIES CONCEPT

5.3.1 To what extent can this be applied to algae?

Mann (1995) stated that 'The biological species concept can perhaps be viewed as a practical guide to the interpretation of the variation pattern and to the equivalence of the species category between different plant and animal groups'. However, this guidance can be used only for biparental, sexually reproducing organisms. The application of the biological species concept to the algae must therefore be very limited, as eight classes are not known to reproduce sexually; sexual reproduction in many of the remainder is a rare and unpredictable event. Some algal lineages show a tendency to have abandoned sexual processes in the course of evolution. Even in sexually reproducing algae, the biological species concept has been explored only for comparatively few groups. The results of algal breeding studies cannot always be reconciled with other data, and as Guiry (1992) has pointed out 'phycologists, while professing to approve of . . . a biological species concept, quickly abandon it when difficulties arise'. The biological species concept is criticised by cladists on the grounds that sexual compatibility is a primitive trait (a 'universal symplesiomorphy') and consequently cannot be used for defining monophyletic taxa (Manhart and McCourt, 1992). Notwithstanding such criticisms, attempts to employ a biological species concept in the algae have provided valuable new insights into algal taxonomy. Here, we present examples of recent research on the biological species concept in three algal classes, the

Rhodophyta (red algae), the Bacillariophyta (diatoms) and the Chlorophyta (green algae).

5.3.2 Rhodophyta

The red algae are ideal for breeding experiments since many readily grow and reproduce sexually in culture. Many hybridization experiments have been performed following the demonstration by Edwards (1970) that two morphologically distinct species of *Polysiphonia* were not interfertile. In a recent review by Guiry (1992), it is concluded that intra- and interpopulation crossability patterns in red algae are equally as complex as those found in other organisms. He recognized four categories of congruence or otherwise of the morphological and biological species concepts (Table 5.2). The first and last of these categories are quite straightforward, showing a good correspondence between the two species concepts. Considerable research effort has been devoted to some examples in the second and third categories, and some progress has been made recently in resolving the apparent incompatibility of the two species concepts. For example, Koch (1986) reported that Danish isolates of *Polysiphonia fibrillosa* (Dillwyn) Sprengel and *Polysiphonia violacea* sensu Harvey were intersterile although morphologically very similar. Maggs and Hommersand (1993) showed that Koch's '*P. fibrillosa*' was a misidentification of *P. harveyi* Bailey, a species readily distinguishable by several morphological characters (e.g. position of plastids within the periaxial cells). Of the cases placed by Guiry (1992) in his third category (see Table 5.2), one can now be clarified. A more extensive breeding study of the *P. harveyi* complex has shown (C.A. Maggs, unpublished data) that two of the Japanese forms currently recognized as separate species are actually elements of a nearly globally distributed interbreeding complex, or single biological species. The morphological differences between field populations, which can be maintained in culture to some extent, presumably indicate infraspecific genetic variability.

5.3.3 Bacillariophyta

Conventionally diatoms have been recognized and classified according to details of the acid-cleaned siliceous wall or frustule revealed on examination of field-collected samples under the light or electron microscope. One of the principal problems in diatom taxonomy has been to decide which discontinuities in the morphological characteristics correspond to boundaries at different levels of classification (genera, species, infraspecific taxa). The major taxonomic revisions of diatoms by Krammer and Lange-Bertalot (1986, 1988, 1991a,b) have adopted a wide morphological concept and abandoned many infraspecific taxa in the belief that minor variations

Table 5.2 Congruence and non-congruence of the morphological and biological species concepts in the red algae (From Guiry, 1992.)

1. Morphologically indistinguishable plants that are capable of interbreeding with each other throughout their geographical ranges: complete congruence between the two species concepts.
2. Morphologically indistinguishable plants that are incapable of crossing with each other. In this case the variation is cryptic and such sibling species are not afforded species status in algae. Cryptic variation is detectable in various ways including the analysis of allozymes and molecular data.
3. Morphologically dissimilar plants that will cross, some having various levels of genetic and ecological distinctness. These are recognized to be very problematical entities with fertility sometimes depending on the nature of the cross.
4. Morphologically dissimilar plants that will not cross. Many of these were poorly known morphologically and, although separated initially by breeding experiments, further morphological studies showed clear differences.

in cell size, shape, stria density, pattern and, possibly, ultrastructural features, are of little or no taxonomic significance. Mann (1989b) considers that a narrower rather than a broader morphological concept is needed, stating there to be 'a general failure to look critically at the variation pattern and a belief that variation is often continuous within, and sometimes also between, traditionally recognized species. Arguing logically from this unsubstantiated, and I believe incorrect premise, various authors have suggested that most infraspecific taxa are worthless and that many species should be combined'. Mann and co-workers have explored the mating patterns in mixed semi-natural populations of raphid diatoms and discovered incompatibility barriers between infraspecific taxa recognized by small morphological differences. For example, Mann and Droop (1996) discovered intrinsic barriers to hybridization existing between six morphologically distinct, sympatric populations of *Sellaphora pupula* (Kützing) Mereschkowsky and recommend species status for these incompatible 'morphotypes'. Other closely related species of *Sellaphora* have proved to be reproductively isolated (Mann, 1989a, 1995), indicating congruence between the morphological and crossability data.

5.3.4 Chlorophyta

The 'mating type' phenomenon is widespread in algae including many unicellular and colonial green algae (e.g. *Chlamydomonas moewusii* Gerloff, *Eudorina* species, *Gonium pectorale* Mueller, *Cosmarium botrytis* Meneghini ex Ralfs, *Micrasterias thomasiana* Archer, *Closterium ehrenbergii* Meneghini ex Ralfs). It involves multiple, genetically controlled mating types existing within a morphologically defined species. These mating units might be regarded as sibling or incipient biological species whose cryptic variation is often only detectable by means of physiological, biochemical or molecular analysis. One alga that has been extensively investigated is the 16-celled volvocine green alga *Pandorina morum* Bory. Coleman (1977) discovered that within this morphological species exist 20 distinct mating complexes, or syngens. These subordinate sibling or biological species are the functional or operational species units. In the case of *Pandorina morum* the morphological species is a much more comprehensive entity than the biological species. Mating groups (biological species) are thought to be of polyploid origin in the desmid *Closterium ehrenbergii* (Ichimura and Kasai, 1990). As soon as a ploidy change occurs, the new morphotypes are normally incapable of interbreeding so may be regarded as incipient new species.

Compatibility (interfertility) is used to reveal relationships between clones or 'species', but caution has to be exercised when interpreting the findings. For example, the identification of strains provisionally identified as *Chlamydonomas reinhardtii* Dangeard is routinely tested by carrying out crossability experiments with an authenticated laboratory clone. Spanier *et al.* (1992) found that despite crossing experiments demonstrating partial fertility between some clones, these differed in a wide range of non-morphological traits including heavy metal tolerance, protein composition, mitochondrial DNA length and nuclear, chloroplast and mitochondrial DNA restriction fragment length polymorphisms (RFLPs). The majority of *Chlamydomonas* species cannot be tested by breeding experiments because reproduction is known in less than 20% of described species (Ettl and Schlösser, 1992).

5.4 THE PHYLOGENETIC SPECIES CONCEPT

This concept is favoured by the cladistic school of systematics with its practitioners continuing to debate it although there seems general acceptance that 'species' are terminal taxa on their cladistic trees (Theriot, 1992). Two types of phylogenetic species concept are recognized by cladists: one identifies species as groups of organisms possessing at least one diagnostic character (terminal taxa on cladistic tree), and the other focuses on species as monophyletic and sharing one or more derived characteristics (for a detailed discussion, see Nelson, 1989; Baum, 1992; Manhart and McCourt, 1992; and references therein). In the context of species the term monophyletic has at least two meanings: all the descendants of a common ancestor together with that ancestor, or organisms more closely related to one another than to any other. The argument among cladists revolves around the question as to whether species are like any other taxa, or are they in some way special. Cladistic studies of intrageneric relationships in algae using morphological data tend to address relationships between species groups rather than focusing on individual species (cf. Williams, 1985; Mrozinska, 1991, 1993; Littler and Littler, 1992). A notable exception to this is the detailed phylogenetic analysis of several genera of green seaweeds by a research group at Groningen in the Netherlands (e.g. Kooistra *et al.*, 1993; Bakker *et al.*, 1995). A problem not unique to algae, but particularly acute in the group, is that the data on potentially important characters are insufficient to warrant rigorous statistical treatment. Organisms with a higher level of organization than algae generally have a greater number of more robust characters for defining species limits.

5.5 ROLE OF MOLECULAR DATA

Various types of molecular data provide a wealth of information with which to address problems at all taxonomic levels in the major algal lineages. Detailed analysis of genetic variation by protein sequencing, immunological comparisons and enzyme (isozyme) electrophoresis have been used to examine phylogenetic relationships and address species-level problems. Enzyme electrophoresis has been used widely to provide species markers, but interpretation remains problematical due to a lack of information on the stability of enzyme loci and the effects of physiological state and environment on their expression. For a brief review of algal species problems addressed by protein-based comparisons, see Manhart and McCourt (1992). These molecular approaches have been largely superseded by recent advances in nucleic acid technology.

To date, the majority of studies involving the analysis of nucleic acids have concentrated upon resolving higher-order relationships rather than those at lower taxonomic levels. In addressing species problems, molecular characters are being used to test established species concepts (principally morphological), at least until 'the development of a "molecular species concept" employing disembodied sequences as the *sine qua non* of species definition' (Manhart and McCourt, 1992). Studies addressing species-level problems have involved the analysis of nuclear, chloroplast and mitochondrial DNA at different levels of sequence resolution using RFLPs, randomly amplified polymorphic DNA (RAPD), DNA sequencing and single copy DNA–DNA hybridization. This last method is useful for testing evolutionary divergence (e.g. *Cladophora*; Bot *et al.*, 1989a,b), but it has been largely replaced by more powerful molecular techniques. Of the four methods, RAPD is potentially the most valuable for population analysis (Patwary *et al.*, 1993) although it is beset with many technical difficulties that affect the reliability of the results (Dutcher and Kapraun, 1994). The methods of choice for algal researchers are RFLPs and sequencing, particularly of rRNA and rDNA genes (nuclear: 18S, 26S; chloroplast: 16S, 23S). The moderately fast-evolving spacer regions (ribosomal internal transcribed spacers; spacer between large and small Rubisco subunits) are becoming important sources of informative sequence variation in algae now that more is known of evolution rates in different coding and non-coding regions (Bakker *et al.*, 1995).

The chloroplast DNA of land plants is highly conserved, whereas in the more ancient algal lineages its astonishing diversity enables it to be used for examining all levels of relationships from populational to those between orders and classes (Freshwater *et al.*, 1994). Chloroplast DNA inheritance, like that of mitochondrial DNA, is usually uniparental and so any boundaries do not necessarily correlate with those obtained by nuclear DNA analysis. Mitochondrial DNA previously received little attention in algae (Coleman and Goff, 1991) but as the genome has now been better characterized (Boyen *et al.*, 1994) it may be more widely used in future.

Nuclear and plastid sequence data are now applied with increasing frequency to resolving taxonomic problems in difficult groups where traditional morphological comparisons appear to have failed. However, nucleic acid data are most informative in taxonomic decision-making when combined with data on other types of character. For example, an unusually complete analysis of the red algal genus *Gymnogongrus* (now split into *Gymnogongrus* and *Ahnfeltiopsis*) involved comparative molecular, morphological, and culture studies (Maggs *et al.*, 1992). In this study plastid DNA RFLP analysis and nucleotide sequences for the Rubisco spacer region revealed the existence of what appeared to be two distinct taxa in the Atlantic, both going under the name *Gymnogongrus devoniensis*. These were distinguished not only by RFLP analysis, probing and nucleotide sequence data, but on vegetative, reproductive and life-history characters. The approach enabled resolution of complex taxonomic problems in the genus, including the relationship between populations with different life-histories, and answered questions concerning its distribution in the Atlantic.

Congruence between molecular, morphological and other data types is often regarded as evidence in support of the original morphospecies concept. Conversely, the absence of congruence between datasets indicates that the original concept was either too narrow or too wide. As might be expected, there tends to be less consensus (although there may be no greater conflict) between molecular and morphological data in relatively simple algae where uncertainty frequently surrounds species and generic boundaries due to the existence of few discrete and easily definable characters. For example, an attempt was made by Kooistra *et al.* (1993) using cladistics to discover whether morphological data supported the findings of the molecular analysis of four genera of green algae belonging to the same order. As the limits of morphologically well-defined species showed little correlation with the results of cladistic analysis, the genera were revealed to be paraphyletic. All the genera were seen to be artificial: none was monophyletic and two were paraphyletic.

There are many examples where molecular data have been used to define the limits of morphologically distinguished species. For example, morphospecies of the green alga *Microthamnion* were tested by rigorous experimentation in culture and revealed to be too narrow, so that several previously recognized species were considered to be mere growth forms or 'ecomorphs' (John and Johnson, 1987). RFLP analysis by John *et al.* (1993) of 18S rRNA for 20 isolates from different habitats and geographical regions showed that there was only one pattern; this supported the culture-based view that *Microthamnion* is monospecific. Similarly Lange *et al.* (1994) compared sequences of the same subunit in three colony-forming species of the haptophyte *Phaeocystis* and discovered congruence between the molecular, morphological, and physiological datasets. In the brown seaweed, *Costaria costata* (Turner) Saunders, morphologically distinct wave-sheltered and wave-exposed populations were compared using nuclear DNA RFLPs probed with an array of anonymous randomly cloned fragments (Druehl and Saunders, 1992). Only one RFLP separated the two populations; the results supported previous hypotheses that morphological variation in this species reflects a plastic phenotypic response to environmental variables.

Genomic relationships in *Chlamydomonas* have been examined by Buchheim *et al.* (1990), who discovered close agreement between morphological data and 18S rRNA partial sequences in 14 species. As in most other algal groups, delimitation of the 500-plus recognized *Chlamydomonas* species is possible only by the experienced specialist. In the past the presence of very specific sporangial wall autolysins was used by Schlösser (1976) to distinguish species. The majority of these 'species' are so similar in terms of cellular morphology, reproduction, development and physiological properties that they are now considered to be clones of the same species (Ettl and Schlösser, 1992).

If breeding and morphometric studies in diatoms continue to lead to the separation of distinctive allopatric or sympatric morphotypes (cf. Mann, 1989b; Mann and Droop, 1995), then the current species concept is too wide or 'coarse-grained'. This view was supported by Medlin *et al.* (1991) when describing a new species, *Skeletonema pseudocostatum*, on the basis of small differences in morphology and nucleotide substitutions in 16S rRNA sequences among four clones originally attributed to *Skeletonema costatum*, two of which were reassigned to the new species. Medlin (1995) considers polymorphisms in rRNA sequence data among morphologically indistinguishable clones to indicate the presence of a 'species complex'. The current view of the diatom species concept is summarized by Cox (1993) as follows: 'While it may be possible to distinguish taxa on certain non-morphological criteria,.... if the taxa cannot be readily and routinely distinguished, formal description of a new species is largely irrelevant.... If there is no significant detectable difference in ecology, physiology, or reproductive behaviour of morphologically alike individuals or populations, they should be considered the same taxon'.

The red alga *Gracilaria* is an example of a genus in which systematic revision required extensive molecular studies. The discreteness of *Gracilaria* species defined on the basis of morphological criteria has been tested by comparing morphological and molecular data (Goff and Coleman, 1988; Bird and Rice, 1990). In several morphological species there was close agreement between morphology and DNA data. For example, Goff and Coleman (1988) demonstrated different RFLP patterns for chloroplast DNA in *Gracilaria pacifica* and *G. robusta.* Rice and Bird (1990) investigated various populations previously identified as *G. verrucosa* (Hudson) Papenfuss, a terete species with few taxonomically useful morphological characters. They were able to demonstrate that a core group of interfertile isolates had similar RFLP patterns, but outside this group, various levels of difference in RFLP patterns were evident. The '*G. verrucosa*-like algae' included representatives of several other species of *Gracilaria* and a separate genus, *Gracilariopsis*. The longstanding taxonomic and nomenclatural confusion in this group has now been largely resolved by use of molecular data in combination with traditional morphological studies (Steentoft *et al.*, 1995). Similarly, relationships within the morphologically very variable red algal genus *Gelidium* have been clarified using rbcL sequences (Freshwater and Rueness, 1994). Various species complexes were recognized and this enabled a problematic species to be correctly identified as *G. attenuatum* (Turner) Bornet.

5.6 CONCLUSIONS

The vast majority of algae are distinguished by morphological discontinuities – the 'morphological species concept' dominates algal systematics. Problems associated with the continued acceptance of traditionally recognized 'morphospecies' or 'morphotypes' relate to disagreements over the weighting of characters, the discovery of cryptic molecular variation and to extreme levels of phenotypic plasticity. Breeding experiments have enabled clusters of reproductively isolated cryptic or sibling species (syngens) to be identified within traditional morphological species. Groups of sibling species are equivalent to what have been termed 'species complexes' when morphologically indistinguishable clones show nucleotide substitutions in a reasonably highly conserved region of the genome. Sibling species have generally not been afforded any taxonomic status: only when characters are discovered that make it possible readily and routinely to distinguish them are they formally diagnosed. The results of these compatibility experiments involving biparental, sexually reproducing algae should perhaps be viewed as a practical guide to assist in interpreting morphological variation and making taxonomic decisions. However, algae are largely asexual so the biological species concept is often not applicable and genomic relationships are untestable through a breeding programme. Sexuality is regarded as a primitive characteristic by cladists who point out that genomically very different clones or 'species' are often capable of interbreeding. There is a considerable body of evidence from algae and other organisms to indicate the occurrence of considerable evolutionary divergence without the development of strong isolating mechanisms. Compatibility data therefore need to be viewed with caution as they are not necessarily always a good indication of the degree of taxonomic relatedness.

Systematists are increasingly applying molecular approaches to examine relationships in algae at different taxonomic levels. One advantage of using molecular data is that what is measured represents differences in the genome rather than the phenotype. Nucleic acid sequences are proving especially useful in testing the limits of morphological species and detecting the existence of 'species complexes' within indistinguishable clones. Sequence data are most powerful when combined with other datasets to test the morphological species hypotheses. Inevitably congruent datasets give more reliable and well-corroborated hypotheses for the recognition of monophyletic taxa such as species. It is generally becoming recognized that cladistics is a valuable tool for rigorously analysing these datasets. It is important to recognize that it still fails to address the classical problem of how to prevent personal bias and *a priori* opinion influencing selection of phylogenetically informative characters and equally parsimonious cladistic trees. Ideally, different morphological and molecular datasets should be treated in as similar a fashion as possible in order to facilitate comparison (Williams, 1993). The 'phylogenetic species concept' is favoured by practising cladists, although to date it has been applied principally to diatoms where many putative natural groupings have been revealed using cladistics (cf. Williams, 1985).

Determining discontinuities in morphological variation will undoubtedly remain the principal practical approach to species-level taxonomy in the algae. In the future the traditional morphological species concept is likely increasingly to operate alongside less formalized concepts involving data from other disciplines (e.g. molecular systematics). These data will enable larger suites of concordant characters to be used for calibrating species concepts. If cladistics is accepted as the analytical method of choice for studying species-level relationships then the 'phylogenetic species concept' will come into more frequent use. Finally, we agree with the statement by Wilmotte and Golubic (1991: 4) that 'for practical reasons, all these results [molecular data] will finally have to feed back to the taxonomy based on the morphology and simple testing methods, so that an improved systematic practice based on morphology will then be able to deliver fast and reliable determinations, as pressing ecological questions demand'.

Acknowledgements

We should like to thank Dr David Williams for bringing us up-to-date on current debate on the 'phylogenetic species concept', Drs David Mann and Stephen Droop for commenting upon the diatom section, and Dr Judith John for reading the manuscript and offering valuable suggestions for its improvement.

5.7 REFERENCES

Andersen, R.A. (1992) Diversity of eukaryotic algae. *Biodiversity and Conservation*, **1**, 267–92.

Bakker, F.T., Olsen, J.L. and Stam, W.T. (1995) Evolution of nuclear rDNA ITS sequences in the *Cladopora albida/sericea* clade (Chlorophyta). *Journal of Molecular Evolution*, **40**, 640–51.

Baum, D. (1992) Phylogenetic species concept. *Trends in Ecology and Evolution*, **2**, 1–3.

Bird, C.J. and Rice, E.L. (1990) Recent approaches to the taxonomy of the Gracilariaceae (Gracilariales, Rhodophyta) and the *Gracilaria verrucosa* problem. *Hydrobiologia*, **204/205**, 111–18.

Blackburn, S.I. and Tyler, P.R. (1987) On the nature of eclectic species – a tiered approach to genetic compatibility in the desmid *Micrasterias thomasiana*. *British Phycological Journal*, **22**, 277–98.

Bliding, C. (1963) A critical survey of European taxa in Ulvales. Part I. *Capsosiphon, Percursaria, Blidingia, Enteromorpha. Opera Botanica Lund*, **8**(3), 1–160.

Bold, H.C. (1974) Phycology, 1947–1972. *Annals of the Missouri Botanic Garden*, **61**, 14–44.

Bot, P.V.M., Holton, R.W., Stam, W.T. and Hoek, C. van den (1989a) Molecular divergences between North Atlantic and Indo-West Pacific *Cladophora albida* (Cladophorales, Chlorophyta) isolates as indicated by DNA–DNA hybridization. *Marine Biology* (Berlin), **102**, 307–33.

Bot, P.V.M., Stam, W.T., Boele-Bos, S.A. Hoek, C. van den and Delden, W. van (1989b) Biogeographic and phylogenetic studies in three North Atlantic species of *Cladophora* (Cladophorales, Chlorophyta) using DNA–DNA hybridization. *Phycologia*, **28**, 159–68.

Boyen, C., Leblanc, C., Kloareg, B. and Loiseaux-de Goer, S. (1994) Physical and gene map organization of the mitochondrial genome of *Chondrus crispus* (Rhodophyta, Gigartinales). *Plant Molecular Biology*, **26**, 651–97.

Brook, A.J. (1997) *A Revision of the West' s British Desmidiaceae*. Volume 1. The Ray Society, London.

Buchheim, M.A., Turmel, M., Zimmer, E.A. and Chapman, R.L. (1990) Phylogeny of *Chlamydomonas* (Chlorophyta) based on cladistic analysis of nuclear 18S rRNA sequence data. *Journal of Phycology*, **26**, 689–99.

Castenholz, R.W. (1992) Species usage, concept, and evolution in the cyanobacteria (blue–green algae). *Journal of Phycology*, **28**, 737–45.

Coleman, A.W. (1977) Sexual and genetic isolation in the cosmopolitan algal species *Pandorina morum*. *American Journal of Botany*, **64**, 361–8.

Coleman, A.W. and Goff, L. (1991) DNA analysis of eukaryotic algal species. *Journal of Phycology*, **27**, 463–73.

Corliss, J.O. (1994) An interim utilitarian ('user-friendly') hierarchical classification and characterization of the protists. *Acta Protozoologica*, **33**, 1–51.

Cox, E.J. (1993) Diatom systematics – a review of past and present practice and a personal vision for future development. *Beiheft Nova Hedwigia*, **106**, 1–20.

Druehl, L.D and Saunders, G.W. (1992) Molecular explorations in kelp evolution. *Progress in Phycological Research*, **8**, 47–78.

Dutcher, J.A. and Kapraun, D.F. (1994) Random amplified polymorphic DNA (RAPD) identification of genetic variation in three species of *Porphyra* (Bangiales, Rhodophyta). *Journal of Applied Phycology*, **6**, 267–73.

Edwards, P. (1970) Attempted hybridization in the red algal genus *Polysiphonia*. *Nature*, **226**, 467–8.

Ettl, H. and Schlösser, U.G. (1992) Towards a revision of the systematics of the genus *Chlamydomonas* (Chlorophyta). 1. *Chlamydomonas applanata* Pringsheim. *Botanica Acta*, **105**, 323–30.

Freshwater, D.W. and Rueness, J. (1994) Phylogenetic relationships of some European *Gelidium* (Gelidiales, Rhodophyta) species, based on rbcL nucleotide sequence analysis. *Phycologia*, **33**, 187–94.

Freshwater, D.W., Fredericq, S., Butler, B.S., Hommersand, M.H. and Chase, M. (1994) A gene phylogeny of the red algae (Rhodophyta) based on plastid *rbcL*. *Proceedings of the National Academy of Science, USA*, **91**, 7281–5.

Friedl, T. and Zeltner, C. (1994) Assessing the relationships of some coccoid green lichen algae and the Microthamniales (Chlorophyta) with 18S ribosomal RNA gene sequence comparisons. *Journal of Phycology*, **30**, 500–6.

Gärtner, G. von (1985) Die Gattung *Trebouxia* Puymaly (Chlorellales, Chlorophyceae). *Archiv für Hydrobiologia*, suppl. **71**, **4**, 495–548.

Goff, L.J. and Coleman, A.W. (1988) The use of plastid DNA restriction endonuclease patterns in delineating red algal species and populations. *Journal of Phycology*, **24**, 357–68.

Grant, V. (1981) *Plant Speciation*, 2nd edn. Columbia University Press, New York.

Greville, R. K. (1821) Description of a new species of Fucus, found in Devonshire. *Memoirs of the Wernerian Natural History Society*, **3**, 396–9.

Guiry, M. (1992) Species concepts in marine red algae. *Progress in Phycological Research*, **8**, 251–78.

Hegewald, E. and Silva, P. (1988) Annotated catalogue of *Scenedesmus* and nomenclaturally related genera, including original descriptions and figures. *Bibliotheca Phycologica*, **80**, 1–587.

Hennig, W. (1966) *Phylogenetic Systematics*. University of Illinois Press, Urbana.

Hoek, C. van den (1963). *Revision of the European Species of Cladophora*. Brill, Leiden.

Hoek, C. van den (1964) Criteria and procedures in present-day taxonomy, in *Algae and Man* (ed. D.F. Jackson), pp. 31–58.

Hoek, C. van den (1982) A taxonomic revision of the American species of *Cladophora* (Chlorophyceae) in the North Atlantic Ocean and their geographic distribution. *Verhandelingen. Koninklijke akademie van wetenschappen Tweede Sectie*, **78**, 1–236.

Hudson, W. (1762) *Flora Anglica*. London.

Ichimura, T. and Kasai, F. (1990) Mating systems and speciation in haplontic unicellular algae, desmids, in *Biological Approaches and Evolutionary Trends in Plants* (ed. H. Hara), Academic Press, London, pp. 195–201.

Irvine, L. M. and Dixon, P. S. (1982) The typification of Hudson' s algae: a taxonomic and nomenclatural reappraisal. *Bulletin of the British Museum of Natural History (Botany)*, **10**, 91–105.

Islam, A.K.M.N. (1963) A revision of the genus of the genus *Stigeoclonium*. *Beiheft Nova Hedwigia*, **10**, 1–164.

John, D.M. (1994) Biodiversity and conservation: an algal perspective. *The Phycologist*, **38**, 15–18.

John, D.M. and Johnson, L.R. (1987) Observations on the developmental morphology, growth rate, and reproduction of *Microthamnion kuetzingianum* Nageli (Pleurastraceae, Pleurastrales) in culture and a taxonomic assessment of the genus. *Nova Hedwigia*, **44**, 25–53.

John, D.M. and Johnson, L.R. (1989) A cultural assessment of the freshwater species of *Pseudendoclonium* Wille (Ulotrichales, Ulvophyceae, Chlorophyta). *Algological Studies*, **54**, 79–112.

John, D.M., Bhoday, R., Russell, S.J., Johnson, L.R., and Gacesa, P. (1993) A molecular and morphological analysis of *Microthamnion* (Chlorophyta, Microthamniales). *Archiv für Protistenkunde*, **143**, 33–9.

Kadlubowska, J.Z. (1984) Conjugatophyceae I Zygnemales = Chlorophyta VIII, in *Süsswasserflora von Mitteleuropa*. G. Fischer, Stuttgart and New York.

Koch, C. (1986) Attempted hybridization between *Polysiphonia fibrillosa* and *P. violacea* (Bangiophyceae) from Denmark; with culture studies primarily on *P. fibrillosa*. *Nordic Journal of Botany*, **6**, 123–8.

Kociolek, J.P., Theriot, E.C. and Williams, D.M. (1989) Inferring diatom phylogeny: a cladistic perspective. *Diatom Research*, **4**, 289–300.

Kooistra, W.H.C.F., Olsen, J.L., Stam, W.T. and Hoek, C. van den (1993) Problems relating to species sampling in phylogenetic studies: an example of non-monophyly in *Cladophoropsis* and *Struvea* (Siphonocladales, Chlorophyta). *Phycologia*, **32**, 419–28.

Krammer, K. and Lange-Bertalot, H. (1986) Bacillariophyceae 1. Teil: Naviculacae, in *Süsswasserflora von Mitteleuropa* (eds H. Ettl, J. Gerloff, H. Heynig and D. Mollenhauer), vol. 2(1), G. Fischer, Stuttgart, New York, pp. 1–876.

Krammer, K. and Lange-Bertalot, H. (1988) Bacillariophyceae 2. Teil: Bacillariaceae, Epithemiaceae, Surirellaceae, in *Süsswasserflora von Mitteleuropa* (eds H. Ettl, J. Gerloff, H. Heynig and D. Mollenhauer), vol. 2(2), G. Fischer, Stuttgart, New York, pp. 1–596.

Krammer, K. and Lange-Bertalot, H. (1991a) Bacillariophyceae 3. Teil: Centrales, Fragilariaceae, Eunotiaceae, in *Süsswasserflora von Mitteleuropa* (eds H. Ettl, J. Gerloff, H. Heynig and D. Mollenhauer), vol. 2(3), G. Fischer, Stuttgart, New York, pp. 1–576.

Krammer, K. and Lange-Bertalot, H. (1991b) Bacillariophyceae 1. Teil: Achnanthaceae. Kritische Ergänzungen zu Navicula (Lineolatae) und *Gomphonema*, in *Süsswasserflora von Mitteleuropa* (eds H. Ettl, J. Gerloff, H. Heynig and D. Mollenhauer), vol. 2(4), G. Fischer, Stuttgart, New York, pp. 1–437.

Lange, M., Medlin, L.K. and Baumann, M.E.M (1994) Genetic differentiation among three colony-forming species of *Phaeocystis*: further evidence for the phylogeny of the Prymnesiophyta. *Phycologia*, **33**, 199–212.

Lewin, R.A. (1975) A *Chlamydomonas* with black zygospores. *Phycologia*, **14**, 71–4.

Linnaeus, C. (1753). *Species Plantarum*. Vols. 1, 2. Laurenti Salvii, Stockholm.

Littler, D.S. and Littler, M.M. (1992) Systematics of *Avrainvillea* (Bryopsidales, Chlorophyta) in the tropical western Atlantic. *Phycologia*, **31**, 375–418.

Maggs, C.A. and Hommersand, M.H. (1993) *Seaweeds of the British Isles*. Volume 1 *Rhodophyta*, Part 3A *Ceramiales*. Natural History Museum/HMSO, London.

Maggs, C.A., Douglas, S.E., Fenety, J. and Bird, C.J. (1992) A molecular and morphological analysis of the *Gymnogongrus devoniensis* (Rhodophyta) complex in the North Atlantic. *Journal of Phycology*, **28**, 214–32.

Manhart, J.R. and McCourt, R.M. (1992) Molecular data and species concepts in the algae. *Journal of Phycology*, **28**, 730–7.

Mann, D.G. (1989a) The diatom genus *Sellaphora*: separation from *Navicula*. *British Phycological Journal*, **24**, 1–20.

Mann, D.G. (1989b) The species concept in diatoms: evidence for morphologically distinct, sympatric gamodemes in four epipelic diatoms. *Plant Systematics and Evolution*, **164**, 215–37.

Mann, D.G. (1995) Gamodemes and genodemes in diatoms: the biological species concept in action. *The Phycologist*, **40**, 24–5.

Mann, D.G. and Droop, S.J.M. (1996) Biodiversity, biogeography and conservation of diatoms. *Hydrobiologia*, **331**, in press.

Mayr, E. (1992) A local flora and the biological species concept. *American Journal of Botany*, **79**, 222–38.

McCourt, R.M. and Hoshaw, R.W. (1990) Noncorrespondence of breeding groups, morphology, and monophyletic groups in *Spirogyra* (Zygnemataceae: Chlorophyta) and the application of species concepts. *Systematic Biology*, **15**, 69–78.

Medlin, L.K. (1995) Can molecular techniques enhance our understanding of a species concept? *The Phycologist*, **40**, 25.

Medlin, L.K., Elwood, H.J., Stickel, S. and Sogin, M.L. (1991) Morphological and genetic variation within the diatom *Skeletonema costatum* (Bacillariophyta): evidence for a new species, *Skeletonema pseudocostatum*. *Journal of Phycology*, **27**, 514–24.

Mrozinska, T. (1991) A preliminary investigation of the taxonomical classification of the genus *Oedogonium* Link (Oedogoniales) based on the phylogenetic relationship. *Archiv für Protistekunde*, **139**, 85–101.

Mrozinska, T. (1993) A preliminary investigation of the taxonomical classification of the genus *Bulbochaete* Agardh (Oedogoniales, Chlorophyta) based on the phylogenetic relationship. *Archiv für Protistenkunde*, **143**, 113–23.

Necchi, O., Jr and Zucchi, M.R. (1995) Systematics and distribution of freshwater *Audouinella* (Acrochaetiaceae, Rhodophyta) in Brazil. *European Journal of Phycology*, **30**, 209–18.

Nelson, G. (1989) Species and taxa: systematics and evolution, in *Speciation and Consequences* (eds D. Otte and J.A. Endler), Sinauer Associates, Sunderland, Massachusetts, pp. 60–81.

Patwary, M.U., MacKay, R.M. and van der Meer, J.P. (1993) Revealing genetic markers in *Gelidium vagum* (Rhodophyta) through the random amplified polymorphic DNA (RAPD) technique. *Journal of Phycology*, **29**, 216–22.

Preisig, H.R. and Hibberd, D.J. (1982) Ultrastructure and taxonomy of *Paraphysomonas* (Chrysophyceae) and related genera 2. *Nordic Journal of Botany*, **2**, 601–38.

Prescott, G.W. (1951) History of phycology, in *Manual of Phycology: An Introduction to the Algae and their Biology* (ed. G.M. Smith), Chronica Botanica, Massachusetts, pp. 1–11.

Pringsheim, E.G. (1967) Phycology in the field and in the laboratory. *Journal of Phycology*, **3**, 93–5.

Rice, E.L. and Bird, C.J. (1990) Relationships among geographically distant populations of *Gracilaria verrucosa* (Gracilariales, Rhodophyta) and related species. *Phycologia*, **29**, 501–10.

Russell, G. (1994) A Baltic variant of *Pilayella littoralis* (Algae, Fucophyceae). *Annales Botanici Fennici*, **31**, 127–38.

Russell, G. and Fielding, A.J. (1981) Individuals, populations and communities, in *The Biology of Seaweeds* (eds C.S. Lobban and M.J. Wynne), Blackwell Scientific, Oxford, pp. 393–420.

Schlösser, U.G. (1976) Entwicklungsstadien -und sippenspezifische Zellwand-Autolysine bei der Freisetzung von Fortpflanzungszellen in der Gattung *Chlamydomonas. Bericht der Deutschen Botanischen Gesellschaft*, **89**, 1–56.

Silva, P. C. (1952) A review of nomenclatural conservation in the algae from the point of view of the type method. *University of California Publications in Botany*, **25**, 241–324.

Simons, J., Beem, A.P. van and Vries, P.J.R. de (1986) Morphology of the prostrate thallus of *Stigeoclonium* (Chlorophyceae, Chaetophorales) and its taxonomic implications. *Phycologia*, **25**, 210–20.

Spanier, J.G., Graham, J.E. and Jarvik, J.W. (1992) Isolation and characterization of three *Chlamydomonas* strains interfertile with *Chlamydomonas reinhardtii*. *Journal of Phycology*, **28**, 822–8.

Steentoft, M., Irvine, L.M. and Farnham, W.F. (1995) Two terete species of *Gracilaria* and *Gracilariopsis* (Gracilariales, Rhodophyta) in Britain. *Phycologia*, **34**, 113–27.

Steussy, T.F. (1990) *Plant Taxonomy The Systematic Evaluation of Comparative Data*. Columbia University Press, New York.

Tehler, A. (1994) Cladistic analysis in ascomycete systematics: theory and practice, in *Ascomycete Systematics: Problems and Perspectives in the Nineties* (ed. D.L. Hawksworth), Plenum Press, New York, pp. 185–97.

Theriot, E. (1989) Phylogenetic systematics for phycology. *Journal of Phycology*, **25**, 407–11.

Theriot, E. (1992) Phylogenetic systematics, the theory and practice of taxonomy, in *Modern Approaches to the Analysis of Algal Morphology and Systematics* (eds D.M. John and D. Garbary), vol. 8, Progress in Phycological Research, pp. 180–207.

Theriot, E., Stoermer, E.F and Håkansson, H. (1987) Taxonomic interpretation of the rimoportula of freshwater genera in the centric family Thalassiosiraceae. *Diatom Research*, **2**, 251–65.

Trainor, F.R. (1991) The format for a *Scenedesmus* monograph. *Algological Studies*, **61**, 47–53.

Trainor, F.R. and Egan, P.F. (1991) Discovering the various ecomorphs of *Scenedesmus*. *Archiv für Protistenkunde*, **139**, 125–32.

Vis, M.L., Sheath, R.G. and Entwistle, T. J. (1995) Morphometric analysis of *Batrachospermum* section *Batrachospermum* (Batrachospermales, Rhodophyta) type specimens. *European Journal of Phycology*, **30**, 35–56.

Williams, D.M. (1985) Morphology, taxonomy and inter-relationships of the ribbed araphid diatoms from the genera *Diatoma* and *Meridion* (Diatomaceae: Bacillariophyta). *Bibliotheca Diatomologia*, **8**, 1–228.

Williams, D.M. (1992) Analytical techniques for systematic molecular data. *Progress in Phycological Research*, **8**, 209–49.

Williams, D.M. (1993) Diatom nomenclature and the future of taxonomic databasing studies. *Beiheft Nova Hedwigia*, **106**, 21–31.

Wilmotte, A. and Golubic, S. (1991) Morphological and genetic criteria in the taxonomy of Cyanophyta/Cyanobacteria. *Algological Studies*, **64**, 1–24.

6

The species concept in lichens

O. W. Purvis

Contacting address: Department of Botany, The Natural History Museum, Cromwell Road, London SW7 5BD, UK

ABSTRACT

Lichens are a combination of at least one fungus and an alga or cyanobacterium (blue–green alga). The formation of the lichen body (thallus), typically morphologically, physiologically and biochemically distinct from either partner in the free living state, is one of the greatest enigmas in biology. The classification of lichens at a species level is discussed on a historical basis and evaluated in relation to the varying importance attached to external morphology and internal anatomy of fungal tissues, the significance of the alga in morphogenesis, and the importance of sexual reproductive structures, asexual vegetative propagules and reproductive strategies. Environmental factors are shown to have a profound influence on phenotypic expression and careful field evaluation must therefore be undertaken when describing new species. The wide range of chemical characters including secondary metabolites are considered as an important tool for defining species. The difficulties in maintaining the composite organism under laboratory conditions have resulted in little experimental attention being applied to the nature of species in lichen fungi, artificial crosses being impossible for technical reasons, although recent attempts have been made to study speciation using chemical markers. Molecular techniques have as yet hardly been applied at the specific level in lichen systematics and species specific markers have yet to be developed. The concept of a lichen as an individual is explored with reference to fusions between individuals within a lichen population, species within a genus or species in different genera, which results in the incorporation of different fungal and algal partners within the same thallus. Although documentations of thallus fusions are rare, these pose important consequences for experimental work and also in classification. Personal experience and intuition continue to play a major role in defining species. The morphological species concept, based on phenetic characters, including secondary chemistry, remains therefore of major importance. This must lead the work of molecular systematists.

Species: The Units of Biodiversity. Edited by M.F. Claridge, H.A. Dawah and M.R. Wilson. Published in 1997 by Chapman & Hall. ISBN 0 412 63120 2

6.1 HISTORICAL ASPECTS

Since lichens were first recognized as members of the genus *Lichen* by Tournefort in 1694, many schemes of classification have been outlined (Smith, 1921; Zahlbruckner, 1926; Lorch, 1988). Linnaeus unfortunately did not understand the dual nature of lichens and regarded cryptogams generally as 'rustici pauperrimi' – the poor little peasants of nature. He described only 109 species of lichens, mostly placed in his genus *Lichen* but some were mistakenly treated within the '*Algae*' (Jørgensen *et al.*, 1994). That this great pioneer of phanerogamic systematics did not understand these organisms no doubt greatly contributed to the study of lichenology and of cryptogams as a whole being greatly held back. It was not until Acharius first made lichens a subject of special study by his scientific system of classification in which he introduced a new terminology for lichenology in his *Methodus Lichenum* (1803), when he raised them to the rank of the classes of the other great classes of plants, describing 906 species (Smith, 1921). He and other authors to the mid-19th century based nearly all taxonomy on externally visible characters, such as general morphology, colour, hairiness, the external shape of apothecia, and the presence of a range of structures specific to lichens including several asexual reproductive structures (Almborn, 1965). An increased value was placed on anatomical characters by particularly Massalongo and Körber towards the mid-19th century. However, Nylander, although describing a very large number of species unwittingly put the clock back by establishing a classification based almost entirely on thalline form overlooking the importance of characters of the fruiting bodies or ascomata (see section 6.3.1(a)), contrary to the situation in mycology where the thallus is usually not visible. Following the proposal of the composite nature of the lichen thallus by Schwendener in 1868, there was a prolonged and often heated debate over the dual nature of lichens which continued for about 50 years (Lorch, 1988). Today it is generally accepted that a lichen consists of at least one fungal (myco-) and one algal (phyto-) biont and that lichenization is an extremely successful life strategy for fungi that is taken up by almost 50% of all ascomycete and 20% of all known fungi (Hawksworth, 1988b). There have been various fashions in assigning a taxon to a specific rank. During the 19th and early part of the 20th century, many genera were based on relatively few characters and many infraspecific taxa were recognized. This trend has been dramatically reversed during the past 30 years, with crustose genera such as *Lecidea* Ach. originally united on a single spore character, having been divided into numerous segregates including the resurrection of several originally described, e.g. *Psilolechia* Massal. (1850). For instance the current checklist of lichens of Great Britain and Ireland accepts 42 species of *Lecidea*, a further 80 species since the previous 1980 checklist being assigned to 30 other genera (Purvis *et al.*, 1994). Conversely, the numbers of infraspecific taxa recognized has dramatically decreased following realization of the importance of environmental factors on phenotypic expression (see section 6.5). Estimates of known lichenized taxa have varied from a conservative 13 500 (Hawksworth *et al.*, 1983) to 17 000 (Hale, 1974). Britain is one of the most intensively surveyed countries in the world for lichens and yet since publication of the recent *Lichen Flora of Great Britain and Ireland* (Purvis *et al.*, 1992) including 1700 taxa, over 50 additional species have been added within 2 years. Galloway (1992) draws attention to the large number of recent monographic revisions where new taxa are discovered in even quite well-known macrolichen genera and suggests the world total may be closer to 20 000. Although for vascular plants the tropics are regarded as major sites of biodiversity, much less is known about tropical lichens whose biodiversity tends to be richest in canopy vegetation, which is still very poorly sampled in many tropical areas. Furthermore, as crustose lichens have been so inadequately studied even in many temperate areas with their wide variations of habitat, geology and climate, a conservative estimate of the total number of species would seem to be 30 000 and this would not encompass chemical variation considered by some to indicate sibling speciation (see section 6.6).

6.2 WHAT ARE LICHENS?

In simple terms a lichen is an association between a fungus and one or more photosynthetic partners (photobiont) which may be an alga or a cyanobacterium (blue–green alga). A precise definition of a lichen, although much debated, has not yet been satisfactorily resolved (Hawksworth, 1988a, 1994; Ahmadjian, 1993a) due in part to the great diversity of fungal–algal associations (Table 1 in Hawksworth, 1988a) exhibiting a range of biological strategies from apparently mutualism to antagonism where up to five bionts or possibly more may be involved. A key feature of the symbiosis is the formation of the lichen body or thallus within which photobionts are housed to best advantage for photosynthesis, although the two organisms behave from a physiological point of view rather as a single organism subjected to natural processes of selection. This successful mode of nutrition for the fungus, taken up by widely different groups of fungi, has enabled these organisms to colonize habitats unfavourable to most plants, contributing a significant component of biodiversity in many of the world's ecosystems. The association, although traditionally considered to be of mutual benefit to both partners, appears to represent more of a controlled parasitism on behalf of the fungus at the level of the individual cells (Smith and Douglas, 1987; Kappen, 1994). The challenge of maintaining the composite organism under laboratory conditions has resulted in little experimental attention being applied to the nature of species in lichenized fungi as compared with free-living fungi which are generally easier to culture, artificial crosses being impossible for technical reasons. However, recent significant advances in tissue culture techniques (Yamomoto *et al.*, 1993; Yoshimura *et al.*, 1994) and resynthesis studies (Yoshimura *et al.*, 1993, 1994; Stocker-Wörgötter and Türk, 1994; Crittenden *et al.*, 1995) provide many exciting opportunities to test the specificity of particular bionts, their combined influence on morphogenesis and role in speciation as currently understood.

Given the great complexities of lichen thalli, on what basis are species defined? Almborn (1965) stated that it is essential that lichen taxonomy be founded on firm principles not deviating from those generally accepted in other plant groups. Firm principles are clearly important, but to what extent should we be examining other plant groups? These aspects are now considered in relation to key developments in modern systematic research.

6.3 CURRENT TAXONOMIC CONCEPTS

6.3.1 Fungal partner (mycobiont)

The mycobiont is an obligate symbiont unable to occur free-living other than as a reproductive propagule; the reasons for the dependence of the fungus on the alga are not known. The mycobiont is typically unique to each lichen and the *Code of Botanical Nomenclature* rules that 'for nomenclatural purposes names given to lichens shall be considered as applying to their fungal component'; [Article 13.1(d), Greuter, 1994]. Thus, it follows that those characters derived from the fungus are regarded as most important taxonomically. Herein lies a problem – what are the characters that are derived from the fungus? In general a lichen thallus receives its shape from a mycobiont (Jahns, 1988) but it is still unclear what processes regulate the formation of the specific thallus which allows us to classify the lichen taxonomically. In pure culture many mycobionts remain relatively featureless, mostly lacking the delimitation of specific layers and structures, so to a certain extent the key to this must be the influence of the photobiont on thallus morphogenesis (see section 6.3.2). Most lichen fungi are ascomycetes; only a few form associations with basidiomycete or deuteromycete fungi. Non-lichenized fungi, however, typically lack a perennial thallus and fruiting bodies (ascomata). During the past 30 years lichen taxonomists have increasingly become aware of the importance of mycological characters, notably anatomical structures of the thallus and reproductive structures such as ascomata (see below) that provide a more natural means of delimiting genera mainly based on spore characters adopted by Zahlbruckner (1926a). In view of the polyphyletic origin of lichen fungi (Gargas *et al.*, 1995a) it is hardly surprising that different characters are diagnostic at a species level according to the group of fungi.

(a) Reproductive fungal characters

The type and basic morphology of ascomata, the sexual fruiting body in the ascomycetes ('sac fungi') containing sac-like asci in which spores are formed, are useful characters at a specific and also at higher taxonomic levels. Generic and also higher level taxonomic concepts in fungi forming crustose lichens are nowadays based on microscopic features of the ascomata and asci, including their ontogeny – characters which have only exceptionally been critically examined in the foliose and fruticose lichens – which partly explains the reasons for recent segregates of several macrolichen genera including foliose and fruticose representatives of Teloschistaceae by Kärnefelt *et al.* (1992), although more critical studies are needed in many cases, especially within the genus *Parmelia* Ach. *sensu lato* (Hawksworth, 1994).

A major problem facing lichenologists is our scant knowledge of many details of lichen reproduction, which has not been directly observed in nature or the laboratory and it is assumed that what happens in free-living ascomycetes has a parallel in lichenized ascomycetes, although this need not necessarily be true. However, indirect evidence suggests that sexual reproduction does occur, as shown by the extreme variability of some crustose lichen species which form interlocking mosaics (Poelt, 1994) and from laboratory investigations including isozyme and molecular studies (see section 6.6). During the early stages of ascomatal development in some species specialized cells (trichogynes) are produced which are assumed to be able to receive conidia (spermatia) produced by special reproductive structures producing conidia (pycnidia) of the same lichen and probably also from other lichens of the same species (Culberson and Culberson, 1994). Formed through the fusion of dikaryotic cells, the ascus is the characteristic cell of ascomycetes which brings about karyogamy (the fusion of two sex nuclei after cell fusion), undergo genetic recombination through meiosis, and during ascosporogenesis divide mitotically to form ascospores, which are subsequently liberated (Bellemère, 1994). Data are sparse concerning the dikaryotization process (where cells having two genetically distinct haploid nuclei fuse), and although spermatia have been observed in contact with trichogynes in *Cladonia furcata* (Huds.) Schrader (Honegger, 1984), the sexual process of dikaryotization has not, however convincingly been seen in nature or the laboratory (Ahmadjian and Jacobs, 1987). There is no evidence for the presence of the parasexual cycle occurring in lichens where plasmogamy, karyogamy and haploidization result in genetic recombination during mitosis, although the presence of thallus fusions (see section 6.7) offer intriguing possibilities that genetic exchange between different mycobionts may occur, surely one of the most challenging questions which molecular biologists need address as an urgent priority.

The relative constancy of ascus types linked to ontogenetic differences in the ascomata are considered to be the result of a long evolution, suggesting that they can be usefully used to define taxa at a higher level (Bellemère, 1994). The special types of ascus apex structures, and particularly their iodine reactions, has recently been used to rearrange the classification of lichens (Hafellner, 1984). There is no doubt that studies of large unwieldy heterogeneous genera in the past 10 years such as *Lecidea* and *Lecanora* Ach. have benefited from the use of this additional character. However, there has not been universal acceptance for its use and Bellemère (1994) concludes that ascus characters are of secondary importance in ascomycete systematics. He also points out technical problems in interpreting observations made only by light microscopy on the I+ blue patterns which may reflect the position of certain polysaccharides. There is clearly here a need for more standardized techniques as well as detailed ontogenetic studies using TEM.

Ascospores, the reproductive cell in ascomycete fungi, are important species characters varying greatly in lichens in shape, size, structure, number per ascus, septation and may be colourless, greenish or brown. Ascospore septation is nowadays considered diagnostic at a specific level thus species in recent genera such as *Vezdaea* Tscherm.-Woess and Poelt may contain simple or varyingly septate ascospores with 8- to 16-spored asci (Purvis *et al.*, 1992).

Conidiomata, specialized multi-hyphal structures bearing conidia occur as mostly flask-shaped pycnidia in lichens. Different types of pycnidia relating to hyphal arrangement are recognized which are useful characters at a generic and also specific levels (Vobis, 1980). For instance, *Ramalina siliquosa* (Huds.) A.L. Sm. and *R. cuspidata* (Ach.) Nyl. have differently coloured pycnidia and thus may be conveniently distinguished in the field on this basis (Purvis *et al.*, 1992). Conidia vary in shape, size and septation and are particularly useful characters at the specific level as in *Opegrapha*. Some species produce more than one type of conidium as in *M. denigrata* (Fr.) Hedl. producing three types (Coppins, 1983) and the function of the different types has yet to be conclusively established (see discussion above). Although conidia may be formed in a range of widely different structures), they have not been recorded in all lichenized fungi. The ability of lichens to produce pycnidia varies widely, even within the same genus. For example, *Lecanactis abietina* (Ach.) Körber produces many pycnidia and rarely fruits, while others such as *L. subabietina* Coppins and P. James have abundant pycnidia but fruiting bodies are unknown. Specialized asexual conidia termed thalloconidia occur in 17 species within the genus *Umbilicaria* Hoffm. as well as in other, including crustose lichen genera (Poelt and Obermayer, 1990; Hestmark, 1991; Poelt, 1991). These have been interpreted as hypothalline, hyphomycetous anamorphs occurring attached to the teleomorph. Differences in structure of these are useful at a specific level.

6.3.2 Lichen algae (photobionts) and photomorphs

Most algae occurring as photobionts belong to the class Chlorophyceae, to a lesser extent the Cyanobacteria (Cyanophyceae or blue–green algae) and rarely the Xanthophyceae (yellow–green algae) with most occurring within three genera: *Nostoc, Trentepohlia* and *Trebouxia* (Tschermak-Woess, 1988). The number of species currently identified as photobionts has been estimated to include about 100 green algae in relatively few (about 23) genera, and an uncertain, but much lower number of cyanobacterial species in some 15 genera (Tschermak-Woess, 1988; Hawksworth and Honegger, 1994). However, given the woefully little attention paid to the taxonomy of these organisms many statements and/or identifications, particularly in the older literature are questionable. As reported in this volume (John and Maggs, 1997: Chapter 5) there are many problems surrounding species concepts in green algae belonging to the order Chlorococcales, e.g. *Trebouxia*, with an additional problem in lichens being the algae are modified through symbiosis resulting, for instance, in a suppression of sexual reproduction and the formation of filaments. The popular perception that the photobiont has little taxonomic value in lichens has no doubt further contributed to the lack of study of these organisms (Ahmadjian, 1993b). Ahmadjian argued that the type of photobiont selected by a mycobiont might legitimately be considered a taxonomic character along with thallus or secondary compound characteristics. However, several studies have shown that mycobionts are not very selective in their choice of photobionts. For example, some crustose lichens may contain different photobionts as *Chaenotheca chlorella* (Ach.) Müll. Arg. lichenized with a trebouxioid alga in Costa Rica rather than *Stichococcus* with which it is associated in more northerly areas of the northern Hemisphere (Tibell, 1982). On the other hand Tschermak-Woess (1995) examined some 150 thalli of *Phlyctis argena* (Sprengel) Flotow from Europe and North America concluding that there was strong selectivity for the photobiont *Dictyochloropsis splendida* Geitler since all cultures were identical in both morphology and development. Many lichen species contain different genera of algae within their thallus. Thus species with green algal photobionts may also contain accessory cyanobacteria normally localized in distinct structures termed cephalodia which are considered valuable species characters in several genera including *Peltigera* (Vitikainen, 1994) and *Stereocaulon* (Lamb, 1977). Other photobiont differences have also been used to distinguish between species and certain genera. For example, *Staurothele* Norman is unusual in having algae within the spore-bearing layer of the fruiting bodies (the hymenium). These hymeneal algae differ morphologically from the algae within their thalli and also between species, ranging from cuboid to cylindrical and are considered diagnostic species characters. The pattern of arrangement of algae within fungal tissues is also an important species character in many lichen genera, including *Acarospora* (Clauzade and Roux, 1981) and *Thelotrema* (Salisbury, 1971). In other cases closely related species containing different photobionts have been assigned to different genera, as for instance the separation of *Ionaspis* Th. Fr. with a *Trentepohlia* photobiont from *Hymenelia* Krempelh. with a *Trebouxia* photobiont; these are currently being evaluated (P.M. Jørgensen, personal communication).

However, while the fungus has traditionally been regarded as playing the major role in thallus formation since the early decades of this century, it has been recognized that the photobiont nevertheless also plays a significant role. The first to describe composite lichen thalli, with two morphologically different structures, attached together, were the pioneering studies of Dughi during the 1930s and 1940s on *Dendriscocaulon* Nyl. (James and Henssen, 1976). In one, the mycobiont was associated with a green alga and had a leaf-like (foliose) morphology; in the other with a cyanobacterium it had a shrubby (fruticose) morphology. The different morphotypes were also found as independent thalli. The differences in morphology, ecology, distribution and occasionally chemistry of the two independent morphotypes can be so striking that taxonomists at times have even ascribed the unattached entities to different genera. In such cases, the true relationship between the two morphs has only been elucidated by the discovery of composite thalli as, for example, in *Sticta felix* (Räusch.) Nyl. and *Dendriscocaulon*. This phenomenon has since been observed in many species, particularly occurring within the Peltigerales and Pannariaceae and are usually termed photosymbiodemes or photobiont morphotypes (Hawksworth, 1994) or, more recently, photomorphs (Laundon, 1995). Although histological observations of hyphal continuity suggests the same fungus to be present (James and Henssen, 1976), several authors have suggested they might not be identical, given the frequency with which taxonomically unrelated lichens associate with one another (Hawksworth, 1988). Artificial resynthesis experiments offer one opportunity to establish whether the same fungus occurs in both green- and cyanobacterial photomorphs; however, this is not always possible in the case of the genera *Lobaria* (Schreber) Hoffm., *Peltigera*, *Pseudocyphellaria* Vainio and *Sticta* (Schreber) Ach. since these mycobionts are difficult to grow in laboratory culture (Ahmadjian, 1993a). Armaleo and Clerc (1991) therefore applied molecular methods to define the relatedness between paired fungal components in the photomorphs. Using Southern hybridization and the polymerase chain reaction they demonstrated a genetic near-identity of the mycobionts in two photomorphs of *Pseudocyphellaria* and *Sticta* (Figure 6.1). While they concluded that paired mycobionts within two photomorphs should be assigned to the same species, they pointed out that detailed quantitative definitions of 'relatedness versus molecular variation' in lichen mycobionts will need the inclusion of additional taxa and data. Further direct evidence that the green thallus and cephalodiate structures are formed by the same mycobiont has recently been obtained by artificial resynthesis of the photosybiodeme *Peltigera leucophlebia* (Nyl.) Gyelnik from the one fungus and two algal partners (Stocker-Wörgötter and Türk, 1994).

According to the *International Code of Nomenclature*, which restricts the name of the lichen to the mycobiont, in those lichens with paired mycobionts, priority is given to the first published name. There is therefore no indication of which photomorph is implied when the name alone is used. The use of a single name can therefore result in a loss of ecological and

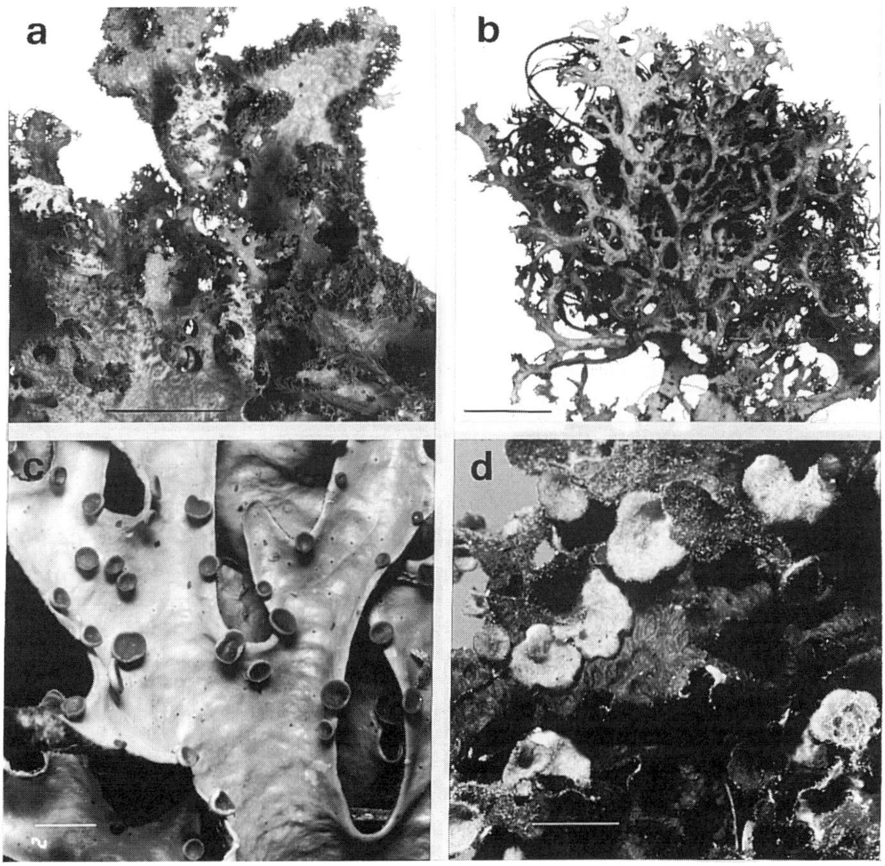

Figure 6.1 Variations in morphology within the lichen species, *Sticta canariensis*, containing the same fungus, but different photobionts. (a) leaf-like growth form containing cyanobacteria; (b) shrub-like growth form containing cyanobacteria; (c) leaf-like growth form containing green alga; (d) composite thalli with lobules containing green algal morph joined to lichen containing cyanobacteria. Scale bars = 0.5 cm. Figures (a–c), Alagadiços, Terceira, Azores; Purvis, James and Smith (BM); (d) from Galloway, SW Scotland, James (BM).

geographical information about the morphs unless qualified by a very cumbersome communication system. Thus, for example, in the Azores the blue–green photomorph of *Sticta canariensis* ('*S. dufourii*') is frequent while the green algal photomorph ('*S. canariensis*') is extremely rare and poorly developed on the island of Pico, though locally abundant on the islands of Terceira and Flores (O.W. Purvis and P.W. James, unpublished data, see Figure 6.1). There is therefore a good case to be made for maintaining such morphs as distinct taxa, which are not only morphologically dissimilar but may also be ecologically, geographically and physiologically distinct (Jørgensen, 1991) perhaps best accomplished at a subspecific level (Laundon, 1995).

Many lichen species contain different algae during different stages of their life-cycles. *Psoroma durietzii* P. James and Henss. is unusual in developing sorediate cephalodia (James and Henssen, 1975) capable of forming small, independent, structured thalli containing only a blue–green photobiont (*Nostoc*) which can secondarily capture a coccoidal green alga. In *Solorina crocea* (L.) Ach. fungal propagules lacking algae form initially an association with cyanobacteria and latterly with coccoidal green algae resulting in a layered thallus (Jahns, 1987). Lichenicolous lichens may also associate with different photobionts at different stages of their life-histories. For example, in *Diploschistes muscorum* (Scop) R. Sant. where the thallus starts as a parasite on squamulose thalli on different species of *Cladonia* and is first associated with the photobiont of the *Cladonia*, (*Trebouxia irregularis*), but later exchanges it for another species of the same green algal genus, *T. showmanii* (Friedl, 1987).

6.4 VEGETATIVE CHARACTERS, ASEXUAL PROPAGULES AND 'SPECIES PAIRS'

Most non-lichenized fungi consist of a diffuse mycelium below ground and seasonal ascomata with usually hardly any vegetative characteristics and hence lacking in important taxonomic characters. In the lichenized ascomycetes vegetative structures are mostly evident, highly complex and it is hardly surprising therefore that great significance has been applied to these, particularly at the species level. Although gross morphology, pigmentation and anatomy have traditionally been used at specific and even generic levels, recently even fruticose and crustose lichens have been recognized as being closely related as, for example, the crustose lichens, *Toninia leucophaeopsis* (Nyl.) Th. Fr. and *Bilimbia tornense* Magnusson which share a similar chemistry and apothecial anatomy with the shrub-like *Stereocaulon* and are now placed in that genus (Purvis and James, 1985). In many macrolichens the great diversity of gross morphological features as compared with crustose lichens, has often led to a lack of critical attention paid to ascomatal anatomy and ontogeny, e.g. in the genus *Parmelia* and its segregates. There exists wide variation in the taxonomic importance given to different characters between genera. In some genera ascospores provide useful characters, whereas in *Usnea* Hill these are apparently remarkably uniform and not useful at a specific level, although detailed studies involving statistical analyses of length : breadth ratios are lacking. However, the wide range of asexual reproductive structures including soredia, isidia and schizidia are routinely used to differentiate between taxa at a species level in many lichen genera, including *Usnea* (Clerc, 1987). Since production and morphology of soralia correlates with populations of *Usnea* having distinct chemical, geographic and other morphological properties, they are considered to be genetically determined. However, caution is required in interpreting such characters as these characters may also be influenced by environmental factors, thus emphasizing the importance of carrying out field investigations when describing new taxa.

Species concepts have traditionally been partly based on sexual strategies and characters associated with reproductive structures. In those lichens where no fruiting bodies are known to exist, ascospores and ascomatal characters cannot be used and morphological and chemical characters are therefore crucially important. Only relatively few lichens are regularly able to simultaneously propagate by ascospores and in addition by some type of sexual diaspore. One example is the well-known *Baeomyces rufus* (Huds.) Rebent., a successful colonizer of eroded soil which develops apothecia, schizidia and soredia at the same time (Jahns, 1987). In many cases, however there exists a continuum between fertile species and sorediate counterparts as in *Catillaria pulverea* (Borrer) Lettau, which may occur as rarely fertile, continuously granular, sorediate thalli (*Catillaria pulverea*) and thin, frequently non-sorediate, richly fertile thalli (*C. albocincta* Degel.), but which as these intergrade are now considered a single species (Purvis and James, 1993). Where lichens have sexual and sorediate phases these entitities have also occasionally been accorded specific status. For example, in the genus *Peltigera* Willd. where *P. spuria* (Ach.) DC lacking soredia, bearing small erect fertile lobes was regarded as distinct from sorediate morphs ('*P. erumpens* (Taylor) Elenkin.') and morphs bearing apothecia and soralia on the same thallus as ('*P. hazslinszkyi* Gyeln.') but which are now collectively regarded as the single species *Peltigera didactyla* (With.) Laundon (Vitikainen, 1994).

Many cases are now known of 'Artenpaare' or species pairs, one of which is fertile with ascomata (a 'primary species') and the other non-fertile, bearing isidia or soredia (a 'secondary species'); these pairs do not intergrade and appear to be genetically isolated, unable to exchange genes (Tehler, 1982). The character sorediate versus non-sorediate is often used as a species criterion, provided that the taxa share different geographic distributions. Vegetative, asexual reproduction by means of soredia, isidia or other vegetative diaspores has the clear advantage of both fungal and algal partners being distributed simultaneously without a need for resynthesis in an environment where there may be few suitable algal hosts. As a rule the sorediate element is more widely distributed, as shown in polluted environments where there is typically a loss of sexuality, and many islands have disproportionately fewer fertile species than on the adjacent mainland. It has been suggested that the sorediate species might be formed in response to unfavourable conditions; thus, species at the edges of their ranges might be expected to be sorediate. Asexual reproduction apparently also has its disadvantages, however, as the organism also presumably loses its genetic flexibility and can no longer adapt to new or changing environments through gene recombination. The situation may well be far more complex than we currently believe with not only sexual and sorediate counterparts, but possibly also various asexual counterparts with alternative asexual propagules. In the genus *Menegazzia*, *M. eperforata* P. James and D.J. Galloway is thought to be the isidiate counterpart of the sorediate *M. nothofagi* (Zahlbr.) P. James and D.J. Galloway which is related to the primary sexual species *M. prototypica* P. James (James and Galloway 1992; see Figure 6.2 for further explanation). Arguments to routinely relegate secondary 'species' to form rank are not now generally accepted (Hawksworth, 1994; Poelt, 1994), but the existence of such pairs or groups of organisms provides a valuable opportunity to study speciation by the application of, for example, molecular techniques and isozyme analysis to determine just how distinct such organisms are and also whether they have evolved on single or multiple occasions.

Jahns and Ott (1994) draw attention to the apparently low number of specialized cells and determined cells and tissues in lichens, which has important consequences for their taxonomy. Lichens might therefore be viewed as a complicated building constructed from few types of stones. Jahns and Ott have shown through various ontogenetic studies a multifunctionality of tissues and considerable plasticity, i.e. old pycnidia can produce apothecia, root-like rhizines may acquire algae and produce layered thalli and out-growing cortical cells in *Peltigera aphthosa* (L.) Willd. might start the development of cephalodia. Therefore, it follows naturally that it is not sufficient to describe the adult organs but that the ontogenetic aspect should also be considered.

6.5 THE IMPORTANCE OF ENVIRONMENTAL FACTORS IN INFLUENCING PHENOTYPIC EXPRESSION

Lichens are intimately associated with their environment, many species having subtly different ecological requirements, e.g. to the extent of preferring particular tree or rock types as substrate. It is those species with a wide ecological amplitude which usually exhibit the greatest range of morphological variation. An understanding of the sources of phenotypic variation in lichens, therefore, is critical to the evaluation of criteria for tax-

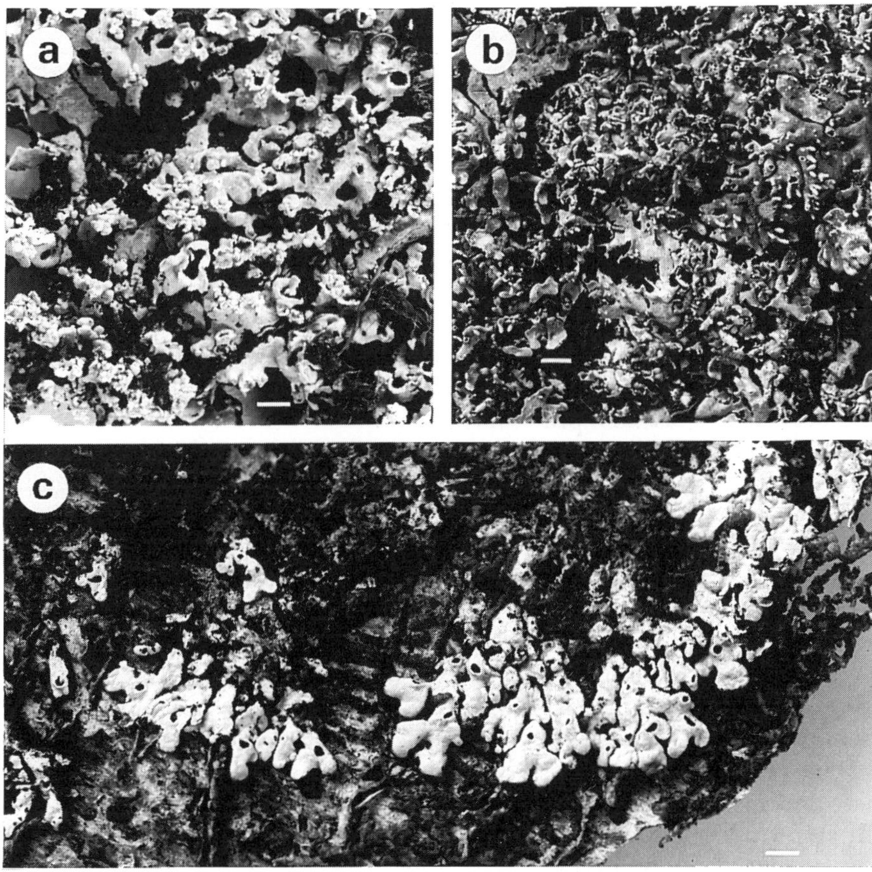

Figure 6.2 Three species of *Menegazzia* found in temperate rainforest distinguished by having different asexual, vegetative propagules. The possible 'primary' fertile species is much rarer than the 'secondary' asexual counterparts, though the evolutionary relationships of these and many other groups of species, characterized essentially by alternative methods of reproduction, are in need of critical evaluation. (a) *M. nothofagi* (sorediate: having powdery soredia, consisting of fungal hyphae intermixed with algae), Tasmania, *G. Kantvilas* 603/80, BM); (b) *M. eperforata* (isidiate: having cylindrical isidia, peg-like outgrowths from the upper fungal layer of the thallus, the cortex), Tasmania, J. 6494, CBG); (c) *Menegazzia prototypica* (lacking soredia and isidia and rarely fertile and bearing disc-shaped fruits as in Figure 6.1(c), Projection Bluff, Tasmania, *G. Kantvilas*, HO). Scale bar = 1 mm.

onomic delimitation. Such variation may reflect either genetic differences between individuals due to mutational and recombinational processes, or result from changes in developmental patterns in response to heterogeneous habitats. The unfortunate practice of taxonomists describing lichens on the basis of fragmentary material without field study has resulted in the description of many superfluous taxa (Weber, 1968; Hawksworth, 1973). This is especially the case in crustose lichens occurring on smooth, hard rocks which are difficult to sample adequately. For instance, several species within the crustose lichen genus *Aspicilia*, which is notoriously lacking in diagnostic characters, are distinctively pigmented when fresh, but these colours are lost in storage in the herbarium. Similarly, extreme morphological variation observed in *Acarospora* has resulted in the description of many superfluous names; numerous species in both *Acarospora* and *Aspicilia* therefore require careful field evaluation.

Lichenized ascomycetes are characterized by the great longevity of their thalli and ascomata. Given their exposure to harsh environmental conditions, it is perhaps hardly surprising that they typically show great variation in morphology. Indeed, two taxa displaying the extreme ranges in morphology of a species may appear more different than distinct species (Poelt, 1994). In the past a large number of taxa were introduced at species and sub-specific level to accommodate such morphotype variation. For instance, Zahlbruckner (1927) recognizes 59 varieties and forms of *Cladonia squamosa* Hoffm. Many lichens are restricted to distinct habitats and it is reasonable to assume that ecological factors have played an important role in speciation. Such is the importance of ecological factors in phenotypic expression that these have been emphasized in definitions of species where these correlate with other features. Thus, Hawksworth (1973) claimed that species concepts in lichens are 'currently based on sharp discontinuities in one or several morphological and anatomical (or) chemical characters, particularly where there is evidence that genotypic differences are involved (e.g. the two entities growing side by side in a uniform ecological situation and retaining their identities) or there are differences in either ecological requirements or geographical distribution, or both.' Some lichens have a relatively wide ecological amplitude and their thalli become modified by the environment in many ways, resulting in changes in growth form, anatomy, pruinosity in those lichens growing on calcareous substrates caused by superficial deposits of calcium oxalate, and alterations in pigment due to differences in light intensity. The variation in morphology observed within taxa in different ecological situations is vast and the problem facing lichenologists is deciding when such variation is important or trivial, especially difficult when examining herbarium material with inadequate field data. A detailed discussion of ecological factors is beyond the scope of this review but has been outlined by Hawksworth (1973).

Among the most extreme environments in which modifications are numerous are metal-rich ones characterized by specialized lichen communities with different rock types possessing distinctive assemblages (Wirth, 1972; Purvis and James, 1985; Purvis, 1993). On copper-rich rocks certain lichens accumulate copper as a chelated Cu–lichen acid complex and turn greenish (Purvis *et al.*, 1987). As colour is normally regarded as an important character at a specific level it is perhaps hardly surprising that this phenomenon has been the cause of taxonomic confusion on several occasions (Purvis, 1984; Purvis *et al.*, 1990). Thus the Cu-rich morph of *Acarospora smaragdula* (Wahlenb.) Massal. has been described on at least three separate occasions previously as *A. undata* G. Clauzade, C. Roux and V. Wirth, *A. isortoquensis* Alstrup and *A. alberti* Tavares (Purvis *et al.*, 1985). A further consequence of the complexing of metals by lichen acids to form stable ligands is that the free acid may be difficult to detect using thin-layer chromatography, thus adding further weight to the unsuspecting researcher that his taxon might be distinct. Additionally, many mine environments are characteristically acidic and under these conditions cyanobacterial containing lichens are absent perhaps owing to the sensitivity of the key enzyme, nitrogenase involved in nitrogen metabolism, to low pH (Farmer *et al.*, 1992). For this reason, Purvis *et al.* (1992) considered the mine 'species' *Placopsis lambii* Hertel and V. Wirth (lacking cephalodia containing cyanobacteria) doubtfully distinct from *P. gelida* (L.) Lindsay (with cephalodia). While several ecotypes inevitably occur in such specialized environments, there are other species which are more or less restricted, such as *Psilolechia leprosa* Coppins and Purvis, which also occurs beneath copper lightning conductors on church walls throughout Britain (Coppins and Purvis, 1987), as well as on sheltered rock faces in cloud forest in the Azores (O.W. Purvis and P.W. James, unpublished data). Such environments might be regarded as analogous to islands and thus provide models for the study of evolution of isolated populations. Several other lichens occur in anthropogenic habitats whose precise wild origin is unknown, e.g. *Lecanora vinetorum* Poelt and Huneck growing on vine supports sprayed with Bordeaux mixture in Austria (Poelt and Huneck, 1968), *Buellia pulverea* Coppins and P. James subjected to fluoride pollution arising from nearby aluminium smelters, and the ubiquitous pollution lichen *Lecanora conizaeoides* Nyl. ex Crombie first recorded in herbarium collections in the early 1860s.

6.6 SECONDARY METABOLITES, LICHEN CHEMOTAXONOMY AND MOLECULAR STUDIES

When colour was accepted as a generic or specific character, chemical discrimination was being inadvertently applied. For example, the grey genus *Physcia* Nyl. (containing the colourless substance atranorin in the cortex) is distinguished from the superficially similar yellow–orange genus *Xanthoria* (Fr) Th. Fr. (containing an orange pigment parietin in the cortex) based on colour and hence chemical characteristics. Similarly at a species level, the yellow *Parmeliopsis ambigua* (Wulfen) Nyl. (containing yellow usnic acid) is separated from the grey *P. hyperopta* (Ach.) Arnold (containing colourless atranorin). Most secondary metabolites are, however, colourless and require analytical procedures for their detection. There has been a long history of using chemical tests for taxonomic purposes dating back to the simple, though unrefined, spot test reactions performed by Nylander in the 1860s followed by a wide range of more sophisticated analytical methods (Santesson, 1973). Indeed, the discipline of 'chemosystematics' was pioneered by lichenologists. Over 400 substances in 6000 lichen species have been characterized, many unique to lichen fungi and new compounds are being continually discovered (Culberson and Elix, 1989). The occurrence of these heterogeneous secondary metabolites has provided the lichenologist with useful additional characters for classifying lichens at the supraspecific, specific and varietal levels and is essential when describing sterile crustose lichens which are unknown fertile. Recent research has also the demonstrated the ability of certain lichenized fungi to produce these characteristic lichen acids in mycobiont culture (Hamada, 1989).

The taxonomic value of chemical criteria in lichen taxonomy has, however, been a matter of considerable controversy for over 130 years (Hawksworth, 1976), although chemical investigations now form an integral part of all serious taxonomic investigations of lichen-forming fungi. There are three common types of chemical pattern studied based upon replacement type compounds, chemosyndromic variation and accessory type compounds. Varying taxonomic significance is applied according to the relatedness of the compounds, although this depends on the group of lichens and the specific monographer. Clearly chemical variation when associated with morphological differences, merit formal taxonomic recognition, but where chemical variation occurs within morphologically similar plants, then the question as at what rank these should be treated remains a cause of considerable debate. The taxonomic interpretation of chemical variation is discussed in detail elsewhere (Hawksworth, 1976; Brodo, 1978, 1986; Culberson and Elix, 1989). Of key importance is to ascertain the genetic basis for the variation in chemistry which will then allow a more informed judgement of how and at what level to differentiate between different taxa.

There are many well-defined lichens having a uniform secondary chemistry throughout their range and in these chemistry is as much a key character as the morphology characterizing the species. This is not always the case with other species exhibiting minor chemical variation not associated with obvious morphological, distributional or ecological differences. Most lichenologists interpret morphologically similar chemotypes as components of single polymorphic species mainly because they believe the genetic basis of chemical differences to be trivial. The Culbersons (Culberson, 1986; Culberson and Culberson, 1994), making a not clearly proven parallel principally with the animal kingdom, have recognized a number of such morphs in the *Ramalina siliquosa* and *Cladonia chlorophaea* (Flörke ex Sommerf.) Sprengel complexes as sibling species where extremely similar and visually indistinguishable cryptic species exist, the individuals of which, when co-existing, do not interbreed with each other but maintain the integrity of their respective gene pools (Mayr, 1992). In many groups of organisms, this question would have been readily resolved by the appropriate experimental crosses, not possible in the lichen fungi. Culberson and Culberson (1994) used lichen secondary products produced in either mycobiont culture (in *Ramalina siliquosa* agg.) or resynthesis experiments (in *Cladonia chlorophaea* agg.) as genetic markers for the indirect analysis of progeny from maternal individuals in nature. These findings, although extremely interesting, require wider testing using complementary analytical procedures including molecular studies to conclusively determine the reasons for the chemical variation observed before the full implications of these results can be assessed.

Molecular techniques have as yet hardly been applied to the systematics of the lichen-forming fungi, especially at the species level, although several studies have been successfully applied at either higher taxonomic levels or at a population level (DePriest, 1994). Analysis of small subunit ribosomal DNA sequences suggests at least five independent origins of the lichen habit in several groups of fungi, thus confirming evidence from morphological studies (Gargas *et al.*, 1995a). At a species level, Armaleo and Clerc (1991) used restriction-fragment polymorphism of anonymous portions of the genomic DNA and ribosomal DNA (rDNA), to show that lichen chimeras are fungi of a single species forming distinct photomorphs with photobionts from different kingdoms (see section 6.3.2). We now know that lichens show extreme variability for rDNA, there being a large amount of length and restriction site variation, even within a species (De Priest, 1993; Gargas *et al.*, 1995b). At the Nato Conference on Ascomycete Systematics held in Paris during 1994, a need for developing species specific markers was identified (see Berbee and Taylor, 1994: 222–3). Studies using enzymes and proteins – the first apparent products of genes – have been used with some success to demonstrate intrapopulational enzyme polymorphisms as in *Umbilicaria* species in central Ellesmere Island (Fahselt, 1989) and at a species level, two lichen species *Ramalina cuspidata* and *R. siliquosa* have also been separated on the basis of banding patterns (Mattsson and Kärnefelt, 1986).

6.7 WHAT IS AN INDIVIDUAL LICHEN? FUSED THALLI AND MECHANICAL HYBRIDS

There is now good evidence that fusions of lichen thalli may occur within populations of the same species, between different species of the same genus or between different genera (Figures 6.3–6.5), although how

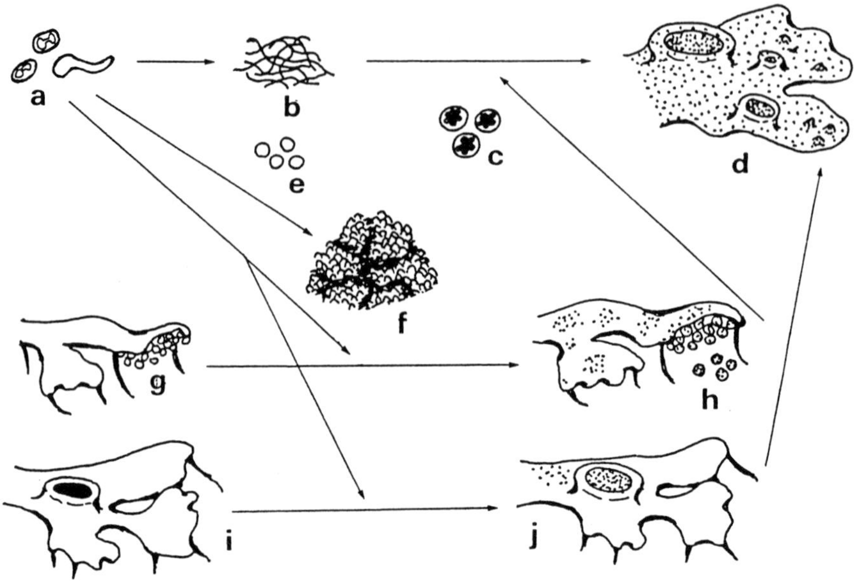

Figure 6.3 Life-cycle of *Xanthoria parietina* (simplified after Ott, 1987a) where the yellow colour due to the substance parietin serves as a marker. (a) Germinating fungal ascospores; (b) developing network of fungal hyphae; (c) free-living lichen alga – *Pseudotrebouxia* normally involved in lichenization; (d) mature *Xanthoria* thallus formed through the interaction between the fungal hyphal network and lichen alga; (e) foreign coccoid green algae not involved in lichenization; (f) undifferentiated, areolated 'lichen' crust containing fungal hyphae and foreign coccoid green algae, enabling the mycobiont to persist until it meets the right algal partner for lichenization to occur; (g) *Physcia* sp. producing powdery soredia consisting of fungal hyphae intermixed with lichen algae; (h) *Physcia* thallus and soredia infected by *Xanthoria* fungal spores or the undifferentiated crust in (f) resulting in *Physcia* disseminating *Xanthoria* through powdery soredia which *Xanthoria* itself does not produce; (i) *Physcia* sp. producing spore-bearing fruiting bodies – apothecia; (j) *Physcia* thallus and apothecia infected by *Xanthoria* resulting in *Physcia* thallus producing *Xanthoria* spores. Yellow colour of *Xanthoria* dotted.

frequently these occur is unknown. It is hardly surprising that this phenomenon has caused some confusion in the past. For instance, *Phaeophyscia orbicularis* is typically grey in colour, but individuals growing adjacent to the bright yellow *Xanthoria parietina* (L.) Th. Fr. containing areas with an atypical yellow pigmentation due to parietin, were described as f. *virella* A.L. Sm. We now know this to be one of a few documented mechanical hybrids between lichens belonging to different genera and demonstrates

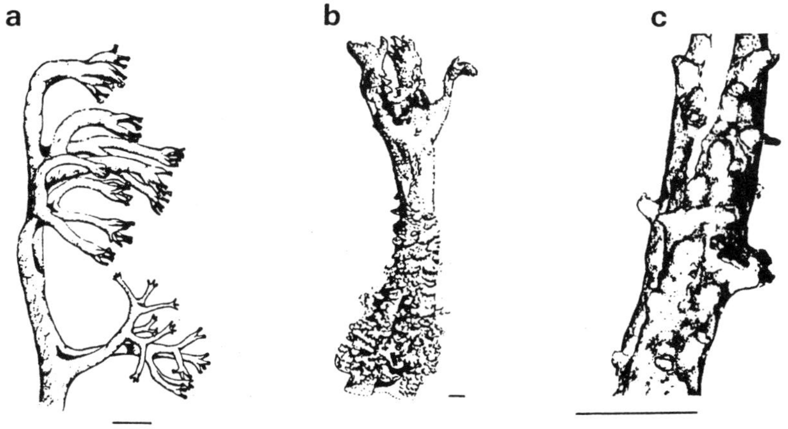

Figure 6.4 Thallus fusion in *Cladonia* (adapted from Jahns, 1987). (a) The 'reindeer lichen', *Cladonia rangiferina* with smooth surface; (b) *C. squamosa* with abundant squamules; (c) *C. rangiferina–C. squamosa* chimera with scattered squamules. Figures (a) and (b) reproduced with kind permission from Société Botanique du Centre Ouest, Le Clos de la Lande, Saint-Suplice-de-Royan, 17200 Royan, France; (c) reproduced with kind permission from J. Cramer, Gebrüder Borntraeger, D-1000 Berlin, D-7000 Stuttgart, Germany.

how easy it is to describe taxa based on phenetic characteristics without proper consideration of other factors. Ott (1987a,b) demonstrated that the mycobiont of *Xanthoria parietina* (L.) Th. Fr. is capable of living together and obtaining nutrition not only from *Trebouxia* but also from other coccoid green algae unable to undergo lichenization in the true sense (see Figure 6.3). However, with foreign algae the fungus does not undergo differentiation into a normal thallus but exists as an areolated crust and is even capable of extracting suitable algae from foreign soredia of *Physcia* (Schreber) Michaux, *Parmelia* and *Evernia* Ach. or else parasitizing mature thalli and taking over the hymenium as in the case of *Physcia tenella* (Scop) DC. that has yellow apothecia containing *Xanthoria* spores (Ott, 1987a,b). These have been considered analogous to graft–chimeras in higher plants (Hawksworth, 1988a). Such combined phenotypes do not apparently represent true or 'sexual' hybrids, as there seems to be no exchange of genetic material although this needs rigorous testing. There are several examples of fusions between species within the same genus, as in *Cladonia* subgenus *Cladina*, which are rarely fertile. Many of these so-called reindeer lichens grow in cushions with the individual plants touching one another and the branches may fuse (Jahns, 1987). Therefore, most hyphae will contain hyphae from different parental plants and it is probable that the phenotype is influenced by the genetic material of all hyphae.

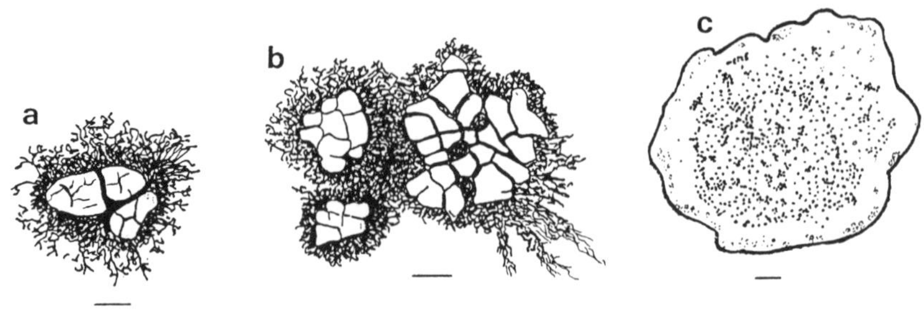

Figure 6.5 Fusion of *Rhizocarpon geographicum* thalli (simplified after Letrouit-Galinou and Asta, 1994). (a) Confluence of areoles on black young fungal hypothallus; (b) fusion of young thalli; (c) fully mature thallus with a continuous margin and whose central yellow part is entirely covered with areoles and apothecia.

Different species rarely fuse, though fusions do occur as, for instance, between *C. rangiferina* (L.) Weber ex F.H. Wigg. and *C. mitis* Sandst. and perhaps, more surprisingly, between *C. rangiferina* (L.) Weber ex F.H. Wigg. and *C. squamosa* Hoffm. (subgen. *Cladonia*) when the former may develop small phyllocladia typical for *C. squamosa* (Figure 6.4(c)). Many crustose lichens often form mosaics. Letrouit-Galinou and Asta (1994) demonstrated that in *Rhizocarpon geographicum* (L.) DC. several thalli may fuse at an early stage of development, resulting in the formation of a single thallus with a well-delimited margin. It is therefore likely that individual thalli will not be genetically uniform and that populations of mycobionts and photobionts may exist within a single thallus, though this remains to be tested. Yet other lichens, such as in species of *Toninia* Massal, commence development on other lichens, usually species containing cyanobacteria (Timdal, 1991). Although some thalli in mosaic-forming lichens may fuse together, there are many other species in which this does not happen even between thalli of the same species, as for instance in the genus *Pyrenula* Massal. where thalli are typically marked by clearly delimited prothalline boundaries. Poelt (1994) draws attention to the variable mosaics formed by several crustose lichen genera typically occurring on siliceous rocks (e.g. *Fuscidea* V. Wirth and Vězda, *Bellemerea* Hafellner and Roux, *Lecidea*) and in *Graphis scripta* (L.) Ach. on trees. Such variation has in the past been accorded subspecific rank, e.g. Zahlbruckner (1923) enumerates no less than 72 varieties and forms of *Graphis scripta*.

The extent to which mixed thalli occur in nature remains unclear. Indeed, the extraordinary specificity of many fungi growing on lichens ('lichenicolous fungi') to particular genera or species of lichen would suggest there are great incompatabilities and that it is therefore unreasonable to suggest a free-for-all. Molecular biologists need to be aware, however, of the possibilities for thallus fusions and to formulate hypotheses to test their frequency of occurrence and influence on lichen biology .

6.8 CONCLUSIONS

Lichen systematists involved in monographic revisions rarely discuss their basis for defining species. Considerably greater effort has recently been focused on defining the far less stable higher levels. Thus, Hafellner (1989) in discussing the 'principles of classification' deals exclusively with genera and higher taxa and does not define a species. Systematics has two principal objectives, namely to communicate the identity of an organism by means of latinized names, and to indicate the probable evolutionary relationships of organisms. The basic assumption of lichenologists that every lichen species consists of a distinct and unique fungus and its specific alga is flawed. We now know that individual thalli can be composed of populations of different mycobionts, as well as different species of algae. The preliminary evidence of sexual outcrossing provided by the Culberson and Culberson (1994) and De Priest (1994) and consequent formations of sibling species, if found to be widespread, will result in a significant increase in taxa, but would we be performing a service if we recognized these at a specific level and re-defined our concepts? The reviewer adopts a pragmatic view in such matters and believes this would cause undue complexity for the user.

Systematics, if considered a mirror of evolution is faced with one major problem – we were not there to watch it happening (Jahns, 1988). We attempt to reconstruct the course of evolution through comparative evaluation of characters from characteristics of present day organisms but the fossil record of fungi is non-existent. This means a subjective component is inevitable. This is also true for cladistics and systematics based on molecular techniques. The cladistic method has forced a more logical way of thinking and molecular data additional characters. Molecular systematists, certainly in ascomycete lichenized and non-lichenized taxa have tended to work with supraordinal taxa to avoid problems of identification at a specific level. We now need to focus attention on particular problem groups and apply a range of techniques to test our methodologies. There are woefully few researchers and results are often accepted without repeated testing in other laboratories.

The phylogenetic species concept has yet to play an important role in defining species, mycologists have been slow to take up cladistics, and only recently have phylogenetic studies been used to study relationships between taxa at higher levels (Tehler, 1994). The biological species concept is inappropriate for lichens owing to technical problems in studying breeding behaviour in culture. It is to be hoped that the many exciting new analytical techniques at our disposal, including molecular and ultrastructural studies, which when combined with morphological, ontogenetic and environmental studies will further contribute to our basic understanding of these intriguing organisms. Only by such a multi-disciplinary approach involving both comparative morphologists as well as other biologists, who have mutually much to learn from each other's disciplines, can we hope to progress beyond the morphological species concept (Mayr, 1992) which presently remains the only practical way of naming species.

In conclusion, lichen species are based on clear discontinuities in one or more unrelated fungal characters. A character which is of fundamental importance in one group may be much less important in another. Indeed, it has been said the 'art of taxonomy is in devising the most appropriate scheme of character weighting' (Brodo, 1986). The lichen symbiosis is complex and its existence poses one of the most fundamental questions in biology, namely how two such distinct organisms can combine to form such distinctive organisms so dissimilar from either component. The time has come for the barriers to come down between molecular and morphological systematists. Morphologists need to become more involved in molecular work and vice versa as they know the taxa and interesting questions to be answered.

Acknowledgements

I am particularly grateful to Mr P.W. James and Professor P.M. Jørgensen for stimulating discussions on all aspects of lichen biology on many occasions. Dr M. Gibby, Professor D.L. Hawksworth, Dr D.M. John and Mr J. Vogel are thanked for their valuable comments on an earlier draft of the paper. Professor M. Blackwell is thanked for advice regarding molecular studies.

6.9 REFERENCES

Ahmadjian, V. (1993a) *The Lichen Symbiosis.* John Wiley and Sons, Inc, New York.

Ahmadjian, V. (1993b) The lichen photobiont – what can it tell us about lichen systematics? *Bryologist*, **96**, 310–13.

Ahmadjian, V. and Jacobs, J.B. (1987) Studies in the development of synthetic lichens. In *Progress and Problems in Lichenology in the Eighties. Bibliotheca lichenologica*, **25**, 47–58.

Almborn, O. (1965) The species concept in lichen taxonomy. *Särtryck ur Botaniska Notiser*, 1965 (**4**), 454–7.

Armaleo, D. and Clerc, P. (1991) Lichen chimeras, DNA analysis suggests that one fungus forms two morphotypes. *Experimental Mycology*, **15**, 1–10.

Bellemère, A. (1994) Asci and ascospores in ascomycete systematics, in *Ascomycete Systematics: Problems and Perspectives in the Nineties* (ed. D.L. Hawksworth), Plenum Press, London, pp. 111–25.

Berbee, M.L. and Taylor, J.W. (1994) 18S Ribosomal DNA, sequence data and dating, classifying and ranking the fungi, in *Ascomycete Systematics: Problems and Perspectives in the Nineties* (ed. D.L. Hawksworth), Plenum Press, London, pp. 213–23.

Brodo, I.M. (1978) Changing concepts regarding chemical diversity in lichens. *Lichenologist*, **10**, 1–11.

Brodo, I.M. (1986) Interpreting chemical variation in lichens for systematic purposes. *Bryologist*, **89**, 132–8.

Clauzade, G. and Roux, C. (1981) Les Acarospora de l'Europe occidentale et de la région Méditerranéenne. *Bulletin du Muséum d'Histoire Naturelle de Marseille*, **41**, 41–93.

Clerc, P. (1987) On the morphology of soralia in the genus *Usnea*, in *Progress and Problems in Lichenology in the Eighties, Bibliotheca lichenologica* 25 (ed. E. Peveling), Cramer, Berlin, pp. 99–102.

Coppins, B.J. (1983) A taxonomic study of the lichen genus *Micarea* in Europe. *Bulletin of the British Museum (Natural History). London*, **11**, 17–214.

Coppins, B.J. and Purvis, O.W. (1987) A review of *Psilolechia. Lichenologist*, **19**, 29–42.

Crittenden, P.D., David, J.C., Hawksworth, D.L. and Campbell, F.S. (1995) Attempted isolation and success in the culturing of a broad spectrum of lichen-forming and lichenicolous fungi. *New Phytologist*, **130**, 267–97.

Culberson, C.F. and Elix, J.A. (1989) Lichen substances, in *Methods in Plant Biochemistry* (eds P.M. Dey and J.B. Harborne), Academic Press, London, pp. 509–35.

Culberson, W.L. (1986) Chemistry and sibling speciation in the lichen-forming fungi: ecological and biological considerations. *Bryologist*, **89**, 121–31.

Culberson, W.L. and Culberson, C.F. (1994) Secondary metabolites as a tool in ascomycete systematics: lichenized fungi, in *Ascomycete Systematics: Problems and Perspectives in the Nineties* (ed. D.L. Hawksworth), Plenum Press, London, pp. 155–63.

DePriest, P.T. (1993) Variation in the *Cladonia chlorophaea* complex I, morphological and chemical variation in Southern Appalachian Populations. *Bryologist*, **96**, 555–63.

DePriest, P.T. (1994) Variation in the *Cladonia chlorophaea* complex II, ribosomal DNA variation in a southern Appalachian Population. *Bryologist*, **97**, 117–26.

Fahselt, D. (1989) Enzyme polymorphism in sexual and asexual umbililcate lichens from Sverdrup Pass, Ellesmere Island, Canada. *Lichenologist*, **21**, 279–85.

Farmer, A.M., Bates, J.W. and Bell, J.N. (1992) Ecophysiological effect of acid rain on bryophytes and lichens, in *Bryophytes and Lichens in Changing Environments* (eds J. Bell and A. Farmer), Oxford Science, Oxford, pp. 284–313.

Friedl, T. (1987) Thallus development and phycobionts of the parasitic lichen *Diploschistes muscorum* (Cop.) Ant, in *Progress and Problems in Lichenology in the Eighties, Bibliotheca Lichenologica* 25 (ed. E. Peveling), Cramer, Berlin, pp. 81–93.

Galloway, D.J. (1992) Biodiversity, a lichenological perspective. *Biodiversity Conservation*, **1**, 312–23.

Gargas, A., DePriest, P.T., Grube, M. and Tehler, A. (1995a) Multiple origins of lichen symbioses in fungi suggested by SSU rDNA phylogeny. *Science*, **268**, 1492–4.

Gargas, A., DePriest, P.T. and Taylor, J. (1995b) Positions of multiple insertions in SSU rDNA of lichen-forming fungi. *Molecular Biology and Evolution*, **12**, 208–18.

Greuter, W. (1994) *International Code of Botanical Nomenclature (Tokyo).* Koeltz, Königstein.

Hafellner, J. (1984) Studien in Richtung einer natürlichen Gliederung der Sammelfamilien *Lecanoraceae* und *Lecideaceae. Bibliotheca Lichenologica*, **79**, 241–371.

Hafellner, J. (1989) Principles of classification and main taxonomic groups, in *Handbook of Lichenology* (ed. M. Galun), Vol. 1, CRC, Boca Raton, pp. 41–52.

Hale, M.E. (1974) *The Biology of Lichens*, 2nd edn, Edward Arnold, London.

Hamada, N. (1989) The effect of various culture conditions on depside production by an isolated lichen mycobiont. *Bryologist*, **92**, 310–13.

Hawksworth, D.L. (1973) Ecological factors and species delimitation in the lichens, in *Taxonomy and Ecology* (ed. V.H. Heywood), Vol. 5, Academic Press, London, pp. 31–69.

Hawksworth, D.L. (1976) Lichen chemotaxonomy, in *Lichenology: Progress and Problems* (eds D.H. Brown, D.L. Hawksworth and R.H. Bailey), Academic Press, London, pp. 139–184.

Hawksworth, D.L. (1988a) The variety of fungal–algal symbioses, their evolutionary significance, and the nature of lichens. *Botanical Journal of the Linnean Society*, **96**, 3–20.

Hawksworth, D.L. (1988b) The fungal partner, in *Handbook of Lichenology* (ed. M. Galun), Vol. 1, CRC, Boca Raton, pp. 35–8.

Hawksworth, D.L. (1994) The recent evolution of lichenology: a science for our times. *Cryptogamic Botany*, **4**, 117–29.

Hawksworth, D.L. and Honegger, R. (1994) The lichen thallus: a symbiotic phenotype of nutritionally specialized fungi and its response to gall producers, in *Plant Galls* (ed. M.A.J. Williams), Clarendon Press, Oxford, pp. 77–98.

Hawksworth, D.L., Sutton, B.C. and Ainsworth, G.C. (1983) *Ainsworth & Bisby's Dictionary of the Fungi.*, 7th edn, Commonwealth Mycological Institute, Kew.

Hestmark, G. (1991) Teleomorph–anamorph relationship in *Umbilicaria* 1. Making the connections. *Lichenologist*, **23**, 343–59.

Honegger, R. (1984) Scanning electron microscopy of the contact site of conidia and trichogynes in *Cladonia furcata. Lichenologist*, **16**, 11–19.

Jahns, H.M. (1987) New trends in developmental morphology of the thallus, in *Progress and Problems in Lichenology in the Eighties. Bibliotheca lichenologica* (ed. E. Peveling), Vol. 25, Cramer, Berlin, pp. 17–33.

Jahns, H.M. (1988) The establishment, individuality and growth of lichen thalli. *Botanical Journal of the Linnean Society*, 96, 21–9.

Jahns, H.M. and Ott, S. (1994) Thallic mycelial and cytological characters in ascomycete systematics, in *Ascomycete Systematics: Problems and Perspectives in the Nineties* (ed. D.L. Hawksworth), Plenum Press, London, pp. 57–62.

James, P.W. and Galloway, D.J. (1992) *Menegazzia, Flora of Australia*, **54**, 213–46.

James, P.W. and Hesssen, A. (1975) A new species of *Psoroma* with sorediate cephalodia. *Lichenologist*, **7**, 143–7.

James, P.W. and Henssen, A. (1976) The morphological and taxonomic significance of cephalodia, in *Lichenology: Progress and Problems* (eds D.H. Brown, D.L. Hawksworth and R.H. Bailey), Academic Press, London, pp. 27–77.

John, D.M. and Maggs, C.A. (1997) Species problems in Eukaryotic algae: a modern perspective, in *Species: the Units of Biodiversity* (eds M.F. Claridge, H.A. Dawah and M.R. Wilson), Chapman & Hall, London, pp. 83–107.

Jørgensen, P.M. (1991) Difficulties in lichen nomenclature. *Mycotaxon*, **40**, 497–501.

Jørgensen, P.M., James, P.W. and Jarvis, C.E. (1994) Linnaean lichen names and their typification. *Botanical Journal of the Linnean Society*, **115**, 261–405.

Kappen, L. (1994) The lichen, a mutualistic system – some mainly ecophysiological aspects. *Cryptogamic Botany*, **4**, 193–202.

Kärnefelt, A., Mattsson, J.-E. and Thell, A. (1992) Evolution and phylogeny of cetrariod lichens. *Plant Systematics and Evolution*, **183**, 113–60.

Lamb, I.M. (1977) A conspectus of the lichen genus *Stereocaulon* (Schreb.) Hoffm. *Journal of the Hattori Botanical Laboratory*, **43**, 191–355.

Laundon, J.R. (1995) On the classification of lichen photomorphs. *Taxon*, **44**, 387–9.

Letrouit-Galinou, M.A. and Asta, J. (1994) Thallus morphogenesis in some lichens. *Cryptogamic Botany*, **4**, 274–82.

Lorch, J. (1988) The true nature of lichens – a historical survey, in *Handbook of Lichenology* (ed. M. Galun), Vol. 1, CRC, Boca Raton, pp. 3–32.

Mattsson, J.E. and Kärnefelt, J. (1986) Protein banding patterns in the *Ramalina siliquosa* group. *Lichenologist*, **18**, 231–40.

Mayr, E. (1992) A local flora and the biological species concept. *American Journal of Botany*, **79**, 222–38.

Ott, S. (1987a) Reproductive strategies in lichens, in *Progress and Problems in Lichenology in the Eighties* (ed. E. Peveling), *Bibliotheca lichenologica*, J. Cramer, Berlin, pp. 81–93.

Ott, S. (1987b) Sexual reproduction and developmental adaptations in *Xanthoria parietina*. *Nordic Journal of Botany*, **7**, 219–28.

Poelt, J. (1991) Homologies and analogies in the evolution of lichens, in *Frontiers in Mycology* (ed. D.L. Hawksworth), CAB International, Wallingford, pp. 85–97.

Poelt, J. (1994) Different species types in lichenized ascomycetes, in *Ascomycete Systematics: Problems and Perspectives in the Nineties* (ed. D.L. Hawksworth), Plenum Press, London, pp. 273–8.

Poelt, J. and Huneck, S. (1968) *Lecanora vinetorum* nova spec., ihre Vergesellschaftung, ihre Ökolgie und ihre Chemie. *Öserreichische Botanische Zeitschrift*, **115**, 411–22.

Poelt, J. and Obermayer, W. (1990) Über Thallosporen bei einigen Krustenflechten. *Herzogia*, **8**, 273–88.

Purvis, O.W. (1984) The occurrence of copper oxalate in lichens growing on copper sulphide-bearing rocks in Scandinavia. *Lichenologist*, **16**, 197–204.

Purvis, O.W. (1993) The botanical interest of mine spoil heaps – the lichen story. *Journal of the Russell Society*, **5**, 45–8.

Purvis, O.W. and James, P.W. (1985) Lichens of the Coniston Copper Mines. *Lichenologist*, **17**, 221–37.

Purvis, O.W. and James, P.W. (1993) Studies on the lichens of the Azores. Part 1. Caldeira do Faial. *Arquipelago*, **1**, 1–15.

Purvis, O.W., Gilbert, O.L. and James, P.W. (1985) The influence of copper mineralization on *Acarospora smaragdula*. *Lichenologist*, **17**, 111–14.

Purvis, O.W., Elix, J.A., Broomhead, J.A. and Jones, G.C. (1987) The occurrence of copper-norstictic acid in lichens from cupriferous substrata. *Lichenologist*, **19**, 193–203.

Purvis, O.W., Elix, J.A. and Gaul, K.L. (1990) The occurrence of copper-psoromic acid in lichens from cupriferous substrata. *Lichenologist*, **22**, 345–54.

Purvis, O.W., Coppins, B.J., Hawksworth, D.L., James, P.W. and Moore, D.M. (1992) *Lichen Flora of Great Britain and Ireland*. Natural History Museum, London.

Purvis, O.W., Coppins, B.J. and James, P.W. (1994) *Checklist of Lichens of Great Britain and Ireland. British Lichen Society*, London. pp. 1–75.

Salisbury, G. (1971) *Thelotrema wightii* (T. Tayl.) Nyl. *Portugaliae Acta Biologica, series B*, **11**, 35–7.

Santesson, J. (1973) Identification and isolation of lichen substances, in *The Lichens* (eds V. Ahmadjian and M.E. Hale), Academic Press, New York, pp. 633–652.

Smith, A.L. (1921) *Lichens.* Cambridge University Press, Cambridge.

Smith, D.C. and Douglas, A.E. (1987) *The Biology of Symbiosis.* Edward Arnold, London.

Stocker-Wörgötter, E. and Türk, R. (1994) Artificial resynthesis of the photosymbiodeme *Peltigera leucophlebia* under laboratory conditions. *Cryptogamic Botany,* **4**, 300–8.

Tehler, A. (1982) The species pair concept in lichenology. *Taxon,* **31**, 708–14.

Tehler, A. (1994) The phylogeny of *Roccellina* (*Rocellaceae*), revisited. *Acta Botanica Fennica,* **150**, 185–95.

Tibell, L. (1982) Caliciales of Costa Rica. *Lichenologist,* **14**, 219–54.

Timdal, E. (1991) A monograph of the genus *Toninia* (*Lecideaceae,* ascomycetes) *Opera Botanica,* **110**, 1–137.

Tschermak-Woess, E. (1988) The algal partner, in *Handbook of Lichenology* (ed. M. Galun), Vol. 1, CRC, Boca Raton, pp. 39–92.

Tschermak-Woess, E. (1995) *Dictyochloropsis splendida* (*Chlorophyta*), the correct phycobiont of *Phlyctis argena* and the high degree of selectivity or specificity involved. *Lichenologist,* **27**, 169–87.

Vitikainen, O. (1994) Taxonomic revision of *Peltigera* (lichenized ascomycotina) in Europe. *Acta Botanica fennica,* **152**, 1–96.

Vobis, G. (1980) Bau und Entwicklung der Flechten-Pycnidien und ihren Conidien. *Bibliotheca lichenologica,* **14**, 1–141.

Weber, W.A. (1968) Environmental modification and the taxonomy of the crustose lichens. *Särtryck ur Svensk Botanisk Tidskrift,* **56**, 293–333.

Wirth, V. (1972) Die Silikatflechten-Gemeinschaften in Ausseralpinen Zentraleuropa. *Dissertationes Botanicae, Lehre,* **17**, 1–305.

Yamomoto, Y., Miura, Y., Higuchi, M., Kinoshita, Y. and Yoshimura, I. (1993) Using lichen tissue cultures in modern biology. *Bryologist,* **96**, 384–93.

Yoshimura, I., Kurokawa, T., Yamamoto, Y. and Kinoshita, Y. (1993) Development of lichen thalli in vitro. *Bryologist,* **96**, 412–21.

Yoshimura, I., Kurokawa, T., Yamamoto, Y. and Kinoshita, Y. (1994) In vitro development of the lichen thallus of some species of *Peltigera. Cryptogamic Botany,* **4**, 314–19.

Zahlbruckner, A. (1922–24) *Catalogus Lichenum Universalis,* 2. Bornträger, Leipzig.

Zahlbruckner, A. (1926–27) *Catalogus Lichenum Universalis,* 4. Bornträger, Leipzig.

Zahlbruckner, A. (1926a) B. Spezieller Teil. Lichenes (Flechten), in *Die Natürlichen Pflanzenfamilien.* Englelmann, Leipzig, pp. 61–270.

7

Fungal species in practice: identifying species units in fungi

C. M. Brasier

Contacting address: Forest Research Station, Alice Holt Lodge, Farnham, Surrey GU10 4LH, UK

ABSTRACT

Through the strong historical association of mycology with botany, taxonomic species in fungi continued to be almost entirely morphologically based up to the middle of this century, despite a paucity of suitable morphological characters in some fungal groups. Even though considerable progress was made in understanding the genetics of fungal breeding systems in the early 1900s, evolutionary biology made little initial impact on fungal systematic concepts. However, the early genetical studies did result in the emergence of the modern fungal genetics that, together with microbial genetics, contributed significantly to recombinant DNA theory.

More evolutionarily based fungal species concepts began to emerge in the 1950s, and an accelerated use of population, genetical and molecular tools to assess variation and species diversity occurred from the 1960s to the 1970s, particularly through the efforts of fungal geneticists, pathologists and ecologists. One important result of these studies is that clusters of biological and sibling species have been identified within many traditional morphological species. For example the well-known basidiomycete pathogen *Armillaria mellea*, or 'honey fungus', has been subdivided into 10 biological species in North America alone. In such cases the original morphological species approximates to a superspecies, the functional or operational species unit occurring at the level of subordinate sibling or biological species (which may or may not subsequently be shown to have useful morphological differences). The extent of fungal biodiversity may therefore have been seriously underestimated in the past.

A great deal of complexity is also being revealed that is as yet little understood. For example, many behaviourally or molecularly distinct, partially reproductively isolated subpopulations are being identified; some will undoubtedly prove to be independent operational

Species: The Units of Biodiversity. Edited by M.F. Claridge, H.A. Dawah and M.R. Wilson. Published in 1997 by Chapman & Hall. ISBN 0 412 63120 2

species units. In addition, operational species are being identified at very different levels of genetic hierarchy within a genus. Other longer-standing complexities are being resolved. Thus, molecular tools are enabling the many asexual fungal taxa to be assigned to sexually defined genera, so throwing new light on the question of their individual status.

Despite these developments, applying a unified species concept within the fungi is likely to prove difficult. Indeed, while such questions are being explored, the fungal species concept must be considered to be in a transitional phase. Currently, traditional morphological species concepts often operate alongside the emerging, largely non-formalized (and sometimes supportive and sometimes contradictory) ecological or molecularly-based species concepts. Ultimately, a new, more flexible, set of hierarchical terms may be needed. The application of genetical, ecological and molecular approaches to species recognition has therefore achieved a belated conjunction of fungal genetics and fungal systematics, and perhaps a much needed reassessment of the fungal species concept, although the outcome is still unclear.

7.1 INTRODUCTION

Current species concepts in fungi, as in most other groups of organisms, are the product of historical, philosophical and technological influences, plus the varied experiences and interests of individual observers. This paper discusses recent progress in the characterization of species units and the development of species concepts in fungi from the standpoint of a practising plant pathologist and fungal population biologist.

An inherent problem, as the author sees it, is that in the past species discrimination in fungi has been too anthropocentric. Too great an emphasis has been put on classification and too little on defining the operational species as it functions in nature. A major challenge for today's mycologist is that of correctly identifying and accurately biologically characterizing these fungal operational species units, or OSUs – the principal population units, sharing a common gene pool and exhibiting a common set of physiological, ecological and morphological attributes, by which fungi have defined themselves making use of all appropriate methods. Since in general OSUs are ecologically distinct entities the methods used to discriminate them should, wherever practicable, extend beyond morphological or molecular criteria to behavioural criteria such as breeding system, habitat and physiological attributes. Indeed, at its most sophisticated, the process of discriminating OSUs can itself be a developmental and evolutionary one. The character and rank of population units becomes apparent to the observer as information accumulates (Brasier and Rayner, 1987), and more gradual progress in their identification may be punctuated by leaps of understanding. The processes involved may require experience and even a degree of intuition on the part of the observer, together with laboratory experimentation and statistical analysis of data.

The need for accuracy in the recognition and labelling of OSUs in fungi extends well beyond the requirements of scientific communication or the assessment of fungal biodiversity to a range of issues of environmental and socio-economic significance (Table 7.1). These include the requirements of international quarantine legislation and the legal specification of organisms used in production of pharmaceuticals or in genetic engineering. Uncritical identification of species units in fungal pathogens, for example, may result in inadequate quarantine protection and so can contribute unwittingly to the spread of highly destructive organisms. A surprising number of important plant pathogens have recently been shown to have been inadequately defined. In particular, distinctive taxa with very different host ranges or geographical distributions are being shown to have been labelled collectively under a single species. Equally, the designation of taxa that are too narrowly defined can create loopholes in quarantine legislation that might be exploited by the unscrupulous. There is, therefore, a hidden environmental and social cost to the species concept debate.

Table 7.1 Accurate species recognition in fungi: applications

Scientific communication	–	reduce ambiguity
Natural history, biodiversity	–	assessment of rain forest resources, etc.
Diagnosis	–	plant and animal pathogens
Quarantine legislation	–	plant and animal pathogens
Patenting	–	pharmaceutical, industrial, genetic engineering, biological control organisms, mycorrhizal symbionts

Table 7.2 Fungal groups (mycelial fungi)

Animal affinity	Chytridomycota	Potato wart disease fungus (*Synchytrium*)
	Zycomycota	Pin moulds (e.g. *Mucor*), VAM mycorrhizae
	Ascomycota	Yeasts, Dutch elm disease pathogens, coral spot fungus, discomycetes (*Peziza*)
	Basidiomycota	Mushrooms, toadstools, bracket fungi, puffballs, truffles, rust pathogens, ectomycorrhizae
	Asexual fungi	'Deuteromycota' – *Penicillium, Aspergillus,* grey mould *Botrytis, Trichoderma*
Algal affinity	Oomycota	Potato blight (*Phytophthora infestans*), Jarrah root-rot fungus (*P. cinnamomi*), *Pythium* (damping off fungi)

A point that may not be appreciated by non-mycologists is that fungi encompass a highly diverse and polyphyletic assemblage of organisms (Table 7.2). Some are of near-animal and others of near-plant affinity (Cavalier-Smith, 1987, 1993; Barr, 1992; Wainwright *et al.*, 1993; Corliss, 1994); they exhibit a wide range of morphological structure from ephemeral unicellular yeasts to large relatively well-differentiated thalli capable of living a thousand years or more (Brasier, 1992; Smith *et al.*, 1992); involve very different genetic systems from haploid, to dikaryotic (two haploid nuclei per cell and functionally diploid), to multinucleate to truly diploid; and have breeding systems that, within most fungal groups, range from obligatorily outcrossing types to largely inbreeding types to totally asexual mitotic forms exhibiting clonal lineages. This organismal diversity and the associated variation in ecological strategies and breeding systems increases the problems of finding a unified species concept appropriate to all fungal groups.

7.2 THE CHANGING FUNGAL SPECIES CONCEPT

Until the middle of the 20th century, fungal species, in practice, were largely morphologically based (cf. Bisby, 1953; Hawksworth, 1974). This derived in part from the necessary and inevitable historical importance of morphological criteria in the development of taxonomic concepts, as in most organisms, and in part from a long tradition of the teaching of mycology as an aspect of botany. The latter is also reflected to this day in the now somewhat anachronistic foundation of terms and naming procedures on the *International Code of Botanical Nomenclature*, and the requirement under the Code for the deposition in a 'herbarium' or culture collection of a single, often fortuitously selected, type specimen (cf. Leonian, 1934; Brasier, 1983, 1990b; Brasier and Rayner, 1987; Hansen, 1990). During the early 1900s considerable progress was made in understanding fungal mating systems, extending in some cases to the demonstration of inter-sterile breeding groups within traditional morphological taxa (Shear and Dodge, 1927; Verrall, 1937; Mounce and Macrae, 1938; reviewed in Burnett, 1983; Brasier, 1987). However, these important early developments and other philosophical milestones of wider evolutionary biology from Darwin and Dobzhansky to Mayr and Stebbins made remarkably little impact on fungal systematic philosophy. Indeed, with insufficient dialogue between them, the fields of fungal systematics and fungal genetics drifted apart (fungal genetics emerging as an advanced discipline that later contributed significantly, along with bacterial genetics, to recombinant DNA theory).

The separation of fungal systematics from fungal genetics and the continued historical pre-eminence of morphological characters in species delimitation inevitably restricted mycological perceptions. While morphological characters are a valuable tool in the fungal systematist's armoury, they can have profound limitations if used as sole criteria, especially when applied to the myriad relatively simple structured microfungi and yeasts. These limitations have been highlighted many times in the past (Leonian, 1934; Nelson, 1965; Brasier, 1983, 1990b; Kurtzman, 1985; Brasier and Rayner, 1987) and are summarized in Table 7.3. For example, emphasis on morphological characters takes little account of the processes of fungal speciation, especially the possibility that morphologically useful differences between taxa may not have accumulated or that parallel or convergent evolution of morphological structures may have occurred (Kemp, 1977; Brasier, 1987). A major practical limitation is that many fungi are isolated into artificial culture as mycelia from a natural substrate before their identification, a state of confinement in which their fruiting structures may be difficult or impossible to obtain.

Following the contributions made by early pioneers such as Shear and Dodge (1927) who took a genetical approach to species identification, more evolutionarily related species concepts began to emerge from around the 1950s (Nelson, 1965) and an accelerated application of statistical, genetical and molecular tools to investigate the ecological status of taxa and to define their limits of variation occurred from the 1970s onwards. Initially, such work was often carried out by plant pathologists, fungal ecologists or fungal geneticists responding to a conceptual problem with a particular organism or group of organisms which demanded an understanding of their population structure, rather than by those with a fundamentally systematic intent. They increasingly applied multivariate analysis to continuous and discontinuous and to behavioural and morphological characters, investigated karyotypes, conducted genetic analysis of mating types and breeding units, and introduced molecular analysis of population structure. Fungi were at last beginning to be considered primarily as field populations responding to a set of environments rather than as a specimen entombed alive in a Petri dish or dead in an herbarium box (Brasier, 1989).

This activity has led to an enormous mass of new information, a process which from the literature appears still to be accelerating. It includes a

Table 7.3 Limitations of morphological characters in fungal systematics. (Based on Brasier, 1991.)

Useful morphological differences between taxa may not have accumulated
Small number of characters available
Plasticity of structures, size range overlaps
Difficulty in finding structures in field (rare, seasonal)
Difficulty in obtaining structures in laboratory
Absence of sexual structures: asexual fungi

range of intriguing, frequently important but not altogether unexpected (Brasier, 1983) developments across the entire range of fungal groups. For example, multiple OSUs are being identified within many traditional morphological species, paralleling recent revelation of multiple sibling species within arthropod taxa (Knowlton, 1993; Claridge and Boddy, 1994). At the same time a great deal of inter- and intra-specific complexity is being revealed that at present remains uninterpreted or poorly understood. To illustrate the diverse approaches used in identifying OSUs in fungi, three case studies each selected from a different major fungal group will now be considered. Emphasis has deliberately been placed on mycelial fungi rather than on the mainly unicellular microbe-like yeasts. For a discussion of similar issues within the yeasts, see especially Kurtzman (1985) and Kurtzman and Robnett (1994).

7.3 IDENTIFYING OPERATIONAL SPECIES UNITS

7.3.1 Identification from evidence of total reproductive isolation: *Armillaria mellea*

A striking example of recent developments is that of the basidiomycete fungus *Armillaria mellea*, a common tree root pathogen known to foresters and horticulturalists as the 'honey fungus' (Figure 7.1).

Until recently, *A. mellea* was viewed in the literature as a single, variable morphospecies of worldwide distribution. In the 1970s, however, a Finnish forest pathologist demonstrated an outcrossing mating system in the fungus (Hintikka, 1973) and, using this system, another Finn identified five totally inter-sterile breeding groups among population samples of the fungus from Europe (Korhonen, 1978a), which came to be known as biological species groups. Similar interfertility tests (Figure 7.2) demonstrated at least ten biological species groups among *A. mellea* isolates in North America (Anderson and Ullrich, 1979). These results have been corroborated by a number of molecular studies which have also revealed the extent of genetic divergence between the groups (Figure 7.3; reviewed in Guillaumin *et al.*, 1991).

At least three of the North American biological species groups are conspecific with a European species group, and several American groups have been named as new taxa (Figure 7.3) (Guillaumin *et al.*, 1991). Extensive field studies have also been stimulated by these observations. These have revealed that the European species groups exhibit different ecological characteristics, e.g. *A. ostoyae* is a primary killer of trees while *A. gallica* is a saprotroph that can also colonize declining or stressed trees. It seems likely that different ecological characteristics will be identified for the North American species as field information accumulates.

These biological species groups are the newly recognized OSUs within what until recently was previously viewed, in practice, as a single mor-

Figure 7.1 Fruit bodies of *Armillaria mellea*. Scale bar = 2 cm.

phospecies, *A. mellea*. Within a continent, the boundaries of the units are clearly defined by total sterility barriers. They therefore represent classical biological species, *sensu* Mayr (1942). The European OSUs correlate with some previously proposed morphological taxa, and also show some potentially useful diagnostic differences in their colony characteristics (reviewed by Watling *et al.*, 1991). However, at present most of these OSUs can only be accurately identified on the basis of an interfertility test in culture. Ultimately a series of molecular protocols to aid their identification seems desirable (Harrington and Wingfield, 1995).

Similar studies are revealing additional *Armillaria* taxa worldwide (Cha *et al.*, 1994). Accurate identification of *Armillaria* taxa is imperative for many of the reasons outlined in Table 7.1, in particular because of their different pathological attributes and ecological characteristics. Indeed much of the enormous literature on the biology and pathology of *A. mellea* before 1980 is now of limited value because of uncertainty over which species was being investigated.

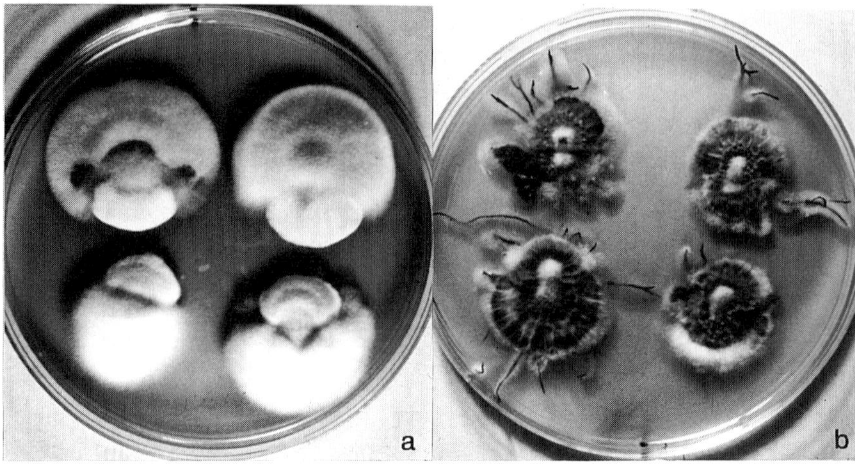

Figure 7.2 Appearance of interfertile and intersterile pairings between *Armillaria* OSUs on culture. (a) Four sets of intersterile pairings, in which the haploid colonies involved belong to different OSUs. A zone of antagonism occurs between the paired colonies. (b) Four sets of interfertile pairings in which the haploid colonies involved belong to the same OSU. Secondary, fruiting competent diploid mycelia develop as a result of fusion between the colonies. These secondary colonies also producing black boot-lace-like rhizomorphs (asexual foraging structures). (Courtesy of S.C. Gregory, Forestry Commission, UK.)

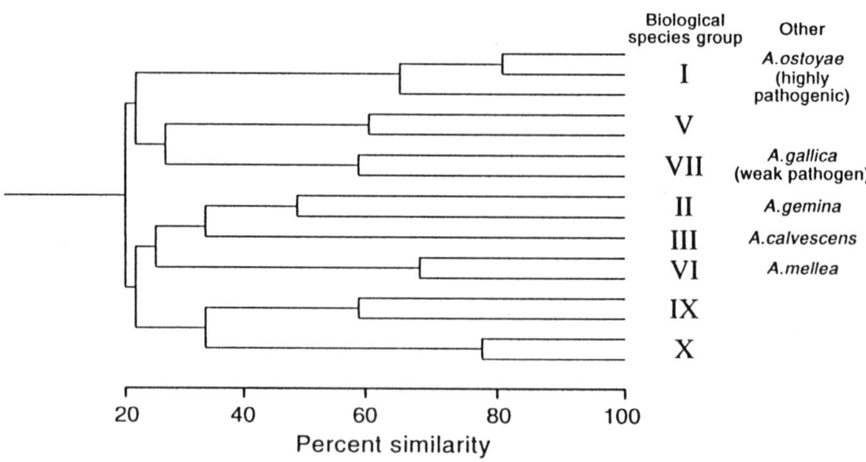

Figure 7.3 Dendrogram of molecular similarity between isolates of *Armillaria* (revealed by *Eco*RI digest of mitochondrial DNAs) showing the relationships between eight of the ten North American biological species groups (revealed by sexual mating tests) and the major molecular groupings. (Redrawn from Smith and Anderson, 1989.)

Discovery of multiple OSUs within traditional taxonomic species via the demonstration of intersterile breeding units is a process that has occurred many times in recent decades, particularly with the culturable ascomycetes and basidiomycetes. Ascomycete examples include *Nectria haematococca, Gibberella fujikoroi, Neurospora* spp., *Ascobolus immersus, Phomopsis oblonga* and *Ophiostoma piceae* (Matuo and Snyder, 1973; Perkins *et al.*, 1976; Kuhlman, 1982; Webber and Gibbs, 1982; Meinhardt *et al.*, 1984; Perkins and Raju, 1986; Brayford, 1990; Brasier, 1993; Brasier and Kirk, 1993). Basidiomycete examples include *Coniophora puteana, Coprinus* spp., *Paxillus involutus, Rhizoctonia solani, Pleurotus ostreatus, Polyporus abietinus, Collybia dryophila* and *Gloeocystidium tenue* (Boidin, 1951; Macrae, 1967; Parmeter *et al.*, 1969; Kemp, 1983, 1985; Fries, 1985; Ainsworth, 1987; Vilgalys and Johnson, 1987).

7.3.2 Identification from cultural and behavioural properties and evidence of a degree of reproductive isolation: the Dutch elm disease pathogens

Where breeding barriers cannot be readily investigated for practical reasons, or where the barriers involved are only partial, the process of identifying OSUs can be a more protracted one. This was so in the recent case of the Dutch elm disease fungi.

Dutch elm disease, a vascular wilt disease of elms, is one of the most catastrophic plant diseases known to man. Until the late 1970s it was believed to be caused by a single morphospecies, the ascomycete, *Ophiostoma ulmi* (Figure 7.4). In the 1970s, population studies carried out in Britain in the context of an unexpected second epidemic of the disease revealed two distinct colony types (Figure 7.5). One proved to be a weak and the other a highly aggressive pathogen of elms. They were initially referred to as the non-aggressive and aggressive strains, and then as subgroups of the fungus. Before long, field samples across the northern hemisphere showed the aggressive and non-aggressive subgroups to be widely geographically distributed, and also recently sympatric as a result of their independent epidemic spread into much of North America and Eurasia from unknown origins (reviewed in Brasier, 1990a, 1991).

Sexual crosses between the aggressive and non-aggressive subgroups, conducted shortly after the aggressive subgroup was discovered in Britain, signalled powerfully that they were separate species. It was shown that a strong but not total pre-zygotic reproductive barrier was operated by the aggressive subgroup against the non-aggressive. Thus, compared with intra-group pairings, perithecial maturation in aggressive × non-aggressive crosses was disrupted and the numbers of viable perithecia (Figure 7.4(d)) were greatly reduced. In addition, the resultant ascospores were of low viability, and the F_1s exhibited a remarkable range of non-parental colony types (Figure 7.6) and showed strong negative

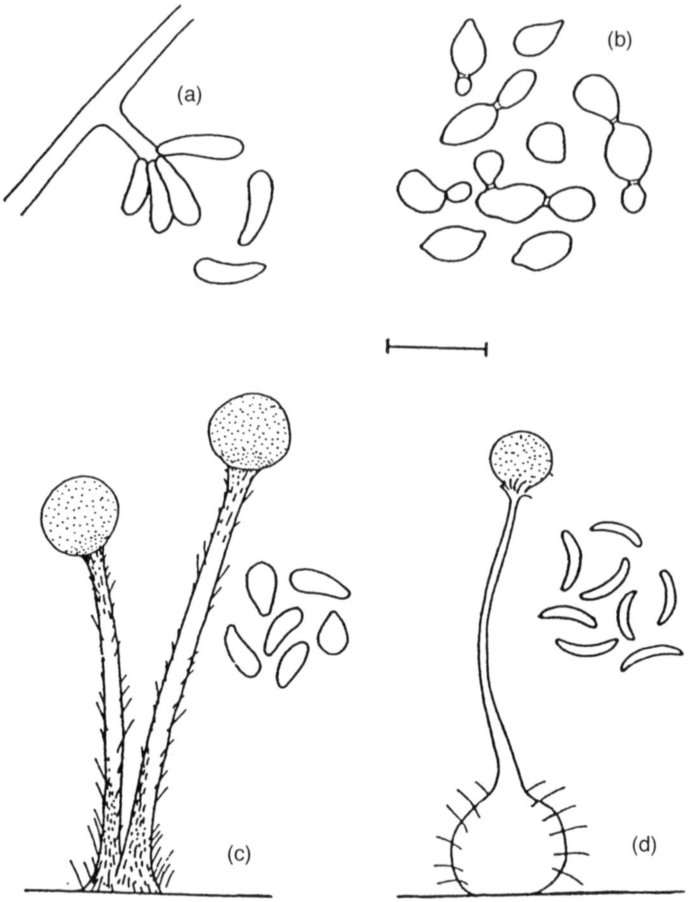

Figure 7.4 Morphological structures of *Ophiostoma ulmi*. (a) Asexual mycelial spores; (b) yeast-like phase; (c) asexual synnemata and synnemiospores; (d) perithecium (sexual stage) and ascospores. Scale bar = 5 mm and refers to spores only. Synnemata and perithecia are approximately 0.5 mm high. (From Brasier, 1981.)

interactions for heritability of continuous characters such as growth-rate and pathogenic ability (Figure 7.7) (Brasier and Gibbs, 1976; Brasier, 1977). Simply put, the genomes of the two subgroups, when recombined, gave rise to F_1s that were unlikely to survive in nature (Kile and Brasier, 1990).

In numerous additional studies, involving both *in vitro* tests and field experiments, the aggressive and non-aggressive subgroups were shown to differ in many important physiological, ecological and morphological attributes. These included characters such as temperature optima for growth (widely different at c. 22° and c. 28°C respectively), pathotoxin production and mean perithecial neck lengths, though there was too much overlap in the latter for this to be diagnostic for many individual fungal isolates. By the late 1970s the two subgroups were considered divergent, probably at the

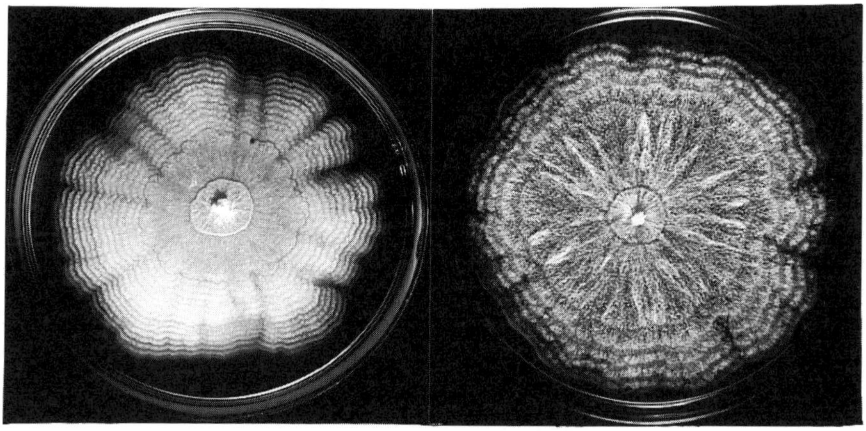

Figure 7.5 Representative colonies of the non-aggressive (left) and aggressive (right) subgroups of *Ophiostoma ulmi*. Note that non-aggressive subgroup colonies are slow-growing and waxy in appearance, the aggressive colonies fibrous-striate petaloid. Subsequently, it was realized that the two subgroups were separate OSUs, the non-aggressive subgroup being retained as *O. ulmi s.s.* and the aggressive being designated *O. novo-ulmi* sp. nov.

level of subspecies. By the early 1980s, they were considered, at least in the author's mind, separate OSUs. Molecular studies in the late 1980s confirmed that they were widely genetically divergent entities (Figure 7.8), and in 1991 they were formally designated as separate species, the non-aggressive being retained as *O. ulmi sensu stricto* and the aggressive recognized as the new species *O. novo-ulmi* (Brasier, 1991).

Turning to the practical relevance of these developments (Table 7.1), *O. ulmi s.s.* was the pathogen responsible for the first pandemic of Dutch elm disease between the 1920s and 1940s, while *O. novo-ulmi* is that responsible for the enormously destructive current pandemic. Had the existence of *O. novo-ulmi* been recognized before 1970, its importation into Britain might have been prevented or perhaps delayed by enactment of appropriate quarantine legislation, or its arrival met by a more immediate sanitation response. However, as with the OSUs recently identified within *A. mellea*, failure to distinguish *O. novo-ulmi* from *O. ulmi s.s.* earlier must be seen in the context of the mycological philosophy of the 1950s and 1960s, when mainstream mycology was relatively isolated from the disciplines of population biology and fungal genetics. Also, the two taxa could not readily have been separated on the basis of traditional morphological criteria, since the morphological differences are limited and their application requires an experimental approach.

In fact, recognition of the aggressive and non-aggressive 'strains' of *O. ulmi s.l.* as separate OSUs was a relatively slow process (some 18 years from start to finish), involving a gradual accumulation of information.

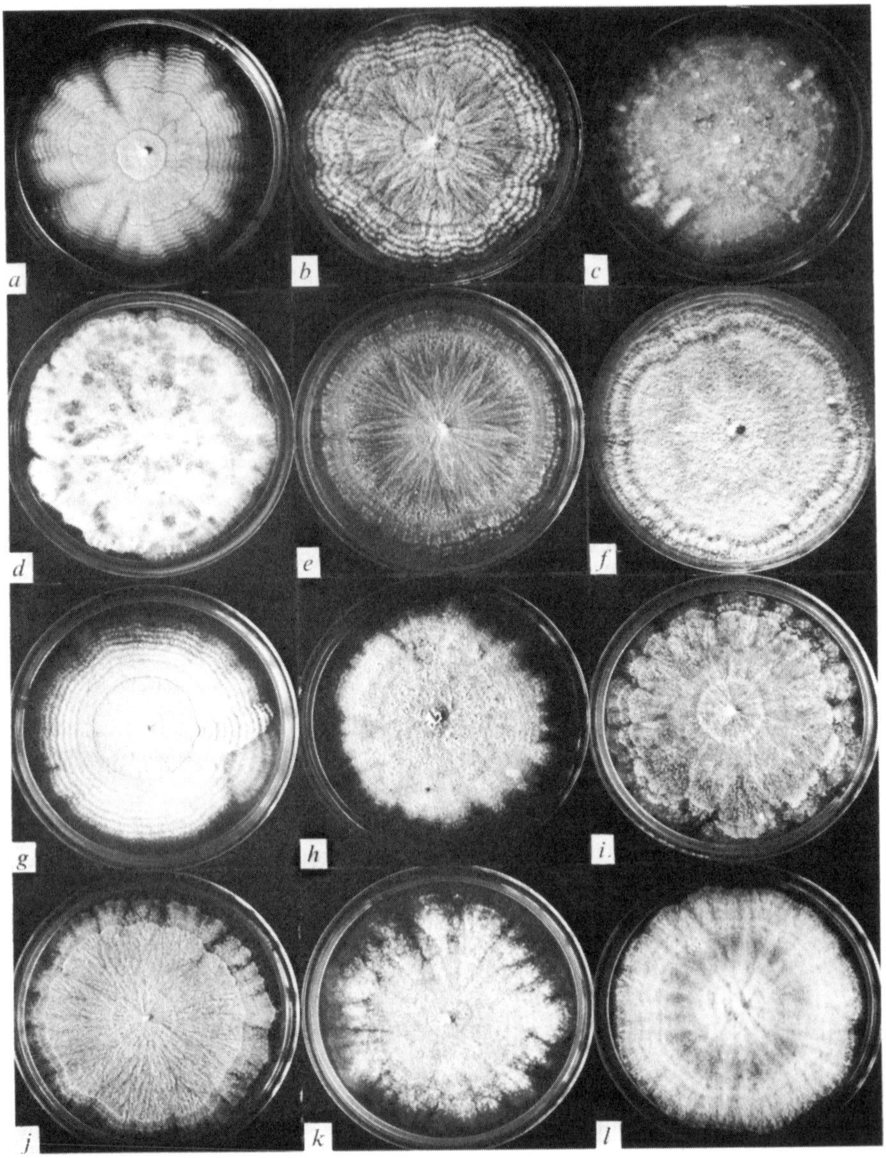

Figure 7.6 Colony characteristics of the progeny of a cross between a non-aggressive subgroup isolate (M35, colony a) and an aggressive subgroup isolate (027, colony b) of *Ophiostoma ulmi* showing the highly unusual non-parental colony types exhibited by the F_1s (colonies c–l). (From Brasier and Gibbs, 1976.)

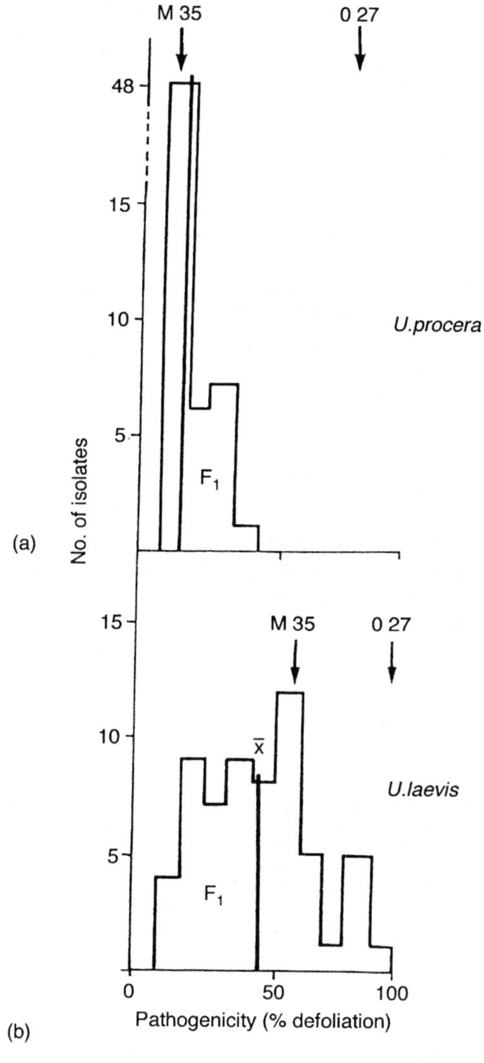

Figure 7.7 Inheritance of pathogenicity (as % defoliation) among 62 progeny of a cross between an aggressive (027) and a non-aggressive (M35) subgroup isolate of *Ophiostoma ulmi*, showing final disease levels: (a) on the moderately susceptible English elm, *Ulmus procera*; (b) on the highly susceptible European white elm *U. laevis*. $\bar{x}$, mean of F_1 progeny. Note the strong negative interaction for inheritance of pathogenicity among the progeny relative to the parents. (From Brasier and Gibbs, 1976.)

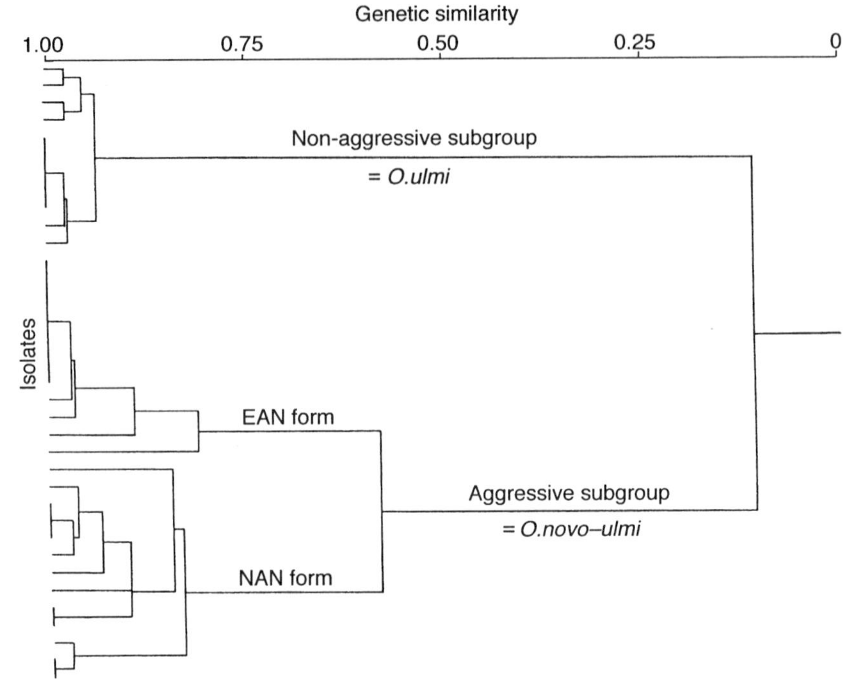

Figure 7.8 Dendrogram of similarity between isolates of the aggressive and non-aggressive subgroups of *Ophiostoma ulmi*, determined by RAPD markers. Also shown is the molecular separation of the two distinct biotypes within the aggressive subgroup, termed the EAN and NAN races (Brasier, 1990a, 1991). (From Pipe *et al.*, 1995.)

This initially amounted to evidence of strong but not total reproductive isolation and of behavioural differences, supported later by evidence of morphological and molecular differences. Similar processes were involved in the identification of OSUs in *Neurospora* (Shear and Dodge, 1927; Perkins *et al.*, 1976). Recently, when yet another Dutch elm disease fungus was discovered in the Himalayas (Brasier, 1994), its characterization as a third OSU (*O. himal-ulmi* sp. nov.) was achieved in only 8 months using many of the techniques already developed for distinguishing *O. ulmi* s.s. and *O. novo-ulmi* (Brasier and Mehrotra, 1995).

7.3.3 Identification from cultural and behavioural properties alone: the case of *Phytophthora megasperma*

Identification of OSUs within morphological taxa does not always involve a direct demonstration of genetic isolation, as can be seen in the case of

Phytophthora megasperma (Figure 7.9), a member of the diploid algal-like Oomycota and a relative of the potato blight fungus, *P. infestans* (Table 7.2). Until the late 1970s, this taxonomic species was viewed as an economically important pathogen on a remarkably wide range of crops from alfalfa and soybean to forest trees worldwide. Some authors considered it to contain two varieties with different oogonial diameters (Fig 7.9), and it was known to contain physiological races specialized in attacking particular crops. Since *P. megasperma* was self-fertile (homothallic), simple sexual mating tests in culture could not be used for defining OSUs. More importantly, it was not even suspected that multiple OSUs existed within this taxon.

However, in the 1980s, two North American pathologists investigated samples of *P. megasperma* isolates from different hosts for a range of morphological and cultural characters and, using multivariate analysis, distinguished distinct host-associated clusters (Hamm and Hansen, 1982; Hansen and Hamm, 1983) (Figure 7.10). Subsequent research demonstrat-

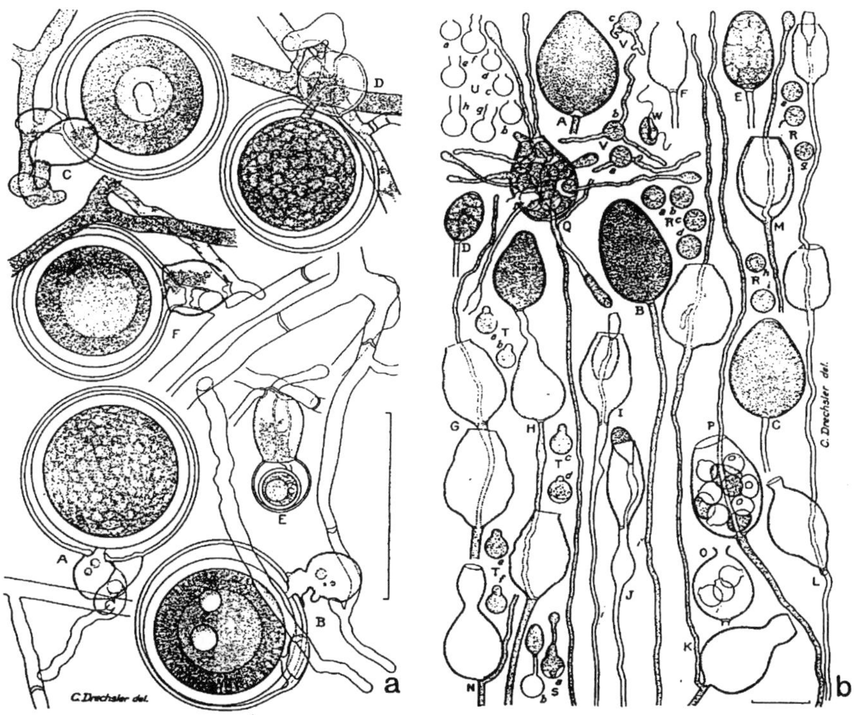

Figure 7.9 Sexual and asexual reproductive structures of *Phytophthora megasperma*: (a) sexual structures, oogonia (♀) and antheridia (♂); (b) asexual sporangia and motile flagellate zoospores. Scale bar (a) = 50 mm; (b) = 30 mm. (From the original description of Drechsler, 1931.)

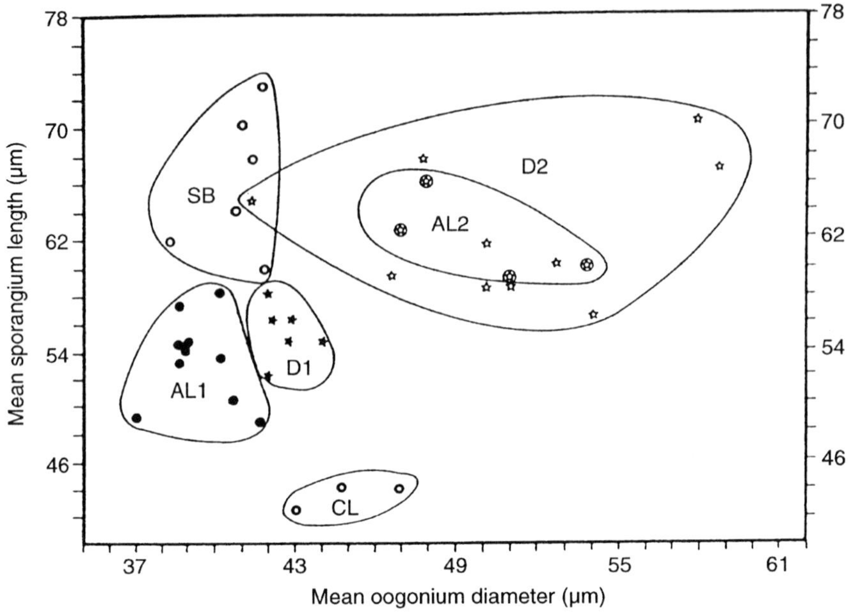

Figure 7.10 Morphometric analysis of host-related subgroups within *Phytophthora megasperma*. SB, soybean group; AL1, alfalfa group 1; AL2, alfalfa group 2; D1, Douglas fir group 1; D2, Douglas fir group 2; CL, clover group. (From Hansen and Hamm, 1983.)

ed additional host-related characters including colony patterns, chromosome number differences and soluble protein banding patterns. This led to the proposal that *P. megasperma* comprised at least six behaviourally different species groups which could be distinguished by their protein patterns (Hansen *et al.*, 1986). A number of molecular polymorphism studies have since confirmed and extended these observations. For example, analysis of mitochondrial DNA and nuclear DNA RFLPs and RAPDs by Förster and Coffey (1993) and Whisson *et al.* (1993) revealed at least seven major groups in *P. megasperma*, six corresponding to those identified by Hansen *et al.* (1986) (Figure 7.11), and a seventh representing isolates from asparagus.

It became increasingly evident from the correlations between their physiological and morphological characteristics and host affinities and later from their wide genetic divergence that the original clusters identified by Hansen and Hamm (1983) were separate OSUs within the old morphospecies '*P. megasperma*'. Several of these groups such as the ALF and SOY 'protein groups' of Hansen *et al.* (1986) have now been formally

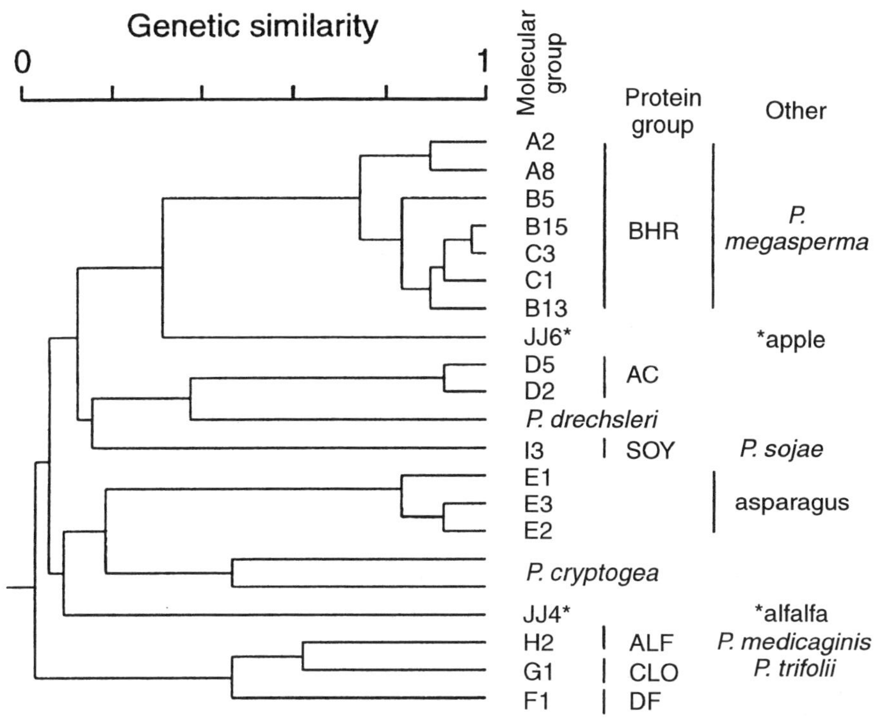

Figure 7.11 Molecular subunits identified within *Phytophthora megasperma* isolates on the basis of a nuclear DNA RFLP analysis, showing the close correspondence with the host-associated protein banding groups of Hansen *et al.* (1986). *P. drechsleri* and *P. cryptogea* were used as outgroups. BHR, broad host range protein group; AC, apple/cherry group; SOY, soybean group; ALF, alfalfa group; CLO, clover group; DF, Douglas fir group. * JJ6, JJ4, representative examples of unique molecular types currently of uncertain status as OSUs. (Redrawn from Förster and Coffey, 1993.)

named as new species, the BHR protein group being redefined as *P. megasperma sensu stricto* (Figure 7.11). The positions of two other traditional morphological taxa, *P. drechsleri* and *P. cryptogea*, in the middle of the same dendrogram when they had been intended as outgroups (Figure 7.11) indicates the extent of the earlier misclassification of '*P. megasperma*' *s.l.* Indeed, this morphological taxon appears to be highly polyphyletic, a situation which is in part a product of the limited morphological criteria available to *Phytophthora* systematists in the past (Brasier, 1990b), and in part a product of convergent evolution.

The initial demonstration of these OSUs was achieved solely by a careful analysis of morphometric, behavioural and cultural data, and was later reinforced by molecular analysis. The latter is also providing the first real measures of genetic distance between the OSUs, which ranges from small to very large (Figure 7.11). Evidence for the extent of reproductive isolation between these OSUs is lacking but these questions are under investigation in a number of laboratories (Förster *et al.*, 1994; Whisson *et al.*, 1994).

Owing to their enormous economic importance, many other traditional *Phytophthora* taxa are under similar scrutiny. In some cases, wide genetic diversity indicating the existence of multiple OSUs has been found (Förster and Coffey, 1991; Oudemans and Coffey, 1991; Brasier and Hansen, 1992). Note here, for example, the wide genetic divergence between the two *P. cryptogea* outgroup isolates shown in Figure 7.11. In other cases such as that of *P. nicotianae*, the behavioural and genetic diversity being revealed is very limited, the species being shown to be a single OSU (Oudemans and Coffey, 1990; Hall, 1993).

7.3.4 The wider perspective: underlying complexities

The above three cases are the tip of a large iceberg. They cannot adequately represent the myriad of other possible examples, each with its special features and circumstances. Fortunately, many traditional morphospecies such as *P. nicotianae*, when subjected to a detailed population biology analysis, are being shown to comprise a single OSU. Such instances need little further elaboration, though it should be emphasized that biological analysis of a suspect morphospecies should be carried out before its status as a single OSU is accepted.

However, there are also many cases where much complexity is surfacing that so far remains unsolved. There are, for example, many instances where newly identified population units show a degree of reproductive isolation, but more information is needed before it is clear whether the units are OSUs or taxa of a lower rank. A well-advanced scientific case is that of the worldwide coniferous tree root pathogen, *Heterobasidion annosum*, a basidiomycete in which mainly pine and spruce attacking subgroups designated as the S and P groups were identified in Europe in the 1970s (Korhonen, 1978b; and see Johansson and Stenlid, 1994). The S and P groups are sympatric, and strongly but not totally reproductively isolated. Host specialization is an important isolating mechanism in sympatric plant pathogens (Brasier, 1987) and is almost certainly involved in sustaining the independence of the S and P groups. Moreover, recent molecular studies indicate that the S and P groups are widely molecularly divergent (Karlsson, 1994). A third group, associated with the true firs (*Abies* sp.) has recently been identified in southern Europe, and is currently referred to as the F group (La Porta *et al.*, 1994). The case for the formal recognition of these groups as separate OSUs seems a powerful one, but they have yet to be presented in these terms in the literature, and continue to be referred to by their trivial names.

While the status of the units within *Heterobasidion* appears to be nearing resolution, it is less clear for those within *Phytophthora capsici*, a major pathogen of perennial crops such as cocoa and pepper across the tropics. In *P. capsici* two ubiquitous molecular sub-taxa termed the CAPA and CAPB groups have been identified that are divergent at about the 40% dissimilarity level on the basis of isozyme analysis. CAPA and CAPB also have slightly differently shaped asexual sporangia, and only CAPB forms chlamydospores (thick-walled asexual resting spores) (Mchau and Coffey, 1995). However, since the host ranges of the two groups are similar, there remains uncertainty as to whether or not they are conspecific.

P. capsici is an outcrossing species and its sexual stage can be obtained in culture. In such circumstances there is a strong case for experimentally investigating the level of reproductive isolation between groups via intergroup crosses. If two groups prove to be totally reproductively isolated at the pre-zygotic level, they should be considered separate OSUs. If partial or no pre-zygotic isolation is found, studies should, if practicable, be extended to examining the relative fitness of the F_1s in order to assess the degree of post-zygotic isolation. The ecological importance of the latter is frequently overlooked in fungal biological species tests (cf. Brasier, 1987). Even where there is full pre-zygotic interfertility, the possibility of post-zygotic isolation should be investigated if there is reason to suspect that two subgroups were once geographically or ecologically isolated.

Even the case of *Armillaria mellea* (Figure 7.3) is less clear-cut when the sympatrically defined OSUs are considered in allopatric terms. Thus, the five sympatric OSUs within Europe or the ten within North America were clearly distinguished on the basis of their total reproductive isolation. However, the European *A. cepistipes* shows full interfertility with the recently discovered North American biological species XI, (not shown in Figure 7.3) and is therefore conspecific with it, yet it also shows partially interfertility with the North American species V (*A. sinapina*) and X (Guillaumin *et al.*, 1991). A similar phenomenon occurs in *Heterobasidion annosum*, in which North American and European S and P groups are more interfertile at the pre-zygotic level than those within North America or Europe alone (Chase and Ullrich, 1990a,b).

Such examples of varying and somewhat unpredictable interfertility between geographically isolated taxa are not new to mycology. The problem was raised during seminal work on fungal breeding units in *Fomitopsis pinicola* by Mounce and Macrae (1938) and similar issues were raised with regard to *Auricularia* (Duncan and MacDonald, 1967) and *Coprinus* (Kemp, 1975, 1977). The phenomenon has since been demonstrated for many closely related allopatric taxa and may reflect the lack of development of isolating mechanisms following geographic separation of populations (Brasier, 1987, 1995). Even with ecologically distinct sympatric taxa, occasional hybrids may form. A probable example has recently been found for the S and P types of *Heterobasidion* in North America (Garbelotto *et al.*,

1995). One way to accommodate such problems, from the point of view of naming and nomenclature, is to formally name the OSUs and to enhance scientific awareness of the potential, if not the real prospects, for interspecific hybridization (cf. Purvis, 1997: Chapter 6). Much better use could also be made of subspecific ranks than at present (cf. Brasier and Rayner, 1987; Brasier, 1992). In *Heterobasidion*, for example, the S and P groups could be named as species and the European versus North American S-types (or the European versus North American P-types) could, if appropriate, be named subspecies.

The complex structure of '*P. megasperma*' *s.l.* has been greatly clarified, but it also remains a case that is still being resolved. In addition to the six well-defined and one partially defined OSUs (Figure 7.11), 14 additional unique molecular types have recently been identified (Förster and Coffey, 1993). Two of these, types JJ6 (from apple) and JJ4 (from alfalfa), are represented in Figure 7.11. The isolates exhibiting these novel molecular types again tend to be associated with particular hosts, but the sample sizes examined so far are small and therefore their status is, as yet, unconfirmed. They may be additional OSUs. A further possibility is that some of these additional molecular types are hybrids between other OSUs. Although the *Phytophthora* taxa involved are mainly inbreeding (homothallic), occasional outcrossing in homothallics is likely (Brasier, 1992). Indeed, outcrossing has recently been demonstrated in *P. sojae* (Förster *et al.*, 1994; Whisson *et al.*, 1994), a homothallic OSU within the old morphospecies '*P. megasperma*' (Figure 7.11). Such outcrossing may sometimes occur between OSUs, particularly if the pre-zygotic isolating mechanisms were weak or the OSUs closely related. The possibility of hybridization is always a consideration where related and once geographically isolated taxa have been subsequently intermixed by human activity (Brasier and Hansen, 1992; Brasier, 1995). Its occurrence, however, may sometimes be difficult to confirm retrospectively, particularly if extensive backcrossing were involved.

In traditional morphological taxonomy, species were often perceived as being of approximately equal weight in the hierarchy (cf. Hawksworth, 1974). The evidence now emerging, however, is that OSUs can occur at widely differing levels of genetic divergence. For example in '*P. megasperma*' (Figure 7.11) OSUs 'ALF', 'CLO' and 'DF' show a much greater molecular relatedness than do, for example, OSUs 'AC' and 'SOY' or indeed the two outgroup isolates of '*P. cryptogea*'. In this instance little is known of the geographical origins of the OSUs, and it is unclear whether the differences in genetic distance relate to differences in their times of appearance or to their exhibiting unequal rates of molecular evolution. It is also worth noting at this point that isozyme differences between traditional *Phytophthora* taxa are often very large (Oudemans and Coffey, 1991) and would be considered nearer the generic level in some organisms such as higher animals. While such knowledge of genetic distances does not affect our ability to identify OSUs as such, it does question the way we have traditionally viewed the status of 'species' in nomenclature and in the systematic hierarchy (cf. Brasier and Rayner, 1987). OSUs are units of evolution, speciation is a continuing process, and species may be subject to near random extinctions over time. The genetic status of modern OSUs will therefore tend to be unequal, reflecting the occurrence of speciation events and extinction events in time and space and the varying environmental processes involved in OSU emergence and survival.

7.4 OPERATIONAL SPECIES UNITS IN ASEXUAL FUNGI

A large number of fungi exist which have 'lost' their sexual stage, possibly as a result of sudden and intense selection events ('episodic selection', Brasier, 1995). The majority of asexual genera, including *Fusarium*, *Verticillium* and *Penicillium* are of ascomycete affinity (Table 7.1). In the absence of sexual structures, morphological characters suitable for systematic or phylogenetic use are often even more limited, though in some asexual fungi complex asexual fruiting structures (e.g. pycnidia), are produced.

As in other asexual organisms, repeated mitotic reproduction in asexual fungi leads to the development of clonal lineages or large genets (Brasier and Rayner, 1987). These lineages may readily become genetically isolated from each other via the fungus's self/non-self recognition system, termed the vegetative incompatibility (vc) system. Vc systems occur in both sexual and asexual fungi and are analogous to tissue incompatibility systems in animals. The vc system allows hyphal fusions to occur between mycelia carrying the same vc loci (which also tend to be genetically similar individuals), but prevents fusions between mycelia carrying different vc loci, thereby restricting gene-flow between 'unlike' genets. Within some traditional asexual taxa that are conventionally assigned the status of species, for example the ubiquitous *Fusarium oxysporum*, numerous vc types occur. These vc types are reproductively isolated micro-units that could in theory, be considered to function as separate OSUs. When they are very numerous, however (as in *F. oxysporum* where some 200 potentially different host-related forms are known), considering each micro-unit as a separate OSU would clearly present problems! When assessing the rank of different vc types or widespread genets, other relevant information, such as behavioural differences and their longer-term genetic stability, must be taken into account. Asexual fungi therefore represent a special case. They may require a different species concept from that applied to most sexually reproducing fungi. Alternatively, they might be included within a general, all-embracing, species concept (see below).

Studies on the molecular structure of asexual fungi have so far often concentrated on resolving subgroupings within the traditional taxa, such as the different host-specialized forms within *Fusarium oxysporum*, rather than on the overall species structure. For example, there is now evidence that the different vc types involved in the group of *F. oxysporum* isolates specific to musk melon ('f.sp. *melonis*') are a polyphyletic assemblage (Jacobson and Gordon, 1990). In *Verticillium*, molecular studies have revealed two widely genetically divergent host-related OSUs within the traditional morphospecies *V. albo-atrum* (Carder and Barbara, 1991) and two other widely divergent units, one haploid, the other diploid, within *V. dahliae* (Morton *et al.*, 1995). The status of these units has yet to be formalized. Within them there are likely to be different vegetative compatibility groups.

An important application of molecular phylogenetics is that asexual fungi are being assigned to their sexual genera, as in the case of asexual penicillia which are being assigned within the sexual genus *Talaromyces* (Figure 7.12; Lobuglio *et al.*, 1993) and fusaria which are being assigned to the genera *Nectria* and *Gibberella* (Guardet *et al.*, 1989). The same two studies have also revealed that loss of the sexual state in these genera has occurred repeatedly over time. The placing of asexual taxa in a natural phylogeny will enhance our understanding of their evolutionary origins (Brasier, 1995) and hence our understanding of their structure, status and ecological significance as OSUs.

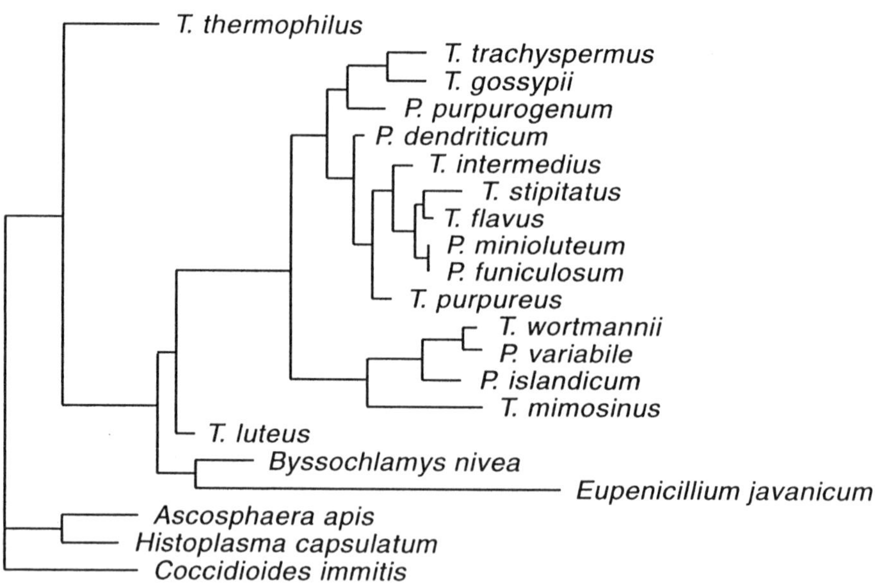

Figure 7.12 Assignment of asexual *Penicillium* species within the sexual genus *Talaromyces* on the basis of combined rDNA domain sequence data. The other genera shown are outgroups. (Redrawn from Lobuglio *et al.*, 1993.)

7.5 OPERATIONAL SPECIES UNITS AMONG STERILE MYCELIA

An assemblage of fungi which have been particularly intractable to morphologically based systematics are the non-sporulating 'sterile mycelia' (Parmeter, 1965). Many sterile mycelia have a basidiomycete affinity, and some show evidence of sexual recognition responses when paired in culture, though it is often uncertain whether the sexual state has been lost, or is simply undiscovered. A well-known sterile fungus of considerable economic importance is the common root pathogen *Rhizoctonia solani* which is morphologically recognized as a basidiomycete. This has been extensively analysed in terms of host-specialization, inter-sterility groups and molecular relationships and up to eleven OSUs have now been identified within it (Parmeter *et al.*, 1969; Vilgalys and Gonzalez, 1990; Sneh *et al.*, 1992).

Another well-known group of sterile mycelia are the so-called 'snow moulds', psychrophilic plant pathogens active in the host at or below freezing point (e.g. under snow cover). By use of combined genetical, behavioural and molecular criteria, considerable progress is being made in the elucidation of taxa within these fungi. A recent molecular study of 23 Canadian isolates of cottony snow mould, previously attributed to the basidiomycete *Coprinus psychromorbidus*, has revealed at least four highly genetically divergent population units (Figure 7.13). The units correspond well to groups with particular host or substrate specificities, and so have identifiable niches. They also correspond with groups based on sexual mating responses between isolates in culture (Laroche *et al.*, 1995).

The processes involved in the discovery of these OSUs in the cottony snow moulds closely parallel the processes involved in the discovery of the OSUs within *Armillaria mellea.* Like so many recent examples, they remain to be formally designated, being for the present referred to by their trivial names.

7.6 CRITERIA AND CONCEPTS FOR THE FUTURE: THE WAY AHEAD

7.6.1 An emerging emphasis on operational species units

In fungal systematics traditional taxa defined mainly on morphological characters form the bulk of the 70 000 or so (Hawksworth, 1991) described species, and new taxa based on these criteria, many representing a single collection, continue to be designated routinely in the literature (c. 1700 per annum; D.L. Hawksworth, personal communication). Where such taxa are based on multiple morphological or other discontinuities, it is likely that they are 'good' taxa in the traditional sense. The increasing revelation of OSUs within such traditional morphological species has, however, brought into question their utility as accurate or natural units. Indeed, the

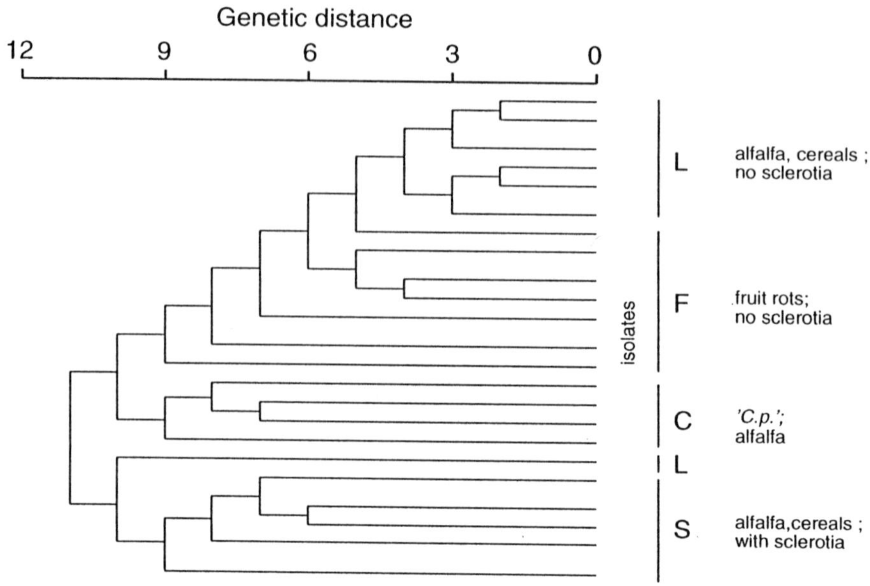

Figure 7.13 Molecular variation among isolates of the cottony snow mould *Coprinus psychromorbidus*, based on RAPDs analysis of total DNA, showing the occurrence of at least four major molecular groups which correspond with groups defined by mating tests. L, low temperature basidiomycete; F, from *Festuca* (fescue); C, *Coprinus psychromorbidus*; S, from wheat stubble. (Redrawn from Laroche *et al.*, 1995.)

status of any morphological taxon must be viewed with a degree of circumspection until it is known whether or not it represents a single OSU. This is especially the case where a morphological species is known to exhibit a wide host or substrate range or a wide geographical and climatic distribution. At the same time, since there is such a large number of described species in the literature, it is improbable that all traditionally described taxa can ever be investigated using modern analytical methods. Most are unlikely to receive a detailed behavioural study, particularly those perceived, rightly or wrongly, as being economically or ecologically unimportant. This being the case, it is inevitable that systems of traditionally described species and OSUs will occur side by side in the literature well into the future.

Nonetheless, with an increasing emphasis on the identification of OSUs, the fungal species concept is in a state of transition. One consequence is that the 'species unit' is, in broad terms, shifting downwards in the phylogenetic hierarchy, with original traditional morphospecies such as *Armillaria mellea* approximating in rank to a superspecies or to a genus (Brasier, 1991). Many traditional 'species' may therefore fall between the species and generic levels. Another consequence of the discovery of multiple OSUs within individual traditional taxa is that, in terms of the numbers of taxa known to exist or remaining to be discovered, the extent of fungal biodiversity on our planet has probably been greatly underestimated.

For the practising plant pathologist, fungal ecologist or fungal geneticist, the logical unit of choice is the operational species unit rather than the traditional morphospecies. OSUs are more precisely biologically defined entities and should most approximate to the principal population units occurring in the fungi themselves. They should therefore best meet the requirements of accurate and unambiguous communication. This requirement is paramount in many fields of applied mycology (Table 7.1), where understanding behavioural and other differences between species units is the key to good ecological management, effective disease control measures and clear quarantine and patent specification. It is OSUs, therefore, that professional systematists must discriminate if they are best to meet the needs of client applied mycologists.

For similar reasons, it is important that, once their status is clear and their variation satisfactorily described, newly identified OSUs be formally designated. The persistence of informal notation and trivial names for OSUs only promotes confusion in the literature. It also devalues the role of taxonomy and nomenclature in scientific communication. The nomenclatural process 'polices' the system. It gives new names a legitimate basis, controls the problem of name duplication and serves as the bases of scientific cataloguing (cf. Hawksworth, 1974). Naming OSUs and other population units under the current system of nomenclature is not, however, ideal. A more flexible system of specific and intra-specific terms fitted to the special requirements of the fungi may eventually be needed (Brasier and Rayner, 1987).

7.6.2 Methodology in a new era

The increasing revelation of OSUs also marks the end of the era in fungal taxonomy in which morphological criteria reigned supreme, and heralds the beginning of an era in which all biological and ecological properties are potentially defining. Where morphological differences exist which can reliably distinguish between OSUs, or where a traditional morphological taxon is shown to comprise a single OSU, as many are, so much the better. The morphological characters can be used as a primary aid in identification, and there may also be less confusion in the literature. Where useful morphological differences are absent, however, as with so many fungi in artificial culture, or where no useful differences have accumulated between OSUs during their evolutionary history, then there is a need to apply methods of diagnosis with a more experimental basis. These range from sexual mating tests to temperature–growth studies and from isozyme profiles to DNA hybridization and sequence studies. To produce an accurate diagnosis of a known OSU several different types of test may sometimes be required.

As already emphasized, the initial discovery of OSUs within traditional taxa is occurring via a number of routes. The demonstration of total pre-zygotic reproductive isolation via a mating test is often the most rapid and the least ambiguous means of revelation. However, the occurrence of strong sterility barriers alone does not in itself fully describe an OSU. It simply defines its reproductive relationship to another unit. To describe it adequately, information on its other biological properties must also be accumulated. Furthermore, many OSUs cannot be revealed by a test for pre-zygotic isolation. In the cases of partially reproductively isolated taxa, inbreeding taxa, asexual taxa, and in fungi not amenable to culture such as the obligate rust parasites, numerous other routes must be undertaken. The examples of the Dutch elm disease Ophiostomas and *Phytophthora megasperma* well illustrate that many of these routes are virtually an evolutionary process in themselves. At some point during the accumulation of data, it becomes apparent which units are functioning as species, sharing a common gene pool. The initial evidence is often of a common set of physiological, morphometric and behavioural characters. This may be supported subsequently by evidence from molecular characters. In other cases molecular evidence is primary to the whole process. Levels of genetic divergence between units can be inferred from the number, type and degree of discontinuity of non-molecular characters, and from the extent of host, substrate or climatic specialization, as well as from the existence of partial pre- or post-zygotic isolating mechanisms or the degree of molecular divergence exhibited.

(a) Molecular criteria

Molecular analysis of fungal taxa has become standard practice. A valuable attribute of a preliminary molecular analysis, or indeed of a preliminary inter-sterility test, is that it can provide a rapid profile of broad relationships within a group, and can therefore quickly reveal anomalies in traditional taxonomic systems where they occur. This is well exemplified by the remarkable complexity, including probable polyphyleticism, being revealed in some traditional taxa such as *P. megasperma*.

The sheer power of molecular techniques makes them both exciting and fascinating. For these very reasons, they can also be seductive! As in cases where morphological criteria only are used to discriminate taxa, or where revelation of OSUs occurs solely via the demonstration of inter-sterility groups, when molecular characters alone are used to discriminate population units and information on the behaviour and ecological properties of the units is not afterwards obtained, the interpretation, biological meaning and taxonomic utility of the units is limited. This problem is more likely to arise where a molecular analysis is undertaken as an end in itself, or where the taxonomic question has not been properly defined in advance. Use of molecular analysis as part of scientific method, for example to test the validity of OSUs hypothesized from non-molecular characters, would seem to be an ideal if not always the most practicable or economically effective approach. Sometimes even the traditional species names attached to isolates in published dendrograms are incorrect due to prior misidentification, a problem which is more likely to arise where the molecular analyst has insufficient background in the biology or traditional taxonomy of the organism being investigated. Where appropriate, it is recommended that isolates widely used in molecular phylogenetic investigations be deposited periodically with a specialist taxonomist or with a major culture collection so their identity may be checked.

Different components of a genome may mutate at different rates (Bruns *et al.*, 1991). It is therefore also desirable that more than one region of the genome (e.g. mitrochondrial versus nuclear DNA) and contrasting methods (e.g. single gene sequencing versus RAPDs analysis of total DNA) be used in order to corroborate resultant molecular relationships, and to reveal both widely divergent and closely related units. An allied aspect of molecular criteria is that they can lead to the development of rapid tools for diagnosis of known OSUs, not only in artificial culture or when no other distinguishing criteria are available, but also even within a host or a substrate. Laboratories involved in routine fungal identification, particularly applied mycology laboratories, therefore need to be equipped for molecular diagnostics. Indeed, at the beginning of the 21st century, the acquisition of molecular methodology by mycology units is likely to be as necessary as the acquisition of a microscope at the beginning of the 20th century.

7.6.3 The biological and phylogenetic species concepts

In systematic biology at large, debate has been focused recently on the value of the biological species concept, encapsulated by Mayr (1942) under the definition 'groups of actually or potentially interbreeding natural populations which are reproductively isolated from other such groups' (see also the modified definition of Mayr, 1982). The fact that fungal intersterility barriers can often be tested in a simple experimental manner in culture, and that totally reproductively isolated units identified in this way form a tidy genetic package, has given added value to the biological species concept so far as mycology is concerned. Indeed, as a result of a flood of studies on sterility barriers and breeding units since the 1970s, the biological species has belatedly become something of a benchmark for a 'species' in fungi (Kemp, 1975; Esser and Hoffman, 1977; Boidin, 1980;

Vilgalys and Miller, 1983; Perkins and Raju, 1986; Brasier and Hansen, 1992).

An obvious limitation of the biological species concept is its inapplicability to asexual fungi. So too is its unsuitability for use with sexually uniparental or mainly inbreeding fungi. Even with outcrossing taxa, however, the concept has certain limitations. That the occurrence of total intersterility between two otherwise similar taxa is a valid criterion for considering them separate species is a point with which few have argued. However, from studies of complexes within basidiomycete genera, Hallenberg and colleagues (Hallenberg and Larsson, 1992; Hallenberg *et. al.*, 1994) have proposed that where the genetic distances between totally intersterile units are very small a sterility barrier may sometimes be a result of a species' propagation strategy. The evolutionary advantages of such a strategy would need to be clearly demonstrated. It would also be necessary, though difficult, to confirm that a given case did not actually represent early evolutionary divergence of a very closely related but adaptively distinct population, i.e. incipient speciation.

A corollary of the widely held viewpoint, that total intersterility implies separate species, is that the ability to interbreed implies conspecificity. This is a far more equivocal matter. Since the genetic basis of intersterility may sometimes be relatively simple (Kemp, 1977; Brasier, 1984; Chase and Ulrich, 1990a,b; Kile and Brasier, 1990), not all freely interbreeding populations will necessarily be conspecific. Inter-sterility groups in fungi, and particularly in basidiomycetes, could sometimes be paraphyletic: geographically distinct populations belonging to the same intersterility group could be of a different molecular evolutionary lineage, and possess different behavioural, ecological and morphological characters. If sufficiently biologically distinct, they could be different OSUs, though shown to interbreed in laboratory tests. A possible example of this problem is given in a study of molecular groups and inter-sterility groups in the 'oyster mushroom' *Pleurotus* by Vilgalys and Sun (1994) (and cf. Hallenberg *et al.*, 1994, 1996). Inter-sterility Group I has North American, European and Asian representatives. These authors' results suggest that these allopatric representatives are actually unique evolutionary lineages and that they may also be paraphyletic to two other inter-sterility groups. The danger of considering interfertility as evidence for conspecificity is also applicable in cases of partial interfertility, as between *O. ulmi s.s.* and *O. novo-ulmi* (see above). While these can be induced to breed in the laboratory, in nature they are strongly post-zygotically isolated. Once again it is apparent that all available characteristics of taxa, including their geography, habitat, and potential for post-zygotic isolation, must be taken into account when assessing their status as OSUs.

In response to the difficulties perceived in the biological species concept, and in an attempt to satisfy a desire for a unified species concept fitted to all organisms, a phylogenetic species concept in which all genealogically monophyletic, geographically distinct clades or lineages are regarded as taxa, has been proposed (Donaghue, 1985; Mishler and Brandon, 1987; Cracraft, 1990). Its possible adoption for the fungi has also been suggested (Vilgalys, 1991; Vilgalys and Sun, 1994). The phylogenetic species concept has the possible virtue that it would cover the wide diversity of OSUs already identified in fungi such as classical biological species, partially reproductively isolated units, and OSUs defined mainly on correlated behavioural and morphological characters. If different ranking criteria were applied (Donaghue, 1985; and *cf.* Brasier and Rayner, 1987), the concept could also be extended to cover widespread genets and vegetatively incompatible micro-units in asexual fungi. The phylogenetic species concept, is, therefore, something of an all-embracing 'catch all', particularly if every terminal taxon in a phenogram is considered of species rank (Donaghue, 1985). It would transcend the current shift in the methodologies used to characterize taxa in fungi, and it would cover the various types of fungal OSU.

So far as mycology is concerned, an implication of the concept present in the author's mind, but not a scientific criticism of it, is that if widely adopted, being so all embracing it might stifle a crucial debate before it has scarcely begun! This is not intended as an altogether flippant statement. It has already been mentioned that in the first half of this century traditional fungal systematics lacked sufficient evolutionary and genetical input. The current application of genetical, ecological and molecular approaches to defining species in fungi, and the resultant demonstration of functional operational species units, is bringing about a much-needed reunification of fungal systematics and fungal genetics. In consequence, fungal systematics is in a state of creative tension. A debate on the species concept is a necessary part of that process, and it would be unfortunate if either a biological or a phylogenetic species concept were accepted by diffusion, without more thorough discussion. Indeed, in view of both the environmental and the nomenclatural implications, the issues involved need to be more widely debated at a formal as well as an informal level. With some exceptions, much of the discussion has proceeded so far at the level of the individual taxon, individual paper and individual scientist rather than being driven at the level of national and international mycological societies and committees. Broadly, there is a need for a greater attempt at synthesis among those most sympathic to the biological species concept, those inclined to the phylogenetic concept, and those whose emphasis has been on morphology.

Whether a species concept can be found that would be acceptable to all practising mycologists remains to be seen. The biological species concept probably appeals most strongly to those with a special interest in genetics. The phylogenetic concept probably appeals most strongly to those involved in phylogenetic grouping and ranking, since it emphasizes genealogy and is itself a product of hierarchical phylogenetic analysis. While the latter might best approximate to a natural system of classification, little has been done so far to suggest that this concept will promote a better understanding of the biological and dynamic aspects of fungal OSUs. OSUs are first and foremost behavioural entities. It is their behavioural differences, and the ecological and genetical processes involved in their emergence and survival, including the splitting or recombining of their lineages, that are their most defining properties. Whatever our concepts and definitions, fungi will continue mainly to exist in such principal functional units, as they have always done. In a changing world, it is important for us to identify as accurately as possible what the boundaries and behavioural properties of these units are, if only to assist in the survival of our own environment.

7.7 REFERENCES

Ainsworth, A.M. (1987) Occurrence and interactions of outcrossing and non-outcrossing populations in *Stereum, Phanerochaetae* and *Coniophora*, in *Evolutionary Biology of the Fungi* (eds A.D.M. Rayner, C.M. Brasier and D. Moore), Cambridge University Press, London, pp. 285–99.

Anderson, J.B. and Ullrich, R.C. (1979) Biological species of *Armillaria mellea* in North America. *Mycologia*, **71**, 402–14.

Barr, D.J.S. (1992) Evolution and kingdoms of organisms from the perspective of a mycologist. *Mycologia*, **84**, 1–11.

Bisby, G.R. (1953) *An Introduction to Taxonomy and Nomenclature of Fungi*. Commonwealth Mycological Institute, Kew, Surrey.

Boidin, J. (1951) Sur l'existence de races intersteriles chez *Gloeocystidium tenue* (Pat.); Étude morphologique et comportement nucléaire de leurs cultures. *Bulletin de la Societé Mycologique de France*, **66**, 204–19.

Boidin, J. (1980) La notion d'espèce. III. Le critère d'infertilité, ou intercompatibilité: resultats et problems. *Bulletin de la Societé Mycologique de France*, **96**, 43–57.

Brasier, C.M. (1977) Inheritance of pathogenicity and cultural characters in *Ceratocystis ulmi*. Hybridization of protoperithecial and non-aggressive stains. *Transactions of the British Mycological Society*, **68**, 45–52.

Brasier, C.M. (1981) The pathogen, in *Compendium of Elm Diseases* (eds R.J. Stipes and R.J. Campana), American Phytopathological Society, St Paul, Minnesota, pp. 11–13.

Brasier, C.M. (1983) Problems and prospects in *Phytophthora* research, in *Phytophthora, Its Biology, Ecology and Pathology* (eds D.C. Erwin, P.H. Tsao and S. Bartnicki-Garcia), American Phytopathological Society, St Paul, Minnesota, pp. 351–64.

Brasier, C.M. (1984) Inter-mycelial interactions in *Ceratocystis ulmi*: their physiological properties and ecological importance, in *The Ecology and Physiology of the Fungal Mycelium* (eds D.H. Jennings and A.D.M. Rayner), Cambridge University Press, London, pp. 451–97.

Brasier, C.M. (1987) The dynamics of fungal speciation, in *Evolutionary Biology of the Fungi* (eds A.D.M. Rayner, C.M. Brasier and D. Moore), Cambridge University Press, London, pp. 231–60.

Brasier, C.M. (1989) Population dynamics of plant pathogenic 'fungi', in *Proceedings of the Fourth International Mycological Congress Regensberg, Germany 1989, (Abstracts)*, University of Regensberg, p.111.

Brasier, C.M. (1990a) China and the origins of Dutch elm disease: an appraisal. *Plant Pathology*, **39**, 5–16.

Brasier, C.M. (1990b) Current questions on *Phytophthora* systematics: the role of the population approach, in *Phytophthora* (eds J.A. Lucas, R.C. Shattock, D.S. Shaw and Louise R. Cooke), Cambridge University Press, London, pp. 105–28.

Brasier, C.M. (1991) *Ophiostoma novo-ulmi* sp. nov., causative agent of the current Dutch elm disease pandemics. *Mycopathologia*, **115**, 151–61.

Brasier, C.M. (1992) A champion thallus. *Nature*, **356**, 382–3.

Brasier, C.M. (1993) The genetic system as a fungal taxonomic tool: gene flow, molecular variation and sibling species in the '*Ophiostoma piceae – Ophiostoma ulmi*' complex and its taxonomic and ecological significance, in *Ceratocystis and Ophiostoma* (eds M.J. Wingfield, K.A. Seifert and J.F. Webber), American Phytopathological Society, St. Paul, Minnesota, pp. 77–92.

Brasier, C.M. (1994) Missing link in tree disease. *Nature*, **372**, 227.

Brasier, C.M. (1995) Episodic selection as a force in fungal microevolution with special reference to clonal speciation and hybrid introgression. *Canadian Journal of Botany*, **73**, 1213–21.

Brasier, C.M. and Gibbs, J.N. (1976) Inheritance of pathogenicity and cultural characters in *C. ulmi*. I. Hybridisation of aggressive and non-aggressive strains. *Annals of Applied Biology*, **83**, 31–7.

Brasier, C.M. and Hansen, E.M. (1992) Evolutionary Biology of Phytophthora. II. Phylogeny, population structure and speciation. *Annual Review of Phytopathology*, **30**, 173–200.

Brasier, C.M. and Kirk, S.A. (1993) Sibling species in *Ophiostoma piceae. Mycological Research*, **97**, 1416–18.

Brasier, C.M. and Mehrotra, M.D. (1995) *Ophiostoma himal-ulmi* sp. nov., a new species of Dutch elm disease fungus endemic to the Himalayas. *Mycological Research*, **99**, 205–15.

Brasier, C.M. and Rayner, A.D.M. (1987) Whither terminology below the species level in the fungi? in *Evolutionary Biology of the Fungi* (eds A.D.M. Rayner, C.M. Brasier and D. Moore), Cambridge University Press, London, pp. 379–88.

Brayford, D. (1990) Variation in *Phomopsis* isolates from *Ulmus* species in the British Isles and Italy. *Mycological Research*, **94**, 691–7.

Bruns, T.D., White, T.J. and Taylor, J.W. (1991) Fungal molecular systematics. *Annual Review of Ecology and Systematics*, **22**, 525–64.

Burnett, J.H. (1983) Speciation in fungi. *Transactions of the British Mycological Society*, **81**, 1–14.

Carder, J.H. and Barbara, D.J. (1991) Molecular variation and restriction fragment length polymorphisms (RFLPs) within and between six species of *Verticillium. Mycological Research*, **95**, 935–42.

Cavalier-Smith, T. (1987) The origin of fungi and pseudofungi, in *Evolutionary Biology of the Fungi* (eds A.D.M. Rayner, C.M. Brasier and D. Moore), Cambridge University Press, London, pp. 339–53.

Cavalier-Smith, T. (1993) Kingdom *Protozoa* and its 18 phyla. *Microbial Reviews*, **57**, 953–94.

Cha, J.Y., Sung, J.M. and Igarashi, T. (1994) Biological species and morphological characteristics of *Armillaria mellea* complex in Hokkaido: *A. sinapina* and two new species, *A. jezoensis* and *A. singula*. *Mycoscience*, **35**, 39–47.

Chase, T.E. and Ullrich, R.C. (1990a) Genetic bases of biological species in *Heterobasidion annosum*: Mendelian determinants. *Mycology*, **82**, 67–72.

Chase, T.E. and Ullrich, R.C. (1990b) Five genes determining intersterility in *Heterobasidion annosum*. *Mycologia*, **82**, 73–81.

Claridge, M.F. and Boddy, L. (1994) Species recognition systems in insects and fungi, in *The Identification and Characterization of Pest Organisms* (ed. D.L. Hawksworth), CAB International, Wallingford, UK, pp. 261–74.

Corliss, A.O. (1994) An interim utilitarian ('user-friendly') hierarchical classification and characterization of the *Protozoa*. *Acta Protozoologia*, **33**, 1–51.

Cracraft, J. (1990) Speciation and its ontology: the empirical consequences of alternative species concepts for understanding patterns and processes of differentiation, in *Speciation and its Consequences* (eds D. Otte and J.A. Endler), Sinauer Associates, Sunderland, Massachusetts, pp. 28–59.

Donaghue, M.J. (1985) A critique of the biological species concept and recommendations for a phylogenetic alternative. *The Bryologist*, **88**, 172–81.

Drechsler, C. (1931) A crown rot of hollyhocks caused by *Phytophthora megasperma* n. sp. *Journal of the Washington Academy of Science*, **21**, 513–26.

Duncan, E.G. and MacDonald, J.A. (1967) Microevolution in *Auricularia auricula*. *Mycologia*, **59**, 803–18.

Esser, K. and Hoffman, P. (1977) Genetic basis for speciation in higher basidiomycetes with special reference to the genus *Polyporus*, in *The Species Concept in Hymenomycetes* (ed. H. Clémencon), J. Cramer, Vaduz, pp. 189–214.

Förster, H. and Coffey, M.D. (1991) Approaches to the taxonomy of *Phytophthora* using polymorphisms in mitochondrial and nuclear DNA, in *Phytophthora* (eds J.A. Lucas, R.C. Shattock, D.S. Shaw and Louise R. Cooke), Cambridge University Press, London, pp. 164–83.

Förster, H. and Coffey, M.D. (1993) Molecular taxonomy of *Phytophthora megasperma* based on mitochondrial and nuclear DNA polymorphisms. *Mycological Research*, **97**, 1101–12.

Förster, H., Tyler, B.M. and Coffey, M.D. (1994) *Phytophthora sojae* races have arisen by clonal evolution and by rare outcrosses. *Molecular Plant–Microbe Interactions*, **7**, 780–91.

Fries, N. (1985) Intersterility groups in *Paxillus involutus*. *Mycotaxon*, **24**, 403–9.

Garbelotto, M., Ratcliff, A., Cobb, F., Otrosina, W. and Bruns, T. (1995) Intersterility groups of *Heterobasidion annosum* in California: a reappraisal of host specificity, hybridisation and intergroup gene flow. *Phytopathology*, **85**, 1141 (abstract 213).

Guardet, J., Julien, J., Lafay, J.F. and Brygoo, Y. (1989) Phylogeny of some *Fusarium* species, as determined by large sub-unit and DNA sequence comparison. *Molecular Biology and Evolution*, **6**, 227–42.

Guillaumin, J.J., Anderson, J.B. and Korhonen, K. (1991) Life cycle, inter-fertility and biological species, in *Armillaria Root Disease* (eds C.G. Shaw and G.A. Kile), *United States Department of Agriculture Handbook* 691. Washington, DC.

Hall, G. (1993) An integrated approach to the analysis of variation in *Phytophthora nicotianae* and a redescription of the species. *Mycological Research*, **97**, 559–74.

Hallenberg, N. and Larsson, E. (1992) Mating biology in *Peniophora cinerea* (Basidiomycetes). *Canadian Journal of Botany*, **70**, 1758–64.

Hallenberg, N., Larsson, K.H. and Larsson, E. (1994) On the *Hyphoderma praetermissum* complex. *Mycological Research*, **98**, 1012–18.

Hallenberg, N., Larsson, E. and Mahlapuu, M. (1996) Phylogenetic studies in *Peniophora*. *Mycological Research*, **100**, 179–87.

Hamm, P.B. and Hansen, E.M. (1982) Single-spore isolate variation: The effect on varietal designation in *Phytophthora megasperma*. *Canadian Journal of Botany*, **60**, 2931–8.

Hansen, E.M. (1990) Variation in the species of the *Phytophthora megasperma* complex, in *Phytophthora* (eds J.A. Lucas, R.C. Shattock, D.S. Shaw and Louise R. Cooke), Cambridge University Press, London, pp. 148–63.

Hansen, E.M. and Hamm, P.B. (1983) Morphological differentiation of host-specialized groups of *Phytophthora megasperma*. *Phytopathology*, **73**, 129–34.

Hansen, E.M., Brasier, C.M., Shaw, D.S. and Hamm, P.B. (1986) The taxonomic structure of *Phytophthora megasperma*: Evidence for emerging biological species groups. *Transactions of the British Mycological Society*, **87**, 557–73.

Harrington, T.C. and Wingfield, B.D. (1995) A PCR-based identification method for species of *Armillaria*. *Mycologia*, **87**, 280–8.

Hawksworth, D.L. (1974) *Mycologists Handbook. An Introduction to the Principles of Taxonomy and Nomenclature in the Fungi and Lichens*. Commonwealth Mycological Institute, Kew, UK.

Hawksworth, D.L. (1991) The fungal dimension of biodiversity magnitude, significance and conservation. *Mycological Research*, **95**, 641–55.

Hintikka, V. (1973) A note on the polarity of *Armillaria mellea*. *Karstenia*, **13**, 32–9.

Jacobson, D.J. and Gordon, T.R. (1990) The variability of the mitochondrial DNA as an indicator of the relationships between populations of *Fusarium oxysporum f. sp. melonis*. *Mycological Research*, **94**, 734–44.

Johansson, M. and Stenlid, J. (ed.) (1994) *Proceedings of the eighth international conference on root and butt rots*. Swedish University of Agricultural Sciences, Uppsala.

Karlsson, J.O. (1994) Studying genetic fingerprinting in *Heterobasidion annosum* using DNA fingerprinting and a mitochondrial DNA ribosomal probe, in *Proceedings of the Eighth International Conference on Root and Butt Rots* (eds M. Johansson and J. Stenlid), Swedish University of Agricultural Sciences, Uppsala, pp. 276–82.

Kemp, R.F.O. (1975) Breeding biology of *Coprinus* species in The Section Lanatuli. *Transactions of The British Mycological Society*, **65**, 375–88.

Kemp, R.F.O. (1977) Oidial homing and the taxonomy and speciation of Basidiomycetes with special reference to the genus *Coprinus*, in *The Species Concept in Hymonomycetes* (ed. H. Clemencon), J. Cramer, Vaduz, pp. 259–73.

Kemp, R.F.O. (1983) Incompatibility and speciation in the genus *Coprinus*. *Revista de Biologia*, **12**, 179–86.

Kemp, R.F.O. (1985) Do fungal species really exist? A study of basidiomycete species with special reference to those in *Coprinus* section *Lanatuli*. *Bulletin of the British Mycological Society*, **19**, 34–9.

Kile, G.A. and Brasier, C.M. (1990) Inheritance and inter-relationship of fitness characters in progeny of an aggressive x non-aggressive cross of *Ophiostoma ulmi*. *Mycological Research*, **94**, 514–22.

Knowlton, N. (1993) Sibling species in the sea. *Annual Review of Ecology and Systematics*, **24**, 189–216.

Korhonen, K. (1978a) Intersterility and clonal size in the *Armillaria mellea* complex. *Karstenia*, **18**, 31–42.

Korhonen, K. (1978b) Intersterility groups of *Heterobasidion annosum*. *Communicationes Instituti Forestalis Fenniae*, **94**, 1–25.

Kuhlman, E.G. (1982) Varieties of *Gibberella fujikuroi* with anamorphs in *Fusarium* section *Liseola*. *Mycologia*, **74**, 759–68.

Kurtzman, C.P. (1985) Molecular taxonomy of fungi, in *Gene Manipulations in Fungi* (eds J.W. Bennett and L.L. Lasure), Academic Press, Orlando, Florida, pp. 35–63.

Kurtzman, C.P. and Robnett, C.J. (1994) Orders and families of ascosporogenous yeasts and yeast-like taxa compared from ribosomal RNA sequence similarities, in *Ascomycete Systematics: the Problems and Prospects in the Nineties* (ed. D.L. Hawksworth), Plenum Press, New York, pp. 249–58.

La Porta, N., Capretti, P., Kammiovirta, K., Korhonen, K. and Karjalainen, R. (1994) Genetic variation in F-group isolates of *Heterobasidion annosum* occurring in Italy, in *Proceedings of the Eighth International Conference on Root and Butt Rots* (eds M. Johansson and J. Stenlid), Swedish University of Agricultural Sciences, Uppsala, pp. 233–7.

Laroche, A., Gaudet, D.A., Schaalje, G.B., Erickson, R.S. and Ginns, J. (1995) Grouping and identification of low temperature basidiomycetes using mating, RAPD and RFLP analyses. *Mycological Research*, **99**, 297–310.

Leonian, L.H. (1934) Identification of *Phytophthora* species. *West Virginia Agricultural Experiment Station Bulletin*, **262**, 1–36.

Lobuglio, K.F., Pitt, J.I. and Taylor, J.W. (1993) Phylogenetic analysis of two ribosomal DNA regions indicates multiple independent loss of a sexual *Talaromyces* state among asexual *Penicillium* species in subgenus *Biverticillium*. *Mycologia*, **85**, 592–604.

Macrae, R. (1967) Pairing incompatibility and other distinctions among *Hirschioporus* (Polyporus) *abietinus*, *H. fusco-violaceous* and *H. laricinus*. *Canadian Journal of Botany*, **45**, 1371–98.

Matuo, T. and Snyder, A. (1973) Use of morphology and mating populations in the identification of formae speciales in *Fusarium solani*. *Phytopathology*, **6**, 562–5.

Mayr, E. (1942) *Systematics and the Origin of Species*. Columbia University Press, New York.

Mayr, E. (1982) *The Growth of Biological Thought*. Harvard University Press, Massachusetts.

Mchau, G.R.A. and Coffey, M.D. (1995) Evidence for the existence of two subpopulations in *Phytophthora capsici* and a redescription of the species. *Mycological Research*, **99**, 89–102.

Meinhardt, F., Koch, H. and Esser, K. (1984) Compatibility groups in *Ascobolus immersus* – an indication of speciation. *Theoretical and Applied Genetics*, **68**, 365–7.

Mishler, B.D. and Brandon, R.N. (1987) Individuality, pluralism, and the phylogenetic species concept. *Biology and Philosophy*, **2**, 397–414.

Morton, A., Tabrett, A.M., Carder, J.H. and Barbara, D.J. (1995) Sub-repeat sequences in the ribosomal RNA intergenic regions of *Verticillium alboatrum* and *V. dahliae*. *Mycological Research*, **99**, 257–66.

Mounce, I. and Macrae, R. (1938) Interfertility phenomena in *Fomes piernicola*. *Canadian Journal of Research*, **16**, 354–76.

Nelson, R.R. (1965) Assessing biological relationships in fungi. *Phytopathology* **55**, 823–6.

Oudemans, P. and Coffey, M.D. (1990) Relationships between *Phytophthora* species: evidence from isozyme analysis, in *Phytophthora* (eds J.A. Lucas, R.C. Shattock, D.S. Shaw and Louise R. Cooke), Cambridge University Press, London, pp. 184–203.

Oudemans, P. and Coffey, M. (1991) A revised systematics of twelve papillate *Phytophthora* species based on isozyme analysis. *Mycological Research*, **95**, 1025–46.

Parmeter, J.R. (1965) The taxonomy of sterile fungi. *Phytopathology*, **55**, 826–8.

Parmeter, J.R., Sherwood, R.T. and Platt, W.D. (1969) Anastomosis groupings among isolates of *Thanatephorus cucumeris*. *Phytopathology*, **59**, 1270–8.

Perkins, D.A. and Raju, N.B. (1986) *Neurospora discreta*, a new heterothallic species defined by its crossing behaviour. *Experimental Mycology*, **10**, 323–38.

Perkins, D.D., Turner, B.C. and Barry, E.G.C. (1976) Strains of *Neurospora* collected from nature. *Evolution*, **30**, 281–313.

Pipe, N.D., Buck, K.W. and Brasier, C.M. (1995) Molecular relationships between *Ophiostoma ulmi* and the EAN and NAN races of *O. novo-ulmi* determined by RAPD markers. *Mycological Research*, **99**, 653–8.

Purvis, O.W. (1997) Species concepts in lichens, in *Species: the Units of Biodiversity* (eds M.F. Claridge, H.A. Dawah and M.R. Wilson), Chapman & Hall, London, pp. 109–34.

Shear, C.L. and Dodge, B.O. (1927) Life histories and heterothallism of the red bread-mould fungi of the *Monilia sitophila* group. *Journal of Agricultural Research*, **34**, 1019–41.

Smith, M.L. and Anderson, J.B. (1989) Restriction fragment length polymorphisms in mitochondrial DNAs of *Armillaria*: identification of North American biological species. *Mycological Research*, **93**, 247–56.

Smith, M.L., Bruhn, J.N. and Anderson, J.B. (1992) The fungus *Armillaria bulbosa* is among the largest and oldest living organisms. *Nature*, **356**, 428–31.

Sneh, B., Burpee, L. and Ogashi, A. (1992) *Identification of* Rhizoctonia *species*. American Phytopathological Society Press: St Paul, MN, USA.

Verrall, A.F. (1937) Variation in *Fomes ignarius*. *Minnesota Technical Bulletin*, **117**, 1–41.

Vilgalys, R. (1991) Speciation and species concepts in the *Collybia dryophila* complex. *Mycologia*, **83**, 758–73.

Vilgalys, R. and Gonzales, D. (1990) Ribosomal DNA restriction fragment length polymorphisms in *Rhizoctonia solani*. *Phytopathology*, **80**, 151–8.

Vilgalys, R. and Johnson, J.L. (1987) Extensive genetic divergence associated with speciation in filamentous fungi. *Proceedings of the National Academy of Sciences, USA*, **84**, 2355–8.

Vilgalys, R. and Miller, O.K. (1983) Biological species in the *Collybia dryophila* group in North America. *Mycologia*, **75**, 707–22.

Vilgalys, R. and Sun, B.L. (1994) Ancient and recent patterns of geographic speciation in the oyster mushroom *Pleurotus* revealed by phylogenetic analysis of ribosomal DNA sequences. *Proceedings of The National Academy of Sciences, USA*, **91**, 4599–603.

Wainwright, P.O., Hinkle, G., Sogin, M.L. and Stickel, S.K. (1993) Monophyletic origins of the metazoa: an evolutionary link with fungi. *Science*, **260**, 340–2.

Watling, R., Kile, G.A. and Burdsall, H.H. (1991) Nomenclature, Taxonomy and Identification, in *Armillaria Root Disease* (eds C.G. Shaw and G.A. Kile), *United States Department of Agriculture Handbook*, 691, Washington, DC.

Webber, J.F. and Gibbs, J.N. (1982) Colonisation of elm bark by *Phomopsis oblonga. Transactions of the British Mycological Society*, **82**, 348–52.

Whisson, S.C., Howlett, B.J., Liew, E.C.Y., Maclean, D.J., Manners, J.M. and Irwin, J.A.G. (1993) An assessment of relationships between members of the *Phytophthora megasperma* complex and *Phytophthora vignae* using molecular markers. *Australian Journal of Systematic Botany*, **6**, 295–308.

Whisson, S.C., Drenth, A., Maclean, D.J. and Irwin, J.A.G. (1994) Evidence for outcrossing in *Phytophthora sojae* and linkage of a DNA marker to two avirulence genes. *Current Genetics*, **27**, 77–82.

8

Practical aspects of the species concept in plants

R. J. Gornall

Contacting address: School of Biological Sciences, Department of Botany, University of Leicester, University Road, Leicester LE1 7RH, UK

ABSTRACT

As a result of its application to a wide variety of different biological situations the universally used taxonomic species concept has evolved from a simple, typological concept based on morphology, into a more complex, pluralistic entity, deriving its ethos from a variety of sources. Owing to its perceived lack of theoretical background and practical difficulties in certain situations, two chief rivals have emerged: the biological and phylogenetic species concepts, respectively. Although much criticised, especially by botanists, features of the biological concept have been absorbed into the taxonomic concept. The relationship between the taxonomic and phylogenetic concepts appears to be close, although differences exist. In particular, an appropriate extension of the concept of monophyly to the species level needs to be formulated, and criteria for the best way to assign rank are needed if a wider acceptance of the phylogenetic concept is to be achieved.

8.1 INTRODUCTION

The debate over what constitutes a species has simmered on and off almost since the start of recorded history. Over the past 50 to 60 years it has occasionally boiled over in flurries of publications. The burgeoning literature is replete with frequently confusing terminology and all sorts of theoretical notions that are untested or even untestable experimentally. The present symposium concerns species concepts in practice. The primary requirement of any practical concept is that the species must be identifiable by the working botanist. Since our interpretations of evolution are

Species: The Units of Biodiversity. Edited by M.F. Claridge, H.A. Dawah and M.R. Wilson. Published in 1997 by Chapman & Hall. ISBN 0 412 63120 2

critically dependent on our operational species concepts, it is also essential that the latter correspond as closely as possible to theoretical models. Matching theory with practice inevitably demands compromise and this is where some of the more apparently intractable problems arise.

As Blackwelder (1967) has pointed out, it is important in any discussion of this subject to clarify the different usages of the word species. In particular, we must distinguish the taxon from the category. A species as a taxon is regarded by many as an individual, an entity made up of component parts (individual plants), with a beginning, an existence in time and space, and an end (Ghiselin, 1974); or it can be seen as a class with defining characteristics (Caplan, 1981; Ruse, 1981). Either way, some method of assembly or grouping is required in order to recognize the taxon. The species as a category is simply a point in the taxonomic hierarchy between genus and subspecies. The task of the taxonomist is to decide what level of variation in the plants at hand corresponds to this particular level in the hierarchy. These two aspects correspond to the assessment of affinity and rank, the twin problems faced every day by practising taxonomists.

In this review I shall concentrate chiefly on the taxonomic species concept, a term I shall apply to the operational construct used almost universally by practising plant taxonomists. In doing so I shall highlight some of the potential problems presented by the plant kingdom, some of which rarely occur, at least to the same degree, in the animal world. Finally, I shall briefly consider two alternative concepts that seek to provide both a working approach and its associated theoretical underpinning. These are the biological and phylogenetic concepts, respectively.

8.2 THE TAXONOMIC SPECIES

The original species concept which forms the basis of our current approach was based essentially on morphological criteria. It has its origin ultimately in folk taxonomy, but can be traced in written form to Aristotle, who recognized groups of organisms on the basis of resemblances in the 'shapes of their parts, or of their whole body'. Successive subdivision of these groups produced smaller units – the species. As taxa these were thus recognized as morphologically well-marked sets of individuals that shared the same essence. An additional criterion, that of reproductive cohesion, was added by Ray (1686), who reckoned that species should breed true within the limits of their variation. This use of the term species was the one adopted by Linnaeus (1753). The concept was typological in the sense that both he and Ray believed in the existence of an underlying, fixed entity created by God, which was overlayed by a certain amount of variation imposed by nature. It was the job of the taxonomist to identify the limits of this variation and thereby deduce the essential nature of each species. However, an appreciation that species were more than types but were made up of natural populations of individuals soon followed.

As a result of the researches of various Russian and German botanists, e.g. Bunge and Wettstein, in the 19th and early 20th centuries, another criterion – that of geographical coherence – was added to those of morphology and breeding behaviour, and together they formed the basis of the morphological–geographical approach. Du Rietz (1930) summarized the situation at the time by offering the following formal definition of a species: 'the smallest natural populations permanently separated from each other by a distinct discontinuity in the series of biotypes'. Since about the 1940s onwards, the similarity and discontinuity has usually been assessed on the basis of evidence from a wide variety of sources, including anatomy, chemistry, geography and ecology, as well as breeding behaviour and cytogenetics, although a key convention has been that any species must be recognisable on morphological grounds (Davis and Heywood, 1963). The inclusion of evidence in addition to that from morphology has blurred considerably the original, simple concept and has led botanists to talk in terms of taxonomic rather than morphological species.

The taxonomic concept of a species is often described as being phenetic, in which recognition of the taxon is on the basis of overall similarity as perceived by the senses, and its rank is attributed on the basis of the degree of its discontinuity from related biotypes. This approach in practice often produces species that are delimited on the basis of certain character correlations involving at least one (or at least one of a very few) key characters, even though many other characters – perhaps all available ones – may have been taken into account. Blackwelder (1964, 1967) has described this practice as 'omnispective'. Numerical methods are also available by which to detect pattern and quantify discontinuity and overall similarity (Sneath and Sokal, 1973).

Despite this well-established approach to the recognition of species, it is a frequently quoted complaint that the taxonomic species concept is highly subjective and incapable of precise formulation (Davis and Heywood, 1963). In particular, it has been claimed that the concept does not deal very well with situations in which there is extensive infra-specific variation, or uniparental inheritance (through inbreeding or asexual reproduction), or hybridization, or the existence of cryptic species. All of these are well-known phenomena in plants, and it is worth considering various aspects of them, if only briefly, to remind ourselves of the particular problems which any species concept, not just a taxonomic one, has to face.

8.3 INFRASPECIFIC VARIATION

Mayr (1942, 1992) claimed that the phenotypic differences between individuals of a taxonomic species can exceed those found between such species, thereby violating the grouping criterion of overall similarity. The phenomenon occurs chiefly in some morphological species complexes in which various infraspecific taxa have been recognized. Detailed analysis (including study of breeding relationships, cytology and morphology), however, can often help to clarify the situation, as it did in the case of the *Turnera ulmifolia* complex (Shore and Barrett, 1985), in which several of the varieties were shown on further investigation not only to be morphologically distinct but also reproductively isolated, i.e. good taxonomic and biological species.

This is not to deny that the occurrence of infraspecific variation can pose problems when regional or local populations display a phenotype that shares the characters diagnostic of the species but also possesses additional distinguishing features. Should these be recognized as separate species? Inconsistency in the attribution of rank by taxonomists in this respect when faced with similar situations can lead to a loss of credibility. This can be remedied to some extent, however, by the adoption of certain limited requirements in terms of numbers of correlated character differences. For example, for specific distinction Hedberg (1958) and Davis (1978) advocated the use of clearly discontinuous variation in at least two independent characters; some go as far as to allow a single morphological character together with a difference in a cryptic character, such as chromosome number. Decisions on whether to lump or split may also be made easier by using appropriate methods of numerical analysis in which morphological variation within and between populations of species can be partitioned and the patterns compared (Gilmartin *et al.*, 1986). Du Rietz (1930) has also reviewed and discussed at length the conventions by which regional variants can be allocated to the rank of subspecies and local variants to the rank of variety. His proposals have been widely accepted, if not always strictly followed. Infraspecific variation is especially hard to handle taxonomically when it is non-hierarchical, as is so often the case, and its treatment has been discussed at a recent symposium (Styles, 1986).

Generally speaking, however, infraspecific variation of the extent complained about by Mayr is not often a problem. That this is so is illustrated firstly by a study by Gilmartin (1980), who analysed the partitioning of morphological variation within and between populations, species and genera in two families, Asclepiadaceae and Apiaceae. She found significant breaks in the pattern of variation at all levels up to but not including the generic and family levels. Important for the present discussion is the finding that the extent of variation within species was consistently much less than it was between species. Secondly, very often the characters distinguishing species are not those that vary significantly within species. This has been amply illustrated by Davis and Gilmartin (1985), who analysed morphological variation in several species of the grass *Puccinellia*. They found that variation within species involved characters (e.g. leaf size and inflorescence branching) that could be interpreted as adaptive, i.e. they found evidence of ecotypic differentiation; in contrast, an entirely different set of characters, whose adaptive value was not clear, was found to distinguish the different species.

8.4 UNIPARENTAL INHERITANCE

The problems surrounding the taxonomic treatment of inbreeders and apomictic complexes have been presented in detail elsewhere (Davis and Heywood, 1963; Grant, 1981), and I shall offer only a short description here. The products of uniparental inheritance are often only slightly different one from another and the term microspecies has gained currency for these particular taxonomic species. The term microspecies is used to cover any morphologically uniform set or population of plants which is slightly different from related uniform sets or populations.

As far as inbreeding is concerned, an array of homozygous pure lines is likely to be produced where persistent selfing occurs. In many cases these lines are morphologically distinguishable and it would be possible to give each a name at specific rank, thereby recognizing them as microspecies. Where morphological recognition is difficult, the terms semi-cryptic, cryptic or sibling species are sometimes used. Jordan (1873) divided a number of inbreeding species in this manner, e.g. in *Erophila*, *Thlaspi* and *Iberis*. The approach can be defended in cases where the reproductive isolation mediated by autogamy is complete, and especially where incompatibility barriers exist between the microspecies, as they do for example among several species of *Gilia* in the *G. inconspicua* complex (Grant, 1981). In some cases, however, it is often found that products of the occasional cross-pollination occur which recombine the diagnostic features of certain of the homozygous lines and consequently do not belong to any of the microspecies. The gene-pools are still evidently linked and, in virtually all such cases, the taxonomic problems caused by inbreeding have been solved more or less satisfactorily by judicious use of infraspecific ranks, e.g. as in *Montia fontana*, or by accepting an undivided, but variable species, e.g. *Capsella bursa-pastoris*.

Apomixis, either through vegetative reproduction or by means of agamospermy, in many respects produces patterns of morphological variation similar to those seen in persistent inbreeders. Large numbers of morphologically pure lines can be identified. Genetically, however, the situation is often different. Thus, recent evidence from isozyme studies of agamospermous variants of *Taraxacum* has shown that the level of heterozygosity is higher than in some sexual, outbreeding species, but it is fixed (Hughes and Richards, 1988). This is much as might be predicted from the belief that many apomicts are of hybrid origin. In terms of genetic variation within agamospecies, it appears that some may be uniform, apparently consisting of a single genotype, but others may exhibit various levels of variation, from the odd plant in a population which differs in one allele at a locus, to regional variation where whole populations in different areas may differ by only one allele or, occasionally, rather more extensively; all three situations occur in species of *Hieracium* sect. *Alpina* (Shi *et al.*, in press). Sometimes this regional variation is accompanied by consistent but minute morphological differences, and the question arises as to whether such populations should be treated as infraspecific taxa or as separate microspecies. The convention that individual morphological lines should be recognized at specific rank is defensible, and indeed the current fashion is for each to be accorded such status, e.g. as in *Hieracium, Taraxacum,* and the *Ranunculus auricomus* group (until recently the various agamospermous variants of the latter in Scandinavia were treated as subspecies). If the spirit is not willing to dissect out the apomictic minutiae, there is the alternative of the species aggregate. Although not of specific rank, use of the aggregate does allow the busy taxonomist to acknowledge the variation and give it a name (Heywood, 1963). This approach has been criticised for somehow perpetuating taxonomic imprecision, but this is only so when the aggregate is allowed to masquerade as an entity comparable with a sexual species; it cannot be criticised for not being a proper species – it is not intended to be one (Heslop-Harrison, 1962).

A situation the taxonomic species concept does not cope well with involves plants with facultative apomixis, whereby new genotypes are produced by occasional sexual events and each of which can form a new apomictic lineage. The taxonomic problems that this entails are probably insoluble, although a modus operandi has been worked out by students of the *Rubus fruticosus* aggregate in which this problem occurs (Newton, 1975).

8.5 HYBRIDIZATION

In many plant groups, particularly herbaceous perennials and woody species, interspecific hybridization can be commonplace. Hybrids may be sterile, as is usually the case in *Potamogeton,* or they may exhibit various degrees of fertility right up to 100%. In the genus *Salix,* a hybrid has been made artificially but sexually involving 13 different parental species. In some families, intergeneric hybrids are not uncommon, e.g. Rosaceae, Poaceae, and a particularly extreme example is the Orchidaceae where one horticultural variant has at least eight genera in its parentage. In the British Isles, 780 hybrids have been reliably reported from a flora of c. 2500 species (Stace, 1989). The phenomenon is widespread in plants and can lead to speciation if the hybrids become stabilized by sexual means (via recombination or polyploidy or permanent structural hybridity) or asexual means (via apomixis or clonal growth) (Grant, 1981).

Hybridization can blur the boundaries of taxonomic species. Nevertheless, it is usually defensible to recognize two species, even though they hybridize in nature, if the frequency, location or viability of such hybrids is such that the parental gene pools do not merge, e.g. in species of

Quercus, Populus and *Geum*. Grant (1981) has used the term 'semi-species' for cases where gene pools are incompletely reproductively isolated.

There are also problems with polyploid pillar complexes, where the taxa belonging to the polyploid superstructure share parts of their genome both among themselves and with their diploid ancestors. Not surprisingly, the patterns of morphological variation can be complex and misleading when identifying the evolutionary patterns involved. Nonetheless, taxonomists have laboured to identify the pathways and have recognized the products as species. In doing this a range of approaches (not just morphology) is often used, including breeding studies, cytogenetics and the use of molecular markers. Good examples of this can be found among the ferns, e.g. in the genera *Asplenium, Dryopteris* and *Polypodium*, in which the complex genome relationships and evolutionary pathways have been worked out and species recognized on the basis of the findings (Lovis, 1977). Reliable identification of several of the species using morphological characters requires use of a microscope and involves features such as the anatomy of the annulus and paraphyses or the pattern on the surface of the spores; in this sense the species may be regarded as semi-cryptic. Other examples of polyploid pillar complexes have been discussed by Stebbins (1971) and Grant (1981).

Recent studies of chloroplast DNA variation in plant species have shed light on another dimension to the phenomenon of hybridization in plants. It involves the capture of the chloroplast genome of one species by another. Following an initial hybridization event the hybrid acts as the maternal parent in repeated backcrosses to one of the species acting as the paternal parent (Rieseberg and Soltis, 1991). The result is the capture of the chloroplast genome of the maternal parent by the male parent. Such transfer of cytoplasm has now been documented in a range of different species, including shrubs, e.g. *Salix* (Brunsfeld *et al.*, 1992), herbaceous perennials, e.g. *Heuchera* (Soltis *et al.*, 1991), and annuals, e.g. *Helianthus* (Rieseberg *et al.*, 1991). Studies of several genera, e.g. *Quercus* (Whittemore and Schaal, 1991) and *Zea* (Doebley, 1989), suggest that cytoplasmic gene flow can occur between species in the absence of significant nuclear gene flow. Such findings present considerable problems to the taxonomist. What is one to do about taxa which appear to be good taxonomic species but which contain an alien cytoplasm, possibly captured in a series of steps via different species from an ultimate 'donor' that is not closely related? Such a case is exemplified by *Heuchera nivalis* which has captured the chloroplast of a species in a different section via hybridization with an intermediary, *H. parvifolia* (Soltis *et al.*, 1991).

8.6 CRYPTIC OR SIBLING SPECIES

Reproductive isolation and genetic differentiation can occur without much associated morphological change. This leads to the formation of cryptic species. They are also often referred to as sibling species, although this term implies a relationship which may not necessarily exist. The phenomenon is clearly a problem for the taxonomic species concept. Grant (1981) cited the example of the *Gilia transmontana* group which experiments have shown to contain five intersterile but morphologically similar, though not identical, species.

A related taxonomic problem is presented by autopolyploidy. There are several well-documented cases now where autoploids that have been diagnosed on taxonomic, cytological and genetical evidence have been shown to occupy a substantially different geographical range and ecological niche from their diploid parent, e.g. *Tolmiea menziesii* and species of *Heuchera* (Soltis and Soltis, 1989). In these cases the diploid and autotetraploid cytodemes appear to be reproductively strongly isolated, with little evidence of any hybridization. At least at first, however, there is no or virtually no differentiation from their diploid parent, either genetically in terms of isozyme loci or, apart from slight differences sometimes in pollen or stomata diameter, morphologically (Ness *et al.*, 1989). Current practice is to accord such autopolyploid populations no formal taxonomic recognition whatsoever, despite the existence of the barrier to gene flow.

In conclusion, the current approach acknowledges that taxonomic species are only equivalent by designation, and not by virtue of the nature or extent of their evolutionary differentiation (Heywood, 1958), and in this respect the concept embodies an important element of pluralism. It is sometimes implied that taxonomic species are not delimited on evidence from cytogenetics or reproductive behaviour, which is usually seen as the preserve of the 'biological' species concept, but in practice, no such distinction is drawn and all sources of data are considered fair game. The practical concept is eclectic in its application. Admittedly, however, whatever other evidence there might be, by convention at least one morphological/anatomical character is required for species recognition. Even at its most refined, therefore, the taxonomic species concept is essentially an operational construct with no explicit reference to any particular speciation theory. Nonetheless, it is clear that many botanists believe that, where possible, the taxa so recognized should represent evolutionary units – witness the taxonomic treatments of allopolyploid reticulate evolution, many inbreeders and apomictic complexes, described above, in which strenuous efforts have been made to delimit putative evolutionary products to be recognized as species (see also Grant, 1981).

It is probably no exaggeration to say that all the c. 300 000 currently accepted plant species have been diagnosed on this basis (Table 8.1). There must therefore be a substantial measure of agreement about the working concept of a plant species (Anderson, 1957). To quote Heslop-Harrison (1962), 'The success of the morphological–geographical method in so many instances indicates with little room for doubt that the variational units it reveals have a degree of cohesion in nature'. It is the causes and nature of this cohesion that have prompted fierce debate in the literature

8.7 ALTERNATIVE CONCEPTS

Aside from the problem areas described earlier, the drive to overturn the traditional taxonomic concept of a species probably owes as much to its lack of any well-argued background in speciation theory as it does to its perceived shortcomings in any particular situation. No causal mechanisms of the species being recognized are offered as part of the concept and this produces a vacuum which, its critics say, leads to subjectivity and muddled thinking. Two alternative concepts, both of which embody hypotheses about the origin of species, have been offered as potential replacements for the taxonomic concept. They are the biological and phylogenetic species concepts, respectively, and are discussed below.

8.7.1 Biological species concept

With the advent in the 1920s and 1930s of the discipline we know as biosystematics, it was realized that at least one of the processes responsi-

Table 8.1 Number of taxonomic species of living green plants. (Data on non-vascular plants are taken from Raven *et al.*, 1986; those on vascular plants are from Mabberley, 1987.)

Group	*No. of species (approximate)*
Chlorophyta (green algae)	7000
Bryophyta	
Anthocerotopsida (hornworts)	100
Hepaticopsida (liverworts)	6000
Bryopsida (mosses)	9500
Tracheophyta (vascular plants)	
Lycophytina (club-mosses)	1150
Psilophytina (whisk-ferns)	5
Sphenophytina (horsetails)	30
Pterophytina (ferns)	8400
Spermatophytina (seed plants)	
Cycadopsida (cycads)	100
Ginkgoopsida (Maidenhair tree)	1
Coniferopsida (conifers)	550
Gnetopsida	70
Angiospermopsida (flowering plants)	240 000

ble for the evolution of species was the development of reproductive isolation (Dobzhansky, 1937). This perception of a causal link between process and product led directly to the so-called biological species concept. Mayr (1942) offered perhaps the most widely-quoted definition as 'groups of actually or potentially interbreeding natural populations which are reproductively isolated from other such groups'. Blackwelder (1967) has justifiably criticised the name of this concept on the grounds that it is no more biological than are any of the other concepts. The idea that plant species should be delimited at least in part on the basis of breeding behaviour in fact has a considerable history, going back, as we have seen, at least to Ray (1686), and, at least in the sense of true-breeding, is an integral (though often unacknowledged) part of the traditional taxonomic species concept. The proposal, however, that reproductive isolation should be considered the *sine qua non* of species status has met with considerable reserve, not to say resistance, by the botanical community. The reasons for this have been set out in detail by others (Davis and Heywood, 1963; Raven, 1976; Levin, 1979; Jonsell, 1984; Donoghue, 1985), and will only briefly be touched on here.

Despite generating much interest, the criterion of actual or potential reproductive isolation was not widely applied to plants as the sole determinant of species status owing primarily to its general inapplicability in a group in which, as we have seen, the breeding unit and the morphological unit are frequently not correlated and in which uniparental inheritance and lack of or reduced sexuality are not uncommon. Consequently, gene flow, or lack of it, is not regarded as the sole criterion for specific status in plants; rather such evidence has been absorbed into the overall taxonomic approach, to be drawn on as and when considered appropriate.

This is not to deny, however, that there can be an intimate relationship between reproductive isolation and speciation in plants. In some cases we find what may be described as the ideal situation (at least for the taxonomist), in which a species is morphologically homogeneous and well-differentiated from its counterparts, and has a sexual, mixed or outbreeding mating system but does not hybridize with other species. Examples of such clear-cut species occur among many annual plants and can also be found in families such as the Apiaceae and Fabaceae, and in the genera *Allium*, *Campanula* and *Sedum* (Stace, 1989); Grant (1981) has listed a selection of others.

There are also now several examples among plants that demonstrate that shifts in the breeding system from outbreeding to inbreeding, often mediated by changes in floral morphology, can cause reproductive isolation and rapid speciation (Barrett, 1989). Molecular studies of recent progenitor–derivative species pairs in *Stephanomeria* and *Clarkia* (Gottlieb, 1973, 1974), however, have revealed that although the new species is morpho-logically distinct it can be no more different at isozyme loci than are populations of the progenitor species from each other. In other words, speciation can occur without much associated genetic change, at least in the genes controlling the fundamental metabolism of the species. Crawford (1990) has listed other examples.

In his most recent response to the vigorous attacks on the biological species concept by botanists, Mayr (1992) analysed the local flora of the township of Concord, Middlesex County, Massachusetts, a well-botanized area of about 27 square miles. Using a non-dimensional version of the concept, i.e. one to be applied in a given place at a given time, he found he had trouble dealing with 54 species (6.44% of the flora), all either apomictic or autopolyploid, and by implication regarded this as a successful outcome. In practice, of course, Mayr used a largely morphological concept on which to base his inferences about gene flow. It can be seriously doubted, however, whether gene flow or its absence was necessarily involved in either causing or preserving any of the phenotypic continuities and discontinuities that he found within and between the species (Levin, 1979).

Despite its unpopularity in botanical circles, the soul of the biological species concept goes marching on in the guise of the recognition species concept (Paterson, 1985). This essentially turns the biological species concept inside out: the recombinational field of the species is delimited not by reproductive isolation from other taxa but by shared fertilization mechanisms acting within. Since it is based on reproductive behaviour, it suffers the same drawbacks in plants as does the biological species concept.

8.7.2 Phylogenetic species concept

Partly as a result of the debate over the biological species concept, and partly as a result of an increasing understanding of the phylogenetic approach to taxonomy, it has become increasingly accepted that the amount and distribution of gene flow is not the only factor influencing the differentiation or stasis of populations. Other factors include selection and drift (Carson, 1985), and descent from a common ancestor (Grant, 1980). The existence of the many cases where morphological discontinuities do not correspond with ecological discontinuities (Mishler and Donoghue, 1982) has diverted attention away from the role of selection as a cohesive force and focused it instead on shared evolutionary history as a factor binding the individuals of a species together (Grant 1980). The mechanisms by which this may occur have been postulated to be of an epistatic nature (Mishler, 1985).

Over the past 30 years or so several versions of an evolutionary or phylogenetic species concept have been proposed (Mishler, 1985). One that has gained some currency is that proposed by Cracraft (1983: 170), according to whom a phylogenetic species 'is the smallest diagnosable cluster of individual organisms within which there is a parental pattern of ancestry and descent'. This is essentially a refinement of the evolutionary species concept proposed by Simpson (1961). One of the advantages of the phylogenetic (and of the taxonomic) species concept is that it can be applied to persistent inbreeders and to asexual as well as sexual organisms, and therefore has a wider applicability among plants than does the biological species concept. According to Cracraft, any measurable feature can be used in identifying 'the smallest diagnosable cluster' of individuals. This is a controversial point because its strict application utilizing, say, DNA data, could result in the segregation from a single taxonomic species of perhaps dozens, even hundreds, of cryptic species, each possibly of very few individuals. The crux of the problem seems to lie in how to delimit the species from among the welter of individual variants, a position which looks like square one. Nixon and Wheeler (1990) and Davis and Nixon (1992) have attempted to resolve this difficulty by suggesting that a phylogenetic species be recognized as the smallest group of populations (sexual) or lineages (asexual) in which a unique combination of character states is present in all comparable individuals, thereby effectively distinguishing intra- from inter-specific variation. The unique character combination is then regarded as an autapomorphy. It would also appear to be a requirement that the phylogenetic species should be monothetic, in contrast to the taxonomic species which can be polythetic. Davis and Nixon (1992) cautioned potential users of this phylogenetic approach about practical problems regarding the assessment of homology and the effects of under sampling either characters or populations. Although the phylogenetic concept makes no reference to gene flow, the way the criteria are used implies that there should be none, although, as with the taxonomic concept, a limited amount may be allowed if hybrid zones are suspected and parental taxa do not become merged.

A recent investigation adopting the phylogenetic approach and using isozyme data showed that populations of the variable western North American grass *Puccinellia nuttalliana* could be divided into six species, each with a unique isozyme profile (Davis and Manos, 1991). Whether there are any morphological features that can be used by field or herbarium botanists to identify these isozyme species remains to be seen; it is possible that some segregate species may correspond. Another test of the phylogenetic concept has been conducted by Freudenstein and Doyle (1994) who examined variation in morphology and plastid DNA in the *Corallorhiza maculata* complex in North America. Three phylogenetic species were recognized by means of unique combinations of morphological characters. Preliminary evidence from variation in plastid DNA, however, showed that one of the species, *C. maculata*, may have acted as the ancestor to the other two, i.e. the second species (new and unnamed) may have evolved from within it, separating Mexican from more northerly populations, and the third species (*C. mertensiana*) may have budded off from the northern populations (Figure 8.1). All four taxa (Mexican and northern *C. maculata*, *C. mertensiana* and the unnamed species) would qualify as phylogenetic species, each being delimited by a unique combination of features, molecular and morphological. The two variants of *C. maculata*, however, appear to lack any autapomorphies and both they and the taxonomic species could clearly be regarded as being paraphyletic.

This brings us to a contentious point among proponents of phylogenetic (and indeed other) species concepts. By what criterion should the species be delimited? It has been argued that all species we recognize should be monophyletic and recognized on the basis of apomorphic characters, e.g. Rosen (1979) and Donoghue (1985). According to Cracraft (1989), however, the phylogenetic concept does not require that the diagnostic features of a species be apomorphic; he nonetheless believed that most phylogenetic species would turn out to be monophyletic. To what extent our present taxonomic species are monophyletic is unknown. On theoretical grounds it may be that only those arising as a result of vicariance processes are of this nature. Other cases, in which phylogenies have been developed from population studies, have shown that many species are paraphyletic, in that not all the descendants of the most recent common ancestor are included in the species. Thus, in studies of cases where a derivative species originates from its progenitor through rapid speciation, the progenitor is regarded by the analysis as paraphyletic because not all of its populations have evolved into the new species. Similarly, when speciation occurs through peripheral isolation and several deriva-

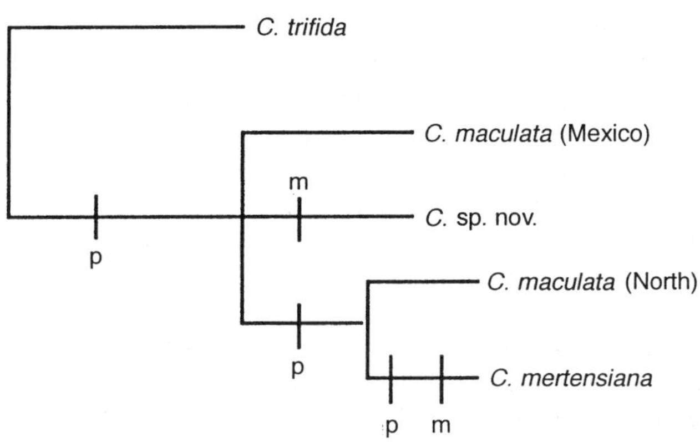

Figure 8.1 Putative phylogenetic tree for American species of the *Corallorrhiza maculata* complex, based on morphological (m) and plastid restriction site (p) apomorphies. This tree is provisional: further sampling is required to confirm that the plastid restriction site mutations are truly fixed. (After Freudenstein and Doyle, 1994.)

tives arise from a single progenitor, the latter is regarded as being paraphyletic, much as in the case of *Corallorrhiza maculata* discussed earlier. According to Davis and Nixon (1992), however, phylogenetic species delimited as the smallest groups that can be recognized on the basis of unique combinations of attributes can never be monophyletic in the sense of Hennig (1966) partly because there is no discoverable subordinate hierarchy, and partly because there are difficulties concerning the concept of the most recent common ancestor at this level. A properly argued extension of the Hennigian definition of monophyly to the species level and below is clearly needed.

Despite the problems with applying the criterion of monophyly to ancestors, the theoretical arguments for adopting a phylogenetic concept of some sort are persuasive, perhaps the more so because they complement and reflect to some extent current practice. For example compare the definition of the phylogenetic species concept enunciated by Cracraft (1983) with the taxonomic or morphological concept offered by Du Rietz (1930). Both require the species to be diagnosable and both attribute species rank to the least inclusive group so diagnosed. Furthermore, just as the phylogenetic concept explicitly recognizes that species are evolutionary products involving parental patterns of ancestry and descent, as already mentioned, so do many practising taxonomists try hard to ensure that the species they recognize are the products of evolution, although this is often not made explicit.

Apart from the need to resolve the problems already alluded to of applying the concept of monophyly at and below the species level, there are several other issues which need to be addressed before wider agreement is achieved. These include the treatment of the products of hybrid speciation, the use of cryptic (and also of quantitative) characters for purposes of species recognition, and the treatment of infraspecific variation.

Although not strictly a problem for the phylogenetic species concept itself, the occurrence of hybridization may present problems in regard to the cladistic methodology and analysis which the phylogenetic approach often involves. There is considerable evidence that very many species may have a hybrid origin and that much plant evolution may be reticulate rather than cladistic. For example, recombinational speciation at the diploid level has been shown to be important in the genus *Helianthus* section *Helianthus*, with three of the stabilized recombinant hybrid derivative species even sharing the same parents (Rieseberg, 1991). Molecular data implicate a similar process of homoploid hybrid speciation in other genera, such as *Gossypium* (Wendel *et al.*, 1991) and *Stephanomeria* (Gallez and Gottlieb, 1982). The role of allopolyploidy, a second process involving lineage fusion as a speciation mechanism in plants, is also well-documented, e.g. as in *Spartina anglica*, *Senecio cambrensis* and *Tragopogon* spp. (Soltis and Soltis, 1993), and a particularly impressive example of reticulate evolution involving allopolyploidy is shown by American members of the

Polypodium vulgare complex (Haufler *et al.*, 1995a,b). It is also notable that many allopolyploids have evolved more than once, sometimes in well-separated places, e.g. *Senecio cambrensis* (Ashton and Abbott, 1992), and therefore may be described as polyphyletic when considered at a population level. Since it has been estimated that 70–80% of angiosperm species are of polyploid origin (Goldblatt, 1980; Lewis, 1980), it is a mode of speciation which is not only sympatric but also possibly represents the most common form in flowering plants. Any phylogenetic species concept must consequently be able to deal the evolutionary pathways and patterns of variation involved.

One practical difference between the phylogenetic and taxonomic approaches appears to lie in the nature of the characters distinguishing the species. In the phylogenetic concept, apparently any character can be used, whereas in the taxonomic concept, by convention, at least one of them must be morphological. At least partly on this basis, Freudenstein and Doyle (1994) proposed to award the Mexican variant of *Corallorrhiza maculata*, a diagnosable phylogenetic species, varietal status under *C. maculata* because it could not be distinguished reliably on morphological grounds but only in terms of its plastid genome. The effect on the number of plant species recognized of abandoning the current morphological requirement can only be guessed at. The example of *Puccinellia nuttalliana* cited earlier may not be a good guide because it is a very variable species, morphological segregates of which have already been described that may correspond to some extent with the isozyme species identified by Davis and Manos (1991). Finally the extent of the problems associated with the use of quantitative characters in a phylogenetic analysis (Stevens, 1991) at and below the species level remains to be evaluated.

Another difference between the taxonomic and phylogenetic approaches lies in the recognition of subspecies and varieties by the taxonomic concept. This is not possible under those phylogenetic concepts in which species are strictly irreducible clusters; no procedures are available for the recognition of infraspecific taxa because variation below the species level (as defined by the concept) is not hierarchical. On one hand, therefore, it may be that many taxonomic subspecies and varieties would not be recognized at all under a phylogenetic approach that involved a monothetic species concept because morphological overlap is allowed, or even required, between them. On the other hand, of course, many infraspecific variants might be elevated to species status as a result of the discovery of diagnostic cryptic characters.

8.8 COMMENTARY

If species are regarded as individuals, they appear to be much like genetic jig-saw puzzles, made up of pieces (individual plants) that fit together to make a whole. The picture on the upper surface is analogous to the operational, largely morphological, species concept, with different species having different pictures of various complexities, corresponding to their pattern and degree of variability. The prongs and notches of the individual pieces are analogous to the mechanisms or processes that lock the individuals of each species into a whole. These processes include not only reproductive isolation and gene flow, but also genetic drift and natural selection (Carson, 1985), and epistatic factors (Mishler, 1985) that result in complex genomic integration and historical (phylogenetic), developmental and ecological constraints which tie the individuals of a species together. Templeton (1989) attempted to integrate all these forces into his cohesion species concept. To what extent this theoretical view of a species, in which phenotypic variation is delimited by genetical, environmental, developmental and phylogenetic cohesion mechanisms, can actually be translated into an operational construct is not clear. Any species concept must address not only the theoretical but also the practical issues posed by the assessment of affinity and rank. What cohesion mechanisms are the most important in delimiting a particular species, and on what basis should the latter be recognized – if by the criterion of monophyly, then how should this be established in practice? Whatever the cohesion factors, the question of assessing species rank arises: how continuous should species be? Perhaps not surprisingly, on the one hand King (1993) regarded the cohesion concept as being no different in its essentials from the biological species concept, and on the other hand Endler (1989) regarded it as being close to degenerating into a phenetic concept.

Where does the analogy of the genetic jig-saw leave us? It is tempting at this point to invoke some form of uncertainty principle and conclude that, electron-like, species exist but are impossible to pin down. The species concept to be adopted must depend on the problem at hand and, within this constraint, be selected so as to provide maximum insight into the biological situation rather as the wave and particle properties of an electron are differentially emphasized depending on context.

Acknowledgement

I should like to thank C.A. Stace for commenting on a draft of this paper.

8.9 REFERENCES

Anderson, E. (1957) An experimental investigation of judgements concerning genera and species. *Evolution*, **11**, 260–1.

Ashton, P. and Abbott, R.J. (1992) Multiple origins and genetic diversity in the newly arisen allopolyploid species, *Senecio cambrensis* (Compositae). *Heredity*, **68**, 25–32.

Barrett, S.C.H. (1989) Mating system evolution and speciation in heterostylous plants, in *Speciation and its Consequences* (eds D. Otte and J.A. Endler), Sinauer Associates, Sunderland, Massachusetts, pp. 257–83.

Blackwelder, R.E. (1964) Phyletic and phenetic versus omnispective classification, in *Phenetic and Phylogenetic Classification* (eds V.H. Heywood and J. McNeill), Systematics Association, London, pp. 17–28.

Blackwelder, R.E. (1967) *Taxonomy. A text and Reference Book.* John Wiley & Sons, New York.

Brunsfeld, S.J., Soltis, D.E. and Soltis, P.S. (1992) Evolutionary patterns and processes in *Salix* section *Longifoliae*: evidence from chloroplast DNA. *Systematic Botany*, **17**, 239–56.

Caplan, A.L. (1981) Back to class: a note on the ontology of species. *Philosophy of Science*, **48**, 130–40.

Carson, H.L. (1985) Unification of speciation theory in plants and animals. *Systematic Botany*, **10**, 380–90.

Cracraft, J. (1983) Species concepts and speciation analysis, in *Current Ornithology*, Plenum Press, New York, vol. 1, pp. 159–87.

Cracraft, J. (1989) Speciation and its ontology: the empirical consequences of alternative species concepts for understanding patterns and processes of differentiation, in *Speciation and its Consequences* (eds D. Otte and J.A. Endler), Sinauer Associates, Sunderland, Massachusetts, pp. 28–59.

Crawford, D.J. (1990) *Plant Molecular Systematics*, John Wiley & Sons, New York.

Davis, J.I. and Gilmartin, A.J. (1985) Morphological variation and speciation. *Systematic Botany*, **10**, 417–25.

Davis, J.I. and Manos, P.S. (1991) Isozyme variation and species delimitation in the *Puccinellia nuttalliana* complex (Poaceae): an application of the phylogenetic species concept. *Systematic Botany*, **16**, 431–45.

Davis, J.I. and Nixon, K.C. (1992) Populations, genetic variation, and the delimitation of phylogenetic species. *Systematic Biology*, **41**, 421–35.

Davis, P.H. (1978) The moving staircase: a discussion on taxonomic rank and affinity. *Notes from the Royal Botanic Garden, Edinburgh*, **36**, 325–40.

Davis, P.H. and Heywood, V.H. (1963) *Principles of Angiosperm Taxonomy*, Edinburgh.

Dobzhansky, T. (1937) *Genetics and the Origin of Species*, Columbia University Press, New York.

Doebley, J.E. (1989) Molecular evidence for a missing wild relative of maize and the introgression of its chloroplast genome into *Zea perennis*. *Evolution*, **43**, 1555–8.

Donoghue, M.J. (1985) A critique of the biological species concept and recommendations for a phylogenetic alternative. *Bryologist*, **88**, 172–81.

Du Rietz, G.E. (1930) The fundamental units of biological taxonomy. *Svensk Botanisk Tidskrift*, **24**, 333–428.

Endler, J.A. (1989) Conceptual and other problems in speciation, in *Speciation and its Consequences* (eds D. Otte and J.A. Endler), Sinauer Associates, Sunderland, Massachusetts, pp. 625–48.

Freudenstein, J.V. and Doyle, J.J. (1994) Plastid DNA, morphological variation, and the phylogenetic species concept: the *Corallorrhiza maculata* (Orchidaceae) complex. *Systematic Botany*, **19**, 273–90.

Gallez, G.P. and Gottlieb, L.D. (1982) Genetic evidence for the hybrid origin of the diploid plant *Stephanomeria diegensis. Evolution*, **36**, 1158–67.

Ghiselin, M.T. (1974) A radical solution to the species problem. *Systematic Zoology*, **23**, 536–44.

Gilmartin, A.J. (1980) Variation within populations and classification. II. Patterns of variation within Asclepiadaceae and Umbelliferae. *Taxon*, **29**, 199–212.

Gilmartin, A.J., Dobrowolski, J., Soltis, D., Kellogg, E. and Harris, G. (1986) Variability within and among populations of four grass species. *Systematic Botany*, **11**, 559–66.

Goldblatt, P. (1980) Polyploidy in angiosperms: monocotyledons, in *Polyploidy: Biological Relevance* (ed. W.H. Lewis), Plenum Press, New York, pp. 219–40.

Gottlieb, L.D. (1973) Genetic differentiation, sympatric speciation, and the origin of a diploid species of *Stephanomeria. American Journal of Botany*, **60**, 545–53.

Gottlieb, L.D. (1974) Genetic confirmation of the origin of *Clarkia lingulata. Evolution*, **28**, 244–50.

Grant, V. (1980) Gene flow and the homogeneity of species populations. *Biologisches Zentralblatt*, **99**, 157–69.

Grant, V. (1981) *Plant Speciation*, 2nd edn, Columbia University Press, New York.

Haufler, C.H., Soltis, D.E. and Soltis, P.S. (1995a) Phylogeny of the *Polypodium vulgare* complex: insights from chloroplast DNA restriction site data. *Systematic Botany*, **20**, 110–19.

Haufler, C.H., Windham, M.D. and Rabe, E.W. (1995b) Reticulate evolution in the *Polypodium vulgare* complex. *Systematic Botany*, **20**, 89–109.

Hedberg, O. (1958) The taxonomic treatment of vicarious taxa. *Uppsala Univ. Arsskr.*, **6**, 186–95.

Hennig, W. (1966) *Phylogenetic Systematics*, University of Illinois Press, Urbana.

Heslop-Harrison, J. (1962) Purposes and procedures in the taxonomic treatment of higher organisms, in *Microbial Classification* (eds G.C. Ainsworth and P.H.A. Sneath), Cambridge University Press, Cambridge, pp. 14–36.

Heywood, V.H. (1958) *The Presentation of Taxonomic Information: A Short Guide for Contributors to Flora Europaea.* Leicester University Press, Leicester.

Heywood, V.H. (1963) The 'species aggregate' in theory and practice. *Regnum Vegetabile*, **27**, 26–37.

Hughes, J. and Richards, A.J. (1988) The genetic structure of populations of sexual and asexual *Taraxacum* (dandelions). *Heredity*, **60**, 161–71.

Jonsell, B. (1984) The biological species concept reexamined, in *Plant Biosystematics* (ed. W.F. Grant), Academic Press, New York, pp. 159–69.

Jordan, A. (1873) Remarques sur le fait de l'existence en societe a l'etat sauvage des especes vegetales affines. *Bulletin Association Française pour l'Avancement des Sciences*, 2.

King, M. (1993) *Species Evolution: the Role of Chromosome Change.* Cambridge University Press, Cambridge.

Levin, D.A. (1979) The nature of plant species. *Science*, **204**, 381–4.

Lewis, W.H. (1980) Polyploidy in angiosperms: dicotyledons, in *Polyploidy: Biological Relevance* (ed. W.H. Lewis), Plenum Press, New York, pp. 241–68.

Linnaeus, C. (1753) *Species Plantarum*, Stockholm.

Lovis, J.D. (1977) Evolutionary patterns and processes in ferns. *Advances in Botanical Research*, **4**, 229–415.

Mabberley, D.J. (1987) *The Plant-Book.* Cambridge University Press, Cambridge.

Mayr, E. (1942) *Systematics and the Origin of Species.* Columbia University Press, New York.

Mayr, E. (1992) A local flora and the biological species concept. *American Journal of Botany*, **79**, 222–38.

Mishler, B.D. (1985) The morphological, developmental, and phylogenetic basis of species concepts in Bryophytes. *Bryologist*, **88**, 207–14.

Mishler, B.D. and Donoghue, M.J. (1982) Species concepts: a case for pluralism. *Systematic Zoology*, **31**, 491–503.

Ness, B.D., Soltis, D.E. and Soltis, P.S. (1989) Autopolyploidy in *Heuchera micrantha* (Saxifragaceae). *American Journal of Botany*, **76**, 614–26.

Newton, A. (1975) *Rubus* L., in *Hybridization and the Flora of the British Isles* (ed. C.A. Stace), Academic Press, London, pp. 200–6.

Nixon, K.C. and Wheeler, Q.D. (1990) An amplification of the phylogenetic species concept. *Cladistics*, **6**, 211–23.

Paterson, H.E.H. (1985) The recognition concept of species, in *Species and Speciation* (ed. E.S. Vrba), Transvaal Museum, Pretoria, pp. 21–9.

Raven, P.H. (1976) Systematics and plant population biology. *Systematic Botany*, **1**, 284–316.

Raven, P.H., Evert, R.F. and Eichhorn, S.E. (1986) *Biology of Plants*, 4th edn, Worth, New York.

Ray, J. (1686) *Historia Plantarum.*

Rieseberg, L.H. (1991) Homoploid reticulate evolution in *Helianthus*: evidence from ribosomal genes. *American Journal of Botany*, **78**, 1218–37.

Rieseberg, L.H., Beckstrom-Sternberg, S.M., Liston, A. and Arias, D.M. (1991) Phylogenetic and systematic inferences from chloroplast DNA and isozyme variation in *Helianthus* sect. *Helianthus* (Asteraceae). *Systematic Botany*, **16**, 50–76.

Rieseberg, L.H. and Soltis, D.E. (1991) Phylogenetic consequences of cytoplasmic gene flow in plants. *Evolutionary Trends in Plants*, **5**, 65–84.

Rosen, D.E. (1979) Fishes from the uplands and intermontane basins of Guatemala: revisionary studies and comparative geography. *Bulletin of the American Museum. Natural History*, **162**, 267–376.

Ruse, M. (1981) Species as individuals: logical, biological, and philosophical problems. *Brain and Behavioural Sciences*, **4**, 299–300.

Shi, Y., Gornall, R. J., Draper, J. and Stace, C.A. (in press) Intraspecific molecular variation in *Hieracium* L. sect. *Alpina* (Asteraceae), an apomictic group. *Folia Geobotanica et Phytotaxonomica.*

Shore, J.S. and Barrett, S.C.H. (1985) Morphological differentiation and crossability among populations of the *Turnera ulmifolia* L. complex (Turneraceae). *Systematic Botany*, **10**, 308–21.

Simpson, G.G. (1961) *Principles of Animal Taxonomy.* Columbia University Press, New York.

Sneath, P.H.A. and Sokal, R.R. (1973) *Numerical Taxonomy, the Principles and Practice of Numerical Classification.* W.H. Freeman, San Francisco.

Soltis, D.E. and Soltis, P.S. (1989) Tetrasomic inheritance in *Heuchera micrantha* (Saxifragaceae). *Journal of Heredity*, **80**, 123–6.

Soltis, D.E. and Soltis, P.S. (1993) Molecular data and the dynamic nature of polyploidy. *Critical Reviews in Plant Sciences*, **12**, 243–73.

Soltis, D.E., Soltis, P.S., Collier, T.G. and Edgerton, M.I. (1991) Chloroplast DNA variation within and among genera of the *Heuchera* group (Saxifragaceae): evidence for chloroplast transfer and paraphyly. *American Journal of Botany*, **78**, 1091–112.

Stace, C.A. (1989) *Plant Taxonomy and Biosystematics*, 2nd edn, Edward Arnold, London.

Stebbins, G.L. (1971) *Chromosomal Evolution in Higher Plants*, Edward Arnold, London.

Stevens, P.F. (1991) Character states, morphological variation, and phylogenetic analysis – a review. *Systematic Botany*, **16**, 553–83.

Styles, B.T. (ed.) (1986) *Infraspecific Classification of Wild and Cultivated Plants*. Clarendon Press, Oxford.

Templeton, A.R. (1989) The meaning of species and speciation: a genetic perspective, in *Speciation and its Consequences* (eds D. Otte and J.A. Endler), Sinauer Associates, Sunderland, Massachusetts, pp. 3–27.

Wendel, J.F., Stewart, J.M. and Rettig, J.H. (1991) Molecular evidence for homoploid reticulate evolution among Australian species of *Gossypium*. *Evolution*, **45**, 694–711.

Whittemore, A.T. and Schaal, B.A. (1991) Interspecific gene flow in oaks. *Proceedings of the National Academy of Sciences, USA*, **88**, 2540–4.

9

Cultivated plant diversity and taxonomy

J. G. Hawkes

Contacting address: School of Continuing Studies, The University of Birmingham, Edgbaston, Birmingham B15 2TT, UK

ABSTRACT

Although species diversity in cultivated plants is roughly similar to that of wild plants, infraspecific diversity is much greater in all but a few of them. This is undoubtedly due to the hand of humans, by selecting and hybridizing them and by moving them into environments far different from their original ones.

The standard taxonomic system used for wild species is also used for cultivated ones in general outline, but various systems have been used for infraspecific categories, bearing in mind that their evolution is largely driven by artificial human-directed selection.

The practice among cultivated plant taxonomists up to the 1940s and 1950s was to use a microspecies concept. This has now given way, under the influence of genetical and cytological thought, to the uniting of such microspecies into broadly based large species, to which breeders and agronomists can more easily relate.

9.1 INTRODUCTION

9.1.1 The basis of diversity and selection

Evolutionary change in cultivated plants is largely due to human selection, both conscious and unconscious. By unconscious selection is meant the pressures caused by particular agricultural systems of sowing, cultivated and harvesting, and the withdrawal of competition from other plants and animals in natural ecosystems.

Species: The Units of Biodiversity. Edited by M.F. Claridge, H.A. Dawah and M.R. Wilson. Published in 1997 by Chapman & Hall. ISBN 0 412 63120 2

Agricultural origins took place some 10 000 years ago, and undoubtedly conscious selection played a role from very early times. This selection, long before the advent of scientific plant breeding, involved:

- A search for greater yields.
- Easier and more efficient harvesting, e.g. non-shattering spikes and spikelets in cereals and non-dehiscing capsules in leguminous crops, flax, etc.
- Shorter stolons or roots in tuber crops, thus rendering harvesting much easier.
- A greater range of adaptation to soil, moisture and general climatic conditions as the crop was moved by man into different environments.
- Control of maturity time – long or short growing seasons, uniform maturity cycle, etc. thus ensuring that all seeds, tubers or fruits matured at the same time.

It must be recognized that the processes of conscious selection were minimal or even non-existent during the earlier years of domestication and farming, but that these gradually became more clearly defined as the centuries passed.

Over and above this pattern of human selection, the general processes of mutation and natural selection in cultivated plants also took place, as it continued to do in their wild ancestors.

As the farmers and their newly domesticated crops moved into other areas, these crops were no doubt brought into contact with different subspecies of their wild ancestors and even related wild species. Natural hybrids and gene flow from such related subspecies and species would have taken place and these would have added to the genetic diversity of the crop itself.

As if these processes were not enough to promote morphological and genetic diversity in the cultivated species, farmers, and later, plant breeders have also been consciously hybridizing and selecting cultivated plants, thus adding to their diversity.

All these processes have tended to provide much more complex patterns of variation in cultivated plants than in wild ones. Species concepts have become even less clear than in some completely wild species, which are themselves often bedevilled with polyploidy, pillar complexes and other untidy situations which often render the task of the cultivated plant taxonomist very difficult.

9.2 POLYPLOIDY

Polyploids occur in cultivated plants under primitive agriculture, and are not by any means freaks of wide artificial hybridization. It has been stated on many occasions that polyploidy is more common in cultivated than in wild species. However, out of 40 widely grown species of field and tree crops, 20 are completely diploid, while the rest are tetraploid, diploid and tetraploid, or possess a range of ploidy levels (Table 9.1). Certain fruit trees, such as apples and pears are basically diploid but possess triploid cultivars. Several crops possess diploid and tetraploid species (cherries, coffee, cotton), while others such as wheat, oats, potatoes and plums, possess a wider range, generally having developed these by ancient natural hybridization and chromosome doubling. Highly variable ploidies occur in sugar cane and yams (Hawkes, 1983: 23), but these are exceptions to the general rule, with the wide ploidy range seeming to be preserved only by means of vegetative reproduction.

9.3 SPECIES CONCEPTS IN CULTIVATED PLANTS

As with wild species there are some taxonomists of cultivated species who see and describe new species for every recognizable variant – the so-called splitters – and on the other hand, there are those who try to group together as many variants as they can – the so-called lumpers. Horticulturalists were certainly more likely in the past to see a new species in every

Table 9.1 Levels of ploidy in certain widely grown species of field and tree crops*

Diploids	*Polypoloids*
Almond	Apple (2×, 3×)
Barley	Banana (2×, 3×, 4×)
Cabbage	Canna (2×, 3x)
Chickpea	Cassava (4×)
Cocoa	Cherry (2×, 4×)
Coconut	Coffee (2×, 4×)
Cocoyam	Cotton (2×, 4×)
Coix	Curcuma (2×, 3×)
Ginger	Oats (2×, 4×, 6×)
Lentil	Peanut (4×)
Maize	Pear (2×, 3×)
Oil palm	Plum (2×, 4×, 6×)
Olive	Potato (2×, 3×, 4×, 5×)
Peach	Rose (2×, 3×, 4×, 5×, 6×)
Phaseolus beans	Sugar cane (variable)
Pineapple	Sweet potato (6×)
Rice	Taro (2×, 3×)
Rye	Tobacco (4×)
Tea	Wheat (2×, 4×, 6×)
Tomato	Yam (variable)

*From *The Diversity of Crop Plants* (Hawkes, 1983)

variant, particularly if it appeared to breed true. This, of course, can also happen with groups of wild species, though the temptation to create new species of horticultural or agricultural plants seems to have been yielded to with greater abandon in the past.

Thus, de Wet *et al.* (1986) show that the cultivated sorghums were once divided by Snowden into 28 species, 156 taxonomic varieties and 521 forms. All of these are now clearly seen to be conspecific, under the single species *Sorghum bicolor*. Similarly, N.I. Vavilov and his co-workers in the 1920s and 1930s divided the known wheat species into one diploid (*Triticum monococcum*, wild and cultivated), eight tetraploids (of which one, *T. dicoccoides*, is wild); and four hexaploids (including *T. aestivum*). All these were allotted innumerable varieties and forms, all with their Latin names (see Vavilov – English translation, 1992, of Russian compilation, 1987). It is now considered best to retain three species *T. monococcum* (2×), *T. dicoccum* (4×), and *T. aestivum* (6×), with all the diverse agricultural variants given no more than vernacular names. The same solution has taken place with other cereals, such as barley, rye and oats.

The cultivated potato has gone through a similar sequence of events. Linnaeus in 1768 named it *Solanum tuberosum*, and thus it was continued to be referred to until 1929 (though *S. esculentum* Neck. was proposed in 1768, and *S. sinense* by Blanco in 1837). However, as a result of the Russian expeditions to South America by S.V. Juzepczuk and S.M. Bukasov in 1927 a polyploid series was discovered, ranging from diploid ($2n = 24$), triploid ($2n = 36$), tetraploid ($2n = 48$) and pentaploid ($2n = 60$) – all cultivated materials. They split the tetraploid *S. tuberosum* into an Andean and a Chilean species, and in addition made six diploid, five triploid and one pentaploid species. After I returned from my first collecting expedition to South America I regret to say, that with an enthusiasm which I have since learned to regret, I added four diploid species, five diploid varieties, one triploid species and one triploid variety, after the fashion of the Russians. The Russians themselves had described among the Chilean tetraploids nine varieties and 28 forms, while they split the Andean tetraploids into two species, 21 varieties and 55 forms (Table 9.2). Ultimately, this sort of classificatory system became of little or no practical use, since natural hybridization between clones can lead to an almost limitless number of genotypes.

The simplified classification shown in Table 9.2 has been helped considerably by the experimental recreation of the polyploid species and even one of the diploids (see Figure 9.1 and Hawkes, 1990). From this it has

Table 9.2 Russian and other infraspecific classifications of cultivated potatoes

Ploidy level	*Species now recognized*	*Previous classification*		
		Species	*Varieties*	*Forms*
	S. ajanhuiri	*S. ajanhuiri*	2	
	S. phureja	*S. phureja*	4	
	(two subspecies)	*S. rybinii*		
		S. boyacense		
2×		*S. kesselbrenneri*		
(2n = 24)		*S. cardenasii*		
		S. ascasabii		
		S. caniarense		
	S. stenotomum	*S. stenotomum*		
	(two subspecies)	*S. goniocalyx*		
		S. churuspi		
		S. yabari		
	S. juzepczukii	*S. juzepczukii*	2	
3×	*S. chaucha*	*S. chaucha*		3
(2n = 36)		*S. tenuifilamentum*		
		S. mamilliferum		
		S. coeruleiflorum		
	S. tuberosum	*S. tuberosum*	17	37
	subsp. *tuberosum*	*S. esculentum*		
		S. sinense		
		S. chiloense		
		S. cultum		
		S. sabinii		
4×		*S. molinae*		
		S. leptostigma		
		S. diemii		
		S. sanmartinense		
		S. oceanicum		
		S. ochoanum		
		S. zykinii		
	S. tuberosum	*S. andigena*	21	55
4×	subsp. *andigena*	*S. herrerae*		
		S. apurimacense		
5×	*S. curtilobum*	*S. curtilobum*		

become fairly clear from experimental studies that the diploid cultivated species *S. stenotomum* was the first to be cultivated and that natural hybridization, followed by unconscious selection gave rise to all the others. Undoubtedly, environmental pressures and isolation gave rise to subspecies *tuberosum* in Chile; and environmental selection for frost resistance at high altitudes 'fixed' the hybridogenic species *S. ajanhuiri, S. juzepczukii* and *S. curtilobum.*

We have now arrived at a stage where the complex series of varieties and forms named and described by Bukasov, Juzepczuk and Lechnovich no longer matter. They played an important role in the 1920s and 1930s by highlighting the value of the immense diversity of potatoes in the Andean mountains and in southern Chile. Now, however, they are no longer of very great importance. In fact, the process of simplification was taken even further by Dodds (1962) who used a group classification (Table 9.3), under a single species, *S. tuberosum* (*sensu lato*).

As one can see, this corresponds very closely with the present author's species, classification, apart from subgroup IIa – Amarilla, which is no more than the northern yellow-fleshed forms of *S. phureja.*

9.4 CONCLUSIONS

What, then, are the units of biodiversity in cultivated plants and how do they differ from those of wild plants? To a large extent they depend on the breeding system. If the species are inbreeding, such as the Old World cereals, Linnean systems may generally be useful, though there is a danger of

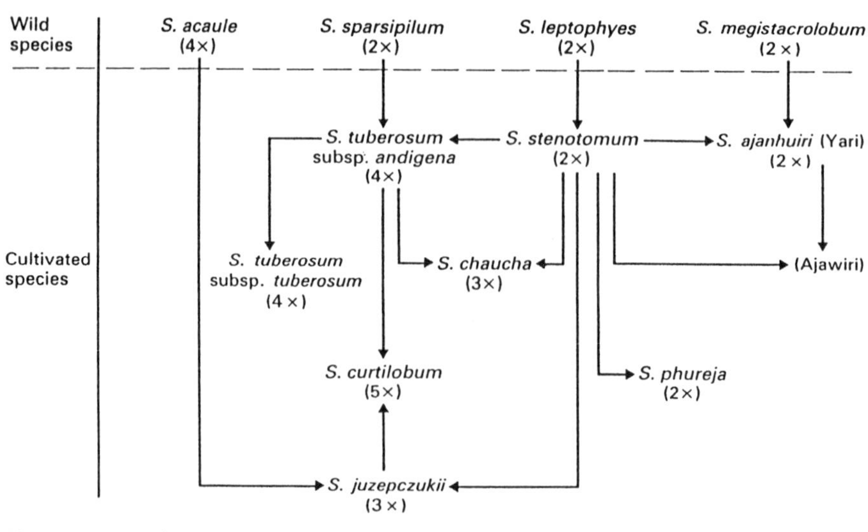

Figure 9.1 Evolutionary relationships of cultivated potatoes and their ploidy levels. (Adapted from Hawkes, 1990, with kind permission of Belhaven Press.)

Table 9.3 K.S. Dodds' classification of *S. tuberosum* L. (*sensu lato*) with ploidy levels and present classification added

*Dodds' groups**		*Ploidy*	*Species equivalent*
I	Group Stenotomum	2×	*S. stenotomum* Juz. *et* Buk. *S. ajanhuiri* Juz. *et* Buk.
I-A	Subgroup Goniocalyx	2×	*S. stenotomum* Juz. *et* Buk. subsp. *goniocalyx* (Juz. *et* Buk.) Hawkes
II	Group Phureja	2×	*S. phureja* Juz. *et* Buk.
IIa	Subgroup Amarilla	2×	Part of *S. phureja* Juz *et* Buk.? (not clarified in text)
III	Group Chaucha	3×	*S. chaucha* Juz. *et* Buk.
IV	Group Andigena	4×	*S. tuberosum* L. subsp. *andigena* Hawkes
V	Group Tuberosum	4×	*S. tuberosum* L. subsp. *tuberosum*
VI	Solanum × juzepczukii Buk.	3×	*S. juzepczukii* Buk.
VII	Solanum × curtilobum Juz. *et* Buk.	5×	*S. curtilobum* Juz. *et* Buk.

*From Dodds (1962: 517–39.)

over-complication when no more than different geographical or altitudinal races are considered to be distinct species. Some attention should be paid to ploidy levels, and presence or absence of crossability barriers should be observed in these cases. Sometimes the wild prototypes (as with *Triticum boeoticum/monococcum*) need to be considered as the same Linnean species, with perhaps subspecies differentiation only. In other cases, as with *Zea mays* and its wild prototype *Zea* (or *Euchlaena*) *mexicana*, specific or even generic distinctions need to be made. With *Solanum tuberosum*, although this author believes it to be a natural hybrid of *Solanum stenotomum* × *S. sparsipilum*, others are less convinced, and consider it to be an autotetraploid of its ancestral diploid, *S. stenotomum*. In this instance it is clearly better to use the Linnean name rather than give it a hybrid designation.

As I have stressed before (Hawkes, 1986), in cultivated plants the lowest taxonomic rank may not be the cultivar as some workers have considered (e.g. Parker, 1978). Under primitive agriculture the cultivar hardly exists, and its more elemental place is taken by the population (or 'pseudocultivar'). This still applies in many parts of the world today, as for instance in Ethiopia (Worede, 1991), where in certain regions cultivars do not exist as such but are preceded by populations, even in predominantly inbreeding species such as barley (*Hordeum*) and tetraploid wheats.

Vegetatively propagated crops such as sweet potatoes, cassava and potatoes are propagated in developed countries as clones and may be classified as such. However, in Third World countries in and around the centres of diversity of these crops there may be considerable outcrossing. Potatoes in any case are mostly allogamous and new seedlings which arise naturally will clearly possess different gene combinations from their parents. Hence, a clonal classification will clearly be unacceptable. Here, as we have seen earlier, a broad classification based on general species delimitations, as with wild species, is the most appropriate one to adopt.

These broadly based species and subspecies units must surely be the generally acceptable unit of diversity in most cultivated plants. Nevertheless, in largely inbreeding or in entirely clonally reproducing apomictic species, then the 'pseudoclone' or inbred population, or even the clones themselves may feature as the most easily recognizable units.

9.5 REFERENCES

Dodds, K.S. (1962) Classification of cultivated potatoes, in *The Potato and its Wild Relatives* (ed. D.S. Correll), Texas Research Foundation, Renner, Texas.

Hawkes, J.G. (1983) *The Diversity of Crop Plants*, Harvard University Press, Cambridge, Massachusetts, USA.

Hawkes, J.G. (1986) Infraspecific classification – the problems, in *Infraspecific Classification of Wild and Cultivated Plants* (ed. B.T. Styles), Clarendon Press, Oxford.

Hawkes, J.G. (1990) *The Potato. Evolution, Biodiversity and Genetic Resources*, Belhaven (Pinter) Press, London.

Parker, P.F. (1978) The classification of crop plants, in *Essays in Plant Taxonomy* (ed. H.E. Street), Academic Press, London.

Vavilov, N.I. (1992) *Origin and Geography of Cultivated Plants*. Translated by Doris Love, Cambridge University Press, from original compilation by V.F. Dorofeyev and A.A. Filatenko, 1987, Nauka.

de Wet, J.M.J., Harlan, J.R. and Brink, D.E. (1986) Reality of infraspecific taxonomic units in domesticated cereals, in *Infraspecific Classification of Wild and Cultivated Plants* (ed. B.T. Styles), Clarendon Press, Oxford.

Worede, M. (1991) An Ethiopian perspective on conservation and utilization of plant genetic resources, in *Plant Genetic Resources of Ethiopia* (eds J.M.M. Engels, J.G. Hawkes and M. Worede), Cambridge University Press, Cambridge.

10

Species of marine invertebrates: a comparison of the biological and phylogenetic species concepts

N. Knowlton and L. A. Weigt

Contacting address: Smithsonian Tropical Research Institute, Apartado 2072, Balboa, Republic of Panama

ABSTRACT

Traditional morphological approaches have substantially underestimated species diversity in living marine invertebrates, as assessed by either the biological or phylogenetic species concepts. The degree to which these concepts succeed individually or agree with each other depends on the group and the biogeographic context. In sympatry, the biological and phylogenetic species concepts should yield the same result, because reproductive incompatibility implies at least one diagnostic difference between isolated forms. Hybridization in sympatry between partially isolated forms may be a problem for both species concepts in some groups, although evidence for this in the field is limited. It is often difficult to find qualitative morphological differences between forms that can be unambiguously recognized by other characters, so that a phylogenetic species concept that depends on morphological characters will miss many reproductively isolated forms. In allopatry, the differences between the two species concepts are potentially much greater, because the phylogenetic species concept has the potential to recognize any diagnosably distinct population at the species level, regardless of its triviality. In groups with extensive dispersal ability, there may be predictable relationships between genetic and reproductive divergence that allow taxonomic decisions to be made using either species concept. In groups with limited dispersal ability and a propensity for founder events and local extinction, substantial and complex patterns of genetic variation will prove challenging for both species concepts. The phylogenetic species concept makes no special distinction between species and higher level

Species: The Units of Biodiversity. Edited by M.F. Claridge, H.A. Dawah and M.R. Wilson. Published in 1997 by Chapman & Hall. ISBN 0 412 63120 2

taxa, while the biological species concept places species at the boundary between reticulation and cladogenesis. This appears to be an important and well-defined boundary for many marine invertebrates, and thus merits special taxonomic recognition.

10.1 INTRODUCTION

Morphologically defined species remain the rule in nearly all groups of marine invertebrates, but recent work has revealed that many so-called species are in fact complexes of taxa that can be most readily distinguished using genetic, behavioural or ecological characters (Knowlton, 1993). Nevertheless, most of these sibling species exhibit subtle (and sometimes not so subtle) morphological differences that were previously ascribed to intraspecific variation, based on a priori assumptions of wide geographic range or extensive non-genetic plasticity.

How much variability should a single species encompass? The taxonomic response to this question will depend upon the nature of the species concept to be employed. Two classes of options currently dominate the literature (Claridge *et al.*, 1997: Chapter 1). The first class is epitomized by the biological species concept (Mayr and Ashlock, 1991). Species are defined in principle on the basis of reproductive compatibility, although indirect evidence for the existence of reproductive barriers marking species boundaries is acceptable with this approach (Avise and Ball, 1990). Templeton's (1989) cohesion species concept is related in accepting the biological species concept for those groups for which it works well, but it also utilizes criteria that can be applied to groups that are asexual, subdivided, or that hybridize extensively. The second class of species concepts consists of a group of cladistically based approaches, all of which have been loosely referred to as the phylogenetic species concept (Mishler and Theriot, 1997). In its least restrictive usage (Cracraft, 1989, 1997: Chapter 16), a phylogenetic species is simply the minimum diagnosable taxonomic unit based on any qualitative character. Some workers, however, insist that minimum units must be strictly monophyletic and thus defined by derived characters (reviewed by Smith, 1994), and others argue that some monophyletic groups are too trivial to merit recognition at the species level (Mishler and Theriot, 1997).

The purpose of this chapter is to compare the implications of these two classes of approaches for the definition of species in marine invertebrates in practice, without getting overwhelmed by the alpha taxonomy of the species concepts themselves. For simplicity, we will focus primarily on the biological species concept of Mayr (Mayr and Ashlock, 1991) and Cracraft's (1989, 1997: Chapter 16) version of the phylogenetic species concept. We shall not consider the special problems raised by obligately clonal or selfing life-histories, since these are rare in marine invertebrates.

The comparison between the biological and phylogenetic approaches is clarified by considering sympatric and allopatric taxa separately. In sympatry, the two approaches make the same recommendations in principle, for taxa that are reproductively isolated in sympatry are also by definition diagnosably distinct in at least the character that generates the isolation. However, as the title of this symposium (*The Units of Biodiversity: Species in Practice*) implied, the translation of theory into practice is not always straightforward. Reproductively isolated, sympatric taxa can generally be distinguished by ecological, behavioural, genetic or morphological differences, but these differences may be quantitative rather than qualitative, and thus not diagnosable in practice by the criteria of phylogenetic species concepts. Moreover, reproductive isolation is itself a quantitative character when isolation is not complete, and can thus be a problem for both species concepts in deciding the status of morphs that are partially interfertile in sympatry. In allopatry, the differences between these approaches are potentially much more marked, since for the phylogenetic species concept any geographically isolated population with a diagnosably distinct characteristic can be recognized at the species level. Founder populations begun by a small number of individuals may often be diagnosably distinct at the genetic level, and thus would be species using the approach favoured by Cracraft (1989), regardless of the triviality of the difference.

10.2 CORALS AND SHRIMPS AS CASE STUDIES

The differences between the biological and phylogenetic species concepts for practical taxonomy are best evaluated by considering real organisms rather than hypothetical situations. Below we discuss the implications of these approaches for corals and snapping shrimp. These two groups share a bad reputation among practising taxonomists trying to define species, but the problems they pose for these alternative species concepts are quite different.

10.2.1 Reef-building corals and the problem of quantitative characters

With few exceptions, alpha taxonomy of corals is entirely based on skeletal characters (but see Lang, 1984). Most reef-building corals are colonial, and characteristics of the skeletal cups secreted by individual polyps, called corallites, play a major role in defining species boundaries in many genera (Budd *et al.*, 1994a). Because corallites often vary considerably within a colony (a genetic individual), species boundaries were traditionally defined to encompass much of this variation (Wood Jones, 1907). Colony growth form is also known in some cases to exhibit phenotypic plasticity in response to wave energy and light level (Willis, 1985), the latter a consequence of the obligate symbiosis between reef-building corals and photosynthetic dinoflagellates. These features have led most coral taxonomists to accept large amounts of intraspecific variation at the level of both the corallite and the colony.

The problems with this approach are well illustrated by the coral *Montastraea annularis*. This 'species' is the most abundant and best studied coral on Caribbean reefs. It exhibits enormous variability in colony morphology, ranging from massive heads to columns to plates. Computer models and transplant experiments designed to evaluate the effect of depth on growth form (Graus and Macintyre, 1976; Dustan, 1979) made it a text-book case of the importance of phenotypic plasticity. Nevertheless, the different morphologies can be found side by side, and intermediates between the major types of colony morphology are rare (Graus, 1977).

Recently, a number of features have been found that are strongly concordant with the different patterns of colony growth. These include corallite morphometrics, skeletal density and growth rate, stable isotope signatures, aggressive behaviour, allozymes, and reproductive biology (Knowlton *et al.*, 1992, and in press; Van Veghel and Bak, 1993; Van Veghel and Kahmann, 1994; Weil and Knowlton, 1994; A.F. Budd and N. Knowlton, in preparation). However, most of these differences, including those likely to be useful in identifying skeletal material, are quantitative rather than qualitative.

The response of coral systematists to these findings has been mixed. Because numerous and presumably unrelated characters were concordant, Weil and Knowlton (1994) resurrected two previously synonomized names to recognize the three common shallow water forms. Van Veghel and Bak (1993), in contrast, referred to them as morphotypes rather than species, based on the absence of qualitative differences in the same set of data. Reviewing this situation, Veron (1995: 33) concluded that there may be no correct answer because species are arbitrary. The species level taxonomy of many other corals is comparably problematic (Veron, 1995).

There are several general and somewhat inter-related explanations for this failure to find reliable qualitative differences. First, reproductively isolated species may be genuinely similar in genetic or morphological characters, either because divergence is slow or because the species are relatively young. Recently or slowly diverging species may still share the same polymorphisms across much of the genome, so that finding qualitative or fixed differences can be equivalent to searching for a genetic needle in a haystack (Avise and Ball, 1990). Romano (1995), for example, found comparatively little genetic divergence even between groups whose fossil record indicates a long history of isolation, suggesting that the molecular 'clock' of corals ticks slowly. Some corals have long generation times, which other things being equal, should act to slow the rate of divergence (Gillespie, 1991).

Second, coral species, like plants, may be particularly prone to hybridization (Veron, 1995). The morphological simplicity of corals may underlie the successful production of hybrids in the laboratory (Wallace and Willis, 1994), and the annual simultaneous spawning of congenerics documented for many corals could provide regular opportunities for interspecific hybridization in the field (Wallace and Willis, 1994; Veron, 1995). Even when hybrids are less fit than the products of intraspecific matings, selection to avoid such matings is ultimately constrained by the fact that even a severely inferior hybrid is fitter than an unfertilized egg.

However, it should be emphasized that hard evidence for rampant natural hybridization in corals is lacking. Indeed, the three described species of the *M. annularis* species complex show considerable overlap in the date of spawning, but appear nevertheless to be reproductively isolated by a combination of barriers to fertilization or development and differences in the hour of peak spawning (Knowlton *et al.*, in press). Moreover, in *Acropora*, there are several examples of incompatibility between morphologically similar colonies that are supposed to be conspecific (Wallace and Willis, 1994), suggesting that the importance of breeding barriers may have been underestimated as well as overestimated.

Molecular techniques have revolutionized the detection of sibling species in many groups, but in corals, apart from protein electrophoresis, they have been little used due to technical difficulties. These difficulties are likely to be resolved in the coming years, and clearly defined species may emerge from these studies. On the other hand, if closely related species of corals diverge slowly and hybridize frequently, there may be no magic molecular bullet for recognizing hybrids or defining species, even in sympatry.

In summary, corals can be problematic for a variety of species concepts, even in sympatry, due to the scarcity of diagnostic differences and a poorly understood potential for hybridization. In the absence of a consensus in sympatry, the situation in allopatry or even in different reef habitats is, of course, more problematic. In the *M. annularis* species complex, for example, deeper water forms converge on a plate-like morphology, making it difficult to distinguish species or evaluate relationships between shallow and deeper water forms. The three basic morphologies can be recognized throughout the Caribbean, but show differences from site to site which are also hard to interpret. Veron (1995) described many comparable situations for Pacific corals.

10.2.2 Snapping shrimp and the problem of allopatric variation

Species of snapping shrimp in the genus *Alpheus* (Alpheidae) are numerous and often difficult to distinguish using conventional morphological characters in preserved material. In sympatry, however, many morphologically similar species can be readily identified by their colour patterns when alive (Knowlton and Mills, 1992), as well as by fixed allozyme differences and substantial divergence in mitochondrial DNA (mtDNA) sequences (Knowlton *et al.*, 1993 and unpublished data). It is also possible directly to assay reproductive cohesion, because males and females are often found paired, and are usually aggressive to all individuals except potential mates. Interspecific pairings are not observed in the field and are almost impossible to establish in the laboratory, suggesting that hybridization is uncommon. Thus, in contrast to corals, both the biological and phylogenetic species concepts have little difficulty in defining species boundaries in sympatry when the appropriate characters are employed.

Neotropical alpheids from opposite sides of the Isthmus of Panama are particularly useful for exploring the nature of species in allopatry. The Isthmus now divides what was once a continuous tropical sea, and the faunas of the eastern Pacific and Caribbean contain numerous examples of transisthmian sister taxa (Knowlton *et al.*, 1993). Detailed geological and palaeontological analyses provide strong support for dating the final closure of the Isthmus at approximately 3 million years ago (Coates *et al.*, 1992). While the biological history of the region is more complex than commonly assumed (see below), it nevertheless represents the best marine example of the classic form of geographic isolation – one large population being divided into two still large populations that subsequently slowly diverge through time.

Seven pairs of transisthmian sister taxa have been examined using a suite of molecular and behavioural techniques (Knowlton *et al.*, 1993). They show considerable but concordant variation in the degree of divergence exhibited, suggesting that some pairs were isolated more recently than others. Although the most similar transisthmian pairs show little aggression when males from one ocean are placed with females from the other ocean, they almost never produce fertile clutches of eggs. This contrasts with the routine production of fertile clutches when conspecifics from the same ocean are paired under identical laboratory conditions. The general conclusion to be drawn from these experiments is that 3 million years is long enough to result in substantial reproductive isolation in these organisms.

Can we use the information obtained from transisthmian pairs to interpret other patterns of divergence in allopatry where it has not been possible to study reproductive compatibility directly? This appears to be a reasonable procedure for two reasons. First, the relatively large number of transisthmian pairs studied strengthens our confidence in the relationship between molecular and reproductive divergence. Second, the classical or dumb-bell mode of geographic isolation is probably the most sluggish way of generating new species, based on theory (Templeton, 1981), empirical observations in other groups (Coyne and Orr, 1989), and limited observations for the alpheids themselves (unpublished data). Thus, we may underestimate reproductive isolation for a given level of molecular divergence by using the transisthmian data as a null model, but we are not likely to overestimate it.

The widely scattered island groups of the Pacific provide an appropriate arena for comparing the biological and phylogenetic species concepts in a more complex pattern of allopatry. To this end, we here summarize preliminary data on *Alpheus lottini*, an obligate symbiont of the corals *Pocillopora* and *Stylophora* that ranges in distribution from the eastern Pacific to the Red Sea. It is a conspicuous and well-studied snapping shrimp, but no attempt has previously been made to evaluate variability across its wide range.

Examination of living material from a variety of sites across the Pacific immediately revealed two clearly distinct colour patterns: animals were characterized by either a solid black stripe down the carapace and paired blotches down the abdomen (type A), or by speckles scattered across both regions (type B). Limited collections from Panama, Pohnpei, Clipperton and Moorea contained only type A, those from Hawaii and Guam only type B, while both types were found in Palau (mated assortatively).

Mitochondrial DNA sequences from types A and B (for the same gene region as was studied for transisthmian taxa; see caption of Table 10.1 for methods) had an average corrected sequence divergence of 10–13% (Table 10.2). This corresponds to an estimated time of divergence of 4–5 million years, based on the transisthmian taxa (Figure 10.1). This degree of sequence divergence was always associated with reproductive isolation across the Isthmus, and the finding of assortatively mated pairs at the one site where they were found together (Palau) also supports an interpretation of strong reproductive isolation. Thus, both the biological and phylogenetic species concepts would recognize types A and B as specifically distinct.

The molecular data also reveal more subtle allopatric structuring within these two types, however. The clearest is that between the Hawaiian and other type B taxa, which differ by about 4.5% in corrected sequence divergence. This is less than that seen between seven transisthmian sister species (Knowlton *et al.*, 1993), but still substantial, particularly in light of the fact that even the least divergent of these (6.6%) never successfully interbred in laboratory experiments. From the perspective of the biological species concept, these taxa are likely to be specifically distinct, but explicit tests of their compatibility to further refine the relationship between genetic and reproductive divergence at the low end of the divergence scale are necessary. For the phylogenetic species concept, however, there is no ambiguity, because the Hawaiian shrimp can be distinguished genetically from the other type B shrimp (20 transitions and three silent transversions

Table 10.1 Aligned sequence haplotypes for variable positions within the 564 bp area of analysis of the CO1 mtDNA gene of *Alpheus lottini* complex (parentheses indicate multiple individuals with same halotype)

	000000000000000000111111111122222222222222222223333333333344444444444444555555555555
Position	011222333445666799900112467890122233445778889900133455677003344567788990011223344 55
	628147039281369503628470421984625814087392581409203245058585814321709581706581403 25
B-PALAU1	AATGCCCTAGTCAAGGACGGTAAGATGTCTCTCACGAGAACCCGCTAGCAGTATATCACGGCCCTACACTCCAAACCCGTGCG
B-PALAU2	....
B-GUAM1	................................T.......T..
B-GUAM2	................................T..
B-GUAM3 (2)	........G...............................T..
B-HAWAII1 (2)	G..A..T.G...G.AA....C..A..A.....TG...A.........AT.A.T.GC.G.CT..........T......A.A..
B-HAWAII2	G..A..T.G.C.G.AA....C..A..A.....TG...A.........AT.A.T.GC.G.CT..........T......A.A..
B-HAWAII3	G..A..T.G...G.AA....C..A..A.....TG...A.........AT.A.T.GC.G.CT..........T......A.A..
A-PALAU1	.GCATAGCCACTGGAAGTA.CGGAGCAATCTG...AG.GG.TTCACGA.GACCC.CTG.CCAT.CG.CGCA.GG...TAC..A
A-PALAU2	.GCATAGCTACTGGA..TA.CGGAGCAATCTG...AG.GG.TTCACGA.GACCC.CTG.CCAT.CGTCGCA.GG...TAC..A

```
A-POHNPEI1      .GCATAGCCACTGGA..TA.CGGAGCAATCTG...AG.GG.TTCACGA.GACCC.CTG.CCAT.CG.CGCA.GG...TAC..A
A-POHNPEI2      .GCATAGCCACTGGA..TA.CGGAGCAATCTG...AG.GG.TTCACGA.GACCC.CTG.CCAT.CG.CGCA.GG...TAC..A
A-POHNPEI3      .GCATAGCCACTGGA.?TAACGGAGCAATCTGT..AG.GG.TTCACGA.GACCC.CTG.CCAT.CG.CGCA.GG...TAC..A
A-MOOREA1       .GCATAG.CACT.....TA.CGGAGCAAT.TG....G..G.TTCAC.A..ACCC.CT..CCAT.CG.CGCA.GG...TAC..A
A-MOOREA2       .GCATAG.CACT.....TA.CGGAGCAAT.TG....G..G.TTCAC.A..ACCC.CT..CCAT.CG.CGCA.GG..T.AC..A
A-MOOREA3       .GCATAG.CACT.....TA.CGGAGCAAT.TG....G..G.TTCAC.A..ACCC.CT..CCAT.CG.CGCA.GGG..TAC..A
A-CLIPPERTON1   .GCATAG.CACT.....TA.CGGAGCAAT.TG....G..G.TTCAC.A..ACCC.CT..CCAT.CG.CGCA.GGG..TAC..A
A-CLIPPERTON2   .GCATAG.CACT.....TA.CGGAGCAAT.TG....G..G.TTCAC.A..ACCC.CT..CCATT.GTCGCA.GGGT.TAC.TA
A-PANAMA1       .GCATAG.CACT..T.GTA.CGGAGCAAT.TG..TAG..G.TTCAC.A..ACCC.CT.TCCAT.CG.CGCA.GGG..TAC..A
A-PANAMA2(2)    .GCATAG.CACT..T.GTA.CGGAGCAAT.TG...AG..G.TTCAC.A..ACCC.CT..CCAT.CG.CGCA.GGG..TAC..A
A-PANAMA3       .GCATAG.CACT..T.GTA.CGGAGCAAT.TG...AG..G.TTCAC.A..ACCC.CT..CCAT.CG.CGCA.GG...TAC..A
```

Dots represent concordance with the top sequence. Mitochondrial DNA methods as in Knowlton *et al.* 1993 with the following modifications: primers (5′–3′ positions in the amplification product) PCR and sequencing = COI (1–20) and *Alpheus* specific primer 1 [5′ CAT TTA GGC CTA AGA AGT GTT G 3′ (619–640)]; internal sequencing: light strand – ALP7 [TGA CTT GGA ACC CTC CAT GG 3′ (304–323)] and ALP5 [5′ ATT GCY CAC TGA TTC CCC YTA TT 3′ (514–536)]; heavy strand – ALP2 [5′ CCR TGG AGG GTT CCR AGT CA 3′ (304–323)]. Double stranded sequencing as in method 2 of Kessing *et al.* (1989). Sequences in Genbank (u76428–u76455).

Table 10.2 Kimura (1980) corrected percent sequence divergence values* (calculated by Sequencer 3.0; Kessing, 1995) averaged by site for two types of *Alpheus lottini*

	B-PALAU	B-GUAM	B-HAWAII	A-PALAU	A-POHNPEI	A-MOOREA	A-CLIPPERTON	A-PANAMA
B-PALAU	1 0.0 XXX–XXX							
B-GUAM	8 0.4 0.2–0.4	6 0.3 0.2–0.5						
B-HAWAII	8 4.7 4.6–4.8	16 4.6 4.4–4.8	6 0.1 0.0–0.2					
A-PALAU	4 12.7 12.6–12.8	8 13.0 12.8–13.2	8 11.8 11.5–12.0	1 0.7 XXX–XXX				
A-POHNPEI	6 12.5 12.4–12.8	12 12.7 12.6–13.0	12 11.7 11.5–11.8	6 0.5 0.4–0.7	3 0.3 0.0–0.4			
A-MOOREA	8 10.3 10.2–10.4	12 10.5 10.4–10.8	12 10.7 10.6–11.1	6 2.4 2.2–2.5	9 2.1 1.8–2.6	3 0.4 0.2–0.5		
A-CLIPPERTON	4 10.8 10.4–11.1	8 11.0 10.6–11.5	8 11.4 10.9–11.7	4 2.8 2.4–3.3	6 2.5 2.0–3.3	6 0.7 0.0–1.4	1 0.9 XXX–XXX	
A-PANAMA	8 11.1 10.8–11.5	16 11.3 11.0–11.9	16 11.5 11.1–11.9	8 2.2 1.8–2.7	12 2.1 1.8–2.5	12 0.8 0.5–1.4	8 1.1 0.5–1.8	6 0.3 0.2–0.5

Top value is number of pairwise comparisons, bottom values indicate range. See text for details.

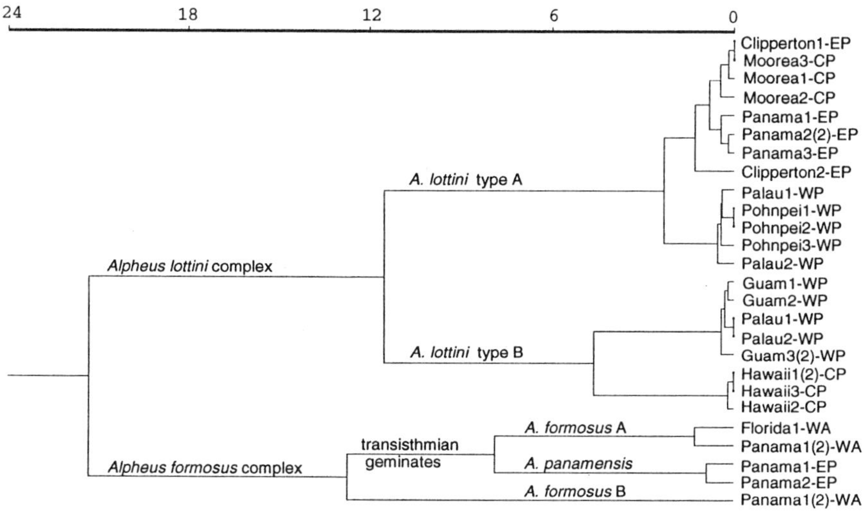

Figure 10.1 UPGMA tree for *Alpheus lottini* and the *A. formosus* sister clade (unpublished data) complex within the genus. The latter includes a pair of geminate (transisthmian) species that apparently diverged at the final closure of the Isthmus approximately three million years ago (Knowlton *et al.*, 1993). EP, eastern Pacific; CP, central Pacific; WP, western Pacific; WA, western Atlantic.

at third position sites; Table 10.1). The Hawaiian individuals also have a distinct egg colour that would permit their recognition on that basis.

The potential conflict between the two species concepts becomes clearer as we consider more similar populations. Within the type A form, maximum corrected sequence divergence between sites is only 2.8% (Table 10.2). This value is substantially less than that exhibited across the Isthmus and is not much more than apparently intraspecific differences between Panama and Florida or even within Panama (Figure 10.1). Nevertheless, within the type A clade, fixed third position genetic differences distinguish a Palau/Pohnpei taxon (nine transitions), a Panama taxon (one silent transversion), and a Clipperton–Moorea taxon (two transitions) (Table 10.1). If confirmed by more extensive sampling, this would result in the elevation of each of these to specific status using the phylogenetic species concept of Cracraft (1989, 1997: Chapter 16), but the slight sequence divergence would argue against such a distinction using the biological species concept in the absence of other evidence.

Thus, the biological and phylogenetic species concepts provided rather different results when applied to variability in allopatric *Alpheus*, although both support recognition of several taxa that had been missed in traditional morphological analyses. Their treatment of the transisthmian taxa is comparable, for Pacific and Caribbean forms are both reproductively isolated and diagnosably distinct. For the pan-Pacific *A. lottini*, however, the biological species concept clearly supports the existence of only two or three species, while the phylogenetic species concept would recognize five based on the samples available (Figure 10.2).

10.3 OTHER MARINE INVERTEBRATES

In the discussion above we focused on two genera and ignored the vast majority of marine invertebrates. While we cannot do justice in a brief review to the diversity represented by over 30 phyla, we can try to assess the generality of the patterns we highlighted.

10.3.1 Quantitative characters in sympatry

The absence of qualitative morphological differences between closely related species is a fact of life for many groups of marine invertebrates, including those that are clearly reproductively isolated (Langan-Cranford and Pearse, 1995). Multivariate analyses of quantitative morphological characters (Budd *et al.*, 1994a; Jackson and Cheetham, 1994) often clearly define taxa, however, and molecular techniques can confirm their status. This approach is compatible with the phylogenetic species concept *sensu* Cracraft (1989) when fixed genetic differences can be found, and concordance of independent characters provides strong indirect evidence for reproductive isolation (Avise and Ball, 1990).

However, as argued above, there are no easy practical solutions for species that are either extremely similar across much of the genome or for species that sporadically hybridize. How common are such cases likely to be? Although natural hybridization is often inferred from the ability to produce viable hybrids in the laboratory and the occurrence of 'intermediates', we generally lack the genetic data to test the inference. Lessios and Pearse (in press) show for Indo-Pacific sea urchins in the genus *Diadema* that genetic data do not support the idea of common hybridization, despite the absence of conspicuous pre- or post-zygotic isolating mechanisms and the presence of morphologically intermediate specimens.

We may be fortunate to be doing systematics now, rather than two million years ago, when many new species emerged during the turnover event that created much of the modern tropical marine invertebrate fauna (Budd *et al.*, 1994b; Jackson, 1994). Since that time, comparatively little appears to have happened in terms of speciation, thereby increasing the probability that lineage sorting will have created qualitative genetic differences between species (Avise and Ball, 1990). Very recently diverged species of marine invertebrates (Palumbi and Metz, 1991; Johannesson *et al.*, in press) seem to be the exception rather than the rule, although the

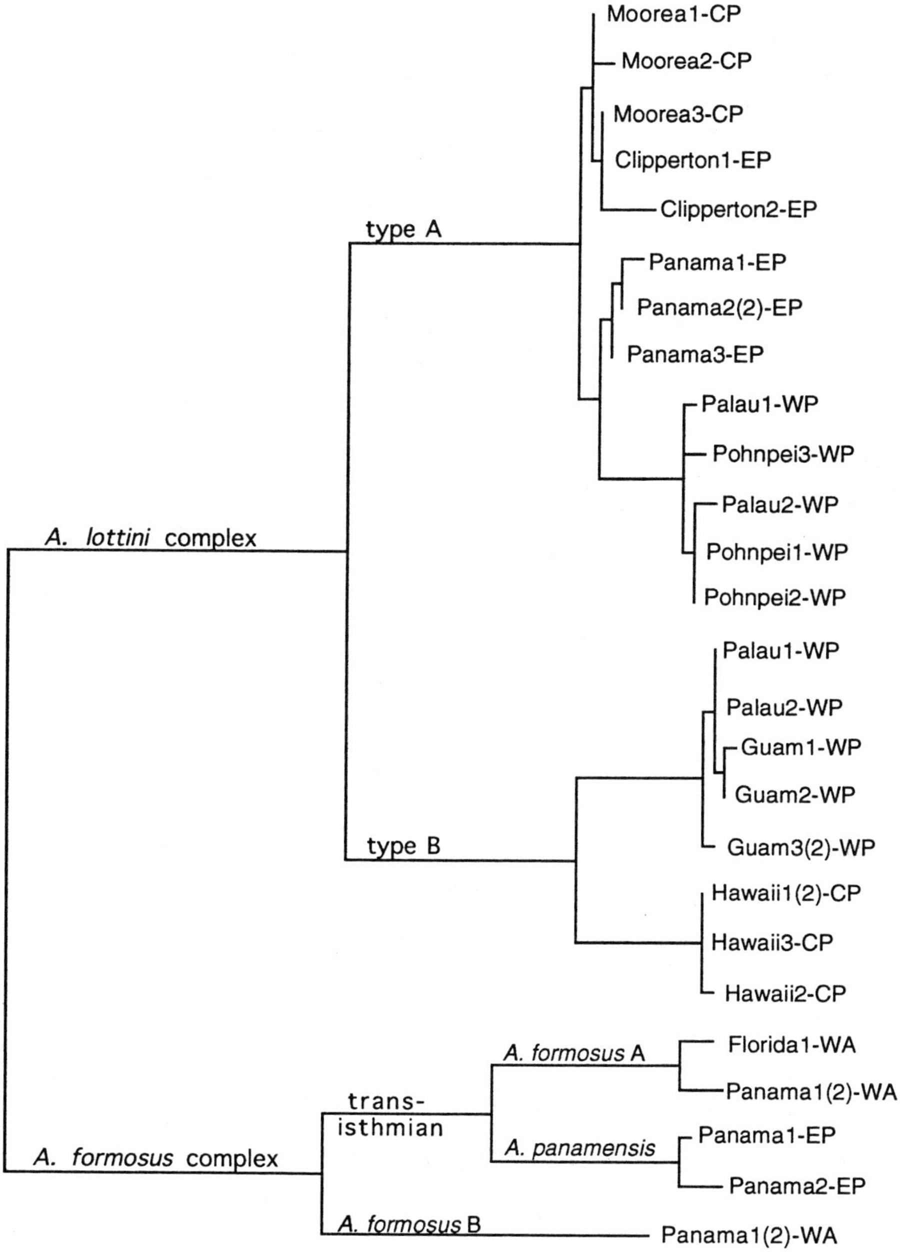

Figure 10.2 One of 144 equally parsimonious trees. (PAUP analysis as in Knowlton *et al.*, 1993, for the same taxa as in Figure 10.1.)

difficulty of identifying recently diverged species may contribute to our perception of their rarity.

10.3.2 Divergence in allopatry

While problems with corals may represent an atypically pessimistic view of species definitions in sympatry, the snapping shrimp case may give an overly optimistic view of the potential for understanding and categorizing allopatric variation. The dispersal of most *Alpheus* larvae is moderate in range (with a development time of several weeks), and rafting is probably not a regular feature of their dispersal biology. While many marine invertebrates share these features, many others do not (Jackson, 1986). Of particular concern are the numerous sessile or sedentary groups with limited larval dispersal. These taxa as currently defined often have very large geographic ranges relative to their dispersal ability, and exhibit extensive and complex patterns of variation.

Several well-studied, small crustaceans provide examples of problems that are likely to plague many such groups. In the tide pool dwelling copepod *Tigriopus californicus*, six of nine study sites analysed in the less than 800 km of coastline between San Francisco and San Diego are diagnosably distinct based on allozymes and DNA sequences (Burton and Lee, 1994; R.S. Burton, personal communication). Some of these differences appear to represent divergence times of 7 million years, but there is no predictable relationship between genetic divergence and post-zygotic compatibility, or between different molecular measures of divergence (Burton, 1990; Burton and Lee, 1994; Ganz and Burton, 1995). Similarly complex patterns have also been reported for the intertidal, beach-dwelling isopod *Excirolana braziliensis*, whose current range, based on traditional morphological criteria, extends from the Gulf of California to southern Chile in the eastern Pacific, and from the Gulf of Mexico to southern Brazil in the Atlantic. Just within Panama, populations from 10 of 11 sites could be distinguished by fixed allozyme differences (Lessios and Weinberg, 1993), and nearby populations sharing the same morphology were typically almost as genetically divergent as populations of different morphologies isolated by the Isthmus of Panama for at least 3 million years (Lessios and Weinberg, 1994). There were also marked genetic changes at one site over a 2-year period.

Organisms like these are going to be a problem for both the biological and phylogenetic species concept. If there are no predictable relationships between potential for interbreeding and other measures of divergence, there are going to be no short-cuts such as the Isthmus for *Alpheus* in defining species boundaries based on the biological species concept. For the phylogenetic species concept, sufficiently detailed genetic information could result in every sample (including samples over time from the same site) being diagnosably distinct. Even the biological species concept may result in a staggering increase in species level diversity in such taxa.

10.4 BROADER IMPLICATIONS

Taxonomy does not operate in an intellectual vacuum, unconnected with other fields of science. Thus, the species concept embraced by students of marine invertebrates is likely to have substantial consequences both within taxonomy and across other disciplines. Here we consider a few examples.

10.4.1 Nomenclature

Rigorous application of either the phylogenetic or the biological species concept will inevitably result in the future recognition of numerous sibling species within marine invertebrates. Regardless of the species concept that is ultimately employed, use of terms such as 'species complex' or 'species group' allows one to recognize informally clusters of similar and apparently related forms. This seems a more useful approach than splitting up speciose genera (see also Mayr and Ashlock, 1991) or designating numerous subgenera without a phylogenetic analysis.

Application of the phylogenetic species concept as defined by Cracraft (1989, 1997: Chapter 16) could result in orders of magnitude increases in species level diversity in some cases (Avise and Ball, 1990), effectively replacing a taxonomic entity with a geographic one. Cracraft (1989) has scorned the 'how many names can you learn' concern, but names that cannot be readily used by non-specialists will not be used at all. There is likely to be little sympathy beyond the ranks of a subset of systematists for a system that gives species names to every genetically distinct population, regardless of its potential transience and the triviality of its divergence. Indeed, Mishler and Theriot (in press), while arguing for their version of the phylogenetic species concept, suggest that monophyletic groups that are evolutionary trivial, cryptic or poorly supported not be given formal recognition. This solution, however, negates the chief virtue of Cracraft's approach, namely its unambiguous universality.

10.4.2 Ecology

Sympatric sibling species, once recognized, generally exhibit ecological differences that have important implications for understanding community structure (Knowlton, 1993). For example, previous conclusions that corals were generalists (Connell, 1978) are suspect in light of recent taxonomic discoveries which suggest that niche divergence may be more important than previously realized (Knowlton and Jackson, 1994).

Diversity recognized by adherents of both the biological and phylogenetic species concepts could result in a substantial change in the taxonomic database upon which ecologists rely.

10.4.3 Biogeography

Both the biological and the phylogenetic species concepts are likely to result in greater estimates of endemism, particularly with the use of genetic data. For example, only 30% of Hawaii's marine invertebrates are thought to be endemic (Kay and Palumbi, 1987), but *Alpheus lottini* is apparently a false member of the remaining 70%, despite its ability to disperse via planktonic larvae. Based on the studies of *Excirolana* and *Tigriopus* summarized above, essentially all of Hawaii's brooding marine invertebrates are likely to be endemic.

10.4.4 Conservation

In the era of the Rio Convention on Biodiversity, species boundaries have important political as well as biological implications. Fine scaled allopatric splitting, as promoted by the phylogenetic species concept, would facilitate national 'ownership' of biodiversity resources. On the other hand, if such splitting were unwarranted, it could inhibit the development of international cooperation in the conservation of interconnected populations whose survival depends on regional approaches (Committee on Biological Diversity in Marine Systems, 1995).

10.4.5 Palaeontology

While this book focuses deliberately on living species, many marine invertebrate species living today have extensive fossil records. Species concepts that work in the present but not in the past are particularly problematic for these groups. Smith (1994), for example, argued that quantitative statistics are an inappropriate tool for recognizing species, yet they are an essential and powerful technique for recognizing living species morphologically for many of the groups with the best fossil records, such as corals, bryozoans and molluscs. Felsenstein (1988) argued that the biological difference between quantitative and qualitative characters was more illusory than real, and that the focus on qualitative characters in parsimony analysis was primarily a reflection of computational constraints. Thus, there seems to be little justification for abandoning these fossil species as unrecognizable. Rather, genetic data for species in the present should be used to evaluate quantitative morphometric methods for application in the past (Jackson and Cheetham, 1994).

10.4.6 Phylogeny

Phylogenetic analyses have largely ignored the problem of species boundaries. A recent study of cheilostome bryozoans, however, showed that accurate resolution at the species level greatly improved the quality of phylogenetic analyses in terms of parsimony, consistency, and concordance between morphological and genetic data (Jackson and Cheetham, 1994). For species that are readily distinguished by qualitative characters, rigorous discrimination of species by either concept will lead to an improved understanding of relationships. However, in these cheilostomes, as with the corals discussed earlier, quantitative statistical methods were required to resolve the species level taxa, so that many of the species discriminated in this study would not be recognized by most supporters of the phylogenetic species concept.

10.5 CONCLUSIONS

The traditional, morphological species concept, as applied to marine invertebrates, has been unduly conservative. In sympatry, application of both the biological species concept and the phylogenetic species concept will result in the recognition of numerous new species that reflect previously undetected reproductive barriers between morphologically similar forms. However, insistence on qualitative morphological characters by some advocates of the phylogenetic species concept will make it impossible to recognize many species, because morphological differences are often quantitative.

In allopatry both approaches would again result in an increase in the number of species, but the differences between the two approaches are more marked because any distinctive population can become a phylogenetic species. There are two problems here. First, the difference between allopatry and sympatry is less clear-cut in the sea because of the variable and difficult to measure dispersal of larvae. Over what spatial scale does one assess reproductive cohesion, an essential step if one is 'to avoid assigning species status to individual organisms, to different sexes and morphs, or to developmental stages' (Cracraft, 1989)? Second, sporadic recruitment events are likely to create trivial and sometimes transient (Lessios and Weinberg, 1994) but diagnosable groups. Mishler and Theriot (in press) recommend ignoring these minor monophyletic units, but then we are left with the question, how minor is minor enough? This is no less subjective than trying to assess how much differentiation is likely to be associated with reproductive barriers were the allopatric taxa to come together, one of the primary objections raised against the biological species concept. Moreover, the subjectivity of the assessment of reproductive compatibility is declining with the appearance of theory (Orr, 1995) and empirical studies (Coyne and Orr, 1989; Knowlton *et al.*, 1993) that relate divergence in allopatry to reproductive isolation (although some groups, like *Tigriopus*, may prove difficult to evaluate in this fashion).

Thus, it would appear that all reasonable species concepts require the use of scientific judgement. If we are going to have to use judgement, what principle should guide it? The warring species concepts give us two alternatives from which to choose. The phylogenetic species concept argues for systematic consistency: species should be defined in a manner analogous to all higher taxa. The biological species concept, on the other hand, suggests that species are special, because they lie at the boundary that divides the realm of reticulation from the realm of cladogenesis. We strongly prefer the latter approach, because we find the reticulate/cladistic boundary, even if fuzzy (Mishler and Theriot, in press), to be too biologically important to ignore. Most cladists, however, strongly disagree. Ironically, it is not at all clear where Willi Hennig would stand were he to participate in today's debate.

Acknowledgements

The symposium to which Nancy Knowlton was invited provided a stimulating environment in which to think carefully about these issues. We thank A. Budd, A. Cheetham, C. Cunningham, J. Jackson, and S. Palumbi for comments on the manuscript; B. Mishler and H. Lessios for sharing their in press manuscripts with us; and E. Gomez for laboratory assistance. P. Glynn and M. Gleason provided the collections of *Alpheus* from Clipperton and Moorea respectively, while J. Jara and M. Gassel assisted us in the field. Various government agencies graciously granted permission to collect in their waters. The Smithsonian Institution and the molecular evolution program of the Smithsonian Tropical Research Institute provided financial support.

10.6 REFERENCES

Avise, J.C. and Ball, R.M., Jr (1990) Principles of genealogical concordance in species concepts and biological taxonomy. *Oxford Surveys in Evolutionary Biology*, **7**, 45–67.

Budd, A.F., Johnson, K.G. and Potts, D.C. (1994a) Recognizing morphospecies in colonial reef corals: 1. Landmark-based methods. *Paleobiology*, **20**, 484–505.

Budd, A.F., Stemann, T.A. and Johnson, K.G. (1994b) Stratigraphic distributions of genera and species of Neogene to Recent Caribbean reef corals. *Journal of Paleontology*, **68**, 951–77.

Burton, R.S. (1990) Hybrid breakdown in developmental time in the copepod *Tigriopus californicus*. *Evolution*, **44**, 1814–22.

Burton, R.S. and Lee, B.N. (1994) Nuclear and mitochondrial gene genealogies and allozyme polymorphism across a major phylogeographic break in the copepod *Tigriopus californicus*. *Proceedings of the National Academy of Sciences, USA*, **91**, 5197–201.

Claridge, M.F., Dawah, H.A. and Wilson, M.R. (1997) Practical Approaches to species concepts, in *Species: the Units of Biodiversity* (eds M.F. Claridge, H.A. Dawah and M.R. Wilson), Chapman & Hall, London, pp. 1–5.

Coates, A.G., Jackson, J.B.C., Collins, L.S. *et al.* (1992) Closure of the Isthmus of Panama: the near-shore marine record of Costa Rica and western Panama. *Geological Society of America Bulletin*, **104**, 814–28.

Committee on Biological Diversity in Marine Systems (1995) *Understanding Marine Biodiversity*. National Academy Press, Washington, DC.

Connell, J.H. (1978) Diversity in tropical rain forests and coral reefs. *Science*, **199**, 1302–10.

Coyne, J.A. and Orr, H.A. (1989) Patterns of speciation in *Drosophila*. *Evolution*, **43**, 362–81.

Cracraft, J. (1989) Speciation and its ontology: the empirical consequences of alternative species concepts for understanding patterns and processes of differentiation, in *Speciation and Its Consequences* (eds D. Otte and J.A. Endler), Sinauer Associates, Sunderland, Massachusetts, pp. 28–59.

Cracraft, J. (1997) Species concepts in systematics and conservation biology: an ornithological viewpoint, in *Species: the Units of Biodiversity* (eds M.F. Claridge, H.A. Dawah and M.R. Wilson), Chapman & Hall, London, pp. 325–39.

Dustan, P. (1979) Distribution of zooxanthellae and photosynthetic chloroplast pigments of the reef-building coral *Montastrea annularis* Ellis and Solander in relation to depth on a West Indian coral reef. *Bulletin of Marine Science*, **29**, 79–95.

Felsenstein, J. (1988) Phylogenies and quantitative characters. *Annual Review of Ecology and Systematics*, **19**, 445–71.

Ganz, H.H. and Burton, R.S. (1995) Genetic differentiation and reproductive incompatibility among Baja California populations of the copepod *Tigriopus californicus*. *Marine Biology*, **123**, 821–7.

Gillespie, J. H. (1991) *The Causes of Molecular Evolution*, Oxford University Press, New York.

Graus, R.R. (1977) Investigation of coral growth adaptations using computer modelling. *Proceedings of the Third International Coral Reef Symposium*, **2**, 463–9.

Graus, R.R. and Macintyre, I.G. (1976) Control of growth form in colonial corals: computer simulation. *Science*, **193**, 895–7.

Jackson, J.B.C. (1986) Modes of dispersal of clonal benthic invertebrates: consequences for species' distributions and genetic structure of local populations. *Bulletin of Marine Science*, **39**, 588–606.

Jackson, J.B.C. (1994) Constancy and change of life in the sea. *Philosophical Transactions of the Royal Society, London, B*, **344**, 55–60.

Jackson, J.B.C. and Cheetham, A.H. (1994) Phylogeny reconstruction and the tempo of speciation in cheilostome Bryozoa. *Paleobiology*, **20**, 407–23.

Johannesson, K., Rolán-Alvarez, E. and Ekendahl, A. Incipient reproductive isolation between two sympatric morphs of the intertidal snail *Littorina saxatilis*. *Evolution* (in press).

Kay, E.A. and Palumbi, S.R. (1987) Endemism and evolution in Hawaiian marine invertebrates. *Trends in Ecology and Evolution*, **2**, 183–7.

Kessing, B., Croom, H., Martin, A. *et al.* (1989) The simple fool's guide to PCR, version 1.0. Unpublished.

Kessing, B. (1995) Sequencer 3.0: DNA sequence manipulation and analysis. Program written for the MacIntosh.

Kimura, M. (1980) A simple method for estimating evolutionary rate of base substitutions through comparative studies of nucleotide sequences. *Journal of Molecular Evolution*, **16**, 111–20.

Knowlton, N. (1993) Sibling species in the sea. *Annual Review of Ecology and Systematics*, **24**, 189–216.

Knowlton, N. and Jackson, J.B.C. (1994) New taxonomy and niche partitioning on coral reefs: jack of all trades or master of some? *Trends in Ecology and Evolution*, **9**, 7–9.

Knowlton, N. and Mills, D.K. (1992) The systematic importance of color and color pattern: evidence for complexes of sibling species of snapping shrimp (Caridea: Alpheidae: *Alpheus*) from the Caribbean and Pacific coasts of Panama. *Proceedings of the San Diego Society of Natural History*, **18**, 1–5.

Knowlton, N., Weil, E., Weigt, L.A., and Guzmán, H.M. (1992) Sibling species in *Montastraea annularis*, coral bleaching, and the coral climate record. *Science*, **255**, 330–3.

Knowlton, N., Weigt, L.A., Solórzano, L.A., Mills, D.K., and Bermingham, E. (1993) Divergence in proteins, mitochondrial DNA, and reproductive compatibility across the Isthmus of Panama. *Science*, **260**, 1629–32.

Knowlton, N., Maté, J. L., Guzmán, H.M., Rowan, R. and Jara, J. Direct evidence for reproductive isolation among the three species of the *Montastraea annularis* complex in Central America (Panama, Honduras). *Marine Biology* (in press).

Lang, J.C. (1984) Whatever works: the variable importance of skeletal and non-skeletal characters in scleractinian taxonomy. *Palaeontographica Americana*, **54**, 18–44.

Langan-Cranford, K.M. and Pearse, J.S. (1995) Breeding experiments confirm species status of two morphologically similar gastropods (*Lacuna* spp.) in central California. *Journal of Experimental Marine Biology and Ecology*, **186**, 17–31.

Lessios, H.A. and Pearse, J.S. Hybridization and introgression between Indo-Pacific species of the sea urchin *Diadema*. *Marine Biology* (in press).

Lessios, H.A. and Weinberg, J.R. (1993) Migration, gene flow and reproductive isolation between and within morphotypes of the isopod *Excirolana* in two oceans. *Heredity*, **71**, 561–73.

Lessios, H.A. and Weinberg, J.R. (1994). Genetic and morphological divergence among morphotypes of the isopod *Excirolana* on the two sides of the Isthmus of Panama. *Evolution*, **48**, 530–48.

Mayr, E. and Ashlock, P.D. (1991). *Principles of Systematic Zoology*, 2nd edn, McGraw-Hill, New York.

Mishler, B.D. and Theriot, E. (1997) Monophyly, apomorphy, and phylogenetic species concepts, in *Species Concepts and Phylogenetic Theory: A Debate* (eds Q. D. Wheeler and R. Meier), Columbia University Press, New York (in press).

Orr, H.A. (1995) The population genetics of speciation: the evolution of hybrid incompatibilities. *Genetics*, **139**, 1805–13.

Palumbi, S.R. and Metz, E.C. (1991) Strong reproductive isolation between closely related tropical sea urchins (genus *Echinometra*). *Molecular Biology and Evolution*, **8**, 227–39.

Romano, S.L. (1995) *A Molecular Phylogenetic Analysis of Reef-Building Corals*, PhD Thesis, University of Hawaii.

Smith, A.B. (1994) *Systematics and the Fossil Record: Documenting Evolutionary Patterns*, Blackwell Scientific, Oxford.

Templeton, A.R. (1981) Mechanisms of speciation – a population genetic approach. *Annual Review of Ecology and Systematics*, **12**, 23–48.

Templeton, A.R. (1989) The meaning of species and speciation: a genetic perspective, in *Speciation and Its Consequences* (eds D. Otte and J. A. Endler), Sinauer Associates, Sunderland, Massachusetts, pp. 3–27.

Van Veghel, M.L.J. and Bak, R.P.M. (1993) Intraspecific variation of a dominant Caribbean reef building coral, *Montastrea annularis*: genetic, behavioral and morphometric aspects. *Marine Ecology Progress Series*, **92**, 255–65.

Van Veghel, M.L.J. and Kahmann, M.E.H. (1994) Reproductive characteristics of the polymorphic Caribbean reef building coral *Montastrea annularis*: II. Fecundity and colony structure. *Marine Ecology Progress Series*, **109**, 221–7.

Veron, J.E.N. (1995) *Corals in Space and Time: The Biogeography and Evolution of the Scleractinia*, UNSW Press, Sydney.

Wallace, C.C. and Willis, B.L. (1994) Systematics of the coral genus *Acropora*: implications of new biological findings for species concepts. *Annual Review of Ecology and Systematics*, **25**, 237–62.

Weil, E. and Knowlton, N. (1994) A multi-character analysis of the Caribbean coral *Montastraea annularis* (Ellis and Solander, 1786) and its two sibling species, *M. faveolata* (Ellis and Solander, 1786) and *M. franksi* (Gregory, 1895). *Bulletin of Marine Science*, **55**, 151–75.

Willis, B.L. (1985) Phenotypic plasticity versus phenotypic stability in the reef corals *Turbinaria mesenterina* and *Pavona cactus*. *Proceedings of the Fifth International Coral Reef Congress, Tahiti*, **4**, 107–12.

Wood Jones, F. (1907) On the growth-forms and supposed species in corals. *Proceedings of the Zoological Society of London*, **1907**, 518–56.

11

Nematode species: concepts and identification strategies exemplified by the Longidoridae, Steinernematidae and Heterorhabditidae

D. J. Hunt

Contacting address: International Institute of Parasitology, 395A, Hatfield Road, St Albans AL4 0XU, UK

ABSTRACT

The differing approaches to the practical identification of phytoparasitic nematodes in the family Longidoridae and entomopathogenic nematodes in the families Steinernematidae and Heterorhabditidae are compared and contrasted. Both groups contain economically important species: the former as parasites of the root systems of plants and, in certain species, with the potential for transmission of plant nepoviruses; the latter as possible biocontrol agents of insect pests in horticulture and agriculture.

The taxonomy of the Longidoridae, a group displaying an abundance of relatively well-defined and stable characters, is almost exclusively concerned with the classical approaches of morphology and morphometrics. However, in one species complex within the genus *Xiphinema*, namely the *X. americanum*-group, molecular approaches are being applied with increasing frequency. This new approach has largely been precipitated by the exceptionally close morphological and morphometric similarity of the members of this substantial group of species and the practical difficulties involved in specific determination; this being of particular relevance bearing in mind the importance of certain members in vectoring economically important plant viruses.

In the Steinernematidae and the Heterorhabditidae, all the members are obligate parasites of the insect haemocoel and exhibit few reli-

Species: The Units of Biodiversity. Edited by M.F. Claridge, H.A. Dawah and M.R. Wilson. Published in 1997 by Chapman & Hall. ISBN 0 412 63120 2

able morphological or morphometric characters. This has led to confusion as to the value and status of the differences which have been observed, together with concomitant problems in establishing the specific identity of isolates. In these families the major progress to date has been via molecular techniques involving PCR products, a technique which can rapidly yield reliable results from a single infective juvenile.

A synthesis of classical and molecular methodologies is advocated as the most productive approach in delimiting and recognizing the species taxon in these families.

11.1 INTRODUCTION

The Phylum Nematoda comprises a diverse assemblage with tens of thousands of nominal species. Estimates of the total number of species are speculative, but range into the hundreds of thousands. Representatives may be parasitic in vertebrates and/or invertebrates, parasitic on plants and/or fungi, predators, or free-living microbivores in soil or water. They are found throughout the world's seas and continents, often in colossal numbers, and may be the dominant biomass in selected habitats. They are even found colonizing such extreme niches as the abyssal ocean depths and the lacunary system on the undersurface of Arctic sea ice (Tchesunov and Rieman, 1995). Despite this ecological diversity they are usually rather similar in their gross morphology, being essentially cuticle-bounded tubes supported by a hydrostatic skeleton. The internal organs essentially conform to a tubular pattern and lie within a fluid-filled space. Those nematodes parasitizing plants or insects tend to be small, typically less than 1 mm in length, although exceptionally over 50 cm in some insect parasites from the Mermithida. They are usually transparent so that the internal organs are readily visible with the light microscope, even at low magnifications. This phenomenon has led to a conflict between features which, although useful as key characters, are not necessarily of equivalent value in systematics. Similarly, some characters which are of great systematic importance are visually obscure and of little practical use in routine identification procedures.

What is a species? This fundamental question continues to spawn a confusing plethora of erudite concepts (biological, evolutionary, morphological, phylogenetic, etc.) which are often championed by vehement protagonists, but a unified theory is lacking, perhaps unattainable. The conflict between theoretical and operational constraints implicit in any concept inevitably results in compromises. Cronquist (1977) expressed the morphological species concept thus: 'Species are the smallest groups that are consistently and persistently distinct, and distinguishable by ordinary means'. Mayr, an ardent exponent of the major alternative viewpoint, has long advocated the use of a biological species concept with its implication of cryptic species (Mayr, 1969). Both concepts are used in plant and insect nematology, the emphasis being on the phenetic approach as nematodes are inadequately known at the biological level to facilitate widespread use of the latter. Philosophical conflict between proponents of the morphological or biological species concept is frequent, but sterile when indulged to excess. The theories are surely complementary in overall purpose and deserve due recognition as viable perspectives with differing strengths and weaknesses. A more pragmatic, less adversarial approach is required, one which recognizes the unpalatable fact that species are not completely definable and will not conform neatly to the tenets of any one concept. Our idea of what constitutes a species demands continual reappraisal; it must evolve with increasing knowledge, but carries the proviso that it is unlikely to be consistently applicable over the whole gamut of Life.

A common artifice is to regard the opinion of a competent taxonomist as the determining factor as to whether an organism qualifies for specific status or not. Despite provoking additional comment on the definition of taxonomic competence, this circular argument has considerable functional appeal, particularly in view of the lack of consensus on a more objective alternative. As the species is the fundamental taxon within systematics, any imprecision in definition inevitably invites problems by allowing subjective interpretation of the degree of plasticity acceptable before the permitted variation of one species crosses some imaginary boundary and becomes another entity, distinct, at least, in name. One could argue, with little difficulty, that a species, as such, does not really exist as a discrete, definable object. It is merely a convenient pigeon hole with which we seek to define and categorize the variability of an expression of life (the organism) in question. A species may exhibit a core or suite of polythetic character states, but the margins of phenotypic variability are usually diffuse and ill-defined and more indicative of our state of knowledge than anything more germane. The specific attribution of an organism displaying a phenotype at the periphery of variation is therefore probabilistic, i.e. taxa are fuzzy, not discrete entities.

The purposes and/or uses of species categorization may be manifold, but the desire to label and catalogue is surely a prime mover (labelling is often more comforting than informative – it is always easier to label than to define!). The need for a species concept is, however, inescapable and cataloguing can be a reasonable compromise, particularly if the 'entities' are then arranged into patterns which shed light on their inter-relationships. We should be careful not to assume too readily that these named life forms are imbued with validity or relevance outside our own philosophy – the species concept is not as plastic as the organisms it purports to define. As we often have little idea of the potential variability within one of our 'species' it follows that the species concept itself, although possessing a set of core criteria, must of necessity be diffuse and imprecise if it is to be anything other than an exercise in semantics. This imprecision is itself not constant and may be influenced by prevailing fashion, type of organism at hand and/or environmental parameters.

The *International Code of Zoological Nomenclature* conveniently defines a species by a nomen attached, in the ultimate resort, to a single specimen, the holotype. Such a nomenspecies concept, although necessary for nomenclatural stability, offers little solace to the person faced with the phenotypic (and genotypic) plasticity of populations. The interpretation of what constitutes a species *sensu lato* is intuitive. In this respect the term 'unit of diversity' expresses rather more of a sense of precision, of being in control, than is warranted. Despite semantic and philosophical quibbling, there is an irreducible requirement to identify and name organisms as a prerequisite to many areas of biological research. A species concept, if interpreted sensibly, provides an acceptable means of categorizing and packaging variability in order to achieve this purpose – even if our convenient units do prove a trifle leaky and bleed characters across boundaries which are often a reflection of ignorance, rather than anything more profound.

Which species concept should be applied? The choice may be eclectic and involve factors such as the predilection of the user; the group being studied and overall purpose. For example, a phylogenetic species concept may be robust in highly visible, well-studied groups, such as birds (Cracraft, 1997: Chapter 16), but suffers overwhelming practical disadvantages in cryptic, poorly known groups such as nematodes where the necessary criteria are more elusive. Although the ultimate (and laudable) purpose of systematists may be to produce an entirely natural scheme reflecting evolutionary lineages, this is not the only raison d'être for classification. Cain (1959) wrote: 'Taxonomists have forgotten for too long that they are the name makers for all zoologists and botanists, pure and applied, and are under an obligation never to impose an unnecessary burden on others'. With the current widespread interest in biodiversity, the user list could be expanded to include politicians and the public at large. The role of the taxonomist in facilitating a practical, reliable and consistent approach to problems inherent in the identification process is crucial. Biosystematists no longer work in an esoteric backwater. Consequently, they need to assume broader responsibilities than previously and perhaps exercise more caution and restraint in the pursuit of their art – internecine squabbles over semantic abstractions risk misinterpretation and ultimately detract from the wider cause of biosystematics.

The problem of correctly naming a plant or insect parasitic nematode species has practical considerations in that a proportion of species are economically important, perhaps causing severe damage to agricultural crops either directly, as a result of their feeding activities, indirectly by facilitating attack by pathogenic fungi or, as in the case of the longidorids and trichodorids, by vectoring plant viruses. Species with biocontrol potential, such as entomopathogenic nematodes, may have a restricted host range or require specific environmental parameters to facilitate invasion and control of the target organism. Correct specific attribution in such instances carries considerable practical implications, not to mention responsibilities.

In modern nematology, the traditional phenetic species concept combining morphology and morphometrics still holds sway, regardless of discipline. Apomixis, a phenomenon of common occurrence in the Nematoda, poses awkward questions for both morphological and biological species concepts and such problems have yet to be satisfactorily resolved. Biochemical and molecular approaches, although late starters, are now well established in many aspects of nematology and show particular potential where classical techniques lack the necessary precision to discriminate closely related taxa or host races. The seminal value of molecular techniques is, in the absence of fossil nematodes of any geologically significant age, likely to prove particularly potent in elucidating phylogenetic relationships.

Two contrasting paradigms of the methodology and exigent problems concerning the concept and characterization of the species taxon within the Nematoda, exemplified here by the phytoparasites in the Longidoridae and the entomopathogenic nematodes in the Steinernematidae and Heterorhabditidae, are now discussed in greater depth.

11.2 PHYTOPARASITIC NEMATODES OF THE FAMILY LONGIDORIDAE (WITH PARTICULAR EMPHASIS ON THE GENUS *XIPHINEMA* COBB, 1913)

11.2.1 Introduction

The Longidoridae, a family of phytoparasitic nematodes in the Order Dorylaimida, comprises a handful of genera, representatives of which are found more or less worldwide. Longidorids are large nematodes, ranging in length from about 1.5 mm to over 12 mm. They are all characterized by the form of the oesophagus and by the possession of an attenuated, needle-like odontostylet with which plant roots are pierced in order to feed on the cell contents. The degree to which the odontostylet is developed, the form of the junction between the distal portion (the odontostyle) and the proximal section (the odontophore) and the location and relative size of the nuclei of the oesophageal glands together allow the family to be subdivided into subfamilial groups. Furthermore, these characters, in combination with others such as the form of the amphid, a sense organ opening on the cephalic region, also serve to define the genera. A brief systematic scheme is as follows:

Order Dorylaimida
Superfamily Dorylaimoidea
Family Longidoridae
Subfamily Longidorinae
Genus *Longidorus* Micoletzky, 1922 (Filipjev, 1934)
Genus *Longidoroides* Khan, Chawla & Saha, 1978
Genus *Paralongidorus* Siddiqi, Hooper & Khan, 1963
Subfamily Xiphidorinae
Genus *Xiphidorus* Monteiro, 1976
Genus *Paraxiphidorus* Coomans & Chaves, 1995
Subfamily Xiphinematidae
Genus *Xiphinema* Cobb, 1913

The Longidoridae currently contains about 400 valid species, over half belonging to *Xiphinema* with *Longidorus* (about 100 species), *Paralongidorus* (about 50 species), *Longidoroides* (about 15 species) and *Paraxiphidorus* and *Xiphidorus* (nine species) making up the total. The two genera of most economic importance are *Longidorus* and *Xiphinema*, both genera being known to contain species capable of vectoring plant viruses in addition to causing direct root damage by their feeding activities.

The morphological and morphometric characters used to define species within the Longidoridae are similar across all genera, although *Xiphinema* exhibits appreciably more intrinsic variation, particularly with regard to the anatomy of the female genital apparatus and the form of the tail. Indeed, attempts, albeit unsuccessful, have been made to use such character set variation as justification for subgenera (Cohn and Sher, 1972). To date, the only genus in the Longidoridae where species have been studied using biochemical or molecular techniques is *Xiphinema*. This is mainly a result of the economic and quarantine importance of an intractable species aggregation containing virus vectors – the so-called *americanum*-group – where accurate specific determination is paramount.

The genus *Xiphinema*, partly because of the comparatively large size (1.5–8 mm) of its members, attracts considerable interest from nematologists wishing to describe a new species. While this carries a positive aspect, it inevitably opens the door to less than adequate species descriptions which a more rigorous approach would ascribe to existing taxa. Geographically widespread taxa, such as *X. brasiliense* Lordello, 1951, *X. elongatum* Schuurmans Stekhoven and Teunissen, 1938, *X. ensiculiferum* (Cobb, 1893) and *X. radicicola* T. Goodey, 1936 are particularly prone to this malaise, a fact reflected in their depressing list of synonyms. *Xiphinema*, species-rich by phytoparasitic nematode standards, currently contains over 200 valid species (Hunt, 1993), all members being ectoparasites of plant roots and feeding deep within the tissues by means of a protrusible odontostylet which may be over 350 μm long in some species. In addition to the plant damage (cell death, galling, disruption of root function) caused by their direct feeding activities, some species achieve considerably greater importance by virtue of their ability to transmit plant nepoviruses such as tobacco ringspot virus, tomato ringspot virus and grapevine fanleaf virus.

Species identity within *Xiphinema* has traditionally relied upon the assessment of phenotypic characters. With the ever-increasing number of nominal species and the close similarity of many of the monosexual species, the situation is becoming more difficult and a group which was once reasonably accessible to the non-expert is now posing considerable problems even to the cognoscenti.

Problems in *Xiphinema* taxonomy are exacerbated by the fact that, as with many soil nematodes, a substantial proportion of the nominal taxa are monosexual. Most of these species appear to reproduce by obligate meiotic parthenogenesis. Apparently functional males do occur rarely in some species, thus implying the existence of a facultative parthenogenetic condition which may be more widespread than the data reflect. Monosexual taxa display a frustrating tendency to radiate into clusters of closely similar forms or morphospecies. The more populations of a form that are studied the better are the chances that the spectrum of variation can be adequately assessed, thus leading to a more representative blend of character states and a stable nomenclature. The ability to delimit such forms in a meaningful way such that other workers can recognize and identify them is therefore crucial. The manifest failure of morphological characters to fulfil this function in perplexingly speciose complexes such as the *X. americanum*-group was pointed out by Heyns (1983). The intervening years have seen this situation become more refractory and an alternative methodology is urgently required to assist in systematically untying, rather than cutting, the Gordian knot. This alternative approach may now be on the threshold of realization with the advent of readily accessible and reliable PCR (polymerase chain reaction) based techniques. The characters used to define species of *Xiphinema* will now be examined in greater detail.

11.2.2 Characterization

(a) Classical morphology and morphometrics

A varied suite of reliable phenetic characters for species determination in *Xiphinema* has been developed over the past 35 years (Luc and Dalmasso, 1975a,b; Loof and Luc, 1990). These mainly concentrate on the female sex because many species are not amphimictic, males being either entirely absent or but rarely recorded. The major morphometric and morphological characters currently used to define species of *Xiphinema* are as follows:

Morphometric characters

- Body length (L)
- Odontostylet length
- Tail length
- Vulva position (V – vulval position from the anterior end of the body expressed as a percentage)
- Ratio a (body length divided by maximum body width)
- Ratio c (body length divided by tail length)
- Ratio c′ (tail length divided by anal body width)

Morphological characters

- Shape of cephalic extremity (continuous, offset or expanded from body contour)
- Habitus of heat-relaxed nematode
- Number and development of the female genital tracts
- Presence or absence of Z-organ
- Presence or absence (and type) of pseudo Z-organ
- Tail shape
- Presence or absence of blind terminal canal on tail
- Shape of juvenile tail
- Monosexual or amphimictic

When applied cautiously, these are all good characters. Problems arise when the material is distorted or shrunken by bad fixation or when the coverslip of the slide mount is not supported properly. This causes the nematode to become squashed and therefore alters the appearance of both head contour and tail shape, not to mention changes in the value of the various morphometric ratios. Even in properly supported mounts, specimens can still flatten over a relatively brief time-span (Heyns, 1983) and thus elicit misleading responses when re-examined.

The presence or absence in the female genital tracts of a Z-organ or pseudo Z-organ is an important and reliable character (Luc and Dalmasso, 1975a,b), but if an old description does not mention such a structure (particularly the pseudo Z-organ) as being present, the assumption, a priori, that it is absent, can be only too false. The original author may have overlooked the feature entirely, may have seen it, yet not considered the fact important, or misinterpreted it as something inconsequential. Re-examination of type material for this type of structure is absolutely essential if mistakes are to be avoided and 'new species' proposed unnecessarily merely on presence versus purported absence of such a character.

Particular problems are caused by the *X. americanum*-group. This bloated complex contains over 40 nominal species, many of which are taxonomically adjacent in their morphology with credence given to differences which would not be entertained or countenanced in other members of the genus. Resolving this chaotic situation is made more urgent by the economic importance of the group as certain species can vector damaging plant viruses. The problem of how to identify taxa with virus vector potential from non-virus vectors is pressing and is particularly pertinent in quarantine procedures where nematodes from this group are regularly intercepted in the rhizosphere of imported plants. The taxonomy of the group is made more intransigent by the predominance of an apparently obligate parthenogenetic reproductive strategy. This phenomenon removes male morphology from the characterization/identification process and also results in genetically isolated populations which effectively operate outside the usual biological species concept as clones or morphospecies. Heyns (1983) commented on these problems and stressed the need for an alternative approach to morphological characterization within the group to satisfactorily distinguish the nominal species. Since his paper, the problems have been exacerbated, not diminished, by many more taxa being proposed.

Problems of species delimitation within the *X. americanum*-group have long been recognized (Tarjan, 1969). It was, however, the paper by Lamberti and Bleve-Zacheo (1979) which precipitated many of the problems, when they proposed and differentiated 15 new species in the complex by utilizing a combination of minute differences in head and tail shape and small differences in morphometric characters such as odontostyle length. Such parameters may well be discriminatory, but unless other taxonomists can see and are prepared to accept such nuances as being reliable and meaningful criteria at the species level, the proposed species will be disputed or risk being dismissed in a perfunctory manner out of sheer exasperation. Indeed, the situation has become so abstruse that the authors of otherwise comprehensive multiple-entry or polytomous keys to the genus (Loof and Luc, 1990, 1993) declined to tackle the *X. americanum*-group and Loof *et al.* (1993) disputed the validity of the simplistic dichotomous key approach to the complex proposed by Lamberti and Carone (1991).

(b) Scanning electron microscopy (SEM)

Although SEM techniques have proved valuable in some groups of phytoparasitic nematodes such as the Heteroderidae, Meloidogynidae and Pratylenchidae, the longidorids have not been extensively studied. There are several good reasons for this. Longidorids, being members of the Order Dorylaimida, do not have the outer body cuticle annulated or otherwise adorned as is the norm in many phytoparasitic genera from the Order Tylenchida. Furthermore, they lack features such as longitudinal cuticular incisures in the lateral field and their cephalic structure is generally less complex, although the recent discovery (Swart and Heyns, 1987) of small, flap-like cephalic lobes in certain species of the genus *Paralongidorus* is a notable exception which may yet validate the related genus *Siddiqia*, currently regarded as a synonym, or at best a subgenus, of *Paralongidorus* (Hunt, 1993). SEM has, however, been used to elucidate the form of dissected-out internal organs such as the cuticular odontostylet with which the nematode pierces the plant root. In conclusion, it must be said that SEM has little immediate value in the determination of longidorids to species level, although its merit in other groups of nematodes sporting more complex cuticular and/or cephalic structures is indisputable.

(c) Biochemical and molecular techniques

Although biochemical and molecular techniques have been successfully applied to other phytoparasitic nematodes, comparatively little work has been done on longidorids. The reason for the lack of biochemical methodologies is probably due to the difficulty in obtaining sufficient nematode material as longidorid species, in marked contrast to members of the Heteroderidae, Meloidogynidae or Pratylenchidae, are difficult to culture and usually very slow to reproduce, life-cycles of several years being reported for some of the taxa where detailed biological studies have been made.

With the availability of relatively cheap and reliable PCR techniques, it is now possible to clone enough DNA from a single specimen to facilitate molecular studies. Vrain and Wakarchuk (1989), using cloned rDNA probes and Vrain *et al.* (1992), using rDNA restriction fragment length polymorphism (RFLP) techniques, studied several species in the *X. americanum*-group from North America. Vrain *et al.* (1992) designed two primers to amplify the ITS region using PCR and compared the 1.5 kb amplified product from each of 16 populations for RFLPs. Fourteen restriction enzymes were used to digest the PCR product. The DNA products were separated by electrophoresis, the data being used to calculate dissimilarity coefficients and to construct dendrograms by cluster analysis. This enabled Vrain *et al.* (1992) to separate the original 16 populations into five clusters and show that two populations of *X. rivesi*, one of the better-defined morphological species, were well separated from other *X. americanum*-group populations. The results also implied the affinity of unidentified populations with existing species. Vrain *et al.* (1992) concluded that the molecular approach could be used to separate species in the *X. americanum*-group and thereby assist classical taxonomic techniques in gauging the usefulness of morphological and morphometric characters for species determination.

De Giorgi *et al.* (1994) attempted to use amplification of a segment of mtDNA from *X. index* as an identification aid. They employed a subunit of the cytochrome oxidase gene and concluded that mtDNA offered possibilities as a species specific identification tool, but cautioned that much more data on the gene organization and nucleotide sequences in plant nematodes were required to assess intraspecific variation in the mitochondrial genome.

11.2.3 Discussion

Although plant nematology has been traditionalist in its approach, there is an increasing awareness that alternative methodologies are necessary, particularly in more problematic groups where the definition of specific or subspecific taxa demands increased precision and objectivity. This is particularly so for those species reproducing via meiotic or mitotic parthenogenesis. A variety of techniques have been tried of which the most promising to date have been biochemical or molecular. Burrows (1990) reviewed some of these approaches, but the intervening years have witnessed rapid progress [see the compilation edited by Lamberti *et al.* (1994) for a recent appraisal].

Most biochemical and molecular studies have concentrated on complex, economically important groups, such as the Heteroderidae and Meloidogynidae, although other damaging genera, such as *Radopholus*, have also been studied (Hahn *et al.*, 1994; Kaplan, 1994) while species in the notoriously difficult genus *Aphelenchoides* have been proposed partly on the results of esterase and PCR techniques (Hooper and Ibrahim, 1994; Ibrahim and Hooper, 1994; Ibrahim *et al.*, 1994a,b). Isozyme electrophoresis has been used successfully in identifying *Meloidogyne* spp. on a routine basis (Esbenshade and Triantaphyllou, 1990; Cenis *et al.*, 1992), yet suffers from the disadvantage that the technique is restrictive in the nematode material required and does not operate satisfactorily for single infective juveniles, a problem which PCR is able to overcome. Powers and Harris (1993) showed that PCR could differentiate juveniles from a number of root knot (*Meloidogyne*) species by employing differences in mtDNA and Williamson *et al.* (1994), using this technique, claimed a high success rate in obtaining an amplified band for some 80% of single juveniles studied.

The randomly amplified polymorphic DNA (RAPD) technique allows a broad spectrum of markers to be developed from single primers and can be a useful tool, if currently less consistent, than PCR products from specific regions of the genome DNA. Williamson *et al.* (1994) commented on the potential of RAPDs in assisting research into nematode genetics and systematics and presented results of studies on *Heterodera* spp. and *Meloidogyne* spp. They concluded that RAPDs could be a useful tool in distinguishing species and populations, but felt that the technique was not yet robust enough for a role in general identification practices. Fargette *et al.* (1994), using both RFLP and RAPD patterns, distinguished various species of *Meloidogyne*, although one species in particular, *M. arenaria*, was polymorphic. The dendrograms produced by these two techniques were reasonably congruent and allowed grouping of resistance breaking lines of the nematodes.

Heterologous cloned DNA probes have been used to distinguish between races of *Meloidogyne incognita*, an important phytoparasitic nematode throughout the tropics and subtropics (Chacon *et al.*, 1995). Such accuracy is fundamental with nematode species displaying host races as non-chemical control strategies often employ non-host or tolerant varieties to minimize crop damage. Here, the molecular approach can facilitate host race determination more conveniently, for example, than differential host tests (Sasser and Carter, 1985).

Although classical taxonomic techniques have served well and justifiably dominate within the Longidoridae, a group enjoying a rich and varied suite of morphological characters, there is little doubt that the application of molecular techniques will prove to be of immense value. The resulting synthesis, if sensibly implemented, should help to resolve problems posed by groups of closely related, parthenogenetic species as exemplified by the *X. americanum*-group, where the molecular approach (Vrain and Wakarchuk, 1989; Vrain *et al.*, 1992) may dismiss or validate emergent species based on minute phenotypic differences (see, for example, Cho and Robbins, 1991; Lamberti and Carone, 1991). From the limited information available, there now seems real hope that the *X. americanum*-group will yield to molecular techniques and that the validity or otherwise of the nominal species will be resolved, thus assisting not only nematode taxonomists, but also quarantine authorities and virologists concerned with the nematode transmission of plant nepoviruses. The use of molecular techniques, not only on other species of *Xiphinema* (De Giorgi *et al.*, 1994), but also on other genera within the family, may also illuminate phylogenetic relationships within the group and thus confirm or refute the current, somewhat speculative, theories concerning the higher systematics.

11.3 ENTOMOPATHOGENIC NEMATODES OF THE FAMILIES STEINERNEMATIDAE AND HETERORHABDITIDAE

11.3.1 Introduction

Entomopathogenic nematodes are characterized by their ability to carry specific pathogenic bacteria which are released into the insect haemocoel after penetration of the insect host has been achieved by the infective stage of the nematode. Such nematodes have been known since the early part of the century (Steiner, 1923) and one species, *Steinernema glaseri*, was used as a biocontrol agent of a scarabaeid grub as early as the 1930s (Glaser and Farrell, 1935). There are two families of major importance, the Steinernematidae and the Heterorhabditidae. Both fall within the Order Rhabditida. Both carry symbiotic entomopathogenic bacteria; *Xenorhabdus* Thomas & Poinar, 1979 in *Steinernema*, and *Photorhabdus* Boemare, Akhurst & Mourant, 1993 in *Heterorhabditis*, and both share a broadly similar life-cycle. Despite these shared attributes, they are otherwise remarkably distinct in the morphology of the male tail and copulatory apparatus and possibly have a diphyletic origin, their similarities arising from convergent evolution (Poinar, 1990, 1993). The Steinernematidae comprises two genera: *Steinernema* (syn. *Neoaplectana*), the type genus, and *Neosteinernema*. The Heterorhabditidae is monotypic, represented by the genus *Heterorhabditis* (syn. *Chromonema*). The systematic problems encountered in this group when applying the phenetic approach arise because the adult nematodes feed and reproduce in the protected environment of the insect haemocoel. Such specialized, but essentially similar *modus operandi*, implies a considerable degree of similarity in the morphological features expressed in combination with excessive morphometric variability attributable to density-dependent nutritional factors. Classical techniques have therefore concentrated on the free-living infective stage which, although lacking considerable gross morphological variation as a result of being a non-feeding stage, does show enhanced morphometric consistency. These problems have resulted in considerable confusion as to the status of the nominal species, a confusion which has enormous practical importance now that the nematodes have attracted commercial interest as potential biocontrol agents. In addition, accurate identification is often demanded by quarantine regulations stipulating that only indigenous species/isolates can be released as part of a biocontrol programme. Partly in response to this stricture, more extensive surveys for the infective, soil-dwelling stage have been carried out both in temperate and tropical regions with the result that a large number of isolates, many of which appear to be new species, have been found and cultured. This explosion of data has further stressed the already ill-defined and unstable systematics at a juncture when precision in attributing identity is most needed to facilitate the exploitation of such potentially useful biocontrol agents.

11.3.2 Bionomics

The typical life-cycle of a steinernematid is as follows. The infective third stage juvenile (IJ3) occurs in the soil and, although non-feeding, can survive for a considerable period due to extensive food reserves. The IJ3 carries a specific bacterium of the genus *Xenorhabdus* in the intestine and either penetrates an insect directly via the cuticle, spiracles or anus, or is ingested. Whichever mode of entry is effected, the IJ3 then penetrates to the haemocoel where it moults and releases the bacteria which rapidly produce a fatal septicaemia within 24 to 48 hours. The nematodes feed on the bacterial soup produced by the breakdown of the host tissues and moult to the first generation adults. These mate and subsequently the second generation adults occur. These are smaller than the first generation and may show morphological differences. Typically, the progeny from the second generation cease development at the third stage juvenile as by this time virtually all the food reserves have been exhausted and the cadaver is reduced to a nematode-filled skin. Eventually the nematodes escape from the remains of the host and migrate into the soil as the IJ3. Heterorhabditids show a basically similar life-cycle except for the crucial difference that the first generation is hermaphroditic and not amphimictic. The IJ3 also carries a different genus of bacterium, namely *Photorhabdus*, so-called because of its ability to bioluminesce.

11.3.3 Systematics

The families Steinernematidae and Heterorhabditidae are regarded as being closely related by some authorities, but rather more distant by others (Poinar, 1990, 1993). Certainly the male tail region is very different – heterorhabditids have a bursa supported by caudal rays whereas the steinernematids lack a bursa completely, have differently shaped copulatory spicules and possess numerous copulatory papillae. Such features are considered fundamental in the Rhabditida. Both, however, share a similar life-cycle and depend on the same symbiotic relationship with related genera of insect pathogenic bacteria. Regardless of the higher systematics, the genera are grouped as follows:

- Family Steinernematidae
  - Genus *Steinernema* Travassos, 1927
  - Genus *Neosteinernema* Nguyen & Smart, 1994
- Family Heterorhabditidae
  - Genus *Heterorhabditis* Poinar, 1976

Steinernema currently contains 15 valid species; *Neosteinernema* is monotypic and *Heterorhabditis* is represented by five recognized species. According to A.P. Reid (personal communication) there are currently at least another 20 isolated species of *Steinernema*, the status of which is clearly indicated by molecular data. These await specific designation and formal description via a combination of phenotypic and genotypic characterization. In contrast to *Steinernema*, the genus *Heterorhabditis* appears to exist as species complexes and a biological species concept is often used.

11.3.4 Characterization

(a) Classical morphology and morphometrics

Initial attempts at characterizing the species within these two families involved the tried and tested approaches used on other soil-dwelling or phytoparasitic nematodes, i.e. morphometrics and morphology (for a review, see Poinar, 1990). The difficulties in employing morphometric techniques are profound, as virtually all the stages in the life-cycle occur bathed in their food supply in the protected environment of the host haemocoel and parameters such as body length, width, etc., depend very largely on extrinsic factors such as host species and nutrient availability. In addition, as there are usually two generations within the host, there are marked differences in nematode size between the generations, differences which do not appear to be purely density-dependent. In an attempt to overcome these difficulties, much attention has been focused on the infective third stage juvenile (the IJ3). This stage is the survival/infective aspect of the life-cycle in both genera and is found in the soil. Being non-feeding, its morphometric characters are more stable. Conversely, being a non-feeding stage also means that its internal morphology is degenerate and does not provide a wealth of useful characters. Some of the measurements widely used in IJ3 identification are listed below:

- Total body length (L)
- Oesophagus length
- Distance from cephalic extremity to the excretory pore
- Tail length
- Ratio E (head to excretory pore divided by tail length)
- Ratio D (head to excretory pore divided by oesophagus length)

Ratios E and D are often represented in percentage form.

Body length, although variable, provides a useful parameter, dividing the species of *Steinernema*, for example, into broad groups. Oesophagus length is often difficult to determine with accuracy due to the degenerate nature of the structure, but excretory pore position relative to the cephalic extremity appears to be a fairly stable character in all the genera and has an acceptable coefficient of variation. For this reason, the ratio E is often employed in distinguishing species, as tail length is also reasonably consistent.

Morphological characters include the tail shape and whether the terminus contains a refractive spine or not, and the number and form of the incisures in the lateral fields (Kozodoi and Spiridonov, 1988; Mracek and Bednarek, 1991). Although the lateral field characters appear to work in a reasonably satisfactory manner with the currently described species (and isolates), their reliability will increasingly come under scrutiny now that the groups potential as biocontrol agents has been realized, the resulting attention throwing up a great many different isolates and not a few new species. The great number of new isolates has caused particular problems by increasing the perceived phenotypic variation in character profile of a 'species'. Attempts to counteract this tendency by inventing new ratios (Nguyen and Smart, 1992) and by turning once again to the parasitic stages for inspiration have yet to be overburdened by success. Male characters, such as the shape (or even colour) of the copulatory spicules and associated gubernaculum have been particularly used in *Steinernema* where the presence or absence of a terminal mucron on the tail is another, if somewhat more variable, character. While the spicule is undeniably a potentially useful character, great care needs to be exercised as the shape varies not only within a generation, but even more markedly between generations in the same host. In addition, the spicule structure is not confined to one plane and under the light microscope it is possible to interpret different shapes which are not substantive – they merely reflect relative observation angle. Variations in number and disposition of copulatory papillae in *Steinernema* (typically numbering 21 or 23) offer another possibility.

(b) Scanning electron microscopy

Use of SEM first contributed to the synonymy of *Neoaplectana* under *Steinernema* on the basis of the number of cephalic papillae on the adults (Mracek and Weiser, 1979, 1981). Subsequent work has shown that the number and spacing of the incisures in the lateral field of the IJ3 can, if characterized in a standard way, be a useful diagnostic character (Kozodoi and Spiridonov, 1988; Mracek and Bednarek, 1991), although the technique has yet to be rigorously tested on the profusion of new isolates and/or new species. It seems likely that, although the technique should prove useful in dividing the species into a number of groups, there will be insufficient variation in the lateral field structure to characterize specifically all of the new material awaiting examination. Other uses of SEM include the morphology of the male copulatory spicules and gubernaculum and the disposition of the copulatory papillae on the tail region of *Steinernema*. These features are often difficult to see or interpret under the light microscope, the spicule shape being particularly difficult to ascertain as the outline of this three-dimensional structure varies considerably with differing angles of view, a factor which is not always appreciated.

(c) Cross-breeding

Cross-breeding putative species using virgin females from one and males from the other (together with the reverse cross) and assessing the viability of the offspring is currently regarded as the acid test of reproductive isolation and hence specific status of isolates. Although this technique works relatively easily with *Steinernema*, an amphimictic genus, *Heterorhabditis* has an hermaphroditic first generation and so the bisexual second generation adults must be used. Two isolates currently grouped as *H. bacteriophora* have been reported not to produce viable offspring (Dix *et al.*, 1991) when crossed.

(d) Host specificity

There is some evidence that certain species, such as *S. kushidai* (Mamiya, 1988) and *S. scapterisci* (Nguyen and Smart, 1990) reproduce much better on specific insect hosts rather than on the commonly employed laboratory insects. Most other species appear to be non-host-specific, at least under laboratory conditions. There is also the complicating factor that our knowledge of host specificity may be incomplete as most isolations from soil samples are done by using a baiting technique employing the larvae of the wax moth, *Galleria mellonella*. Thus, host specific isolates could well be discriminated against in the collection/survey process and consequently go unrecorded.

(e) Biochemical and molecular techniques

Attempts to characterize these taxa reliably have involved a range of techniques including allozyme electrophoresis (Akhurst, 1987), isoelectric focusing (Kozodoi *et al.*, 1986; Griffin *et al.*, 1994; Joyce *et al.*, 1994b) and various techniques for analysing DNA (Curran and Webster, 1989; Smits *et al.*, 1991; Curran and Driver, 1994; Joyce *et al.*, 1994a,b; Reid, 1994). Molecular techniques involving DNA banding patterns appear particularly promising in the Steinernematidae (Curran, 1990; Reid, 1994) and probably represent the most rapid, accurate and reliable first step with any unknown isolate. Reid and Hominick (1992, 1993, 1994) and Reid (1994) reported on the use of RFLPs in the taxonomy of *Steinernema*. The technique involves oligonucleotide primers flanking the ITS region of the ribosomal DNA (rDNA) repeat unit and allows the amplification of the DNA of a single IJ3 via PCR. Subsequently, the PCR products are subjected to restriction digestion by a suite of enzymes and the fragments separated via agarose gel electrophoresis. The resulting library of restriction maps based on the rDNA repeat unit, together with the comparative location of enzyme sites, allows phylogenetic comparisons and relationships between individual species within a single family and also between the Steinernematidae and Heterorhabditidae to be constructed. The topology of the resulting phylogenetic tree (Reid, 1994) implied that the Steinernematidae and Heterorhabditidae were more closely related than had been suggested by the diphyletic theory of Poinar (1990, 1993) and showed that the maps of the rDNA repeat units of *Steinernema* revealed greater heterogeneity than those of *Heterorhabditis*. Interestingly, two species (*S. anomali* and *S. glaseri*) differing only by their failure to interbreed (Poinar, 1990) were linked by the tree as were two other morphologically similar species (*S. affinis* and *S. intermedia*), the IJ3 of which is differentiated by the presence or absence of a refractive spine in the tail tip. Joyce *et al.* (1994c) discussed a RAPD technique able to identify entomopathogenic nematodes from either purified DNA or crude lysates of single IJ3s and it would appear that this technique will be increasingly employed.

In the Heterorhabditidae, several workers report reliable results with isoelectric focusing (IEF) of proteins (Griffin *et al.*, 1994; Joyce *et al.*, 1994b) whereas others (Curran and Driver, 1994; Joyce *et al.*, 1994a) employ RFLPs in the rDNA intergenic spacer region (IGS) of *Heterorhabditis* isolates. The latter technique was successful in allocating the isolates to broad groupings concordant with the phenotypic species. Intra-group variation was identified by the use of various restriction endonucleases and the authors concluded that molecular markers would be useful in species recognition, although simultaneously advocating caution in the interpretation of such preliminary data and acknowledging the need for additional studies to complement their findings (Curran and Driver, 1994).

11.3.5 Discussion

In contrast to the Longidoridae, the most operational approach to the entomopathogenic nematodes seems to be to use molecular methodology as the preliminary investigative technique on new isolates. The precision of this more objective approach provides an indispensable tool not only at the subspecific level where accurate identification of isolates can be critical for biocontrol purposes, but also as a first approach at the species level. If significant genomic differences are detected by reference to a restriction map library, phenotype characterization of the putative new species can then proceed with cross-breeding studies to determine genetic isolation as the final arbiter of specific status. To date, *Heterorhabditis* species and isolates appear to show much less variability in their banding profiles than *Steinernema* species. This phenomenon may be related to the life-cycle where the first generation of *Heterorhabditis* is hermaphroditic, not amphimictic, thus resulting in a less variable genotype. Molecular techniques seem set to dominate in the initial characterization of entomopathogenic nematodes because of their ability to bypass phenotypic variability by direct analysis of the genotype and to discriminate at all systematic levels.

11.3.6 Conclusion

Any identification system must not only be capable of working with the existing species, but must be sufficiently soundly based to accommodate new material without risk of undue compromise or, worse still, collapse. One of the major problems in delimiting similar species on minute differences in phenotypic characters is the difficulty of incorporating a new population into the scheme. Such a population, particularly in monosexual species where non-lethal mutations can more readily increase in frequency and become fixed, is likely to be subtly different from the other nominal species and so the researcher is faced with the option of expanding the range of an existing species, a process which, if oft repeated, leads to the obliteration of supposed differences and synonymy of adjacent species. This reinvokes the broad species concept which precipitated the dilemma in the first place. The alternative approach of creating yet another name for the new, intermediate, population risks virtually every population being attributed to a different species – a less than enticing prospect, if one which may have to be faced. In botany, this concept results in several hundred nominal microspecies of, for example, the genus *Rubus* or the apomictic hawkweeds and dandelions. Such zealous partitioning, even when well-founded, has considerable practical implications in denying access to any but the most determined esotericist.

The Longidoridae and the genus *Xiphinema* have, with the notable exception of the *X. americanum*-group, been mostly well served by the classical approach of judiciously combining morphometrics with morphology. The main problems in the taxonomy of the genus have arisen with the parthenogenetic species, an area where the 'traditional' species concept is always likely to be in difficulties. Heyns (1983) pointed out some of the problems of using morphological characters in longidorid taxonomy. Such factors include: subjective terminology when describing head or tail shapes; differing interpretations of the same structure; post-mortem changes due to fixation, processing techniques and temporal changes in mounted specimens due to a gradual flattening allied with concomitant changes in body shape, morphometric parameters, such as width and length, and calculated ratios. Such problems are not restricted to the Longidoridae – fixation and mounting methods may have profound effects on morphology in other groups (Curran and Hominick, 1981).

Subtle differences may well be readily apparent and have great significance to a specialist intimately familiar with a group of species, but tend to be overlooked, viewed with suspicion or disregarded by others. It is partly a result of relying on such potentially subjective characters, that the delimitation of species within the *X. americanum*-group has proved so controversial. Yet there are clearly valid species within the group matrix and there are undoubtedly others, perhaps many others, which are also valid, but which may require the support of a non-morphological technique to unambiguously express and categorize their distinctness. Miniscule phenotypic characters are not necessarily invalid, but there are very real problems in communicating such subtle differences via words and/or conventional illustrations to a broader, and (perhaps rightly) cynical audience. Taxonomy advances by consensus and muddying the water is not a step to be taken lightly. What is needed is an alternative methodology with a different perspective; one which can provide a more profound view of species and their relatedness one to another. Such an approach could be, as with the entomopathogenic nematodes, DNA profiling based on PCR methodologies. Vrain *et al.* (1992) showed the potential of such techniques, although many more of the nominal species in the *X. americanum*-group need to be scrutinized before substantive conclusions can be drawn.

By way of contrast, the classical approach has stumbled when faced with the confusingly variable morphometrics and morphologically conserved entomopathogenic nematodes of the Steinernematidae and Heterorhabditidae. Although the classical approach is still employed in describing new species, there is a fundamental requirement for an alternative, independent diagnostic technique which can readily handle and categorize the influx of new isolates from around the world. Such a technique is molecular taxonomy. The ramifications and ultimate potential of, for example, PCR-based techniques as a taxonomic tool can only be guessed at, but molecular taxonomy has gone a long way in a relatively short time span and its full potential has yet to be realized. Molecular techniques have an intrinsic elegance; they must surely have a secure future in helping to resolve problems in nematode identification and phylogenetic relationships. As such, they should be welcomed and judged by classical taxonomists on their merits rather than being viewed with suspicion.

This is neither to advocate nor endorse a total reliance on molecular characterization – the methodology offers an alternative insight into the problem, a view which then needs to be tested with data from other approaches before a consensus on speciation is reached. Molecular taxonomy is not a panacea; it is a powerful tool, but one which needs to be handled with awareness and understanding. While DNA studies are undeniably efficient at producing differing genomic profiles, we are entitled to ask what do these mean? We need to clarify whether such differences reflect the evolution of a particular gene, the organism, or both. In the biosystematics of phytoparasitic and entomopathogenic nematodes, both the classical and molecular methodologies are valid approaches. The two are complementary, offering different perspectives on the same problem. The onus is on biosystematists to integrate the methodologies and to achieve a synthesis, a synthesis with the ability to delimit, recognize and identify an organism as its ultimate purpose. As advocated by Hyman and Powers (1991) and Ferris and Ferris (1992), a judicious combination of classical and molecular approaches can only enhance our understanding of nematode speciation. In conclusion, the following quote from Powers and Adams (1994) is particularly apt:

> '...when we make the conceptual step from genetic marker to genetic relatedness, we must be careful not to bias our view of nematode relationships with expectations of well-defined species boundaries. It would be convenient if nematode species existed at the terminal branches of a well-pruned phylogenetic tree. The reality may be that

some nematode groups exist at the terminal branches of a tangled hedge, with rather indistinct boundaries'.

Acknowledgements

I thank Alex Reid and Bernie Briscoe for helpful discussions on the characterization of the Steinernematidae and Heterorhabditidae and Julie Oliver and Janice Sheldon for assistance.

11.4 REFERENCES

Akhurst, R.J. (1987) Use of starch gel electrophoresis in the taxonomy of the genus *Heterorhabditis* (Nematoda: Heterorhabditidae). *Nematologica*, **33**, 1–9.

Burrows, P.R. (1990) The use of DNA to identify plant parasitic nematodes. *Nematological Abstracts*, **59**, 1–8.

Cain, A.J. (1959) The post-Linnaean development of taxonomy. *Proceedings of the Linnaean Society of London*, **170**, 234–44.

Cenis, J.L., Opperman, C.H. and Triantaphyllou, A.C. (1992) Cytogenetic, enzymatic and restriction fragment length polymorphism variation of *Meloidogyne* spp. from Spain. *Phytopathology*, **82**, 527–31.

Chacon, M.R., Parkhouse, R.M.E., Burrows, P.R. and Garate, T. (1995) The use of heterologous cloned DNA probes to distinguish between races of *Meloidogyne incognita*. *Nematologica*, **41**, 251–7.

Cho, M.R. and Robbins, R.T. (1991) Morphological variation among 23 *Xiphinema americanum* populations. *Journal of Nematology*, **23**, 134–44.

Cohn, E. and Sher, S.A. (1972) A contribution to the taxonomy of the genus *Xiphinema* Cobb, 1913. *Journal of Nematology*, **4**, 36–65.

Coomans, A. and Chaves, E. (1995) *Paraxiphidorus michelluci* n. g., n. sp. from Argentina (Nematoda: Longidoridae). *Fundamental and Applied Nematology*, **18**, 303–6.

Cracraft, J. (1997) Species concepts in systematics and conservation biology – an ornithological viewpoint, in *Species: the Units of Biodiversity* (eds M.F. Claridge, H.A. Dawah and M.R. Wilson), Chapman & Hall, London, pp. 325–39.

Cronquist, R. (1977) Once again, what is a species? in *Beltsville Symposia in Agricultural Research. 2. Biosystematics in Agriculture* (ed. J.A. Romberger), Wiley, New York, pp. 3–20.

Curran, J. (1990) Molecular techniques in taxonomy, in *Entomopathogenic Nematodes in Biological Control* (eds R. Gaugler and H.K. Kaya), CRC Press, Boca Raton, pp. 63–74.

Curran, J. (1992) Molecular taxonomy of nematodes, in *Nematology from Molecule to Ecosystem* (eds F.J. Gommers and P.W.Th. Maas), Proceedings, Second International Nematology Congress, 11–17 August, 1990, Veldhoven, The Netherlands, European Society of Nematologists, pp. 83–91.

Curran, J. and Driver, F. (1994) Molecular taxonomy of *Heterorhabditis*, in *COST 812 Biotechnology: Genetics of entomopathogenic nematode-bacterium complexes, Proceedings of Symposium and Workshop, St. Patrick's College, Maynooth, Co. Kildare, Ireland* (eds A.M. Burnell, R.-U. Ehlers, and J.P. Masson), Luxembourg, European Commission, DG XII, pp. 41–8.

Curran, J. and Hominick, W.M. (1981) Effect of mounting methods on taxonomic characters of adult male mermithids (Nematoda: Mermithidae). *Nematologica*, **26** (1980), 455–66.

Curran, J. and Webster, J.M. (1989) Genotypic analysis of *Heterorhabditis* isolates from North Carolina. *Journal of Nematology*, **21**, 140–5.

De Giorgi, C., Sialer, M.F. and Lamberti, F. (1994) The use of mtDNA for the identification of plant parasitic nematodes, in *Advances in Molecular Plant Nematology* (eds F. Lamberti, C. De Giorgi and D. Mck. Bird), vol. **268**, NATO ASI Series, Series A: Life Sciences, Plenum Press, pp. 109–18.

Dix, I., Burnell, A.M., Griffin, C.T., Joyce, S.A., Nugent, M.J. and Downes, M.J. (1991) The identification of biological species in the genus *Heterorhabditis* (Nematoda: Heterorhabditidae) by cross-breeding second-generation amphimictic adults. *Parasitology*, **104**, 509–18.

Esbenshade, P.R. and Triantaphyllou, A.C. (1990) Isozyme phenotypes for the identification of *Meloidogyne* species. *Journal of Nematology*, **22**, 10–15.

Fargette, M., Blok, V.C., Phillips, M.S. and Trudgill, D.L. (1994) Genetic variation in tropical *Meloidogyne* species, in *Advances in Molecular Plant Nematology* (eds F. Lamberti, C. De Giorgi and D. Mck. Bird), vol. **268**, NATO ASI Series, Series A: Life Sciences, Plenum Press, pp. 91–6.

Ferris, V.R. and Ferris, J.M. (1992) Integration of classical and molecular approaches in nematode systematics, in *Nematology from Molecule to Ecosystem* (eds F.J. Gommers and P. Th. Maas), Proceedings, Second International Nematology Congress, 11–17 August, 1990, Veldhoven, The Netherlands. European Society of Nematologists, pp. 92–100.

Glaser, R.W. and Farrell, C.C. (1935) Field experiments with the Japanese beetle and its nematode parasite. *Journal of the New York Entomological Society*, **43**, 345–71.

Griffin, C.T., Joyce, S.A., Dix, I., Burnell, A.M. and Downes, M.J. (1994) Characterization of the entomopathogenic nematode *Heterorhabditis* (Nematoda: Heterorhabditidae) from Ireland and Britain by molecular and cross-breeding techniques and the occurrence of the genus in these islands. *Fundamental and Applied Nematology*, **17**, 245–53.

Hahn, M.L., Burrows, P.R., Gnanapragasam, N.C., Bridge, J., Vines, N. and Wright, D.J. (1994) Molecular diversity amongst *Radopholus similis* populations from Sri Lanka detected by RAPD analysis. *Fundamental and Applied Nematology*, **17**, 275–81.

Heyns, J. (1983) Problems of species delimitation in the genus *Xiphinema*, with special reference to monosexual species, in *Concepts in Nematode Systematics* (eds A.R. Stone, H.M. Platt and L.F. Khalil), vol. **22**, Systematics Association Special Volume, pp. 163–74.

Hooper, D.J. and Ibrahim, S.K. (1994) *Aphelenchoides nechaleos* n. sp. and *A. paranechaleos* n. sp. (Nematoda: Aphelenchoididae) from rice plants. *Fundamental and Applied Nematology*, **17**, 153–60.

Hunt, D. J. (1993) *Aphelenchida, Longidoridae and Trichodoridae: Their Systematics and Bionomics.* CAB International, Wallingford, UK.

Hyman, B. and Powers, T. (1991) Integration of classical and molecular approaches in nematode taxonomy. *Annual Review of Phytopathology*, **29**, 89–107.

Ibrahim, S.K. and Hooper, D.J. (1994) *Aphelenchoides varicaudatus* sp. n. (Nematoda: Aphelenchoididae). *Afro-Asian Journal of Nematology*, **4**, 210–14.

Ibrahim, S.K., Perry, R.N. and Hooper, D.J. (1994a) Use of esterase and protein patterns to differentiate two new species of *Aphelenchoides* on rice from other species of *Aphelenchoides* and *Ditylenchus angustus* and *D. myceliophagus*. *Nematologica*, **40**, 267–75.

Ibrahim, S.K., Perry, R.N., Burrows, P.R. and Hooper, D.J. (1994b) Differentiation of species and populations of *Aphelenchoides* and of *Ditylenchus angustus* using a fragment of ribosomal DNA. *Journal of Nematology*, **26**, 412–21.

Joyce, S.A., Burnell, A.M. and Powers, T.O. (1994a) Characterization of *Heterorhabditis* isolates by PCR amplification of segments of mtDNA and rDNA genes. *Journal of Nematology*, **26**, 260–70.

Joyce, S.A., Griffin, C.T. and Burnell, A.M. (1994b) The use of isoelectric focusing and polyacrylamide gel electrophoresis of soluble proteins in the taxonomy of *Heterorhabditis* (Nematoda: Heterorhabditidae). *Nematologica*, **40**, 601–12.

Joyce, S.A., Reid, A., Driver, F. and Curran, J. (1994c) Application of polymerase chain reaction (PCR) methods to the identification of entomopathogenic nematodes, in *COST 812 Biotechnology: Genetics of entomopathogenic nematode-bacterium complexes, Proceedings of Symposium and Workshop, St. Patrick's College, Maynooth, Co. Kildare, Ireland* (eds A.M. Burnell, R.-U. Ehlers and J.P. Masson), Luxembourg, European Commission, DG XII, pp. 178–87.

Kaplan, D.T. (1994) Molecular characterization of the burrowing nematode sibling species, *Radopholus citrophilus* and *R. similis*, in *Advances in Molecular Plant Nematology* (eds F. Lamberti, C. De Giorgi and D. Mck. Bird), vol. **268**, NATO ASI Series, Series A: Life Sciences, Plenum Press, pp. 77–83.

Kozodoi, E.M. and Spiridonov, S.E. (1988) Cuticular ridges on lateral fields of invasive larvae of *Neoaplectana* (Nematoda: Steinernematidae). *Folia Parasitologica*, **35**, 359–62.

Kozodoi, E.M., Voronov, D.A. and Spiridonov, S.E. (1986) [The use of an electrophoretic method to determine the species association of *Neoaplectana* specimens]. *Trudy Gel' minthologicheskoi laboratorii (Voprody biotsenologii gel' mintov)*, **34**, 55–9.

Lamberti, F. and Bleve-Zacheo, T. (1979) Studies on *Xiphinema americanum sensu lato* with descriptions of fifteen new species (Nematoda, Longidoridae). *Nematologia mediterranea*, **7**, 51–106.

Lamberti, F. and Carone, M. (1991) A dichotomous key for the identification of species of *Xiphinema* (Nematoda: Dorylaimida) within the *X. americanum*-group. *Nematologia Mediterranea*, **19**, 341–8.

Lamberti, F., De Giorgi, C. and Bird, D. Mck. (eds) (1994) *Advances in Molecular Plant Nematology*. NATO ASI Series, Series A: Life Sciences, vol. **268**, Plenum Press.

Loof, P.A.A. and Luc, M. (1990) A revised polytomous key for the identification of species of the genus *Xiphinema* Cobb, 1913 (Nematoda: Longidoridae) with exclusion of the *X. americanum* group. *Systematic Parasitology*, **16**, 35–66.

Loof, P.A.A. and Luc, M. (1993) A revised polytomous key for the identification of species of the genus *Xiphinema* Cobb, 1913 (Nematoda: Longidoridae) with exclusion of the *X. americanum* group. Supplement 1. *Systematic Parasitology*, **24**, 185–9.

Loof, P.A.A., Luc, M. and Coomans, A. (1993) The *Xiphinema americanum* group (Nematoda: Dorylaimida). 1. Comments upon the key to species published by Lamberti and Carone (1992). *Fundamental and Applied Nematology*, **16**, 355–88.

Luc, M. and Dalmasso, A. (1975a) A 'lattice' for the identification of species of *Xiphinema* Cobb, 1913, in *Nematode Vectors of Plant Viruses* (eds F. Lamberti, C.E. Taylor and J.W. Seinhorst), Plenum Press, London, New York, pp. 53–70.

Luc, M. and Dalmasso, A. (1975b) Considerations on the genus *Xiphinema* Cobb, 1913 (Nematoda: Longidoridae) and a 'lattice' for the identification of species. *Cahiers de l' Office de la Recherche Scientifique et Technique Outre-Mer, Biologie*, **10**, 303–27.

Mamiya, Y. (1988) *Steinernema kushidai* n. sp. (Nematoda: Steinernematidae) associated with scarabaeid beetle larvae from Schizuoka, Japan. *Applied Entomology and Zoology*, **23**, 313–20.

Mayr, E. (1969) *Principles of Systematic Zoology*, McGraw-Hill, New York.

Mracek, Z. and Bednarek, A. (1991) The morphology of the lateral fields of infective juveniles of entomogenous nematodes of the family Steinernematidae (Nematoda). *Nematologica*, **37**, 63–71.

Mracek, Z. and Weiser, J. (1979) The head papillae of the Steinernematidae. *Journal of Invertebrate Pathology*, **34**, 310–11.

Mracek, Z. and Weiser, J. (1981) Head and cuticular structures of some species in the family Steinernematidae (Nematoda). *Nematologica*, **27**, 443–8.

Nguyen, K. B. and Smart, G. C. (1990) *Steinernema scapterisci* n. sp. (Rhabditida: Steinernematidae). *Journal of Nematology*, **22**, 187–99.

Nguyen, K. B. and Smart, G. C. (1992) *Steinernema neocurtillis* n. sp. (Rhabditida: Steinernematidae) and a key to the species of the genus *Steinernema*. *Journal of Nematology*, **24**, 463–77.

Poinar, G. O. (1990) Taxonomy and biology of Steinernematidae and Heterorhabditidae, in *Entomopathogenic Nematodes in Biological Control* (eds R. Gaugler and H.K. Kaya), CRC Press, Boca Raton, pp. 23–61.

Poinar, G. O. (1993) Origins and phylogenetic relationships of the entomophilic rhabditids, *Heterorhabditis* and *Steinernema*. *Fundamental and Applied Nematology*, **16**, 333–8.

Powers, T.O. and Adams, B. (1994) Nucleotide sequences in nematode systematics, in *Advances in Molecular Plant Nematology* (ed. F. Lamberti, C. De Giorgi and D. Mck. Bird), NATO ASI Series, Series A: Life Sciences, vol. **268**, Plenum Press, pp. 99–108.

Powers, T. O. and Harris, T. S. (1993) A polymerase chain reaction method for identification of five major *Meloidogyne* species. *Journal of Nematology*, **25**, 1–6.

Reid, A.P. (1994) Molecular Taxonomy of *Steinernema*, in *COST 812 Biotechnology: Genetics of entomopathogenic nematode-bacterium complexes, Proceedings of Symposium and Workshop, St. Patrick' s College, Maynooth, Co. Kildare, Ireland* (ed. A.M. Burnell, R.-U. Ehlers and J.P. Masson), Luxembourg, European Commission, DG XII, pp. 49–58.

Reid, A.P. and Hominick, W.M. (1992) Restriction fragment length polymorphisms within the ribosomal DNA repeat unit of British entomopathogenic nematodes (Rhabditida: Steinernematidae). *Parasitology*, **105**, 317–23.

Reid, A.P. and Hominick, W.M. (1993) Cloning of the rDNA repeat unit from a British entomopathogenic nematode (Steinernematidae) and its potential for species identification. *Parasitology*, **107**, 529–36.

Reid, A.P. and Hominick, W.M. (1994) Molecular taxonomy and phylogeny of entomopathogenic nematodes. *Paper presented at the 22nd International Nematology Symposium, Gent, Belgium, 7–12 August, 1994*; abstract p. 107.

Sasser, J.N. and Carter, C.C. (1985) Overview of the International *Meloidogyne* Project 1975–1984, in *An Advanced Treatise on* Meloidogyne. *Biology and Control* (eds J.N. Sasser and C.C. Carter), North Carolina State University, pp. 19–24.

Smits, P.H., Groenen, J.T.M. and De Raay, G. (1991) Characterization of *Heterorhabditis* isolates using DNA restriction fragment length polymorphism. *Revue de Nématologie*, **14**, 445–53.

Steiner, G. (1923) *Aplectana kraussei* n. sp., einer in der Blattwespe *Lyda* sp. parasitierende Nematodenform, nebst Bemerkungen ber das Seitenorgan der parasitischen Nematoden. *Zentralbl. Bakteriol. Parasitenkd. Infektionskr. Hyg. Abt.* 2, **59**, 14.

Swart, A. and Heyns, J. (1987) Comparative morphology of the head regions of some longidorid nematodes from South Africa using the SEM. *Phytophylactica*, **19**, 99–101.

Tarjan, A.C. (1969) Variation within the *Xiphinema americanum* group (Nematoda: Longidoridae). *Nematologica*, **15**, 241–52.

Tchesunov, A.V. and Rieman, F. (1995) Arctic sea ice nematodes (Monhysteroidea), with descriptions of *Cryonema crassum* gen. n., sp. n. and *C. tenue* sp. n. *Nematologica*, **41**, 35–50.

Vrain, T.C. and Wakarchuk, D.A. (1989) Construction of a set of cloned rDNA probes for the specific differentiation of the *Xiphinema americanum* group. *Journal of Nematology*, **21**, 593–4.

Vrain, T.C., Wakarchuk, D.A., Lévesque, A.C. and Hamilton, R.I. (1992) Intraspecific rDNA restriction fragment length polymorphism in the *Xiphinema americanum* group. *Fundamental and Applied Nematology*, **15**, 563–73.

Williamson, V.M., Caswell-Chen, E.P., Wu, F.F. and Hanson, D. (1994) PCR for nematode identification, in *Advances in Molecular Plant Nematology* (eds F. Lamberti, C. De Giorgi and D. Mck. Bird), NATO ASI Series, Series A: Life Sciences, vol. **268**, Plenum Press, pp. 119–27.

12

Species in insect herbivores and parasitoids – sibling species, host races and biotypes

M. F. Claridge , H. A. Dawah and M. R. Wilson

Contacting address: School of Pure and Applied Biology, University of Wales Cardiff, P.O. Box 915, Cardiff CF1 3TL, Wales, UK

ABSTRACT

The species diversity of terrestrial ecosystems is dominated by insects. Of these, the most speciose groups are those that feed either on green plants as herbivores, or as parasitoids on other insects, including the herbivores. Generally these insects are characterized by extreme levels of host specificity and low levels of morphological differentiation. The genetic and taxonomic status of populations from different hosts is often very difficult to determine. Species taxonomy is thus a major problem. Particular difficulties with host-associated populations are shared with all parasitic organisms. The terms host race, biotype, etc. have been widely, but often uncritically, used.

In many families differentiation of the male external genitalia provides morphological markers widely used by taxonomists to discriminate species. Most groups of Hymenoptera Parasitica, which include the greatest diversity of insect parasitoids, do not exhibit such diversification of male genitalia so that species characterization is even more difficult.

So-called biological concepts have been used widely for sexually reproducing species. The investigation of specific mate recognition systems as factors maintaining reproductive isolation have revealed sibling species in a variety of taxa. Enzyme electrophoresis and DNA technologies are being used widely to provide species markers.

In general, biological and phylogenetic concepts will provide similar solutions to species problems in these insects.

Species: The Units of Biodiversity. Edited by M.F. Claridge, H.A. Dawah and M.R. Wilson. Published in 1997 by Chapman & Hall. ISBN 0 412 63120 2

12.1 INTRODUCTION

Insects dominate the diversity of life on earth. They comprise more than 56% of the described species of all living organisms and best recent estimates suggest that about 64% of all existing species are insects (Hammond, 1992). Paradoxically, in morphology the insects are a relatively conservative group so that species and higher taxon differentiation has to be based on relatively small differences.

Of living insect species it is estimated that about 50% are herbivores, mostly associated with vascular plants (Strong *et al.*, 1984). Of the remaining 50%, most are thought to be parasitoids, largely attacking insect herbivores as their primary hosts. Thus, the diversity of modern insects is dominated by these two essentially parasitic life-styles (Price, 1980). Both insect herbivores and parasitoids are characterized by extreme specialization. In particular most species are extremely specific in host utilization, typically being restricted to one, or a few related, host species. More polyphagous species are unusual and typically fewer in number (Figure 12.1). Clearly, a precise understanding of the nature of species and the species status of particular populations is important in understanding such phenomena.

Since the first descriptions by Linnaeus (1759), the species taxonomy of insects has been almost entirely dependent on external morphological characters. The generally hardened nature of insect cuticle means that samples are still most frequently preserved dry in museum collections and this dependence on external morphology still continues. Indeed, the history of insect taxonomy may be seen as one of seeking for more and more refined methods for the morphological differentiation of species. In the early years of the 20th century entomologists realized that the structure of the external genitalia, particularly but not exclusively of males, provided a new and very diverse array of characters that were extremely valuable in discriminating between otherwise very similar, or indeed apparently identical, species. The result is that for very many groups of insects from most orders, characters of the genitalia are essential for species delimitation. For example, in the large leafhopper subfamily Typhlocybinae (Cicadellidae), species are mostly small and rather uniform in appearance. The discovery of distinctive features of the male genitalia very greatly increased the number of recognized species. For example, the genus *Edwardsiana* (=*Typhlocyba* in part) in Europe includes small tree-living leafhoppers, mostly of a uniform pale yellow colour. Ribaut (1936) first made detailed studies widely on male genitalia of these insects. He found in particular that the terminal branching of the aedeagus was very variable (Figure 12.2) and then went on to recognize 19 species from France on this basis. Ossiannillsson (1981), using the same methodology, recognized 24 species from Scandinavia. With few exceptions, *Edwardsiana* females show no obvious morphological differentiation. However, some

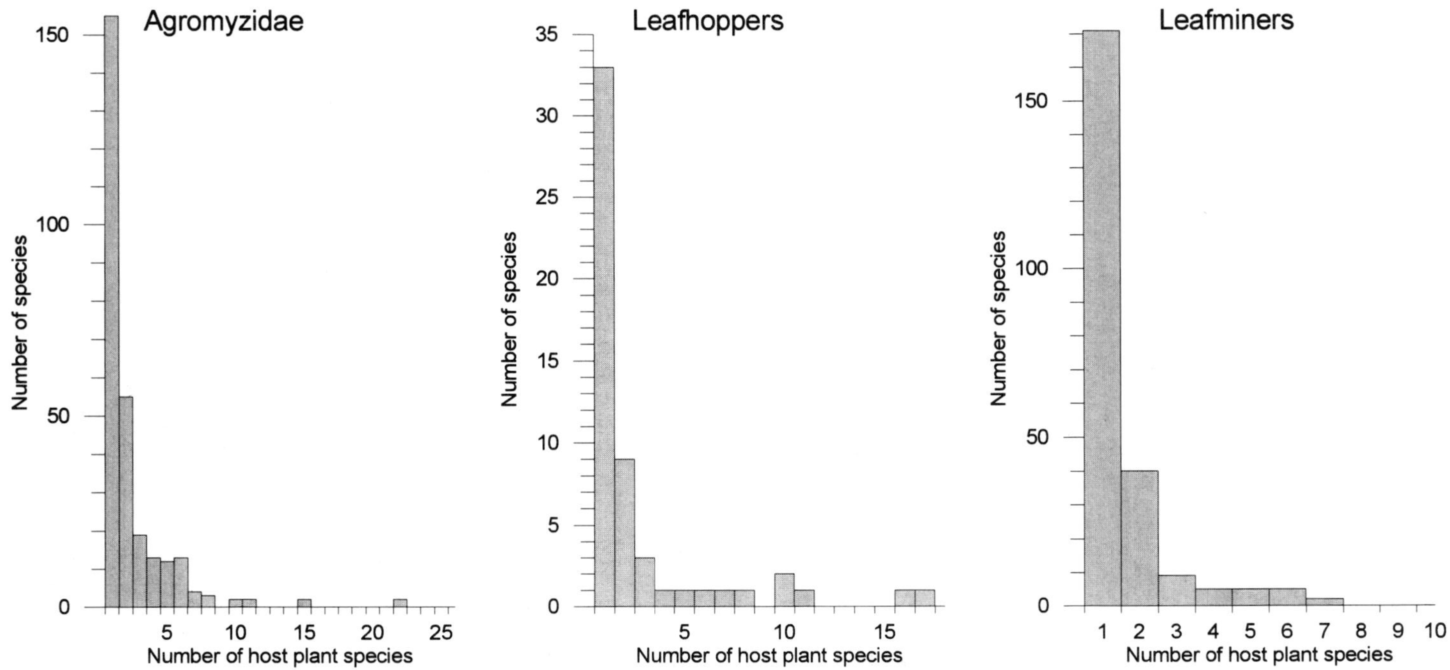

Figure 12.1 Histograms to show spectrum of host specificity in three different groups of insect herbivores in Britain: Diptera, Agromyzidae (after Strong *et al.*, 1984), mesophyll feeding leafhoppers (Cicadellidae, Typhlocybinae) associated with trees (after Claridge and Wilson, 1981), and leafmining insect larvae associated with trees (after Claridge and Wilson, 1982).

confirmation of the biological reality of the male-based differentiation is provided by analyses of host plant-associated distributions of the species so recognized (Claridge and Wilson, 1976, 1981). The realization of the importance of genitalial characters in species taxonomy has been repeated for most major groups of insect herbivores, including many Hemiptera, Lepidoptera, Diptera, Coleoptera, Hymenoptera and most smaller orders. The biological significance of such species characteristics has been reviewed by Eberhard (1985). Most insect parasitoids are Hymenoptera Parasitica. In general, male genitalia in this group are relatively simple in structure and do not provide reliable morphological species markers; consequently, species taxonomy in these families is notoriously difficult. There are some notable, but rare, exceptions to this generalization, for example egg parasites of the genus *Trichogramma* (Nagarkatti and Nagaraja, 1971).

Most insect herbivores and parasitoids are biparental, sexually reproducing organisms. Thus, a biological species concept may be applied to host-associated populations and the determination of species status will ultimately depend on the demonstration of high levels of reproductive isolation in the field. The morphological characters generally used for convenience in the study of these insects are thus markers indicating lack of gene flow. In many groups, refined biochemical markers, primarily obtained as a result of using enzyme gel electrophoresis, but also more often now also by DNA technology, have been used to establish levels of gene flow and reproductive isolation. In some cases, reviewed below, attempts have been made to study the specific mate recognition systems which result in the observed reproductive isolation between species. Also such studies may help in the determination of the genetic status of allopatric populations. In addition, mate choice experiments – if adequately controlled – can provide essential information on the species status of related populations (Claridge and Morgan, 1987; Claridge, 1988).

12.2 STATUS OF HOST-ASSOCIATED POPULATIONS – HOST RACES AND BIOTYPES

In all parasitic organisms a major factor, if not the most important one, is the relationship with their hosts which provide not only a food source but also frequently the bulk of their environment. This is certainly true of insect herbivores and parasitoids. The interpretation of the taxonomic and genetic status of populations from different, often related, hosts has long been an area of controversy. Where such populations differ clearly in morphology there has normally been no problem about recognizing them as distinct species, whatever the author's preferred species concept. However, where host-related populations show no clear-cut diagnostic differences in morphology, but do show some differences in survival and development on a

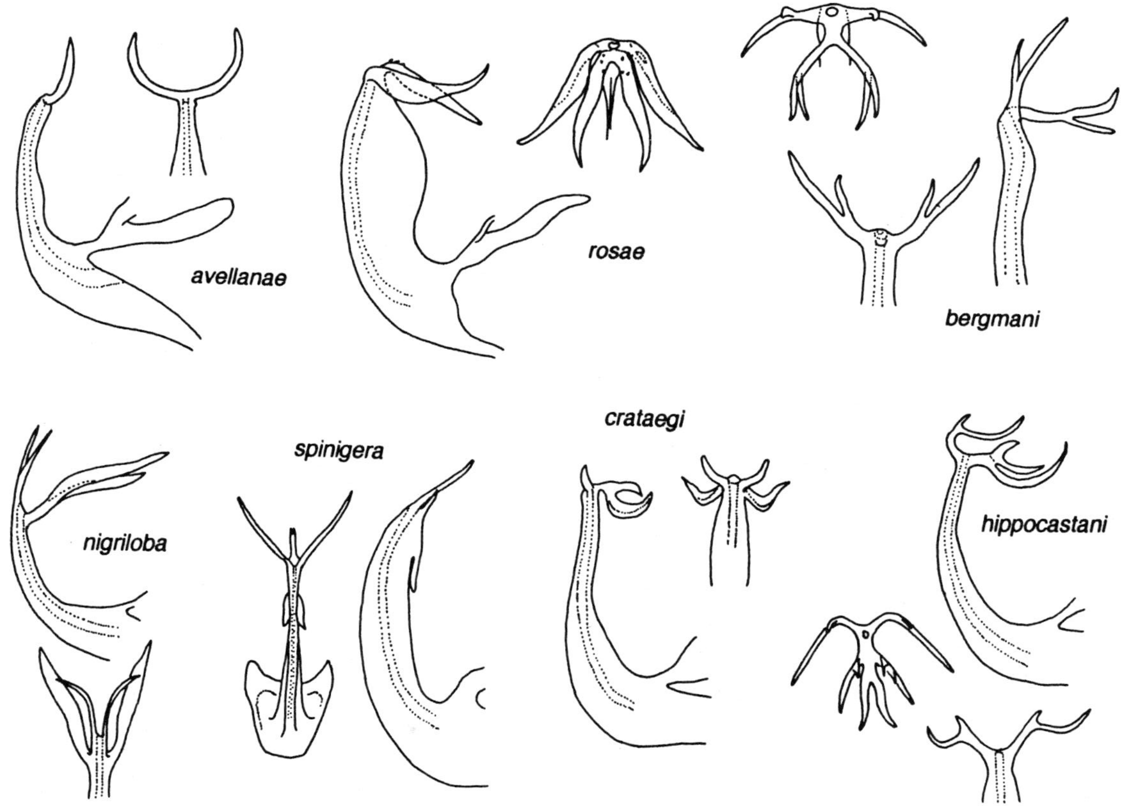

Figure 12.2 Variation in the male aedeagus as shown in seven species of *Edwardsiana* (Cicadellidae, Typhlocybinae). (After Ribaut, 1936.)

particular host, difficulties have arisen. This is a longstanding problem first clearly enunciated by Benjamin Walsh (1864) in his classic paper on 'phytophagic varieties' and 'phytophagic species' (see also Bush, 1995). Not surprisingly, such problems have been most widely studied in pests of agriculture and forestry. The phytophagic races of Walsh have more usually been termed biological races in recent years (see Thorpe 1930 for early review; and Jaenike, 1981) or biotypes (Claridge and den Hollander, 1983; Diehl and Bush, 1984). The interpretation of such populations, which apparently differ primarily in adaptation to survival and reproduction on particular hosts, is difficult, but has often been made more so by exclusive support for particular theories of speciation. Thus, strict adherence to theories of allopatric speciation (Mayr, 1963; Paterson, 1985) would suggest that host-associated differentiation and speciation will occur only when populations are isolated in space. Intermediate, partially differentiated host races or biotypes should then only be found in allopatry. However, if sympatric speciation is prevalent (Bush, 1975, 1993, 1994) then all intermediate host race stages between freely interbreeding parasitic populations and fully isolated biological species should be found reasonably frequently in sympatry. The arguments between the two general schools of thought continue. However, for parasitic organisms the determination of allopatry and sympatry may be difficult. Ultimately, differentiating spatial from host isolation may itself be impossible and have little meaning; indeed, perhaps this is no longer a useful approach. More important is an understanding of the nature of the genetic differences between host-associated populations and the degree and regularity of gene flow between them.

In practice, morphological differentiation is still most widely used in attempts to characterize host associated populations of insects. The continuing development of modern multivariate techniques for handling quantitative data derived from morphometric studies have greatly expanded the possibilities for analysis (Foottit and Sorenson, 1992). However refined the statistical methods available, it is essential to differentiate between those characteristics of a population or individual that are a direct result of induced responses to feeding and living on particular hosts and those that represent real genetic differences between the insect populations. The ideal way to achieve this is to transfer insects between hosts and repeat the original morphometric measurements (Claridge and Gillham, 1992). Unfortunately this is often not easy for purely practical reasons, including difficulties in growing suitable host plants and culturing particular herbivores and parasitoids. However, a few detailed examples are available.

12.2.1 The leafhopper *Alnetoidia alneti*

Alnetoidia alneti (Dahlbom) is a common and easily recognized species in Europe. In Britain it feeds and reproduces in the field on 17 species from

13 different genera of trees and shrubs (Claridge and Wilson, 1981). Individuals reared from some different hosts differ in size and colour. This prompted Ribaut (1936) to recognize two species – *A. alneti* found on alder and elm, and a smaller and paler form, *A. coryli* from hazel. Claridge and Gillham (1992) and Gillham and Claridge (1994) reported multivariate analyses of populations of *A. alneti* from 13 different host plants in South Wales and southern England based on 27 different body measurements. Canonical variates analyses (CVA) were performed on these data and demonstrated significant differences between most populations. The most extreme were those from alder and hazel. Populations sampled in different years and from different parts of South Wales and southern England showed very little within host-plant group variation. First and second instar nymphs collected from a single alder tree in South Wales were reared on the foliage of three alder trees and similar nymphs collected from one hazel tree were reared on three hazels. In addition, a sample of the same nymphs from alder were reared on hazel foliage. CVAs of adult measurement data for these samples show very clearly that those reared on alder and hazel respectively were all similar among themselves, but quite distinct from each other. That transferred from alder to hazel, however, grouped with the hazel insects (Figure 12.3). Thus, the clear differences between these samples, which included features of both size and shape, must be interpreted as being host plant-induced. There is then no support in these data to suggest that the host plant associated populations of *A. alneti* are genetically distinct and certainly not that they are separate biological species.

12.2.2 Biotypes of the brown planthopper, *Nilaparvata lugens*

Nilaparvata lugens (Stål) is a major pest of cultivated rice, *Oryza sativa*, throughout Asia. Populations of *N. lugens* identified by screening against cultivars of rice which incorporate different genes for resistance to feeding have been termed biotypes. They are most obviously categorized by their abilities in mass screening trials to damage particular rice cultivars characterized by different genotypes (Table 12.1). Saxena and co-workers suggested that these biotypes were equivalent to host races and were therefore intermediate stages in a speciation process. They used CVA techniques to demonstrate significant differences between the biotypes numbered 1, 2 and 3 reared on the rice cultivars to which they were adapted (Saxena and Rueda, 1982). However, when the same biotypes were reared on the same susceptible cultivar the significant differences between them disappeared (Claridge *et al.*, 1984) (Figure 12.4). These findings, together with the ease with which these biotypes hybridize and the viability of F_1 and F_2 hybrids (Claridge *et al.*, 1985a), suggest that genetic differentiation between them is slight.

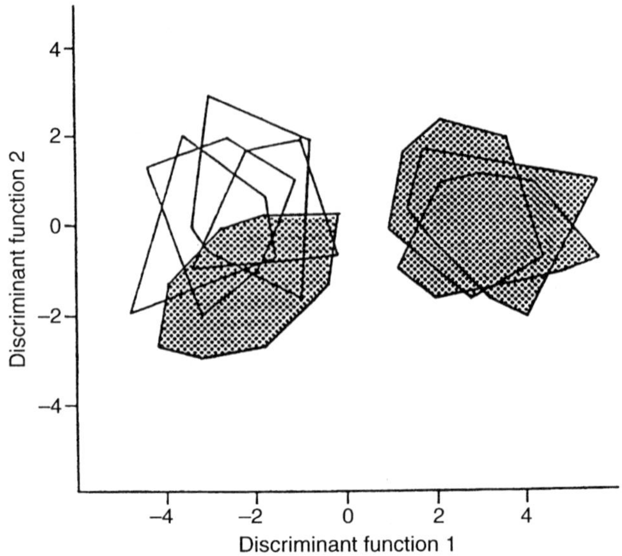

Figure 12.3 Canonical variate plots of space circumscribed for adults of the leafhopper *Alnetoidia alneti* reared on hazel, *Corylus avellana*, (grouped to the left) and alder, *Alnus glutinosa* (grouped to the right). Shaded areas represent adults derived from early instar nymphs originally collected from alder. (After Claridge and Gillham, 1992.)

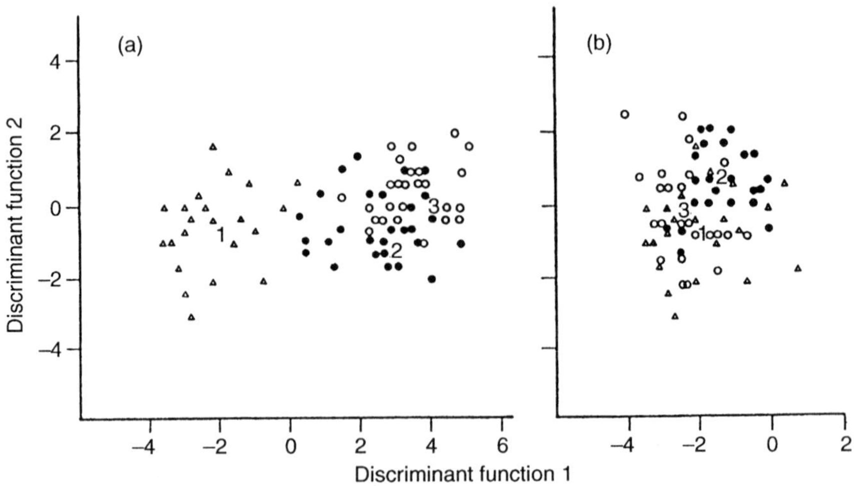

Figure 12.4 Canonical variate plots for biotypes 1, 2 and 3 of *Nilaparvata lugens*. (a) reared respectively on cultivars TNI, Mudgo and ASD7; (b) all reared on TN1. (After Claridge *et al.*, 1984.)

Table 12.1 The possible 'biotypes' of *Nilaparvata lugens* on rice showing some traditional and modern high-yielding cultivars susceptible to each and associated nomenclature. The dominance status of different genes in the plants is also indicated. (After Claridge and den Hollander, 1982.)

Biotype	*Resistance gene*	*Rice variety*	
		Traditional	*Improved*
1	None identified	TB1	IR8
2	Bph 1 (dominant)	Mudgo	IR26
3	bph 2 (recessive)	ASD7	IR36 IR42
4	Bph 3 (dominant)	Rathuheenati	IR56
5	bph4 (recessive)	Babawee	–

Populations attributable morphologically to *N. lugens*, but feeding exclusively on the grass *Leersia hexandra*, show no greater morphometric differentiation than do the rice-associated biotypes and were indeed originally designated as a further biotype by Saxena *et al.* (1983). However, we now know them to represent different biological species (Claridge *et al.*, 1985b, 1988, and see below).

These examples demonstrate the need for care in interpreting differentiation between host-associated populations in the absence of experimental and genetic data.

12.3 BIOLOGICAL SPECIES, SPECIFIC MATE RECOGNITION AND SIBLING SPECIES

A solution to the problems of interpreting the different degrees of host-associated variation discussed above in both insect herbivores and parasitoids is to use the biological species concept with its emphasis on reproductive isolation in the field, achieved through distinct mate recognition systems, and thus lack of significant gene flow between species (Claridge *et al.*, 1997: Chapter 1). This is possible since most insects are biparental. The morphological characters normally used in taxonomy are markers for presumed genetic differentiation of the species under study. More direct methods involve the use of a variety of biochemical and molecular techniques (Loxdale and den Hollander, 1989; Hawksworth, 1994; Symondson and Liddell, 1996). In particular, enzyme gel electrophoresis has provided simple techniques for the estimation of gene flow between populations. The identification of diagnostic and unique enzyme gene loci in host-associated populations clearly establishes their biological species status (Menken and Ulenberg, 1987; Menken, 1989). These techniques have now been applied widely in many groups of insect herbivores and parasitoids.

A particularly thorough study has been made of the small ermine moths, *Yponomeuta* species, in Europe (summary in Menken *et al.*, 1992). Most differ by small, but clear morphological characters, particularly of the male genitalia. The larvae of these insects feed on the foliage of a variety of mostly trees and shrubs. Some species show minor host plant differentiation but no clear-cut morphological characterization. For example, *Y. malinellus* attacks *Malus* and *Pyrus* species and *Y. padellus* a variety of other Rosaceae, including *Crataegus* and *Prunus*, but not *Malus* or *Pyrus*. These two were originally cited as examples of host races of one species (Thorpe, 1929), but Menken (1980) showed that, despite high levels of genetic identity that allow laboratory hybridization between them, interbreeding does not occur in the field and at least one completely diagnostic locus has been identified.

Similar techniques have been used to determine species status in a wide variety of insect groups. Reviews of many such examples are provided in the volume edited by Loxdale and den Hollander (1989). The diversity of genetic techniques available now provides a refined set of markers by which we may establish the existence of reproductive isolation between sympatric populations. However, biological species are characterized by specific mate recognition systems (Paterson, 1985) which result in the observed reproductive isolation between species. Thus, the study of specific mate recognition systems (SMRS) will enable us to identify biological species with greater certainty.

Mate finding and courtship in insects is usually very complex and involves a sequence of signal exchanges between males and females before mating and successful fertilization take place. This complete sequence of signal exchange forms the SMRS of any species. Insects have a wide diversity of sense organs and are capable of discriminating signals in different modalities. Any complete SMRS is likely to involve different signals and receptors, including the chemical, visual, mechanical and auditory senses, and will be unique for each species. Unfortunately, rather few studies have been made of complete SMRS sequences. However, since the chemical senses predominate in many aspects of insect life it is probable that specific chemical signals – pheromones – are usually important. These are particularly well known in the Lepidoptera where in many species specific distance communication is achieved in mate finding by the release of a sex pheromone by virgin females. This, known as 'calling' behaviour, brings conspecific receptive males to the stationary females, often over very long distances. In many species further pheromones are released by both sexes at closer range as further elements of the SMRS sequence. At all stages the potential for chemical discrimination is very great. The pheromones in many species consist of cocktails of several compounds with specificity being conferred by the precise proportions of the different compounds involved. Similar chemical systems of communication dominate mate finding and courtship in many major groups of herbivores and parasitoids, including many Coleoptera, Diptera and Hymenoptera (Phelan, 1992).

Few detailed studies on pheromone systems have been made on groups of related insect herbivores or parasitoids. Probably the best worked are the nine species of small ermine moths, *Yponomeuta*, in Europe (Menken *et al.*, 1992). Virgin females 'call' as described above. Males respond maximally to the female pheromone of their own species (Hendrikse, 1979, 1986). When in close proximity males emit their own pheromone, accompanied usually by characteristic wing fanning behaviour (Hendrikse *et al.*, 1984). The result is that attempts at interspecific copulations are very rare. These chemical interactions are clearly central to the SMRS of *Yponomeuta* species.

Unfortunately, chemical signals and their role in courtship are difficult to study and at least require complex analytical facilities. Difficulties of chemical analysis mean that such systems have rarely been used to determine the nature of SMRS sequences. Most of what we know derives from studies on individual pest species where pheromones may be used in traps for monitoring populations and even in systems of control by disrupting normal mating in the field. However, there is little doubt that further analyses of pheromone systems will lead to the recognition of groups of sibling species presently thought to be one. Published examples of supposed pheromone polymorphisms should be treated with suspicion in this context.

Visual and acoustic systems of communication and mate finding are rarer than chemical ones, but easier for the human observer to study. In particular, acoustic systems have been widely studied, mainly among the Orthoptera and Hemiptera. The Homoptera Auchenorrhyncha is a major group exclusively of herbivores. All species so far as known use acoustic signals in the initial stages of mate finding and courtship (Claridge, 1985a,b). There is no evidence of chemical communication in a sexual context before physical contact is made between potential mates. In the larger cicadas (Cicadoidea) long-distance communication is achieved by loud, high-intensity male calling to which virgin females are attracted. In close contact other calls may be made and pheromones may also be important. In all other groups, including the species-rich Delphacidae and Cicadellidae, low-intensity acoustic signals are produced by males and usually also by females. These calls are transmitted through the plant substrate on which the insects live. Calls of both sexes are usually species specific and form essential elements of the SMRS sequence and have now been used to establish biological species status in a number of different genera (Claridge, 1985a,b; Claridge and de Vrijer, 1994).

One of the best examples of the analysis of a group of sibling species is provided by the work of Wood (1993) on the complex of morphologically almost identical species of treehoppers (Membracidae) in North America, known as *Enchenopa binotata*. In a series of elegant and detailed electrophoretic studies Wood has shown that *E. binotata* consists of at least nine different host-specific and reproductively isolated biological species. Recently, Hunt (1994) has shown that these insects also use acoustic signals in courtship and mating behaviour. Further studies on all of the species will provide exciting evidence in the debate about species and speciation in these insects.

12.3.1 Species of *Nilaparvata* – a herbivore case study

The planthopper genus *Nilaparvata* includes 16 morphological species, differentiated primarily on features of the male genitalia and mostly distributed in the tropical and subtropical regions of the world. *N. lugens* (Stål), the Brown Planthopper, is a major pest of rice in Asia and thus well studied (Claridge and Morgan, 1987; Wilson and Claridge, 1991).

(a) Nilaparvata lugens

N. lugens (Figure 12.5) is clearly differentiated from all other species on a basis particularly of the male aedeagus and paramere structure. It is widely distributed through much of Asia and northern Australia where it feeds only on wild and cultivated rices, *Oryza* (Figure 12.5). Over this enormous geographical range it shows no obvious variation in morphology.

Like other planthoppers, *N. lugens* uses primarily acoustic signals transmitted through the plants on which they live in mate finding and courtship (Claridge and de Vrijer, 1994). Receptive adult males move actively from plant to plant producing characteristic amplitude-modulated signals (Figure 12.6). If a receptive virgin female is within the acoustic field she may respond with her own simple call consisting of a train of regularly repeated pulses (Figure 12.6). In a typical sequence the male and female continue to exchange calls. While the female remains stationary, the calling male moves actively over the plant surface until contact is made. Calling continues and copulation usually ensues. There is no doubt that acoustic signals are vitally important in species recognition in these insects. Indeed, appropriate behaviour from both males and females may be elicited simply by the use of pre-recorded signals in the absence of an insect (Claridge and de Vrijer, 1994).

Populations of *N. lugens* from different parts of Asia and Australia show geographical variation in quantitative features of both male and female calls, in particular the rates at which pulses are produced – pulse repetition frequency (PRF) – in main sections of their calls. Most extreme are the differences between populations from Australia and those collectively from Asia. Correlated with these quantitative differences in calls is the

Figure 12.5 Adult *Nilaparvata lugens* on rice plants. Brachypterous female (left) and macropterous male (right).

relative difficulty of obtaining laboratory hybrids between such populations (Claridge *et al.*, 1985a). Mate choice experiments between Asian and Australian populations show significant preferences for homogametic matings, though not total reproductive isolation. The relative status of populations from Asia and Australia is thus doubtful.

Populations morphologically attributable to *N. lugens* living and feeding on the grass *Leersia hexandra* are now known to occur widely over Asia and Australia where they are frequently sympatric with rice-associated ones (Figure 12.5) (Claridge *et al.*, 1985b, 1988). Calls of both males and females from sympatric populations differ significantly in PRF. However, as in the rice-feeding populations, those from *Leersia* show similar geographical variation (Claridge *et al.*, 1985b, 1988). Mate choice experiments for sympatric populations from both the Philippines and Australia showed very significant preferences for homogametic matings. Laboratory-produced hybrids between the two populations from the Philippines were intermediate in call characters between the two parental

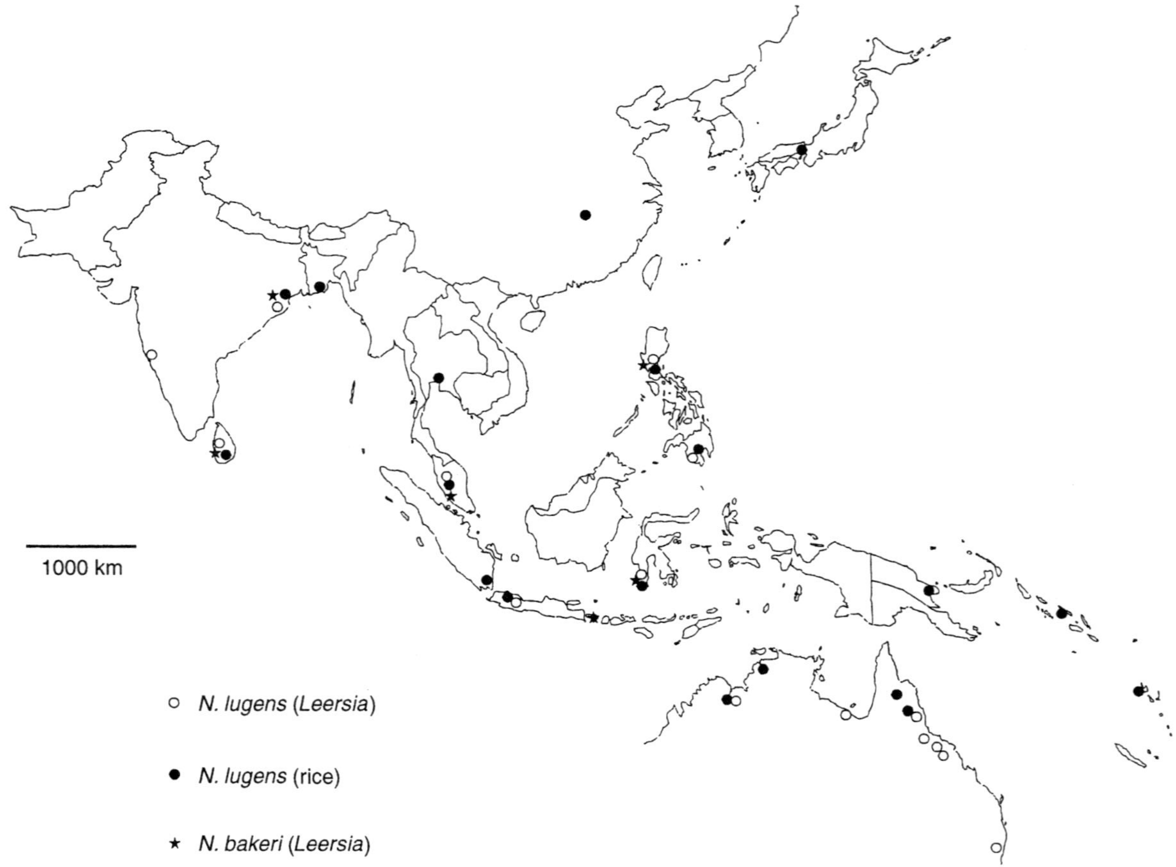

Figure 12.6 Sketch map of Asia to show localities from which Claridge and co-workers have sampled *Nilaparvata lugens* from rice and *Leersia hexandra*, and *N. bakeri* from *Leersia hexandra* only.

types and significantly different from both. Thus, if hybrids were commonly produced in field samples they would be detectable. However, no field-caught individuals with such intermediate calls were found by Claridge *et al.* (1985b, 1988) from any of the many regions in Asia and Australia that were sampled (Figure 12.5).

Playback of pre-recorded calls to both males and females of each host-associated population showed very significant preferential responses for the calls of their own type (Table 12.2). Thus, it is clear that the populations of *N. lugens* from rice and *Leersia* respectively in both Asia and Australia represent different biological species and not just different biotypes or host races, as suggested by Saxena and Barrion (1985). The status of the populations in Australia by comparison with those in Asia is difficult to determine. However, some recent preliminary molecular data suggest that Australian populations should be regarded as specifically distinct from those in Asia (Jones *et al.*, 1996).

(b) Nilaparvata bakeri

Of the remaining morphological species of *Nilaparvata* only *N. bakeri* has been studied intensively. It is a distinct morphological species differing from *N. lugens* in obvious features of both the male aedeagus and parameres. In Asia, *N. bakeri* is widely sympatric with both the rice- and *Leersia*-associated species of *N. lugens*, but it has not yet been found with certainty in Australia. It is a specific feeder associated only with *Leersia*.

The calls of *N. bakeri* males and females show some similarity to those of *N. lugens*, but differ significantly, particularly in the males (Figure 12.7) (Claridge and Morgan, 1993). Neither males nor females respond to the calls of *N. lugens* from either rice or *Leersia*. Despite many efforts neither have we been able to obtain laboratory hybrids between them. They

Table 12.2 Percentages of males and females of rice- and *Leersia*-associated populations of *Nilaparvata lugens* responding to playback of prerecorded calls of female and male calls respectively. Actual numbers of insects responding out of total numbers given in parentheses. (After Claridge *et al.*, 1985b.)

	Pecentage responding to	
	Rice male call	Leersia *male call*
Female response		
Rice females	93 (14/15)	13 (2/15)
Leersia females	27 (4/15)	87 (13/15)
	Rice female call	Leersia *female call*
Male response		
Rice males	77 (23.30)	30 (6/20)
Leersia males	17 (5/30)	90 (18/20)

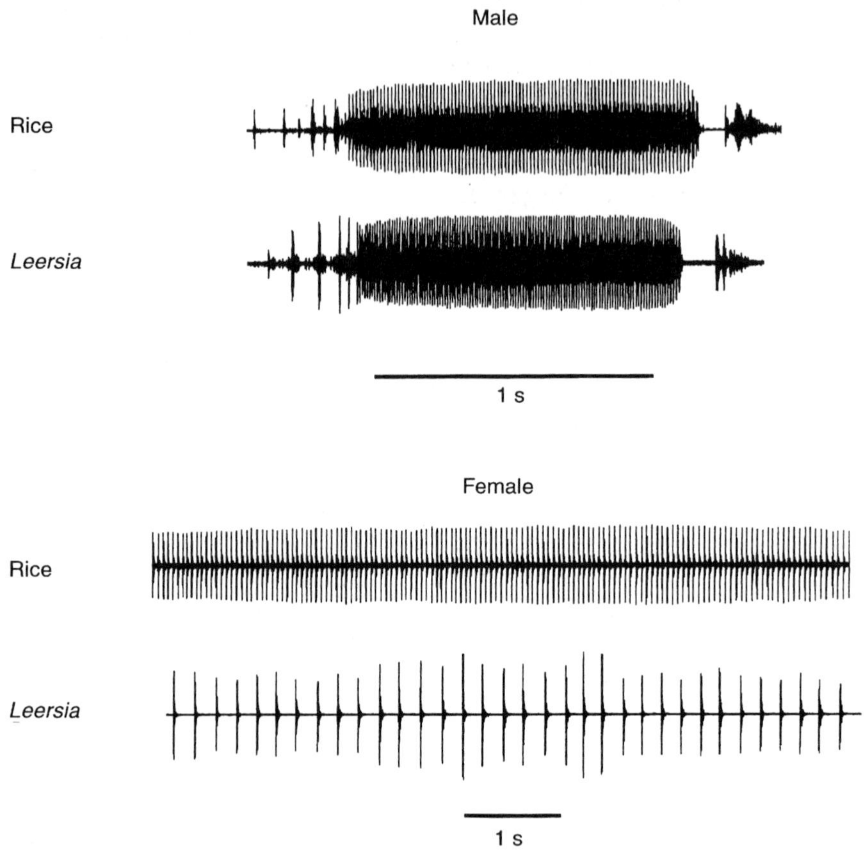

Figure 12.7 Oscillograms of sections of individual male and female calls from rice- and *Leersia*-associated species of *Nilaparvata lugens* from the Philippines. (After Claridge *et al.*, 1985b.)

undoubtedly represent quite distinct species, whatever species concept is employed.

Like both species attributed to *N. lugens*, *N. bakeri* shows no obvious geographical variation in morphology. However, populations from Luzon (Philippines), Bali (Indonesia), West Bengal (India) and Kandy (Sri Lanka) show dramatic differences in male calls (Claridge and Morgan, 1993). In particular, calls from Philippine males are strikingly different to those from Bali to the human observer (Figure 12.8). However, mate choice experiments showed no preference for homogametic matings between these two very distinct call types. Indeed, there is no indication that the striking call differences to the human observer between these allopatric populations would have any role in maintaining reproductive isolation between them. Thus, despite clearly diagnostic differences in calls they cannot be regarded as different species.

Unfortunately, most of the other morphological species of *Nilaparvata* are little known. Distinctive calls have been recorded for *N. muiri* from Japan and *N. maeander* from West Africa, but little more information is available (Claridge and Morgan, 1987).

12.4 HERBIVORE/PARASITOID FOOD WEBS – GRASS-FEEDING CHALCID WASPS

It is estimated that between 20% and 25% of all insect species are parasitoids of other insects (Godfray, 1994: 16). These are dominated by the Hymenoptera Parasitica and the large Dipterous family Tachinidae. Feeding interactions between parasitoid and host species are often specialized, with several parasitoids attacking the same hosts. Similar problems concerning difficulties of determining the status of host-associated populations occur for parasitoids as for herbivores. Because of the difficulties of Hymenoptera taxonomy, there has been a tendency to restrict species in these groups to morphologically diagnosable entities. Indeed, the frustration with the often rudimentary state of taxonomic knowledge has often led ecologists to use artificial minimally recognizable units. For example, Memmott and Godfray (1993: 229) suggested that morphological species or 'morphotype' determination rather than detailed biological species analyses are adequate to determine the structure of parasitoid dominated food webs. However, such procedures undoubtedly underestimate the degree of host specificity and thus of compartmentation in such

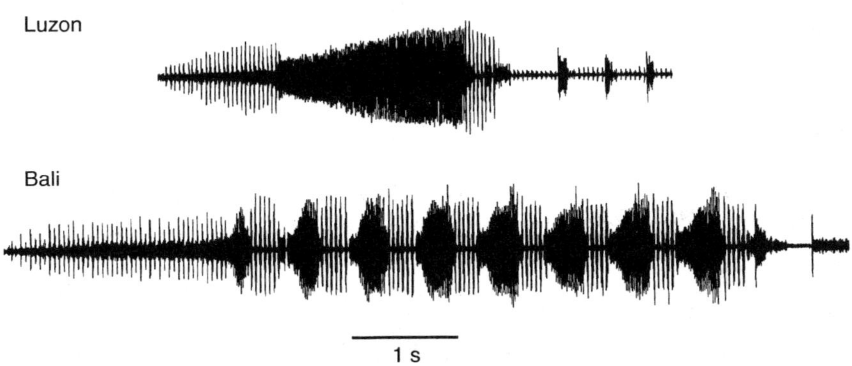

Figure 12.8 Oscillograms of sections of individual male calls of *Nilaparvata bakeri* from Luzon, Philippines, and Bali, Indonesia. (After Claridge and Morgan, 1993.)

webs. Sometimes the scale of such underestimation may be dramatic (Claridge and Dawah, 1994; Dawah *et al.*, 1995).

Relatively few studies have been made in an attempt seriously to apply biological species concepts to groups of parasitoids. However, where it has been done, morphological species have been shown to consist of two or more sibling species, usually with more limited host ranges and other significant differences in behaviour and life histories (Table 12.3).

A good example is provided by the herbivorous chalcid wasps of the genus *Tetramesa* (Eurytomidae) and their mostly also chalcid parasitoids associated with grasses and cereals (Claridge and Dawah, 1994; Dawah *et al.*, 1995). Species discrimination in *Tetramesa* has been based on small,

Table 12.3 Some examples of the application of biological species concepts in Hymenoptera Parasitica using various techniques, including behavioural, pheromones, electrophoresis and molecular methods

Species complex	*Host*	*Reference*
Braconidae		
Aphidius spp.	Various aphid host	Pungerl, 1986; Unruh *et al.*, 1986; Holler, 1991; Castanera *et al.*, 1983
Asobara spp.	Drosophilidae	Vet and Janse, 1984; Vet *et al.*, 1984
Eulophidae		
Pediobius eubis complex	*Tetramesa* sp.	Dawah, 1988a
Tetrastichus spp.	*Crioceris* spp.	Alphen, 1980
Eurytomidae		
Tetramesa sp.	Phytophagous	Dawah, 1987
Eurytoma appedigaster group	*Tetramesa* sp. *Eurytoma* sp.	Dawah, 1988b
Trichogrammatidae		
Trichogramma sp.	Various Lepidoptera	Hung, 1982; Pintureau and Voegele, 1980; Pinto *et al.*, 1992, 1993; Kostadinov and Pintureau, 1991; Landry *et al.*, 1993
Mymaridae		
Anaphes sp.	*Listronotus* spp.	Landry *et al.*, 1993
Aphelinidae		
Aphytis spp.	Diasipididae	Khasimuddin and DeBach, 1976
Pteromalidie		
Chlorocytus spp.	*Tetramesa* sp. *Eurytoma* sp.	Dawah, 1989
Spalangia spp.	*Musca domestica*	Propp, 1986
Muscidifurax spp.	*Stomoxys calcitrans*	Propp, 1986; Assem and Povel, 1973
Nasonia spp.	Muscidae	Assem and Werren, 1994

often difficult to appreciate, and variable characters of the adult females (Walker, 1832; Thomson, 1875; Hedicke, 1920; Phillips, 1920; Phillips and Poos, 1922; Claridge, 1961; Zerova, 1976). In fact, great emphasis has usually been placed on host plant records and features of life-histories. It is only recently that biological species boundaries have been determined more precisely by mate choice experiments and gel electrophoresis to determine evidence for lack of gene flow and therefore of reproductive isolation between supposed species (Dawah, 1987). These studies have confirmed the previously suspected generally extreme host specificity of *Tetramesa* species. Only one species is known to attack host plants from more than one genus of grasses.

Tetramesa are attacked by a series of characteristic parasitoids dominated particularly by the Eurytomidae – *Eurytoma* (Figure 12.9) and *Sycophila*, Eulophidae – *Pediobius*, and Pteromalidae – *Chlorocytus* and *Homoporus*. All of these show extreme host specificity and very limited morphological differentiation, as confirmed particularly by enzyme gel electrophoresis (Dawah 1988a,b, 1989). For *Pediobius*, Dawah (1988a) was able to identify nine different biological species in Britain alone on a basis of mate choice experiments and electrophoresis in the *P. eubius* (Walker) complex. Previously these had been variously regarded as either only one (Bouček, 1965) or three (Graham, 1963) species. Once real species limits were determined, careful and detailed microscopic work revealed very small differences which make it possible normally to identify dead adult females (Dawah, 1988a).

Thus, it is clear that the application of biological species approaches to these groups of Hymenoptera reveals much more significant biological variation than would otherwise have been suspected. The real structure of food webs can only be determined following such studies.

12.5 DISCUSSION AND CONCLUSIONS

Most insect herbivores and parasitoids are biparental, sexually reproducing forms and it is clear that the application of a broadly biological species concept leads to the recognition of more species than a traditional purely morphological approach. Such species represent important ecological entities about which useful generalizations concerning host ranges and patterns of exploitation may be made. In particular, the widespread occurrence of groups of sibling species can only be demonstrated by a biological approach.

For insect herbivores and parasitoids in practice, application of either the biological species or the phylogenetic species (in the sense of Cracraft, 1997: Chapter 16) will almost always produce the same results, though the example of *Nilaparvata bakeri* illustrates the more precise resolving power of the biological concept. Only in very well-worked groups is the problem of

Figure 12.9 Courting male and female of *Eurytoma pollux*, a parasitoid of *Tetramesa calamagrostidis* in *Calamagrostis epigejos*.

determining the status of allopatric populations an important one. On the whole if allopatric populations are clearly diagnosably distinct and, in the absence of contrary evidence, then it is preferable to give them specific rank.

Our major concern about the phylogenetic species as applied to the insects under consideration here is a procedural one. The biological species approach lays emphasis on specific mate recognition systems and reproductive isolation so that sibling species should be readily detected. The phylogenetic species approach gives no incentive to expose the existence of sibling species if no obvious diagnosable differences are apparent between the populations under investigation.

Thus, in practice, for specialist insect herbivores and parasitoids a biological species approach has the best potential for revealing species diversity and is thus to be preferred. However, we see no difficulty in reconciling the two approaches for these organisms.

Acknowledgements

We thank all those colleagues who have helped us in discussion of species problems in insects. In particular our special thanks are due to Arthur Cain for reading a draft of this chapter, John Morgan for preparing some of the figures, David Windsor for the photograph in Figure 12.9, and Rosemary Jones for preparing the manuscript.

12.6 REFERENCES

Alphen, J.J.M. van (1980) Aspect of the foraging behaviour of *Tetrastichus asparagi* Crawford and *Tetrastichus* spec. (Eulophidae), gregarious egg parasitoids of the asparagus beetles *Crioceris asparagi* L. and *C. duodecimpunctata* L. (Chrysomelidae). *Netherlands Journal of Zoology*, **30**, 307–25.

Assem, J. van den and Povel, G.D.E. (1973) Courtship behaviour of some *Muscidifurax* species (Hymenoptera: Pteromalidae): a possible example of a recently evolved ethological isolating mechanism. *Netherlands Journal of Zoology*, **23**, 465–87.

Assem, J. van den and Werren, J.H. (1994) A comparison of the courtship and mating behaviour of three species of *Nasonia* (Hymenoptera: Pteromalidae). *Journal of Insect Behaviour*, **7**, 53–66.

Bouček, Z. (1965) Studies of European Eulophidae IV: *Pediobius* Walk. and two allied genera (Hymenoptera). *Acta Entomologica Musei Nationalis Prague*, **36**, 5–90.

Bush, G.L. (1975) Modes of animal speciation. *Annual Review of Ecology and Systematics*, **6**, 339–64.

Bush, G.L. (1993) A reaffirmation of Santa Rosalia, or why are there so many kinds of small animals? in *Evolutionary Patterns and Processes* (eds D.R. Lees and D. Edwards), Academic Press, London, pp. 229–49.

Bush, G.L. (1994) Sympatric speciation. *Trends in Ecology and Evolution*, **9**, 285–8.

Bush, G.L. (1995) Species and speciation. *Trends in Ecology and Evolution*, **10**, 38

Castanera, P., Loxdale, H.D. and Nowak, K. (1983) Electrophoretic study of enzymes from cereal aphid populations. 11. Use of electrophoresis for identifying aphidiid parasitoids (Hymenoptera) of *Sitobion avenae* (F.) (Hemiptera: Aphididae). *Bulletin of Entomological Research*, **73**, 659–65.

Claridge, M.F. (1961) A contribution to the biology and taxonomy of some Palaearctic species of *Tetramesa* Walker (=*Isosoma* Walk.,=*Harmolita* Motsch)(Hymenoptera: Eurytomidae), with particular reference to the British fauna. *Transactions of the Royal Entomological Society of London*, **113**, 175–216.

Claridge, M.F. (1985a) Acoustic signals in the Homoptera: behaviour, taxonomy and evolution. *Annual Review of Entomology*, **30**, 297–317.

Claridge, M.F. (1985b) Acoustic behaviour of Leafhoppers and Planthoppers: species problems and speciation, in *The Leafhoppers and Planthoppers* (eds L.R. Nault and J.G. Rodriquez), John Wiley, New York, pp. 103–25.

Claridge, M.F. (1988) Species concepts and speciation in parasites, in *Prospects in Systematics* (ed. D.L. Hawksworth), Clarendon Press, Oxford, pp. 92–111.

Claridge, M.F. and Dawah H.A. (1994) Assemblages of herbivorous chalcid wasps and their parasitoids associated with grasses-problems of species and specificity, in *Plant Galls* (ed. M.A.J. Williams), Clarendon Press, Oxford, pp. 313–28.

Claridge, M.F. and den Hollander, J. (1982) Virulence to rice cultivars and selection for virulence in populations of the brown planthopper *Nilaparvata lugens*. *Entomologia experimenatalis et Applicata*, **32**, 213–21.

Claridge, M.F. and den Hollander, J. (1983) The biotype concept and its application to insect pests of agriculture. *Crop Protection*, **2**, 85–95.

Claridge, M.F. and de Vrijer, P. (1994) Reproductive behaviour: the role of acoustic signals in species recognition and speciation, in *Planthoppers Their Ecology and Management* (eds R.F. Denno and T.J. Perfect), Chapman & Hall, New York, pp. 216–33.

Claridge, M.F. and Gillham, M. (1992) Variation in populations of leafhoppers and planthoppers (Auchenorrhyncha): biotypes and biological species, in *Ordinations in the study of Morphology, Evolution and Systematics of Insects: Application and Quantitative Genetic Rationales* (eds R. Foottit and J.T. Sorenson), Elsevier, Amsterdam, pp. 241–59.

Claridge, M.F. and Morgan, J.C. (1987) The Brown Planthopper, *Nilaparvata lugens* (Stål), and some related species: a biotaxonomic approach, in *Proceedings of the 2nd International Workshop on Leafhoppers and Planthoppers of Economic Importance* (eds M.R. Wilson and L.R. Nault), CAB Institute of Entomology, London, pp. 19–32.

Claridge, M.F. and Morgan, J.C. (1993) Geographical variation in acoustic signals of the planthopper, *Nilaparvata bakeri* (Muir), in Asia: species recognition and sexual selection. *Biological Journal of the Linnean Society*, **48**, 267–81.

Claridge, M.F. and Wilson, M.R. (1976) Diversity and distribution patterns of some mesophyll-feeding leafhoppers of temperate woodland canopy. *Ecological Entomology*, **1**, 231–50.

Claridge, M.F. and Wilson, M.R. (1981) Host plant associations, diversity and species–area relationships of mesophyll-feeding leafhoppers of trees and shrubs in Britain. *Ecological Entomology*, **6**, 217–38.

Claridge, M.F. and Wilson, M.R. (1982) Insect herbivore guilds and species–area relationships: leafminers on British treas. *Ecological Entomology*, **7**, 19–30.

Claridge, M.F., den Hollander, J. and Haslam, D. (1984) The significance of morphometric and fecundity differences between the ' biotypes' of the brown planthopper, *Nilaparvata lugens*. *Entomologia Experimentalis et Applicata*, **36**, 107–14.

Claridge, M.F., Den Hollander, J. and Morgan, J.C., (1985a) Variation in courtship signals and hybridization between geographically definable populations of the rice brown planthopper *Nilaparvata lugens* (Stål). *Biological Journal of the Linnean Society*, **24**, 35–49.

Claridge, M.F., den Hollander, J. and Morgan, J.C. (1985b) The status of weed-associated populations of the brown planthopper, *Nilaparvata lugens*(Stål) – host race or biological species? *Zoological Journal of the Linnean Society*, **84**, 77–90.

Claridge, M.F., den Hollander, J. and Morgan, J.C. (1988). Variation in hostplant relations and courtship signals of weed-associated populations of the brown planthopper, *Nilaparvata lugens* (Stål), from Australia and Asia: a test of the recognition species concept. *Biological Journal of the Linnean Society*, **35**, 79–93.

Claridge, M.F., Dawah, H.A. and Wilson, M.R. (1997) Practical approaches to species concepts for living organisms, in *Species: the Units of Biodiversity* (eds M.F. Claridge, H.A. Dawah and M.R. Wilson), Chapman & Hall, London, pp. 1–15.

Cracraft, J. (1997) Species concepts in systematics and conservation–an ornithological viewpoint, in *Species: the Units of Biodiversity* (eds M.F. Claridge, H.A. Dawah and M.R. Wilson), Chapman & Hall, London, pp. 325–39

Dawah, H.A. (1987) Biological species problems in some *Tetramesa* (Hymenoptera: Eurytomidae). *Biological Journal of the Linnean Society*, **32**, 237–45.

Dawah, H.A. (1988a) Taxonomic studies on the *Pediobius eubius* complex (Hymenoptera: Chalcidoidea; Eulophidae) in Britain, parasitoids of Eurytomidae in Gramineae. *Journal of Natural History*, **22**, 1147–71.

Dawah, H.A. (1988b) Differentiation between *Eurytoma appendigaster* group (Hymenoptera: Eurytomidae) using electrophoretic esterase patterns. *Journal of Applied Entomology*, **105**, 144–8.

Dawah, H.A. (1989) Separation of four *Chlorocytus* species (Hymenoptera: Pteromalidae), parasitoids of stem-boring Hymenoptera (Eurytomidae and Cephidae) using enzyme electrophoresis. *The Entomologist*, **108**, 216–22.

Dawah, H.A., Hawkins, B.A. and Claridge, M.F. (1995) Structure of the parasitoid communities of grass-feeding chalcid wasps. *Journal of Animal Ecology*, **64**, 708–20.

Diehl, S.R. and Bush, G.L. (1984) An evolutionary and applied perspective of insect biotypes. *Annual Review of Entomology*, **29**, 471–504.

Eberhardt, W.G. (1985) *Sexual Selection and Animal Genitalia*, Harvard University Press, Cambridge, Massachusetts.

Foottit, R. and Sorenson, J.T. (eds) (1992) *Ordinations in the Study of Morphology, Evolution and Systematics of Insects: Application and Quantitative Genetic Rationales*. Elsevier, Amsterdam.

Gillham, M.C. and Claridge, M.F. (1994) A multivariate approach to host plant associated morphological variation in the polyphagous leafhopper, *Alnetoidia alneti* (Dahlbom). *Biological Journal of the Linnean Society*, **53**, 127–51.

Godfray, H.C.J. (1994) *Parasitoids: Behavioral and Evolutionary Ecology*, Princeton University Press, Princeton, New Jersey.

Graham, M.W.R. de V. (1963) Key to the British genera and species of Elachertinae, Eulophinae, Eutodontinae and Enderinae (Hymenoptera: Chalcidoidea). *Transactions of the Society for British Entomology*, **13**, 169–204.

Hammond, P.M. (1992) Species inventory, in *Global Biodiversity: Status of the Earth's Living Resources* (ed. B. Groombridge), Chapman & Hall, London, pp. 17–39.

Hawksworth, D.L. (ed.) (1994) *Identification and Characterization of Pest Organisms*, CAB International, Wallingford.

Hedicke, H. (1920) Beitrage zu eine Monographie der palaarktischen Isosominen (Hymenoptera: Chalcidoidea). *Archives Naturgeschichtes*, **86**, 1–165.

Hendrikse, A. (1979) Activity patterns and sex pheromone specificity as isolating mechanisms in eight species of *Yponomeuta* (Lepidoptera: Yponomeutidae). *Entomologia Experimentalis et Applicata*, **25**, 172–80.

Hendrikse, A. (1986) Intra- and interspecific sex-pheromone communication in the genus *Yponomeuta*. *Physiological Entomology*, **11**, 159–69.

Hendrikse, A., van der Laan, C.E. and Kerkhof, L. (1984) The role of male abdominal brushes in the sexual behaviour of small ermine moths (*Yponomeuta* Latr., Lepidoptera). *Mededelingen van de Faculteit Landbouwwetenschappen Rijksuniversiteit Gent*, **49**, 719–29.

Holler, C. (1991) Evidence for the existence of a species closely related to the cereal aphid parasitoid *Aphidius rhopalosiphi* De stefani Perez based on host ranges, morphological characters, isoelectric focusing, banding patterns, cross-breeding experiments, sex pheromones specificities (Hymenoptera: Braconidae: Aphidiinea). *Systematic Entomology*, **16**, 15–28.

Hung, A.C.F. (1982) Chromosome and isozyme studies in *Trichogramma* (Hymenoptera: Trichogrammatidea). *Proceedings of the Entomological Society of Washington*, **4**, 791–6.

Hunt, R.E. (1994) Vibrational signals associated with mating behaviour in the tree-hopper, *Enchenopa binotata* Say (Hemiptera: Homoptera: Membracidae) *Journal of the New York Entomological Society*, **102**, 266–70.

Jaenike, L. (1981) Criteria for ascertaining the existence of host races. *The American Naturalist*, **117**, 830–4.

Jones, P., Gacesa, P. and Butlin, R. (1996) Systematics of brown planthopper and related species using nuclear and mitochondrial DNA, in *The Ecology of Agricultural Pests: Biochemical Approaches* (ed. W.O.C. Symondson and J.E. Liddell), Chapman & Hall, London, pp. 133–48.

Khasimuddin, S. and DeBach, P. (1976) Biosystematic studies of three allopatric population of *Aphytis maculicornis* (Hymenoptera: Aphelinidae). *Entomophaga*, **21**, 81–92.

Kostadinov, D. and Pintureau, B. (1991) A possibility to discriminate females of three closely related sp. of *Trichogramma* (Hymenoptera: Trichogrammatidae) with special purpose analysis of the type *T. evanescens* Westwood. *Annales Société Entomologique de France*, **27**, 393–400.

Landry, B.S., Dextraze, L. and Boivin, G. (1993) Random amplified polymorphic DNA markers for DNA fingerprinting and genetic-variability assessment of minute parasitic wasp species (Hymenoptera, Mymaridae and Trichogrammartidae) used in biological-control programs of phytophagous insects. *Genome*, **36**, 580–7.

Linnaeus, C. (1759) *Systema Naturae*, 10th edn, Stockholm.

Loxdale, H.D. and Den Hollander, J. (eds) (1989) *Electrophoretic Studies on Agricultural Pests*, Clarendon Press, Oxford.

Mayr, E. (1963) *Animal Species and Evolution*, Harvard University Press, Cambridge, Massachusetts.

Memmott, J. and Godfray, H.C.J. (1993) Parasitoids' webs, in *Hymenoptera and Biodiversity* (eds J. LaSalle and I.D. Gauld), CAB International, Wallingford, pp. 217–34.

Menken, S.B.J. (1980) Inheritance of allozymes in *Yponomeuta*. II. Interspecific crosses within the *padellus*-complex and reproductive isolation. *Proceedings of the Koninklyke Nederlandse Akademic van Weterschapper*, Series C, **83**, 424–31.

Menken, S.B.J. and Ulenberg, S.A. (1987) Biochemical characters in agricultural entomology. *Agricultural Zoology Reviews*, **2**, 305–60.

Menken, S.B.J., Herrebout, W.M. and Wiebes, J.T. (1992) Small Ermine moths (Yponomeuta): their host relations and Evolution. *Annual Review of Entomology*, **37**, 41–66.

Nagarkatti, S. and Nagaraja, H. (1971) Redescription of some known species of *Trichogramma* (Hymenoptera: Trichogrammatidae) showing the importance of the male genitalia as a diagnostic character. *Bulletin of Entomological Research*, **61**, 13–31.

Ossiannilsson, F. (1981) *The Auchenorrhyncha (Homoptera) of Fennoscandia and Denmark. Part 2: The families Cicadidae, Membracidae and Cicadellidae (excluding Deltocephalinae). Fauna Entomologica Scandinavica*, 7. Scandinavian Science Press Ltd, Klampenborg.

Paterson, H.E.H. (1985) The recognition concept of species, in *Species and Speciation* (ed. E.S. Vrba), Transvaal Museum, Pretoria, pp. 21–9.

Phelan, P.L. (1992) Evolution of sex pheromones and the role of asymmetric tracking, in *Insect Chemical Ecology: An Evolutionary Approach* (eds B.D. Roitberg and M.B. Ismam), Chapman & Hall, London, pp. 265–314.

Phillips, W.J. (1920) Studies on the life history and habits of the jointworm flies of the genus *Harmolita* (*Isosoma*), with recommendation for control. *United States Department of Agriculture Bulletin*, **808**, 1–27.

Phillips, W.J. and Poos, F. (1922) Five new species belonging to the genus *Harmolita* Motschulsky (*Isosoma* Walker *et auctt.*). *Kansas University Science Bulletin*, **14**, 349–59.

Pinto, J.D., Kazmer, D.J., Platner, G.R. and Sassaman, C.A. (1992) Taxonomy of the *Trichogramma minutum* complex (Hymenoptera: Trichogrammatidae): Allozymic variation and its relationship to reproductive and geographic data. *Annals of the Entomological Society of America*, **85**, 413–22.

Pinto, J.D., Platner, G.R. and Sassaman, C.A. (1993) Electrophoretic study of two closely related sp. of North American *Trichogramma*: *T. Pretiosum* and *T. deion* (Hymenoptera: Trichogrammatidae). *Annals of the Entomological Society of America*, **86**, 702–9.

Pintureau, B. and Voegele, J. (1980) Une nouvelle espece proche de *Trichogramma evanescens* et de *T. maidis* (Hymenoptera: Trichogrammatidae). *Entomophaga*, **25**, 431–40.

Price, P.W. (1980) *Evolutionary Biology of Parasites*, Princeton University, New Jersey.

Propp, G.D. (1986) Characterization of intra- and interspecific isoenzyme in *Spalangia* spp. and *Muscidifurax* spp. (Chalcidoidea: Pteromalidae), parasitoids of synathropic Diptera, in *Biological Control of Muscoid Flies* (eds R.S. Patterson and D.A. Rutz), Miscellaneous Publications of the Entomological Society of America, No. 61, pp. 164–74.

Pungerl, N.B. (1986) Morphometric and electrophoresis study of *Aphidius* (Hymenoptera: Aphidiidae) reared from a variety of Aphid hosts. *Systematic Entomology*, **11**, 327–54.

Ribaut, H. (1936) *Faune de France: Homoptères Auchenorrhynques, 1: (Typhlocybidae)*. Paul Lechevalier, Paris.

Saxena, R.C. and Barrion, A.A. (1985) Biotypes of the brown planthopper *Nilaparvata lugens* (Stål) and strategies in deployment of host plant resistance. *Insect Science and its Application*, **6**, 271–89.

Saxena, R.C. and Rueda, L.M. (1982) Morphological variation among three biotypes of the brown planthopper *Nilaparvata lugens* in the Philippines. *Insect Science and its Application*, **6**, 193–210.

Saxena, R.C., Velasco, M.V. and Barrion, A.A. (1983) Morphological variations between brown planthopper biotypes on *Leersia hexandra* and rice in the Philippines. *International Rice Research Newsletter*, **18**, 3.

Strong, D.R., Lawton, J.H. and Southwood, R. (1984) *Insects on Plants: Community Patterns and Mechanisms*, Blackwell, Oxford.

Symondson, W.O.C. and Liddell, J.E. (eds) (1996) *The Ecology of Agricultural Pests: Biochemical Approaches*, Chapman & Hall, London.

Thomson, C.G. (1875) *Hymenoptera Scandinaviae*, **4** (1). Lund, Sweden.

Thorpe, W.H. (1929) Biological races in *Hyponomeuta padella* L. *Zoological Journal of the Linnean Society*, **36**, 621–34.

Thorpe, W.H. (1930) Biological races in insects and allied groups. *Biological Reviews*, **5**, 177–212.

Unruh, T.R., White, W., Gonzalez, D. and Luck, R.F. (1986) Electrophoresis studies of parasitic Hymenoptera and implications for biological control, in *Biological Control of muscoid flies* (eds R.S. Patterson and D.A. Rutz), Miscellaneous Publications of the Entomological Society of America, No. 61, pp. 150–63.

Vet, L.E.M. and Janse, C.J. (1984) Fitness of two sibling species of *Asobara* (Braconodae: Alysiinae), larval parasitoids of Drosophilidae in different microhabitats. *Ecological Entomology*, **9**, 345–54.

Vet, L.E.M., Janse, C., Achterberg, C. van and Alphen, J.J.M. van (1984) Microhabitat location and niche segregation in two sibling species of Drosophilidae parasitoids: *Asobara tabida* (Nees) and *A. rufescens* (Foerster) (Braconidae: Alysiinae). *Oecologia*, **61**, 182–8.

Walker, F. (1832) Monographia Chalcidum. *Entomological Magazine*, **1**, 12–29.

Walsh, B.J. (1864) On phytophagic varieties and phytophagic species. *Proceedings of the Entomological Society of Philadelphia*, **3**, 403–30.

Wilson, M.R. and Claridge, M.F. (1991) *Handbook for the Identification of Leafhoppers and Planthoppers of Rice*, CAB International, Wallingford.

Wood, T. K. (1993) Speciation of the *Enchenopa binotata* complex (Insecta: Homoptera: Membracidae), in *Evolutionary Patterns and Processes* (eds D.R. Lees and D. Edwards), Academic Press, London, pp. 289–317.

Zerova, M.D. (1976) Hymenoptera Chalcidiodea family Eurytomidae, Subfamilies Rileyinae and Harmolitinae. (in Russian). *Fauna SSSR*, **7**, 1–23, Leningrad.

13

The species concept in blood-sucking vectors of human diseases

R. Lane

Contacting address: Department of Entomology, The Natural History Museum, Cromwell Road, London SW7 5BD, UK

ABSTRACT

Blood-sucking insects have attracted much attention from biologists because some species are vectors of devastating human and animal diseases. The need to determine which species transmit pathogens or parasites in any one location has put great demands on systematics on the one hand, but made considerable resources available on the other.

The classical biological species concept underpins the systematics of these insects, although most of the 14 500 or so species of blood-sucking insect are still defined on morphological criteria alone, i.e. discontinuities in morphological variation.

Species complexes of morphologically indistinguishable but biologically distinct and often sympatric species have been discovered in mosquitoes, simuliids and sandflies but not yet in other groups. Surprisingly, there is less proof than might be expected for the existence of species complexes from the obvious route – experimental genetics. Much evidence is inferential rather than experimentally proved and has necessitated the use of many characters and tests of reproductive distinctness. Chromosome markers, isozymes and laboratory cross-mating tests are the most commonly used criteria for establishing the specific status of taxa in species complexes. The chemistry of sex pheromones has even been used in one sandfly complex.

The techniques of molecular biology are only just beginning to make an impact on defining species although DNA probes have been very successful in the identification of wild-caught mosquito sibling species.

Species: The Units of Biodiversity. Edited by M.F. Claridge, H.A. Dawah and M.R. Wilson. Published in 1997 by Chapman & Hall. ISBN 0 412 63120 2

13.1 INTRODUCTION

Insects are the most species-rich group of organisms, with more species known already than all other organisms combined. However, within this vast group only a relatively few – some 14 500 described species – are members of families which suck mammalian blood and are therefore potential vectors of parasites and pathogens to humans. These blood-sucking insects do not form a 'natural' but a polyphyletic assemblage sharing one major life-history characteristic – feeding on vertebrate blood. Blood-feeding on homiotherms is found throughout the Insecta, but it occurs mainly in the Diptera (Lane and Crosskey, 1993).

Although, as a group, haematophagous insects are structurally diverse, they are not as genetically diverse as the parasites they transmit: viruses, bacteria, spirochetes, rickettsia, protozoa and nematodes. What they lack in relative diversity they make up in sheer numbers, so that the number of proven or suspected vector species is always much greater than the number of parasite species they transmit; for example, there are four human malaria parasites (*Plasmodium* spp.) but more than 70 species of *Anopheles* are implicated in their transmission. As a general principle, one species of parasite can be transmitted by several vector species but rarely the other way around. This relationship between the number of parasites and vectors is one reason why there is a considerable difference in the practical criteria used to define species.

An important feature distinguishing the systematics of blood-sucking insects from most other insect groups is the considerable interest it attracts from non specialists – interest shown because of the insects' ability to transmit disease. It is often recognized that accurate delimitation and subsequent identification of species is essential for effective vector control. Perhaps this recognition is not as often as the systematists would like, but in comparison with many other areas of applied entomology the importance of systematics has become axiomatic for effective vector control. To this end, the development of biting-insect taxonomy has been driven by a quest to explore the 'structure' of species at an ever-increasing resolution and this has had an effect not only on the theoretical concepts of species but on the practical application of these ideas.

13.2 THEORETICAL SPECIES CONCEPTS

Several species concepts are currently debated in the systematics literature, although relatively little debate on the broader theoretical aspects has taken place in the medical or veterinary entomology literature. The most well-known of the modern species concepts, is the biological species concept attributed to Dobzhansky (1937) subsequently promulgated by Mayr (1966) in which species are considered 'groups of actually or potentially interbreeding natural populations which are reproductively isolated from other such groups'. The concept is biological because the means of differentiating species is biological (reproductive ability) in contrast to the earlier concepts based on intrinsic or essential qualities used by Linnaeus and his contemporaries.

There are a number of problems, mainly operational but some theoretical, associated with the application of the biological species concept and this has generated a number of alternative concepts. One of the earliest was the evolutionary species concept of Simpson (1951) in which species are 'a lineage evolving separately from others and with its own unitary evolutionary role and tendencies'. This idea emphasized the notion that populations rather than whole species evolve, as is generally accepted, but there are significant difficulties in the practical application of this concept. In response to the emphasis on isolating mechanisms inherent in the BSP, Paterson (1985) focused attention on the unifying reproductive elements of species and therefore in his concept species share a common mate recognition system. It is of note that Paterson is one of the few medical entomologists to have engaged seriously in the debate on species concepts. While the differences between species in mate recognition mechanisms are known and can be characterized accurately in a few species, for example, species-specific sex pheromones of tsetse flies (Carlson *et al.*, 1984) or sandflies (Ward *et al.*, 1988; Ward, 1989), these characters have not been used primarily as the criteria for recognizing species. In the examples given species were defined originally by morphological discontinuities or cross-mating studies rather than by the mate recognition signals.

One of the most recent species concepts to be introduced has been the phylogenetic or diagnostic species concept (Cracraft, 1983) in which species have been defined as the 'smallest diagnosable cluster of individual organisms within which there is a parental pattern of ancestry and decent'. While this concept mirrors the pragmatic emphasis many taxonomists have on detecting the features which diagnose species and is useful for those primarily interested in resolving historical relationships, it has been rejected by others because of its lack of theoretical depth. This concept has not been used yet in an overt manner in medical entomology.

Although not always explicitly stated, the biological species concept of Mayr and Dobzhansky underlies the vast majority of studies in blood-sucking insect systematics. However, the criteria used by systematists to distinguish most species are substantially different from those that directly test adherence to such a theoretical concept (Lane and Crosskey, 1993). As in most groups of organisms, very few species of blood-sucking insects have been tested in relation to this or any other theoretical species concept. Even though all species reproduce sexually, parthenogenicity is known only in the bat bugs (Polyctenidae), thus making it at least theoretically possible to test the degree of reproductive distinctness between taxa.

While recognizing its weaknesses, the strength of the biological species concept (BSP) for many biologists interested in biting insects is that it is closely tied to the process of speciation and to genetic differences between populations. These genetic differences, including differences in parasite or pathogen susceptibility, or host preferences are important factors in assessing the vectorial competence of species, or even in the vectorial capacity (a quantitative measure of disease transmission; Dye, 1990). For example, in the mosquito *Aedes aegypti*, vector competence varies between mosquito populations for yellow fever and dengue viruses and hence the transmission of these pathogens (Gubler *et al.*, 1982; Tabachnick, 1991). There are many factors which influence the ability of a vector to transmit a parasite or pathogen and often there is ambiguity over exactly what is meant by a vector. On one hand it represents vector competence (the physiological ability to transmit) and on the other vectorial capacity (the epidemiological ability to transmit). The above examples notwithstanding, it is not surprising that given the complexity of transmission dynamics there are few really good data on this aspect for many of the important vector complexes, including the *Anopheles gambiae* complex.

13.3 SPECIES COMPLEXES

If there is one characteristic of biting-insect systematics that stands out from the rest of insect systematics it is the notion of species complexes. Species complexes were originally discovered when ecological inconsistencies were observed in the field between populations of the same morphologically defined species. They have received much attention in the medical entomology literature because several major vectors of human disease are members of species complexes and a full understanding of the variation of parasites, vectors and the human population is essential to the correct interpretation of the epidemiology of disease; this is especially the case for malaria and onchocerciasis in Africa (Wright and Pal, 1967; World Health Organization, 1977; Steiner *et al.*, 1982; Service, 1988).

Species complexes consist of sibling species, which are defined as morphologically indistinguishable but reproductively isolated species and frequently are sympatric. Among insects, sibling species are particularly common in the Diptera of medical importance (Simuliidae, Culicidae) but they are not confined to the group, or even to insects (Knowlton, 1993). Once it became known that species complexes existed, the genetic structure of many morphospecies (species defined on morphological criteria only) were examined and further complexes discovered over the past 25 years.

Sibling species are, by definition, isomorphic or virtually so, and other means than morphology have therefore been used to detect the existence of a complex. In practice, this usually means using cytological attributes of chromosomes (cytotaxonomy) or isozyme variation (chemotaxonomy).

The Diptera are cytologically unusual in that they have fewer pairs of chromosomes than is usual in insects and often have massive polytenized chromosomes, of a type unknown in other insect orders and formed by lengthwise multiplication of DNA. Polytene chromosomes are often most developed (or at least easy to visualize) in cells undergoing high levels of protein synthesis, e.g. adult ovarian nurse cells in mosquitoes and larval salivary glands in blackflies. When stained they show conspicuous bands and other micromorphological landmarks characteristic of each species. Although these banding sequences can be used to reconstruct phylogenies (cytophylogenies), in practice their prime significance in systematics is to detect speciation events within a morphospecies. The term 'cytospecies' is frequently used for species, usually sibling species within a complex, and is recognized primarily on chromosomal characters. A cytotype or cytoform is a chromosomally recognized constituent of a morphospecies whose status is still uncertain: it might be a sibling species or only a polymorphic variant. The principal chromosomal differences found between sibling species (and for other species) include fixed sequential differences in inversions and interchanges within populations, sex chromosome differences and polymorphic (floating) inversions in which alternative banding sequences are seen in one of a complementary chromosome pair. If the putative species being compared are sympatric the absence of hybrids is prima facie evidence for the absence of cross-mating.

Similarly, in isozyme studies an absence of hybrids or a deficiency in the proportions of heterozygotes expected by the Hardy–Weinberg Equilibrium indicates the absence of gene flow. The analysis of isozymes, with its very direct genetic interpretation, has been particularly useful in those vector groups (*Culex* and *Aedes* mosquitoes, tsetse flies, phlebotomine sandflies and ticks) which do not have well-developed polytene chromosomes (Miles and Paterson, 1979; Hunt and Hilburn, 1985; Caillard *et al.*, 1986; Munstermann, 1988; Kreutzer *et al.*, 1990; Lanzaro *et al.*, 1993).

Tabachnick and Black (1995) have argued that the tremendous interest in recognizing sibling species might well lead to a revival of the old typological concept of species. Given the broad biological framework in which many systematic studies are now made this seems unlikely; however, the risk of narrow-mindedness remains, as it does for any approach to systematics.

It is important to stress that sibling species within species complexes are not conceptually different from other species – they are only different on operational grounds, i.e. in the sort of characters used to distinguish them. Species complexes are an anthropocentric concept in which human perceptions of readily observable morphological variation are used. Sibling species are frequently presumed to be more recently evolved than morphologically more distinct species but this need not necessarily be the case, morphological distinctness might well be a function of the adaptive significance of the features being examined. However, this caveat notwithstanding, it is likely that species complexes are indeed groups of species of recent, common ancestry. They are probably still in the process of evolving into separate species by reduction of gene flow between populations through both intrinsic and extrinsic mechanisms. If members of species complexes in particular are currently separating into new species it is probably impossible to determine which of the taxa recognized are species in an unambiguous way. There is, theoretically at least, no fundamental difference between sibling species and morphologically discrete species in this respect; it is only a matter of differences in the nature of the characters being assessed and the population scale at which many of these sibling species studies are made.

One important aspect in using genetic data to define species is that there are no hard and fast rules on the minimum threshold for how much, or little, gene flow occurs (or is permitted in the definition) between species. The crux of the issue is estimating gene flow – the genetic variation within a species should be considerably less than the variation between species. Interestingly, this is the genetic analogy of the pheneticist's approach inherent in multiple discriminant function analysis.

13.4 MOLECULAR DATA AND SPECIES CONCEPTS

All reliable taxonomic characters are genetically controlled (environmentally induced variation being of little taxonomic use) and therefore DNA studies should provide some of the most important evidence for species recognition. However, molecular techniques have only just begun to have an impact on recognizing and identifying sibling species, and for determining genetic relationships between species.

Much of the impetus for the development of molecular tools has been the need to find a simple and effective means of identifying field-caught samples using DNA probes (Gale and Crampton, 1988; Hill and Crampton, 1994 for *Anopheles*; Post and Crampton, 1988; Brockhouse *et al.*, 1993 for *Simulium*, Ready *et al.*, 1988 for *Phlebotomus*; Adamson *et al.*, 1991; Ready *et al.*, 1991 for *Lutzomyia*) or the polymerase chain reaction (PCR) (Paskewitz and Collins, 1990; Adamson *et al.*, 1993; Paskewitz *et al.*, 1993; Scott *et al.*, 1993). In the development of DNA probes, most studies do not use sequences of known function, and certainly not genes which are known to be involved in species differentiation. Thus, in many cases, random discriminating sequences are used which give rather limited information on the relatedness of taxa (including populations) by their patterns of hybridization to probes. If they contain sequences that are present in both sexes, DNA probes will identify all stages, adults and immatures, which is a considerable advantage over other techniques.

These tools are particularly useful where the currently used techniques for identifying species are technically limiting (e.g. isozyme electrophoresis requires material frozen from the field), or the technique is confined to one life-stage only (e.g. chromosomes from the salivary glands of mature larvae in *Simulium*). Adult females have polytenized chromosomes but they are frequently difficult to interpret and are often only accessible to analysis within a few hours of a blood meal.

To date, relatively few studies have used DNA methods to explore the genetic variation within species and their constituent populations although the potential is enormous. The deployment of the Randomly Amplified Polymorphic DNA techniques (Kambhamptai *et al.*, 1992; Adamson *et al.*, 1993; Wilkerson *et al.*, 1993; Favia *et al.*, 1994) is likely to be particularly useful in this respect.

Some molecular studies have used sequence data, usually of mitochondrial DNA in reconstructing phylogenies of blood-sucking insects, which often confirm hypotheses based on morphological data (Xiong and Kocher, 1991).

It is unlikely that the introduction of molecular techniques will substantially alter our theoretical concepts of species but they will greatly enhance our means of exploring the nature of species.

13.5 NOMENCLATURE AND SPECIES CONCEPTS

One complication – but operationally a quite different aspect of studying species at fine levels of resolution – is nomenclature, or the naming of taxa. At present, the binomial system of Linneaus is used for naming an organism in which the basic assumption is made that all individuals belong to discrete and mutually exclusive sets. Even though Linneaus and his contemporaries were essentialists it has been possible for most cases to reconcile this now outmoded philosophical standpoint and the pragmatic rules of nomenclature with modern biology.

However, the fundamental problem with the binomial system is that it does not take account of populations (or species) in the process of speciation; many populations might not be unambiguously assignable to one species or another simply because the two species are not yet distinct. In attempting to come to terms with the conflict between the needs of communication which nomenclature seeks to serve on the one hand, and biological reality on the other, many taxonomists have effectively dropped out of the binomial system by developing an informal nomenclature using categories particular to the taxonomic group being studied, e.g. biotypes, populations, cytoforms. The relationship between some of these informal terms and formal nomenclatural categories are given in Lane and Marshall (1981).

The essential need to have a unified and universally agreed set of nomenclatural rules which conflict at times with biological reality is a pragmatic compromise, but unfortunately it adds further confusion to the complicated debate on species concepts. For some viewpoints at least, the discussion of complicated issues is not possible without the appropriate language.

13.6 PRACTICAL SPECIES CONCEPTS

While there are potentially several theoretical species concepts which could be applied to the study of blood-sucking insects, in reality there is only one regularly used – the biological species concept. As in other areas of systematics, there is a substantial gap between the a theoretical model followed by a practising systematist and the criteria that the systematist uses on a day-to-day basis for recognizing and defining species.

Before embarking on a review of the way in which species concepts are approached in blood-sucking insects, it is important to distinguish between the systematics of a group as a whole and that of the taxa containing vector species – the level of sophistication can be very different. For example, contrary to popular belief the higher classification of the mosquitoes is not well resolved, or even of the genus *Anopheles* (Harbach, 1994) but some species groups containing important vector species are very well studied, e.g. *Anopheles gambiae* complex.

To review the working concepts used in the systematics of medically important insects the following scheme has been developed in which practical concepts have been arranged in a sequence of levels of increasing resolution and closeness to testing the Biological Species Concept (Table 13.1). It is important to note that all the techniques and criteria used to distinguish species give relative data, i.e. non-absolute data, and in this way they mirror the relational nature of the biological species concept.

The sequence of criteria for recognizing species given below, do not necessarily reflect the chronological order in which studies are made; i.e. systematists do not work up the levels in an orderly way. However, level 1, in which qualitative morphological data are used, always precedes the others. For example, in the early studies on the *Anopheles gambiae* complex the primary method of recognizing constituent sibling species was by cross-mating tests (Davidson, 1964). This labour-demanding technique was the impetus to discover new methods of discriminating species such as analysing the banding patterns on chromosomes. The various contributions to Lane and Crosskey (1993) give an overview of the characters used to distinguish taxa in the various groups of medically important insects and arachnids.

The criteria levels for recognizing species do not simply describe an increasing sophistication in the technology or characters used, although superficially it looks that way. The techniques employed are a consequence

Table 13.1 Criteria for recognizing species of blood-sucking insects

	Total no. of species	*Level 1 Morphology qualitative*	*Level 2 Morphology quantitative*	*Level 3 Non-morphological*	*Level 4 Genetic*	*Level 5 Cross-mating*	*Level 6 Measuring gene-flow*	*Species complexes known*	*Species described on non-morphological characters*
Culicidae (mosquitoes)	3450	✔	✔	✔	✔	✔	✔	✔	✔
Simuliidae (blackflies)	1580	✔	✔	✔	✔			✔	✔
Phlebotominae (sandflies)	700	✔	✔	✔	✔	✔		✔	
Ceratopogonidae (biting midgs)	1400 (5000 total)	✔	✔		✔				
Tabanidae (horse-flies)	4000	✔		✔	✔				
Glossinidae (tsetse-flies)	23	✔		✔		✔			
Blood-sucking muscids (stable-flies)	50 (5000)	✔							
Triatomine bugs	118	✔			✔				
Cimicidae (bed bugs)	91	✔							
Anoplura (sucking lice)	490	✔	✔			✔			
Siphonaptera (fleas)	2500	✔							

of attempting to base conclusions on ever-more biologically informative characters. Often, experimental design is more important than the sophistication of the techniques used in delimiting and identifying species (Lane, 1994).

13.6.1 Criteria for recognizing species

Level 1: Qualitative morphology

Like most insects, the majority of blood-sucking insects are distinguished at the species level by morphological means, i.e. by recognizing discontinuities in the variation of morphological characters. Many discontinuities can be very overt, such as the patterns on tabanid wings, patterns on mosquito legs and wings or variations in the shape of parts of the genitalia. In this context, morphology is being used as a surrogate for genetic differences and in the vast majority of cases gives quite acceptable results.

Sometimes, special techniques are required to prepare the specimens for examination, especially for microscopical characters such as sensilla or internal structures, or the conditions for examination need to be specified, e.g. the patterns on the thorax of *Simulium* adults are generated by the reflectance from microtrichia arrays and the direction of the incident light greatly affects the pattern (Lowry and Shelley, 1990).

Level 2: Quantitative morphology

Often, the need to resolve species at a finer level (perhaps based on ecological or geographical evidence) has meant that there is a need to look for very fine quantitative differences between species. Discontinuities are still being sought, as at Level 1, but samples are extensive and the concept of a population, at least in the statistical sense, is introduced. There are numerous examples of this approach to defining species scattered throughout the medically important groups (Busvine, 1978; Lane and Ready, 1985). At this level there are no experimental or genetic data to confirm these working concepts of species beyond maintaining biological homogeneity of samples.

Level 3: Non-morphological data

The use of non-morphological data, usually chemical data, to replace morphological characters is usually used where the evidence from morphology is ambiguous or it conflicts with field observations. Often there is no underlying theoretical reason to justify the use of these characters or why chemical data should be intrinsically more predictive that morphology. The majority of the occasions on which cuticular hydrocarbons have been used would fit this description (Carlson and Service, 1979; Hamilton and

Service, 1983; Milligan *et al.*, 1986; Kamhawi *et al.*, 1992), although it is now known that some of these hydrocarbons are contact pheromones (Carlson *et al.*, 1984). If these pheromones are in fact important factors in pre-mating mate choice (whether defined in terms of a mating barrier in the Mayr/Dobzhansky concept, or common mate recognition systems in the Paterson concept) they could be much more effective in defining species than the whole-body extracts accompanied by multivariate analysis that is the usual approach to the use of hydrocarbons. In other taxa where cuticular hydrocarbons have been used, in phlebotomine sandflies, simuliids, and mosquitoes there is no such underlying evidence for their species-specific function.

The recent use of the randomly amplified polymorphic DNA (RAPD) technique (Adamson *et al.*, 1993; Wilkerson *et al.*, 1993) would fit into this category since the genetic basis for these characters is not yet known, although for RAPDs most are thought to be dominant. It is likely that with the rapid development of this technique a much clearer understanding of its genetic significance will be forthcoming and therefore its power increased.

Level 4: Field data to test genetic models

Studies of this type involve sampling from the field and laboratory analysis to test differences according to a predetermined genetic model. This is the level at which most species complexes have been investigated, some classic studies being those of Davidson and Hunt (1973), Mahon *et al.* (1976), Miles (1979) and Coluzzi *et al.* (1985) on *An. gambiae* and Vajime and Dunbar (1975) on *Simulium.*

The usual model is the Hardy–Weinberg Equilibrium to test for panmixis (random mating). Data can be from either isozymes or frequencies of chromosomal inversions, although it is usually the former. Usually, cross-sectional data collected from several sites or localities are used, with the sample pooled within a locality (rather than collected in each locality over a period of time and the changing gene frequencies measured) (Green *et al.*, 1992).

Frequently, this approach is used where there are difficulties in assessing the significance of morphological variation (e.g. in *Lutzomyia yucumensis*, Caillard *et al.*, 1986; or *Anopheles culicifaecies*, Green and Miles, 1980).

There are substantial theoretical difficulties in testing the differences in gene frequencies between allopatric populations since the geographical separation invalidates the prerequisite of free access between individuals.

Level 5: Experimental tests

At this level, data are generated, usually in the form of cross-mating studies to indicate reproductive distinctness between populations or species

(Davidson, 1964; Davidson and Hunt, 1973; Ward *et al.*, 1988). This usually requires colonization of the samples – a major constraint both biologically and in the resources required. These experiments, while appearing to be the 'gold standard', can be difficult to interpret – some species will not mate in laboratory cages (stenogamy) or the offspring cannot be reared easily to determine F_1 sterility. The usual method for conducting these experiments is to compare the fertility of reciprocal crosses to within-sample crosses. Usually, the insects are given the choice between an experimental insect or nothing, rather than an experimental mate and a putative conspecific mate. Sometimes these experiments produce crossing artefacts not present, or rarely found in the field and therefore they express what is physiologically possible rather than what happens in nature.

Level 6: Measurement of gene flow

This is the level at which species can be most comprehensively tested for their integrity and adherence to the BSC. It requires detailed genetic data from extensive field studies. Some of the most comprehensive studies have been those of Coluzzi and colleagues on *Anopheles gambiae* in Africa, in which both geographic and seasonal shifts in gene frequencies have been found (Coluzzi *et al.*, 1979), and other tropical species such as the *Anopheles dirus* complex (Green *et al.*, 1992). The gene flow between populations in relation to dispersal has been investigated by several studies, e.g. Tabachnick (1991) for *Aedes aegypti*, Munstermann (1985) for *Ae. triseriatus*, Narang *et al.* (1991) for *An. albimanus*, Cheng *et al.* (1982) for *Culex pipiens*, and Tabachnick (1992) for *Culicoides variipennis*.

13.7 IDENTIFICATION – A REVERSE LOOK AT SPECIES

It is essential to distinguish between initial recognition of a species and the subsequent identification of samples. Often the same characters are used for both these functions. However, in the case of many cryptic or sibling species, much more detailed data might be used to establish the status of species and then simpler, more economical methods developed for subsequent identification are found. For example, recognition of species by chromosomes or electrophoresis and the subsequent identification by morphometrics (Wilson *et al.*, 1993), establishing species status by breeding then identification by simple morphology (Gebre-Michael and Lane, 1993), defining species by isozymes and chromosomes then identification by DNA probes (Post and Crampton, 1988).

13.8 CONCLUSIONS

The pressure from non-systematists to have an ever-more refined classification of blood-sucking insects which is both a succinct summary of the diversity of these insects as well as providing a predictive framework, has meant that a variety of techniques have been deployed to understand the nature of species. In the future, the ability to measure gene flow in the field with molecular tools to a greater degree of sensitivity and convenience than is currently available will undoubtedly push the theoretical concepts to the limit. Whether the biological species concept, the most widely used concept in medical entomology, will withstand this onslaught remains to be seen.

13.9 REFERENCES

Adamson, R.E., Chance, M.L., Ward. R.D., Feliciangeli, D. and Maingon, R.D.C. (1991) Molecular approaches to the analysis of sympatric sandfly populations in endemic areas of western Venezuela. *Parassitologia*, **33**, 45–53.

Adamson, R.E., Ward, R.D., Feliciangeli, D. and Maingon, R. (1993) The application of random amplified polymorphic DNA for sandfly species identification. *Medical and Veterinary Entomology*, **7**, 1–5.

Brockhouse,C.L., Vajime, C.G., Marin, R. and Tanguay, R.M. (1993) Molecular identification of onchocerciasis vector sibling species in black flies (Diptera: Simuliidae). *Biochemical and Biophysical Research Communications*, **194**, 628–34.

Busvine, J.R. (1978) Evidence from double infestations for the specific status of human head and body lice (Anoplura). *Systematic Entomology*, **3**, 1–8.

Caillard, T., Tibayrenc, M., Le Pont, F., Dujardin, J-P., Desjeux, P. and Ayala, F.J. (1986) Diagnosis by isoenzyme methods of two cryptic species, *Psychodopygus carrerai* and *P. yucumensis* (Diptera: Psychodidae). *Journal of Medical Entomology*, **23**, 489–92.

Carlson, D.A. and Service, M.W. (1979) Differentiation between species of the *Anopheles gambiae* Giles complex (Diptera: Culicidae) by analysis of cuticular hydrocarbons. *Annals of Tropical Medicine and Parasitology*, **73**, 589–92.

Carlson, D.A., Nelson, D.R., Langley, P.A., Coates, T.W., Davis, T.L. and Leegwater-van der Linden, M.E. (1984) Contact sex pheromones in the tsetse fly *Glossina pallidipes* (Austen): identification and synthesis. *Journal of Chemical Ecology*, **10**, 429–50.

Cheng, M.L., Hacket, C.S., Pryor, S.C., Ferrell, R.E. and Kitto, G.B. (1982) The ecological genetics of the *Culex pipiens* complex in North America, in *Recent Developments in the Genetics of Insect Disease Vectors* (eds W.W.M. Steiner, W.J. Tabachnick, K.S. Rai and S. Narang), A Symposium Proceedings, Stipes, Champaign, Illinois, pp. 581–627.

Coluzzi, J., Petrarca, V. and Di Deco, M.H. (1985) Chromosomal inversion intergradation and incipient speciation in *Anopheles gambiae*. *Bolletino di Zoologica*, **52**, 45–63.

Coluzzi, M., Sabatini, A., Petrarea, V. and Di Deco, M.A. (1979) Chromosomal differentiation and adaptation to human environments in the *Anopheles gambiae* complex. *Transactions of the Royal Society of Tropical Medicine and Hygiene*, **73**, 483–97.

Cracraft, J. (1983) Species concepts and speciation analysis, in *Current Ornithology*, vol. 1, Plenum Press, New York, pp. 159–87.

Davidson, G. (1964) The five mating types of the *Anopheles gambiae* complex. *Rivista Malariologica*, **13**, 167–218.

Davidson, G. and Hunt, R.H. (1973) The crossing and chromosome characteristics of a sixth species in the *Anopheles gambiae* complex. *Parassitologia*, **15**, 121–8.

Dobzhansky, T. (1937) *Genetics and the Origin of Species*, Colombia University Press, New York.

Dye, C. (1990) Epidemiological significance of vector parasite interactions. *Parasitology*, **101**, 409–15.

Favia, G., Dimopoulos, G. and Louis, C. (1994) Analysis of the *Anopheles gambiae* genome using RAPD markers. *Insect Molecular Biology*, **3**, 149–57.

Gale, K. and Crampton, J.M. (1988) DNA probes for species identification of mosquitoes in the *Anopheles gambiae* complex. *Medical and Veterinary Entomology*, **1**, 127–36.

Gebre-Michael, T. and Lane, R.P. (1993) Distinguishing the females of *Phlebotomus (Synphlebotomus) martini* and *P. (S) celiae* (Diptera: Phlebotominae), vectors of visceral leishmaniasis in southern Ethiopia. *Bulletin of Entomological Research*, **83**, 353–60.

Green, C.A. and Miles, S.J. (1980) Chromosomal evidence for sibling species of the malaria vector *Anopheles (Cellia) culicifaecies*. *Journal of Tropical Medicine and Hygiene*, **83**, 75–8.

Green, C.A., Munstermann, L.E., Tan, S.G., Panyim, S. and Baimai, V. (1992) Population genetic evidence for species A, B, C and D of the *Anopheles dirus* complex in Thailand and electromorphs for their identification. *Medical and Veterinary Entomology*, **6**, 29–36.

Gubler, D.J., Novak, R. and Mitchell, C.J. (1982) Arthropod vector competence – epidemiological, cytogenetic and biological considerations, in *Recent Developments in the Genetics of Insect Disease Vectors* (eds W.W.M. Steiner, W.J. Tabachnick, K.S. Rai and S. Narang), A Symposium Proceedings, Stipes, Champaign, Illinois, pp. 343–78.

Hamilton, R.J. and Service, M.W. (1983) The value of cuticular and internal hydrocarbons for the identification of larvae of *Anopheles gambiae* Giles, *An. arabiensis* Patton and *An. melas* Theobald. *Annals of Tropical Medicine and Parasitology*, **77**, 203–10.

Harbach, R. (1994) Review of the internal classification of the genus *Anopheles* (Diptera: Culicidae): the foundation for comparative systematics and phylogenetics research. *Bulletin of Entomological Research*, **84**, 331–42.

Hill, S.M. and Crampton, J.M. (1994) Synthetic DNA probes to identify members of the *Anopheles gambiae* complex and to distinguish the two major vectors of malaria within the complex, *An. gambiae* s.s. and *An. arabiensis*. *American Journal of Tropical Medicine and Hygiene*, **50**, 312–21.

Hunt, L.M. and Hilburn, L.R. (1985) Biochemical differentiation between species of tick (Acari: Ixodidae). *Annals of Tropical Medicine and Parasitology*, **79**, 525–32.

Kambhamptai, S., Black, W.C. and Rai, K.S. (1992) Random amplified polymorphic DNA of mosquito species and populations (Diptera: Culicidae). Techniques, statistical analysis and applications. *Journal of Medical Entomology*, **29**, 939–45.

Kamhawi, S., Lane, R.P., Cameron, M.M., Phillips, A., Milligan, P. and Molyneux, D. (1992) Variation in the cuticular hydrocarbons of *Phlebotomus argentipes* from two field populations and a subsequent laboratory colony. *Bulletin of Entomological Research*, **82**, 209–12.

Knowlton, N. (1993) Sibling species in the sea. *Annual Review of Ecology and Systematics*, **24**, 189–216.

Kreutzer, R.D., Palau, M.T., Morales, A., Ferro, C., Feliciangeli, D. and Young, D.G. (1990) Genetic relationships among phlebotomine sandflies (Diptera: Psychodidae) in the *verrucarum* series. *Journal of Medical Entomology*, **27**, 1–8.

Lane, R.P. (1994) Morphometric discrimination of pest species: experimental design, in *Identification and Characterisation of Pest Organisms* (ed. D.L. Hawksworth), CAB International and Systematics Association, pp. 195–205.

Lane, R.P. and Crosskey, R.W. (eds) (1993) *Medical Insects and Arachnids*, Chapman & Hall, London.

Lane, R.P. and Marshall, J. (1981) Geographic variation, races and subspecies, in *The Evolving Biosphere* (ed. P. Forey), British Museum (Natural History), London, pp. 9–19.

Lane, R.P. and Ready, P.D. (1985) Multivariate discrimination between *Lutzomyia wellcomei*, a vector of mucocutaneous leishmaniasis, and *Lu. complexus*. *Annals of Tropical Medicine and Parasitology*, **78**, 469–72.

Lanzaro, G.C., Ostrovska, K., Herrero, M.V., Lawyer, P.G. and Warburg. A. (1993) *Lutzomyia longipalpis* is a species complex: genetic divergence and interspecific sterility among three populations. *American Journal of Tropical Medicine and Hygiene*, **48**, 839–47.

Lowry, C.A. and Shelley, A.J. (1990) Studies on the scutal patterns of three South American *Simulium* species (Diptera: Simuliidae). *Canadian Journal of Zoology*, **68**, 956–61.

Mahon, R.J., Green, C.A. and Hunt, R.H. (1976) Diagnostic allozymes for routine identification of adults of the *Anopheles gambiae* complex (Diptera: Culicidae). *Bulletin of Entomological Research*, **66**, 25–31.

Mayr, E. (1966) *Animal Species and Evolution*, Harvard University Press, Cambridge, Massachusetts.

Miles, S.J. (1979) The *Anopheles gambiae* complex: a biochemical key. *Journal of Medical Entomology*, **15**, 297–9.

Miles, S.J. and Paterson, H.E. (1979) Protein variation and systematics in the *Culex pipiens* group of species. *Mosquito Systematics*, **11**, 187–202.

Milligan, P.J.M., Phillips, A., Molyneux, D.H., Subbarao, S.K. and White, G.B. (1986) Differentiation of *Anopheles culicifaecies* Giles (Diptera: Culicidae) sibling species by analysis of cuticular components. *Bulletin of Entomological Research*, **76**, 529–37.

Munstermann, L.E. (1985) Geographic patterns of genetic variation in the treehole mosquito *Aedes triseriatus*, in *Ecology of Mosquitoes: Proceedings of a Workshop* (eds L.P. Lounibos, J.R. Rey and J.H. Frank), Florida Medical Entomology Laboratory, Vero Beach, Florida, pp. 327–43.

Munstermann, L.E. (1988) Biochemical systematics of nine Nearctic *Aedes* mosquitoes (subgenus *Ochelerotatus annulipes* group B), in *Biosystematics of Haematophagous Insects* (ed. M.W. Service), Clarendon Press, Oxford, pp. 133–47.

Narang, S.K., Seawright, J.A. and Svarez, M.F. (1991) Genetic structure of natural populations of *Anopheles albimanus* in Colombia. *Journal of American Mosquito Control Association*, **7**, 437–45.

Paskewitz, S.M. and Collins, F.H. (1990) Use of polymerase chain reaction to identify mosquito species of the *Anopheles gambiae* complex. *Medical and Veterinary Entomology*, **4**, 367–73.

Paskewitz, S.M., Wesson, D.M. and Collins, F.H. (1993) The internal transcribed spacers of ribosomal DNA in five members of the *Anopheles gambiae* complex. *Insect Molecular Biology*, **2**, 247–57.

Paterson, H.E.H. (1985) The recognition concept of species, in *Species and Speciation* (ed. E.S. Vrba), Transvaal Museum Monograph No. **4**, Pretoria, South Africa, pp. 21–9.

Post, R.J. and Crampton, J.M. (1988) The taxonomic use of variation in repetitive DNA sequences in the *Simulium damnosum* complex, in *Biosystematics of Haematophagous Insects* (ed. M.W. Service), Clarendon Press, Oxford, pp. 245–55.

Ready, P.D., Smith, D.F. and Killick-Kendrick, R. (1988) DNA hybridisations on squash-blotted sandflies to identify both *Phlebotomus papatasi* and infecting *Leishmania major*. *Medical and Veterinary Entomology*, **2**, 109–16.

Ready, P.D., Lainson, R., Shaw, J.J. and Souza, A.A. (1991) DNA probes for distinguishing *Psychodopygus wellcomei* from *Psychodopygus complexus* (Diptera: Psychodidae). *Memorias Instituto Oswaldo Cruz*, **86**, 41–9.

Scott, J.A., Brogdon, W.G. and Collins, F.H. (1993) Identification of single specimens of the *Anopheles gambiae* complex by the polymerase chain reaction. *American Journal of Tropical Medicine and Hygiene*, **49**, 520–9.

Service, M.W. (1988) New tools for old taxonomic problems in bloodsucking insects, in *Biosystematics of Haematophagous Insects* (ed. M.W. Service), Clarendon Press, Oxford, pp. 325–45.

Simpson, G.G. (1951) The species concept. *Evolution*, **5**, 285–98.

Steiner, W.W.M., Tabachnick, W.J., Rai, K.S. and Narang, S. (eds) (1982) *Recent Developments in the Genetics of Insect Disease Vectors*. A Symposium Proceedings, Stipes, Champaign, Illinois.

Tabachnick, W.J. (1991) Evolutionary genetics and arthropod-borne disease: the yellow fever mosquito. *American Entomologist*, **37**, 14–24.

Tabachnick, W.J. (1992) Genetic differentiation among populations of *Culicoides variipennis* (Diptera: Ceratopogonidae), the North American vector of bluetongue virus. *Annals of the Entomological Society of America*, **85**, 140–7.

Tabachnick, W.J. and Black, W.C. (1995) Making a case for molecular population genetic studies of arthropod vectors. *Parasitology Today*, **11**, 27–30.

Vajime, G.C. and Dunbar, R.W. (1975) Chromosomal identification of eight species of the subgenus *Edwardsellum* near and including *Simulium (Edwardsellum) damnosum* Theobald (Diptera: Simuliidae). *Tropenmedizin und Parasitologie*, **26**, 131–8.

Ward, R.D. (1989) Some aspects of the biology of phlebotomine sandflies. *Advances in Disease Vector Research*, 91–126.

Ward, R.D., Phillips, A., Burnet, B. and Marcondes, C.M. (1988) The *Lutzomyia longipalpis* complex: reproduction and distribution, in *Biosystematics of Haematophagous Insects* (ed. M.W. Service), Clarendon Press, Oxford, pp. 257–69.

Wilson, M.D., Post, R.J. and Gomulski, L.M. (1993) Multivariate morphotaxonomy in the identification of adult females of the *Simulium damnosum* Theobold complex (Diptera: Simuliidae) in the Onchocerciasis Control Programme area of West Africa. *Annals of Tropical Medicine and Parasitology*, **87**, 65–82.

Wilkerson, R.C., Parsons, T.J., Albright, D.G., Klein, T.A. and Braun, M.J. (1993) Random amplified polymorphic DNA (RAPD) markers readily distinguish cryptic mosquito species. *Insect Molecular Biology*, **4**, 205–11.

World Health Organization (1977) Species complexes in insect vectors of disease (blackflies, mosquitoes, tsetse flies). Mimeographed document WHO/VBC/77.656, World Health Organization, Geneva.

Wright, J.W. and Pal, R. (eds) (1967) *Genetics of Insect Vectors of Disease*, Elsevier, Amsterdam.

Xiong, B. and Kocher, T.D. (1991) Comparison of mitochondrial DNA sequences of seven morphospecies of black flies (Diptera: Simuliidae). *Genome*, **34**, 306–11.

14

Recognition of parthenogenetic insect species

R. G. Foottit

Contacting address: Eastern Cereal and Oilseed Research Centre, Research Branch, Agriculture and Agri-Food Canada, K.W. Neatby Bldg., Central Experimental Farm, Ottawa, Ontario, K1A 0C6, Canada

ABSTRACT

Parthenogenesis is a common phenomenon in the Animal Kingdom. It is estimated that there are over 1000 obligately parthenogenetic species situated over a broad range of taxa and over 15 000 species reproducing by cyclic parthenogenesis. In such insect groups as aphids, thrips, and gall midges, such factors as complex alternation of generations (with numerous different morphs) and the presence of persistent clonal, parthenogenetic populations, make the recognition of species a difficult task. Parthenogenetic insects require a practical species definition for proper taxonomic and biological handling of species. This paper examines the nature and extent of parthenogenetic insects, their evolution and diversity, and makes recommendations for the taxonomic treatment of species. Topics such as methodologies, degree of discreteness among species, patterns of variation and past treatment under different species concepts are discussed. It is suggested that there is no one comprehensive definition for parthenogenetic species. They can best be handled by concepts and practices that interpret pattern along with biological reality and which incorporate a genealogical perspective at the clonal, population and species level.

14.1 INTRODUCTION

Whether one is a taxonomist or evolutionary biologist, describing or analysing biodiversity, or one who studies pest or beneficial organisms for practical purposes, the species is the fundamental unit of diversity (Wilson, 1992). There have been many proposals for the conceptual and practical handling of species and among these, the biological species concept (BSC; Mayr, 1942) has been the most thoroughly promoted and

Species: The Units of Biodiversity. Edited by M.F. Claridge, H.A. Dawah and M.R. Wilson. Published in 1997 by Chapman & Hall. ISBN 0 412 63120 2

most widely accepted. One of the more frequently stated deficiencies of the BSC is its inapplicability to uniparental organisms. This fact has been considered at length by Mayr, the main proponent of the BSC (Mayr, 1963, 1982, 1987) and by many critics of the BSC and advocates of other viewpoints and solutions to the taxonomic handling of asexual and parthenogenetic organisms.

In terms of sheer numbers, parthenogenesis is a common phenomenon in the Animal Kingdom. It has been estimated that there are over 1000 obligately parthenogenetic species situated over a broad range of taxa and over 15 000 species which reproduce by cyclic parthenogenesis (White, 1978; Bell, 1982). However, obligate parthenogenesis can be considered to be rare when viewed in the relation to the total number of animal species. This phenomenon is estimated to occur in approximately 1% of insect species (Bell, 1982) and 0.1% of the total animal kingdom (White, 1978). The BSC, with its condition of reproductive isolation between interbreeding populations of different species, produces difficulties when one is dealing with completely asexual or parthenogenetic organisms; that is, where no interbreeding takes place. However, these asexual organisms show a level of integrity that is commonly recognized by specialist taxonomists.

Workers differ in their opinions about the need to resolve the issue of species concepts and practical taxonomic handling in asexual and parthenogenetic organisms. Mayr (1963: 27) has stated that 'It is too early for a definitive proposal concerning the application of the species concept to asexually or uniparentally reproducing organisms'. Scudder (1974) has pointed out that this stance hardly satisfies those who study groups where the BSC does not apply. Hull (1970) noted that uniparental organisms adapt, invade new ecological niches and evolve; thus, they form species and criteria are needed to delimit these species. Nevertheless, the task of determining species in completely, partially or cyclically parthenogenetic animals presents a practical challenge. Characteristics associated with the parthenogenetic mode of reproduction, such as the complex alternation of generations with numerous different morphological forms, and the presence of persistent clonal, parthenogenetic populations among cyclically parthenogenetic populations, can make the practical recognition of species operationally difficult. A recent workshop (Hawksworth, 1994) identified again the pressing need to resolve the disorder of differing or even obscure species concepts associated with asexual and parthenogenetic organisms.

This chapter will review the nature of parthenogenesis and examine the theoretical and practical aspects of species recognition in parthenogenetic insects. As will be shown, parthenogenetic insects are diverse, important to society and require a practical definition for proper taxonomic and biological consideration. The problem of parthenogenetic species is not unique to any particular species concept; the impact of various species concepts, from phenetic to phylogenetic interpretations, will be considered.

14.2 NATURE AND EXTENT OF PARTHENOGENETIC REPRODUCTION

General reviews of the nature, taxonomic distribution and ecological and evolutionary significance of asexual and parthenogenetic reproduction are provided by Bell (1982), Hughes (1989), Oliver (1971) and Suomalainen *et al.* (1987). The following is a brief survey of asexual and parthenogenetic reproduction with an emphasis on insects.

Many taxonomic groups of organisms include species that reproduce uniparentally or clonally, that is, the result is an assemblage of individuals that is genetically identical by descent (Bell, 1982). This occurs in animals in two different ways. Asexual (agametic) reproduction involves fragmentation, fission or budding and occurs in organisms of rather simple structure. Parthenogenesis, which includes both the meiotic and the ameiotic formation of reproductive cells, includes all breeding systems in which the eggs develop into new individuals without fertilization by gametes produced by another individual (Bell, 1982).

Parthenogenesis is a varied mechanism with a complex terminology associated with it. There are two main categories of parthenogenesis that are of interest, namely, arrhenotoky and thelytoky. In arrhenotoky, a common mode of reproduction in rotifers, some mite groups and several insect orders, the female undergoes meiosis, resulting in haploid eggs. The egg may develop with (producing a female) or without fertilization (producing a male). There are two types of thelytoky, the development of diploid females from unfertilized eggs. Automictic (meiotic) thelytoky involves meiosis and some mechanism of fusion of meiotic products. Depending on how diploidy is restored, this mechanism could bring about some degree of genetic mixing and result in the formation of diverse, distinct clones with subsequent adaptive radiation over time. In apomictic (ameiotic) thelytoky, a mechanism of cloning, the eggs develop without meiosis; this does not result in genetic mixing. It is typically found in groups with cyclical parthenogenesis, such as the aphids, and obligate parthenogenetic groups that have lost the capacity for bisexual reproduction.

Extensive surveys of the distribution of parthenogenesis in insects are provided by Bell (1982), Suomalainen *et al.* (1976, 1987) and White (1978). The occurrence of parthenogenesis in insects is sporadic throughout the order; it involves a wide range of cytological mechanisms (Suomalainen *et al.*, 1976), but is, in general, concentrated in certain taxa and is entirely lacking in others. Parthenogenetic species often occur sporadically among bisexual species.

Facultative or obligate thelytoky is present in over 200 genera from over 80 families of insects (Bell, 1982; Suomalainen *et al.*, 1987). For example, it is common in the Collembola, Psocoptera, curculionid beetles and the sternorhynchous Homoptera (aphids, scales and whiteflies). Arrhenotoky is found in the order Thysanoptera; there are also some secondarily thelytokous thrips. Parthenogenesis, including both arrhenotoky and thelytoky, is the characteristic mode of reproduction in the Hymenoptera.

Cyclical parthenogenesis is a common reproductive pattern found in groups such as trematodes, cladocerans, rotifers, aphids, cynipid gall wasps, cecidomyiids and the beetle, *Micromalthus*. There is alternation between parthenogenesis (usually apomictic) and sexual reproduction, thus combining the ecological and evolutionary advantages of both modes of reproduction. There are from one to many parthenogenetic generations followed by a sexual generation. Among these cyclical species there may also be purely thelytokous (anholocyclic) populations that do not return to the sexual phase of the life-cycle. Through biogeographical analysis, it has been shown that in one instance an aphid–host plant association and cyclical life history has persisted for 48 million years (Moran, 1989).

14.3 PAST TREATMENT OF ASEXUAL AND PARTHENOGENETIC SPECIES

Species concepts have long been an important topic of discussion in systematics and evolutionary biology. Often, in these many treatments, there has been discussion of asexual and parthenogenetic species. Many viewpoints and solutions to the conceptual and practical handling of these organisms have been proposed. The botanist G. E. Du Rietz (1930), one of the early workers to discuss comprehensively the nature of species concepts, acknowledged the problems for species concepts presented by asexual populations and noted the difficulties in determining the correspondence of rank between sexual species and units of populations of asexual organisms. He cited Turesson's (1929) use of the term agamospecies, 'An apomict-population the constituents of which, for morphological, cytological or other reasons, are to be considered as having common origin', but dismissed this definition as too vague to be of practical value, noting that it could be applicable at a number of taxonomic ranks. In fact, Du Rietz stated (1930) that it was a rather hopeless task to apply a species concept to groups of what he termed apomictic microspecies. He advocated distinguishing sexual and asexual species and recommended applying the corresponding sexual species rank to the asexual microspecies (populations of apomicts) as such, noting that the presence of discontinuity is the most important consideration when determining species, not the relative size of the populations. As will be seen, use of a number of terms such as agamospecies and microspecies, have arisen in discussions by zoologists on a number of occasions.

Dobzhansky (1937) took issue with Du Rietz's conclusions, claiming that agamic complexes do not comprise species. He defined species by the ability actually or potentially to exchange genes. He noted that asexual groups are aggregations of distinct, constantly reproducing genotypes, clustered around adaptive peaks and that these aggregates are arranged in an hierarchical order analogous to that in sexual forms (i.e. species, sub-genera, genera, etc.). Dobzhansky felt that assigning rank was arbitrary, that is, there is no potential ability to develop a species category in asexual organisms that is more fixed and less arbitrary than other ranks; all criteria of species distinction break down. Later, Mayr (1957, 1963) also discussed a similar concept. Dobzhansky (1972) suggested that species in asexual or parthenogenetic organisms should be labelled as pseudospecies, to distinguish their biological differences from sexual species. He noted that, as there are different strategies of evolutionary adaptation, there are different kinds of species.

In his treatment of animal species, Cain (1954) again used the term agamospecies to account for those forms to which the biological species concept cannot apply because there is no sexual reproduction. He regarded agamospecies as being relatively recent phenomena, exhibiting little variability or long-term adaptability, not part of the main evolutionary pathway, although they may be successful in the relative short term. Unlike Dobzhansky, Cain felt that a morphological species concept could be applied equally well to agamospecies as to sexually reproducing species, and that agamospecies could be placed in a natural hierarchy and that they should be recognized.

Meglitsch (1954) proposed an evolutionary concept of the species and argued that it could be applied to both uniparental and biparental organisms and that, while the genetic processes are different, the outcome in both cases is the evolution of species. This was a view that was later developed by Simpson (1961) who also noted that while the evolution of uniparental and biparental populations is very different, this does not mean that species are not formed. However, as argued by some (Sokal and Crovello, 1970), the evolutionary species definition will fit most situations but is so vague as to be operationally impossible, particularly in terms of the criterion of evolutionary role.

While taking an operational approach and noting that thelytokous clones of parasitic Hymenoptera are of importance in biological control and that these entities need to be named for purposes of information retrieval, De Bach (1969) proposed an ethological–ecological species concept for the recognition of species of uniparental organisms. He pointed out that the proposed criteria of an operational approach to species definition would delimit parthenogenetic species, even if these forms were morphologically identical to other bisexual forms. He provided a series of taxonomic tests for comparison with closely related morphological biparental species which would determine if the uniparental group behaved like a biological species; if it had a significantly different ethological or ecological role it should be considered as a valid species.

The basic tenet of the biological species concept is discontinuity among species due to reproductive isolation (Mayr, 1957). In many of his works (1957, 1963, 1969, 1982, 1987) Mayr has dealt with asexuality as the fundamental impediment of the BSC. This obstacle is due to the fact that in these organisms, reproductive isolation cannot be tested and that asexual organisms do not exist in populations in the same sense as sexual organisms, that is, as interbreeding groups. Mayr (1957, 1963, 1969) has argued that the most satisfactory solution in taxonomic terms is to use the biological concept of species in sexual organisms and a morphological concept in asexual organisms. This is justified biologically because of the relationship between reproductive isolation and morphological differences, that is, they are the products of the same degree of genetic difference and are thus correlated. Mayr suggested the same adaptive peak situation as had Dobzhansky, and later (Mayr, 1982) extended this concept to include the fact that morphological discontinuities give indications of the specific ecological niches that asexual organisms occupy and thus provide inferences about their species status. Mayr (1957) also argued that while many previous workers would refer to asexual or uniparental populations, every individual and its descendants are reproductively isolated and that the term population does not apply. However, it would appear that workers, such as Simpson (1961), are really discussing the diversity, the spread and the differential selection of clones.

Realizing that the BSC does not reflect the range of kinds of species and speciation processes that exist, some have argued for a wide, pluralistic species concept which would apply to organisms with different modes of reproduction (Mishler and Donoghue, 1982; Mishler and Brandon, 1987). Some have attempted to resolve the issue by arguing that asexual organisms, by definition, cannot form species (Ghiselin, 1987). A counter argument, presented by Mishler and Brandon (1987) is that the inability of the BSC to apply to uniparental organisms is not a reason to deny that asexual species exist, but that it is simply a fault of the BSC. In several papers, Mayr (1963, 1992) has simply minimized the impact of asexual and parthenogenetic reproduction on the BSC.

In many cases, the empirical evidence of species analysis reveals complex patterns of variation in morphological, genetic and ecological variation while the species concepts are too generalized to represent these situations. The criteria recommended for determining meaningful discontinuities within biological variation are very much dependent upon the biology of specific groups (Mishler and Donoghue, 1982). Scudder (1974) noted that it is an impossible task to develop a single, precise, meaningful species definition that is applicable to all organisms and acceptable to both taxonomists and evolutionists. While there are many published arguments to the contrary, some of which have been discussed above, particularly from an operational viewpoint, I feel that it is both practical and biologically accurate to recognize that there are different kinds of species as a result of different kinds of speciation processes.

14.4 EVOLUTIONARY CONSIDERATIONS

14.4.1 Variability

Many writers, while discussing the difficulties of systematically handling groups of parthenogenetic organisms within existing species concepts, often on the same occasion dismissed the importance of these groups in ecological and evolutionary terms, perhaps in an attempt to reduce the problem. Due to presumed limited variability, parthenogenesis has often been thought to be restricted to narrow, specific environmental conditions in space and time. Many workers (Du Rietz, 1930; Dobzhansky, 1937; Cain, 1954; Mayr 1957, 1963) have generalized that parthenogenetic groups are relatively recent, secondarily derived phenomena and thus are impossible to designate as species in a non-arbitrary way, are too similar to their sexual relatives to distinguish as separate species, or simply do not require an essentially different species category.

Parthenogenesis has also been considered by many theoreticians to be an evolutionary dead-end due to a lack of genetic variability and the steady accumulation of deleterious mutations in clonal lineages (Muller's Ratchet; Felsenstein, 1974) (White, 1978; Bell, 1982; Maynard-Smith, 1986). Clones are believed to be unable to diversify rapidly enough to meet changing environmental conditions and to overcome average rates of extinction. However, there are indications that these traditional views are now considered less general and acceptable (Suomalainen *et al.*, 1976; Ghiselin, 1988). Studies by Suomalainen (1961) showed a large amount of morphological variation among populations of parthenogenetic weevils. Subsequently, allozyme analyses provided further evidence of substantial clonal variation in many instances. For example, Saura *et al.* (1976) revealed 76 clones of a weevil and Mitter *et al.* (1979) identified 36 clones of a parthenogenetic moth. This diversity could be the result of different origins or the result of within-clone evolution.

While there have been evolutionary arguments about just how relatively long- or short-term the evolution of parthenogens is, the genetic variability that does exist in parthenogens is usually considered to be a relatively finite phenomenon; variation is ultimately selected out through clonal selection or driven to homozygosity by automixis. There is growing evidence that clones do become extinct and only persist successfully in situations where the conditions are very favourable for them (Hughes, 1989).

Single gene mutations may result in the development of parthenogenetic lineages within sexual populations. However, the majority of parthenogenetic taxa are the result of hybridization (White, 1978). Polyploidy is a frequent and important correlate of parthenogenesis (Bell, 1982); it is particularly common in those groups with apomictic parthenogenesis. As a result, in the short term, parthenogenetic lineages are able to occupy broad niches and be ecologically successful.

Increasingly, data show that the genetic structure of parthenogenetic populations can be complex (Hebert, 1987) and that the evolutionary potential of these groups in not as limited as has been thought. Finston *et al.* (1995), in a study of genome size variation in aphids, have postulated that shifts in small genome sizes, associated with short generation times in parthenogenetic aphids, provide a means for saltational change in character states and may be important in the evolutionary diversification of this group.

14.4.2 Diversity

Many consider that the taxonomic pattern where most parthenogens have close sexual relatives also indicates that they are successful only in the evolutionary short term. There is taxonomic and phylogenetic evidence that, in some groups, there has been considerable evolutionary radiation in some parthenogenetic lineages. The most widely cited example is the bdelloid rotifers, a thelytokous, species-rich group of over 300 relatively easily recognized species in four families (Hutchinson, 1967). As an explanation for this diversity, it has been postulated that these discrete species represent scattered adaptive peaks that are the survivors from a larger array of produced clonal forms and that, due to their protected aquatic habitats, they have low rates of extinction (Stanley, 1975).

In the oribatid mites, parthenogenesis is estimated to occur in 8–9% of the known species (Norton and Palmer, 1991). Included is the largest group of animals that reproduce solely by thelytokous parthenogenesis, the Desmonomata, an early derivative taxon comprising about 400 species in more than 30 genera and seven families (Palmer and Norton, 1992). It has been proposed (Norton and Palmer, 1991) that many families of oribatid mites have speciated in the absence of sexual reproduction through meiotic thelytoky from ancestors who were also parthenogenetic.

Among the insects, another possible example of parthenogenetic radiation can be found in the aphid genus *Trama*. This group, in which no males have been found, consists of about 30 species that feed exclusively on the roots of Compositae and which have an extensive distribution primarily in Europe and Asia. Again, their comparatively protected environment may have resulted in a low rate of extinction (Eastop, 1953).

14.4.3 Cyclical parthenogenesis

Cyclical parthenogenesis has been a source of evolutionary change in parthenogens. For example, the Aphidoidea exhibit a wide range of variation in cyclical parthenogenesis (Moran, 1992). The complete life-cycle (holocycle) consists of a single sexual generation which occurs during the colder part of the year and series of summer, parthenogenetic generations. There are many previously holocyclic aphids, that are now confined to a single host plant and that are without sexual reproduction in their life-cycle (anholocycle). These obligately parthenogenetic lineages could be the result of accumulated mutations within clones or could have come about through the multiple evolution of thelytoky from different sexual genotypes.

Among the aphids, complex life-cycles involving cyclical parthenogenesis and winter and summer host plant alternation occur frequently. For example, the agricultural pest, *Myzus persicae,* now spread throughout the world, is likely to have originated in China on its primary, winter host plant, *Prunus persica.* In cold climates *M. persicae* undergoes the complete holocycle from winter, primary to summer, secondary hosts while in tropical climates it loses the bisexual part of the life-cycle, remaining completely parthenogenetic on secondary hosts (anholocycly). In intermediate climates, this aphid can be holocyclic, anholocyclic or even androcyclic, where the parthenogenetic females eventually produce functional males (Blackman, 1981, 1985). In addition to heritable life-cycle variation, some clones of *M. persicae* exhibit insecticide resistance, variable ability to transmit plant viruses, general and specific and geographic host plant preferences and interclonal variation in the ability to produce sexual morphs. Related species, such as *M. dianthicola* and *M. ascalonicus* appear to consist entirely of anholocyclic forms.

There are a number of examples of incipient speciation among parthenogens. The Spotted Alfalfa Aphid, *Therioaphis trifolii,* first appeared as anholocyclic populations on alfalfa in the southern United States in the 1950s. It was subsequently determined that this aphid had been introduced into eastern North America over 70 years before and named the Yellow Clover Aphid. The Spotted Alfalfa Aphid subsequently exhibited insecticide resistance which spread to other areas. As its range extended into the northern United States, the aphid started to produce the sexual stage and overwintering eggs. It appears that these two entities represent separate introductions from rather different genotypes of Old World *T. trifolii* (Blackman, 1981).

14.4.4 Geographical variation

Parthenogens show considerable geographic variation which indicates a capacity to evolve. Even in situations where parthenogenesis is a relatively recent, secondarily derived situation, there is evidence of evolutionary change. Geographical analyses have shown that parthenogenetic organisms have distributions that are different from those of their bisexual relatives. It has also been noted that if there is a parthenogenetic and a bisexual form of the same species, the successful parthenogenetic form usually has a wider distribution. These situations have been called geographical parthenogenesis (Lynch, 1984). There are many examples in Europe where the bisexual form is confined to a limited area and the parthenogenetic form has spread extensively (Suomalainen *et al.*, 1976). This situation can be explained by the existence of general-purpose genotypes which occupy a wide range of habitats, that is, there is strong selection pressure on obligate parthenogens to produce general-purpose genotypes, able to tolerate a wide range of conditions (Lynch, 1984). The alternate explanation would be that species consist of a set of distinct genotypes, each specialized for a narrow subset of the habitat range.

Suomalainen (1962) and Suomalainen *et al.* (1976) have explained geographical parthenogenesis in weevils by invoking a glacial refugium model. Glaciation during the last Ice Age has been the selective force involved in determining the distribution of species of parthenogenetic insects. Bisexual forms of the weevil, *Otiorrhynchus scaber*, survived in isolated, ice-free areas. As the glaciers retreated, these populations spread and interbred, producing increased heterozygosity, which became fixed in the parthenogenetic forms. Subsequent polyploid clones were able to adapt to new habitats which were created as the glacier retreated (Saura *et al.*, 1976).

Cyclically parthenogenetic Adelgidae alternate between a primary host which is always a species of spruce (*Picea*) and a secondary host which is another species of conifer. As is the case with the closely related true aphids (Aphididae), the adelgids often have a truncated life-cycle with associated anholocyclic forms. In the Adelgidae these may occur on either the primary or the secondary host plant. Many of these persistent parthenogenetic lineages are considered to be distinct species by taxonomic specialists. Anholocycly has come about either through mutational elimination of the male line or by glacial displacement of the primary host plant. Steffan (1961, 1963) has, for example, shown that the anholocyclic species, *Sacciphantes abietis*, on spruce and *S. segregis* on larch have split off from the holocyclic *S. viridis* which migrates from spruce to larch. Steffan (1964) has used a concept of agamospecies for these parthenogenetic lineages. He has also used the term microspecies, as components of a larger artenkreis or superspecies, in this case the *A. abietis–A. segregis–A. viridis* combination.

A number of adelgid agamospecies are important forest pests; the ability to produce overwintering eggs enables them to persist in otherwise harsh climates. The balsam woolly aphid, *Adelges piceae*, is indigenous to Europe, where it has lost its primary host and is found only on species of *Abies*. This adelgid has been introduced into North America on nursery stock on a number of occasions and into different regions. This completely parthenogenetic species has shown considerable biological variability in its physiological effects on its host. Using multivariate morphometrics, morphologically distinct groupings of population samples, associated with three different areas of introduction in North America have been demonstrated (Foottit and Mackauer, 1980). These groupings were subsequently distinguished as geographical subspecies (Foottit and Mackauer, 1983).

14.5 PRACTICAL CONSIDERATIONS

Claridge (1995) has rightly pointed out that species concepts are necessary in order to describe patterns of biological diversity and that different species concepts will have different consequences for the analysis of speciation processes. Parthenogenetic insects may not exist as populations or form species in the same manner as do sexual species. However, they do adapt and evolve and they have unique ecological and evolutionary roles. Clonal diversity in obligate and cyclical parthenogens is real, generated by life-history dynamics, vicariant processes and selection pressure. In biological terms, it is necessary to recognize parthenogenetic species. A category for parthenogenetic insects is necessary for practical purposes, such as the needs of applied biologists.

14.5.1 Populations

The use of the term population, in the context of parthenogenetic organisms, needs clarification. It is often stated that obligate parthenogens only form clones, not populations in the sense of the interbreeding group of evolutionary biology. As has been outlined above, parthenogenetic species are far from uniform; they are genetically variable and clonally diverse. Mutations that are selectively advantageous spread throughout a species. Parthenogenetic species do form populations of co-adapted clones and it is in that sense that the term population should be used.

14.5.2 Analytical methods

As discussed above, there has been extensive consideration of how to accommodate asexual and parthenogenetic organisms within the various species concepts. Nevertheless, at the operational level, the main approach has been that of the descriptive morphospecies and comparison with closely related sexual species; biological data are incorporated into the taxonomic process, if the data are available. In fact, many workers (Mayr, 1963; De Bach, 1969) have essentially recommended a holomorphological approach to delimiting parthenogenetic species (Enghoff, 1976). Morphological gaps are viewed as representative of the degree of genetic isolation resulting from natural selection and are indicative of the species-specific ecological niche that is occupied (Mayr, 1982).

Recognition of the extent of parthenogenetic clones and species can be a relatively local problem or it may require a broad perspective over wide species, host and geographical ranges. Theory has provided much towards the understanding of the evolution and systematics of parthenogenetic organisms. Further understanding and the practical recognition of parthenogenetic taxa will come about through the empirical comparison of clonal lineages in a phylogenetic framework. A number of approaches, particularly DNA sequencing techniques, allozyme analysis and multivariate morphometrics, will prove useful in parthenogenetic species resolution as these methods not only delimit taxa they also provide inferential information about the ecology and evolution of these species.

Mitochondrial DNA, which is maternally inherited, and which exhibits a rapid rate of sequence divergence, will provide the data necessary to associate parthenogenetic lineages and species with their most closely related sexual species. Allozyme analysis, particularly indications of fixed gene differences (Hales, 1991), provides additional evidence in support of morphological differences and may provide indications of the existence of lineages or species in cases where differences are morphologically cryptic. Parthenogenesis, particularly cyclical parthenogenesis and associated progenetic development, results in a reduction in available morphological characters. Multivariate morphometrics is necessary for the resolution of subtle population and species limits in these cases. Additionally, and based on the assumption that multivariate ordinations evaluate partitioned variance and reflect the polarity of character states, these methods can be used to provide phylogenetic resolution (Foottit and Sorensen, 1992).

14.5.3 Recommendations: handling parthenogenetic lineages

Many absolute statements have been made about what is a variable phenomenon. I doubt that we will obtain a single, comprehensive species concept or species definition for all parthenogenetic insects, given the variety of genetic and evolutionary mechanisms involved. However, species of parthenogenetic insects can be best handled taxonomically by concepts and practices that interpret pattern along with biological reality and which incorporate a genealogical perspective, at the clonal, population and species level, where possible. Sound taxonomic judgement, as always, is an important requirement.

The BSC should be supplemented through a pluralistic approach (Mishler and Donoghue, 1982). The recognition of agamospecies (asexual or parthenogenetic species; species without sexual reproduction) as proposed by a number of workers, mentioned above, is a minimally pluralistic solution to the problem of parthenogenetic species (Mayr, 1992). This recognizes that parthenogenetic species have an integrity that is maintained by selection for gene combinations that are adaptive and that these mechanisms are different from those of sexual species. Cyclical parthenogens can be accommodated in a strict sense with the BSC, although even in these cases, geographically limited, infrequent sexual reproduction will result in patterns similar to obligate parthenogenesis (Mishler and Budd, 1990; Hebert *et al.*, 1991).

Blackman (1995) has recently pointed out that some of the difficulty of the species problem is in fact confusion between the theoretical issues of species concepts and the practical issues of working definitions of species. The biological diversity of parthenogenetic organisms, as reviewed above, clearly exists; these parthenogenetic species exist, evolve and require taxonomic handling. The practical issues of a working definition can best be addressed through the use of the agamospecies definition, as above. Clonal populations in various circumstances, if clearly distinctive and diagnosable, can be designated, in a taxonomically neutral fashion, as uniparental clonal lineages, that is, as parthenogenetic ancestral–descendant sequences of clones that are unique historical entities.

As was recommended by Enghoff (1976) and Suomalainen *et al.* (1987), when a parthenogenetic lineage is derived from a known bisexual species, it should not be designated as a separate species, it should be designated as the parthenogenetic form of species *Aus bus*. Known hybrid forms should be identified by a hybrid designation (e.g. *Aus bus* × *Bus cus*). These measures will prevent unnecessary naming of species where there is inadequate data. However, if there is evidence that clonal populations are evolving and occupying a habitat and perpetuating, they constitute a historically unique species and should be distinguished as agamospecies. Criteria for agamospecies status would be differences from bisexual relatives in terms of host plant preferences, chromosome number, etc., that is, in meaningful differences, not just taxonomic convenience.

While the phylogenetic species concept (Mishler and Brandon, 1987) is an attractive conceptual framework for emphasizing phylogenetic reconstruction, strict adherence to this concept would result in the recognition of all monophyletic, geographically distinct, uniparental clonal lineages as formal taxa. This increase in the number of binomials would not be practically advantageous. Nevertheless, each historical group of organism lines (tokogenetic arrays; Hennig, 1966) could be handled in a phylogenetic framework, assuming appropriate cladistic evidence, but only using informal names for purposes of communication, unless there was convincing, biologically significant evidence for species status. A phylogenetic framework would serve the important function of providing a hierarchical means of grouping clonal lineages.

14.6 CONCLUSIONS

There is increasing evidence that while the distribution of parthenogenesis is sporadic it is not random (Bell, 1982). In many taxa, parthenogens are not evolutionary dead-ends, but they are genetically diverse and ecologically widely adapted and have the ability to adapt to changing environments (Suomalainen *et al.*, 1987). Parthenogenesis is most common in situations of disturbance, such as fire, drought, glaciation and grazing and ephemeral or fluctuating habitats, as are prevalent in agricultural situations. For example, there is considerable evidence for host plant-associated genotypes in parthenogenetic insects; host plant adaptations play an important role in the pest status of agriculturally important insects. It is therefore very important from the perspective of effective pest management that we have operational species definitions which allow workers to handle parthenogenetic insect species.

14.7 REFERENCES

Bell, G. (1982) *The Masterpiece of Nature. The Evolution and Genetics of Sexuality*, University of California Press, Berkeley and Los Angeles.

Blackman, R.L. (1981) Species, sex and parthenogenesis in aphids, in *The Evolving Biosphere* (ed. P. Forey), Cambridge University Press, Cambridge, pp. 75–85.

Blackman, R.L. (1985) Aphid cytology and genetics, in *Evolution and Biosystematics of Aphids, Proceedings of the International Aphidological Symposium at Jablonna*, 1981, Wydawnictowa Polska Akademia Nauk, Warsaw, pp. 171–237.

Blackman, R.L. (1995) What's in a name? Species concepts and realities. *Bulletin of Entomological Research*, **85**, 1–4.

Cain, A.J. (1954) *Animal Species and their Evolution*, Hutchinson, London.

Claridge, M.F. (1995) Species and speciation. *Trends in Ecology and Evolution*, **10**, 38.

De Bach, P. (1969) Uniparental, sibling and semi-species in relation to taxonomy and biological control. *Israel Journal of Entomology*, **4**, 11–28.

Dobzhansky, Th. (1937) *Genetics and the Origin of Species*, Columbia University Press, New York.

Dobzhansky, Th. (1972) Species of *Drosophila*. New excitement in an old field. *Science*, **177**, 664–9.

Du Rietz, G.E. (1930) The fundamental units of biological taxonomy. *Svensk Botanisk Tidskrift*, **24**, 333–428.

Eastop, V. (1953) A study of the Tramini (Homoptera–Aphididae). *Transactions of the Royal Entomological Society of London*, **104**, 385–413.

Enghoff, H. (1976) Taxonomic problems in parthenogenetic animals. *Zoologica Scripta*, **5**, 103–4.

Felsenstein, J. (1974) The evolutionary advantage of recombination. *Genetics*, **78**, 737–56.

Finston, T.L., Hebert, P.D.N. and Foottit, R.G. (1995) Genome size variation in aphids. *Insect Biochemistry and Molecular Biology*, **25**, 189–96.

Foottit, R.G. and Mackauer, M. (1980) Morphometric variation between populations of the balsam woolly aphid, *Adelges piceae* (Ratzeburg) (Homoptera: Adelgidae), in North America. *Canadian Journal of Zoology*, **58**, 1494–503.

Foottit, R.G. and Mackauer, M. (1983) Subspecies of the balsam woolly aphid, *Adelges piceae* (Homoptera: Adelgidae), in North America. *Annals of the Entomological Society of America*, **76**, 299–304.

Foottit, R.G. and Sorensen, J.T. (1992) Ordination methods: Their contrast to clustering and cladistic techniques, in *Ordination in the Study of Morphology, Evolution and Systematics of Insects: Applications and Quantitative Genetic Rationales* (eds J. T. Sorensen and R.G. Foottit), Elsevier, Amsterdam, pp. 1–10.

Ghiselin, M.T. (1987) Species concepts, individuality, and objectivity. *Biology and Philosophy*, **2**, 127–43.

Ghiselin, M.T. (1988) The evolution of sex: A history of competing points of view, in *The Evolution of Sex. An Examination of Current Ideas* (eds R. E. Michod and R. B. Levin), Sinauer Associates, Sunderland, Massachusetts, pp. 7–23.

Hales, D. (1991) Use of allozyme electrophoresis in distinguishing morphologically similar species of Aphididae (Homoptera: Aphidinea). *Entomologia Generalis*, **16**, 193–200.

Hawksworth, D. (ed.) (1994) *The Identification and Characterization of Pest Organisms*, CAB International, Wallingford, UK.

Hebert, P.D.N. (1987) Genotypic characteristics of cyclic parthenogens and their obligately asexual derivatives, in *Evolution of Sex and its Consequences* (ed. S.J. Stearns), Birkhauser, Berlin, pp. 175–95.

Hebert, P.D.N., Finston, T.L. and Foottit, R. (1991) Patterns of genetic diversity in the sumac gall aphid, *Melaphis rhois*. *Genome*, **34**, 757–62.

Hennig, E. (1966) *Phylogenetic Systematics*, University of Illinois Press, Urbana.

Hughes, R.N. (1989) *A Functional Biology of Clonal Animals*, Chapman & Hall, New York.

Hull, D. (1970) Contemporary systematic philosophies. *Annual Review of Ecology and Systematics*, **1**, 19–54.

Hutchinson, G.E. (1967) When are species necessary? in *Population Biology and Evolution* (ed. R.C. Lewontin), Syracuse University Press, pp. 177–86.

Lynch, M. (1984) Destabilizing hybridization, general-purpose genotypes and geographic parthenogenesis. *The Quarterly Review of Biology*, **39**, 257–90.

Maynard-Smith, J. (1986) Contemplating life without sex. *Nature*, **324**, 300–1.

Mayr, E. (1942) *Systematics and the Origin of Species from the Viewpoint of a Zoologist*, Columbia University Press, New York.

Mayr, E. (1957) Difficulties and importance of the biological species, in *The Species Problem* (ed. E. Mayr), American Association For the Advancement of Science, Washington, DC, pp. 371–88.

Mayr, E. (1963) *Animal Species and Evolution*, Harvard University Press, Cambridge, Massachusetts.

Mayr, E. (1969) *Principles of Systematic Zoology*, McGraw-Hill, New York.

Mayr, E. (1982) *The Growth of Biological Thought. Diversity, Evolution, and Inheritance*, Harvard University Press, Cambridge, Massachusetts.

Mayr, E. (1987) The ontological status of species: Scientific progress and philosophical terminology. *Biology and Philosophy*, **2**, 145–66

Mayr, E. (1992) A local flora and the biological species concept. *American Journal of Botany*, **79**, 222–38.

Meglitsch, P.A. (1954) On the nature of the species. *Systematic Zoology*, **3**, 49–65.

Mishler, B.D. and Brandon, R.N. (1987) Individuality, pluralism, and the phylogenetic species concept. *Biology and Philosophy*, **2**, 397–414.

Mishler, B.D. and Budd, A.F. (1990) Species and evolution in clonal organisms – Introduction. *Systematic Biology*, **15**, 79–85.

Mishler, B.D. and Donoghue, M.J. (1982) Species concepts: a case for pluralism. *Systematic Zoology*, **31**, 491–503.

Mitter, C., Futuyma, D.J., Schneider, J.C. and Hare, J.D. (1979) Genetic variation and host plant relations in a parthenogenetic moth. *Evolution*, **33**, 777–90.

Moran, N.A. (1989) A 48-million-year-old aphid–host plant association and complex life cycle: biogeographic evidence. *Science*, **245**, 173–5.

Moran, N.A. (1992) The evolution of aphid life cycles. *Annual Review of Entomology*, **37**, 321–48.

Norton, R.A. and Palmer, S.C. (1991) The distribution, mechanisms, and evolutionary significance of parthenogenesis in oribatid mites, in *The Acari: Reproduction, Development and Life-History Strategies* (eds R. Schuster and P.W. Murphy), Chapman & Hall, London, pp. 107–36.

Oliver, J.H. (ed.) (1971) Parthenogenesis. *American Zoologist*, **11**, 239–398.

Palmer, S.C. and Norton, R.A. (1992) Genetic diversity in thelytokous oribatid mites. *Biochemical Systematics and Ecology*, **20**, 219–31.

Saura, A., Lokki, J., Lankinen, P. and Suomalainen, E. (1976) Genetic polymorphism and evolution in parthenogenetic animals. III. Tetraploid *Otiorrhynchus scaber* (Coleoptera: Curculionidae). *Hereditas*, **82**, 79–100.

Scudder, G.G.E. (1974) Species concepts and speciation. *Canadian Journal of Zoology*, **52**, 1121–34.

Simpson, G.G. (1961) *Principles of Animal Taxonomy*, Columbia University Press, New York and London.

Sokal, R.R. and Crovello, T.J. (1970) The biological species concept: a critical evaluation. *American Naturalist*, **104**, 127–53.

Stanley, S.M. (1975) Clades versus clones in evolution: why we have sex. *Science*, **190**, 382–3.

Steffan, A.W. (1961) Die Stammes- und Siedlungsgschichte des Artenkreis *Sacciphantes viridis* (Ratzeburg 1843) (Adelgidae: Aphidoidea). *Zoologica* (Stuttgart), **109**, 1–113.

Steffan, A.W. (1963) Zur systematischen und phylogenetischen Stellung der Agamospecies in den Adelgidae- Genera (Homoptera). *Verhandlungen der Deutschen Zoologischen. Gesellschaft in Wein*, 1962, pp. 640–55.

Steffan, A.W. (1964) Problems of evolution and speciation in Adelgidae (Homoptera: Aphidoidea). *Canadian Entomologist*, **96**, 155–7.

Suomalainen, E. (1961) On morphological differences and evolution of different polyploid parthenogenetic weevil populations. *Hereditas*, **47**, 309–41.

Suomalainen, E. (1962) Significance of parthenogenesis in the evolution of insects. *Annual Review of Entomology*, **7**, 349–66.

Suomalainen, E., Saura, A. and Lokki, J. (1976) Evolution of parthenogenetic insects. *Evolutionary Biology*, **9**, 209–57.

Suomalainen, E., Saura. A. and Lokki, J. (1987) *Cytology and Evolution in Parthenogenesis*, CRC Press, Boca Raton.

Turesson, G. (1929) Zur Natur und Begrenzung der Arteinheiten. *Hereditas*, **12**, 323–34.

White, M.J.D. (1978) *Modes of Speciation*, W.H. Freeman and Company, San Francisco.

Wilson, E.O. (1992) *The Diversity of Life*, Harvard University Press, Cambridge, Massachusetts.

15

The species in terrestrial non-insect invertebrates (earthworms, arachnids, myriapods, woodlice and snails)

A. Minelli and D. Foddai

Contacting address: Universita di Padova, Dipartimento di Biologia, Via Trieste 75, I35121 Padova, Italy

ABSTRACT

In current practice, species discrimination in these diverse groups often rests on showy, complex morphological traits, such as the reproductive apparatus of gastropods, the complex gonopods of males in juliform millipedes, and the not less complex palps of many spider males. Problems with species delimitation are often caused by parthenogenesis, well studied in oligochaetes, but common in most groups of invertebrates.

Perceptions of species barriers are affected by patterns of geographic distribution, in turn reflecting dispersal power. Allopatric taxa are usually easy to recognize, but their species status is usually less obvious. Easily distinguishable and easily dispersed species include anthropochorous isopods, centipedes and millipedes, well-documented in the faunas of islands. Very interesting patterns of intensive speciation characterize insular biotas. Large species swarms are known among Macaronesian millipedes (*Cylindroiulus*, *Dolichoiulus*, *Acipes*) and Hawaiian spiders (Tetragnathidae). More widespread, however, are the patterns of continental insularity, for example in the extensive cave systems along the Southern margin of the Alps, but also on mountain tops.

Most taxonomic methods have been used, including karyology (good tradition and good results in earthworms), allozymes (relatively few, but for some snails and earthworms), isozymes, and more recently also mtDNA. Allozymes are sometimes better than morphol-

Species: The Units of Biodiversity. Edited by M.F. Claridge, H.A. Dawah and M.R. Wilson. Published in 1997 by Chapman & Hall. ISBN 0 412 63120 2

ogy (earthworms such as *Hormogaster* species; sibling species of *Macrocheles* mites), but quite often they do not correlate with morphology. Attitudes of specialists towards the value of Nei's D estimates as a cue to species status are very divergent.

For this sample of more than 100 000 named species from a diversity of phyla, species are neither easier nor more difficult to recognize than in most other animal groups.

15.1 INTRODUCTION

The animal groups surveyed in this paper are very diverse and the taxonomic traditions are very different. Also some of the key biological features of these animals, such as dispersal power and reproductive systems, are very different. Thus, these groups offer a variegated sample, both in terms of species circumscription and identification. Some groups, such as centipedes, millipedes and many arachnid taxa, are among the less-studied groups of terrestrial invertebrates. Others, however, have often been the subject of critical discussion of species boundaries and a few have even played the role of textbook examples in discussions about species and speciation. Obvious examples include snails of the genera *Cepaea*, *Partula*, *Albinaria* and *Cerion*. To date, very few critical reviews of 'the species in practice' have been devoted to these groups, exceptions being snails (Giusti and Manganelli, 1992) and spiders (Blandin, 1977; Brignoli, 1988).

Overall, more than 100 000 living species from these groups have been described: in crude figures, 12 000 myriapods, some 3500 woodlice, 1000 terrestrial oligochaetes, 35 000 spiders, nearly 40 000 non-aquatic mites, 9000 further arachnids and some 35 000 land snails.

A first reflection of the current state of affairs may be gleaned from a few examples of the levels of synonymy in these groups. We may expect that groups where species are more difficult to identify will have a higher average level of synonymy than groups where species are easier to circumscribe (Holman, 1987). Recent estimates of the number of known living gastropod species range between 40 000 and over 100 000 (Bieler, 1992). To this uncertainty, marine, freshwater and terrestrial forms are likely to contribute in similar ways. It is not too difficult to find species with a few dozen synonyms each, when browsing through recent revisionary works, especially those dealing with land snails from the European/Mediterranean area.

For spiders, very striking are the numbers of nominal taxa recently put into synonymy under other species as the effect of revisionary works on some genera of orb-web spiders (Araneidae). For example, Levi (1983) synonymized 17 nominal species with *Argiope aetherea* (Walckenaer), 14 of them described between 1911 and 1915; and Grasshoff (1986), in revising the African species of *Neoscona*, transferred 19 nominal taxa to *N. subfusca*

(C.L. Koch), 13 to *N. penicillipes* (Karsch) and 11 to *N. triangula* (Keyserling). The most impressive case, however, is that of the pisaurid spider, *Thalassius spinosissimus* (Karsch), of which Sierwald (1987) identified not less than 38 synonyms, 23 among them having been described in a single paper by Roewer in 1954!

Just to add one example from earthworms, let us cite *Aporrectodea caliginosa caliginosa* (Savigny) with 18 synonyms and *A. c. trapezoides* (Dugès) with seven synonyms (Mrsic, 1991).

Some groups, especially land snails, experienced waves of excessive splitting, then followed by strong, if not equally excessive, lumping. However, these very high levels of synonymy are not simply a reflection of the sloppy taxonomy of the past. One must seriously ask, whether the pattern of variability we discover in nature actually allows, today, an easy circumscription of species. As we shall try to demonstrate, animals do actively contribute to taxonomists problems.

15.2 DIAGNOSTIC CHARACTERS VERSUS SPECIFIC MATE RECOGNITION SYSTEMS (SMRS)

In the current practice, species discrimination in these groups often rests on showy, complex morphological traits, such as the reproductive apparatus of gastropods, the gonopods of male helminthomorph millipedes and the palps of many male spiders. In most examples, however, there is no direct evidence as to a function of these traits as specific mate recognition systems (SMRS). One of the few studies demonstrating the specificity of millipede gonopod structure is that of Barnett and Telford (1996). In a laboratory experiment, Costa and Francescoli (1991) found anomalous mating behaviour between interspecific couples of two sympatric sibling species of the wolf spiders, *Lycosa thorelli* (Keyserling) and *L. carbonelli* Costa and Capocasale, probably involving mechanical incompatibility between the genitalia of the two species. Males of *L. carbonelli* proved to be more discriminative between potential partners than male *L. thorelli*. Apparently no post-copulatory isolating mechanism exists, however, as demonstrated by the easy production of F_1 hybrids.

In the recent malacological literature, there has been discussion over the possible role of chirality (the left- or right-handedness of the shell) in the process of speciation. This role has been affirmed by Gittenberger (1988), but rejected by Johnson *et al.* (1990). Field evidence is not clear-cut. In *Partula suturalis* (Pfeiffer) populations respectively monomorphic for right- and left-handed shells are separated by steep clines (Johnson, 1987). Johnson *et al.* (1987) studied geographic variation of allozymes at 23 loci through the range of this species. Some correlation was found between allozymic variation and direction of coil, but the geographic patterns of allozymic variability did not closely overlap the distribution of the two shell types.

In the animal groups we are discussing here, chemical and behavioural SMRS are probably widespread, but have been very seldom exploited by taxonomists, except for spiders (see below).

Sex pheromones have been demonstrated at work in spiders (Krafft and Roland, 1980). In ixodid ticks, contact sex pheromones produced by *Dermacentor variabilis* (Say) and *D. andersoni* (Stiles) are species-specific and enable the courting males to discriminate conspecific females, with which they mate, from heterospecific, which they ignore (Sonenshine, 1984).

Takeda and Tsuruoka (1979) described a head gland of a terrestrial snail (*Euhadra*), consisting of fleshy tissue lying close to the base of the eye stalks. This gland increases in size during the breeding season and mating snails touch each other's gland during courtship. Tests with ethanol extracts succeeded in eliciting sexual behaviour in conspecifics. Further circumstantial evidence concerning the possible role of sex pheromones in land snails was discussed by Croll (1983). A study of competitive interactions between two introduced European land snails in South Australia (Smallridge and Kirby, 1988) suggests that species-specific inhibitory substances in the mucus may aid in keeping the species apart.

Discriminating characters and potential SMRS are often differently developed in otherwise similar, related groups. Within Opilionida, for example, the species of Trogulidae are character-poor, even lacking those conspicuous glandular organs that are present on the chelicerae in other families of harvestmen, as in Ischyropsalididae and Nemastomatidae. In these families these organs seem to play a key role in mating behaviour, and hence in the delimitation of biological species (Martens, 1969a,b). Other groups with uniform, less informative morphology include scorpions, pseudoscorpions, centipedes, symphylids, pauropods, and even two groups of diplopods, i.e. Pselaphognatha and Pentazonia. Sometimes, sizeable differences in usually showy traits are lacking in some species groups, diagnostic characters being provided by other, usually more trivial and less reliable characters, such as colour patterns and even size. Examples are found for example in millipedes.

In spiders, behavioural traits are sometimes accorded prominent importance in the delimitation of species (Blanke, 1986). Den Hollander and Dijkstra (1974), for example, described a new species of wolf spider, *Pardosa vlijmi*, on the basis of its courtship display. This behavioural character is, in practice, the only ground for distinguishing this ethospecies from its closest relative, *P. proxima* (C.L. Koch).

Jumping spiders (Salticidae) are a good group in which to test the importance of behavioural characters in species discrimination. Their showy displays are well known and extensively documented. Other spiders, however, also merit close attention. In Ontario, the spider *Philodromus rufus* Walckenaer (Philodromidae) is represented by two sympatric forms, differing in the way their males use their legs when approaching females. Only one of these forms exhibits conspicuous waving movements of the legs. An attentive morphometric study revealed subtle differences between the two forms in the size of the cephalothorax, in the relative size of the second pair of femora, in colour and pattern, as well as in genitalia. When given a choice, these spiders mate preferentially with individuals of the same form. According to Dondale (1964) these differences in mating behaviour justify recognizing a separate species, *Ph. vibrans* Dondale, but matters are probably more intricate, as shown by subsequent papers of the same author.

Stridulatory organs used in courtship have been described in spiders (Uetz and Stratton, 1982; Grabner and Thaler, 1986). Finally, constructional features of spider webs have been sometimes regarded as taxonomic characters, but these seem to be useful only within local species-poor faunas, not in a comprehensive survey of a whole genus (Eberhard, 1990).

15.3 PITFALLS OF MORPHOLOGICAL EVIDENCE: SEXUAL DIMORPHISM, FUZZY DEVELOPMENTAL SCHEDULES

Practical problems sometimes arise because of sexual dimorphism. Quite often, only one sex, generally the male, offers good taxonomic characters for species identification. Examples are among the juliform millipedes with their very complex male gonopods. In other examples, both sexes exhibit useful diagnostic characters, but the sexual dimorphism is so strong, that putting together male and female of the same species may prove difficult for the museum taxonomist. Examples in spiders are referred to by Coddington and Levi (1991).

Another problem, which does not occur in some other major animal groups, such as pterygote insects, derives from the difficulty of differentiating juveniles from adults. Sometimes, especially in groups with rampant heterochrony, such as lithobiomorph centipedes (Minelli *et al.*, 1996) the differences between juveniles and adults of one and the same species are comparable to those differentiating closely related species. Therefore, many nominal species have been created for the juveniles of other already-named species.

In still other examples, mature specimens of the same species are sometimes intrinsically inhomogeneous because of the plasticity of developmental schedules. Coddington and Levi (1991) have stressed the propensity of spiders to mature in any of several moults. Also julidan millipedes often may mature in any of several moults (e.g. in Enghoff *et al.*, 1993). However, this problem is lessened by the developmental constraints acting in the morphogenesis of adult genitalia, whose dimensions vary, within a given population, much less than overall body dimensions, or proportions, despite the occurrence of allometry.

This problem of the taxonomic implications of relative developmental control has been discussed in centipedes by Minelli and Bortoletto (1990) in respect to a character widely used in the taxonomy of geophilomorph centipedes, that is the (modal) number of body segments.

15.4 MORPHOLOGICAL AND MOLECULAR EVIDENCE: CHROMOSOMES, ALLOZYMES, ISOZYMES AND mtDNA

Most kinds of taxonomic approaches have been tried, including karyology (good tradition and good results in earthworms), allozymes (relatively few, but for snails and earthworms), isozymes; and more recently also mtDNA.

The usefulness of mtDNA for species discrimination is open to discussion. On the one hand, coherent geographic trends in variation of mtDNA polymorphisms have been confirmed at least for some species in one group of snails, for example by Murray *et al.* (1991) for *Partula suturalis* and *P. taeniata* from the Society Islands. On the other hand, the diagnostic value of mtDNA data is lessened by the confirmed presence of common genotypes in what we would plainly regard as different species. One case is discussed by Murray *et al.* (1991) in the same paper. One wonders how far this overlapping of mtDNA variants between closely related species is due to hybridization or introgression and how often it simply documents an ancestral similarity surviving after cladogenesis.

Allozymes are sometimes better than morphology in discriminating species, for example in earthworms of the genus *Hormogaster* (Cobolli Sbordoni *et al.*, 1992), in the lithobiomorph centipedes of the *Eupolybothrus fasciatus* (Newport) group (Zanazzo *et al.*, 1994) and among the sibling species of *Macrocheles* mites (Pomponi *et al.*, 1988). Quite often, however, the biochemical evidence does not correlate with morphology. Examples of more or less extensive mismatch between these different kinds of evidence are provided by the land snails *Medora* (Giusti *et al.*, 1986), *Solatopupa* (Boato, 1988) and *Albinaria* (Kemperman and Degenaars, 1992). We should be cautious, however, that different authors also have different attitudes towards the data, sometimes according electrophoretic data undue preference in assessing taxonomic relationships.

Speaking of genetic distance as determined from protein electrophoretic data, attitudes of specialists towards the value of Nei's D estimates as a clue to species status differ greatly. Some still regard as conspecific two earthworm populations with a D distance up to 0.7, whereas $D = 0.4$ is adequate for others to discriminate species of oribatid mites, and much less is sometimes accepted as discriminating for snail species. Some examples follow:

1. For Barros *et. al.* (1992), the most distantly related strains within the *Allolobophora molleri* Rosa complex (earthworms) belong to one and same species, in spite of D values larger than 0.88!
2. According to Bernini *et al.* (1988), in the oribatid genus *Steganacarus*, D is about 0.4 in comparisons between related species, but only 0.00 to 0.036 between populations of *S. magnus* (Nicolet).
3. The values of D are very low (at most 0.03) between species of the land snails, *Samoana*, from the Society Islands (Johnson *et al.*, 1986).

Such figures can be profitably discussed in relation to Thorpe's (1983) overall averages of D of about 0.40 (range 0.03 to >1.00) in interspecific congeneric comparisons (900 cases) and D <0.10 in 98% of 7000 intraspecific comparisons.

The study of allozymic variation in some species of the Australian genus *Cristilabrum* of camaenid snails by Woodruff and Solem (1990) is very illuminating. These species have extreme, narrowly parapatric ranges. Nei's D, as calculated between *C. primum* Solem and its southern neighbour *C. grossum* Solem, was not significantly different from zero (range 0.026–0.051). Nevertheless, the authors rejected considering them conspecific because of strong morphological differences (shell and genitalia) and the absolute lack of intermediates. Moreover, they point to the lack of a necessary link between speciation and genetic differentiation. It is worth noting here that *C. primum* is more differentiated genetically from the other, northern neighbour, *C. monodon* Solem, with D = 0.17.

Even more impressive is evidence of genetic differentiation among semi-terrestrial (amphibious) snails of the genus *Novisuccinea* by Hoagland and Davis (1987). The widespread *N. ovalis* and the localized *N. chittenangoensis* (Pilsbry) cannot be distinguished electrophoretically, but they are clearly distinct in terms of shell morphology, anatomy and ecology. On the contrary, within what is a morphologically uniform *N. ovalis*, two population groups can be separated easily by electrophoresis.

In the rock-dwelling clausiliid snail, *Solatopupa*, allozymic data match morphology in distinguishing four out of the five conventional species, whereas the fifth, *S. similis* (Bruguière), splits into at least two cryptic, but genetically distinguishable species (Boato, 1988).

In the Ionian island of Kephallinia (Greece) there are several different forms of the clausiliid land snail *Albinaria*. Traditional taxonomy, based on morphological and biogeographical evidence, distinguishes four species with a total of 12 subspecies. Two of the species are endemic to Kephallinia. Kemperman and Degenaars (1992) studied allozymes in 42 populations representative of 11 of the 12 subspecies. The subspecies distinguished within the two non-endemic species are hardly differentiated genetically. However, allozymic differences were found between subspecies of the two endemic species. Moreover, each subspecies of these endemic species showed unexpectedly conspicuous genetic resemblance to different non-endemic species. These authors obviously question the validity of the traditional taxonomic arrangement, but do not feel confident, for the time being, to suggest an alternative systematization.

Allozymes sometimes suggest that a morphologically distinguishable species is nested within the diversity of its morphologically uniform but cladistically paraphyletic closest relative. An example is the large Italian earthworm, *Hormogaster samnitica* Cognetti, clearly derived from within what is traditionally called *H. pretiosa* Michaelsen (Cobolli Sbordoni *et. al.*, 1992). This species could be better regarded as a cluster of cryptic species, estimates of divergence times within it being up to 3 800 000 years!

In the land snail *Partula*, the evolution of the morphological, electrophoretic and mitochondrial phenotypes occur at variable rates, and independently of one another (Murray *et al.*, 1991). That is possibly a widespread phenomenon.

15.5 HYBRIDS

Good evidence for natural hybrids is not extensive. For example, for all myriapod taxa studied to date, the only substantiated case involves *Rhymogona cervina* (Verhoeff) and the related *Rh. silvatica* (Rothenbuehler) (Diplopoda). Hybrids between them have been demonstrated morphologically, and verified electrophoretically, by Pedroli-Christen and Scholl (1990). However, it is important to note that it was in a study of two naturally hybridizing land snails (*Cerion stevensoni* and *C. fernandina*) that Woodruff (1989) first noted the unique electromorphic variants (hybrizymes) which sometimes characterize hybrid populations. A further interesting example has been recently illustrated by Schilthuizen and Gittenberger (1994).

15.6 DISPERSAL, DISTRIBUTION AND SPECIES

Geography, i.e. the spatial distribution of variability, offers perhaps the biggest challenge to the (in principle, non-dimensional) biological species concept. Our groups of terrestrial non-insect invertebrates do not escape from this stricture. Sometimes, what is a couple of well-differentiated and distinguishable species in a part of the common range turns out to be an undifferentiated continuum within a polymorphic species in another part of the same range. That is, at least, what seems to emerge from years of investigations on the European land snails known as *Cochlicopa lubrica* (O.F. Müller), *C. lubricella* (Porro), *C. nitens* (Gallenstein) and *C. repentina* Hudec. Apparently only one species of this group occurs in Italy (Giusti and Manganelli, 1992), but it splits into two or more divergent phenotypes without intermediates in Central Europe.

Our perception of species barriers is affected by patterns of geographic distribution, itself reflecting dispersal power. Critical taxa in different groups are often co-occurring in the same areas (e.g. the Western and Eastern ends of the Alps). Non-insect terrestrial invertebrates include extremely low dispersers, for example many with a subterranean lifestyle. Among earthworms there are a few deep-burrowing species (e.g. *Eophila tellinii* Rosa). There are many subterranean species-rich taxa of isopods, a few spiders, many pseudoscorpions, centipedes and millipedes. These allopatric taxa are usually easy to recognize, but their species status is often not so obvious. At the opposite end of the spectrum, easily dispersed and seemingly coherent species include anthropochorous isopods, centipedes and millipedes, well-documented in the faunas of islands, e.g. Iceland, St. Helena, Hawaii, etc.

Very interesting patterns of species multiplication are revealed by insular biotas. Insular species swarms are known among Macaronesian millipedes of the genera *Cylindroiulus*, *Dolichoiulus* and *Acipes* (Enghoff 1982, 1983a,b, 1992a,b; Enghoff and Baez, 1993) and Hawaiian spiders of the family Tetragnathidae (Okuma, 1988; Gillespie, 1991, 1992, 1993, 1994). Some 29 species of *Cylindroiulus* occur on Madeira; 27 of them are endemic to that island, one is common to the Azores and another to the Desertas. Together with one species from Porto Santo and another from the Canaries, these species form a monophyletic group within this large genus. As many as 15 species occur at a single location, but with clear microhabitat differentiation. Most microhabitats, such as logs or the interface between soil and litter, have more than one characteristic species but, in most cases, such narrowly coexisting species are of different sizes. Also in terms of cladogenesis, size is the most conspicuous trait differentiating pairs of sister species. In particular, each of the pairs *gemellus/madeirae* and *julipes/lundbladi* includes two species essentially similar except for size.

Dolichoiulus is a millipede genus with 52 known species. Total lengths of mature specimens range between 7 and 49 mm, with diameter of their cylindrical body between 0.56 and 3.65 mm. The species differ in details of eye number, leg number and size, and sculpture, but particularly in the shape of the complex male gonopods. Of these 52 species, 46 live in the Canary Islands: 20 on Tenerife, 10 on Gran Canaria, nine on La Gomera and one to three species each on the remaining islands of the archipelago. Nearly all species are confined to one island. On certain islands, the distribution of the individual species is all but random: closely related species occur in micro-allopatry. The so-called archipelago effect (cf. Carlquist, 1974) does not seem to have helped in generating this large number of endemics.

From the Canary Islands large numbers of congeneric species are also known for three spider genera – *Dysdera* (Dysderidae) with 46 species, *Spermophorides* (Pholcidae) with 23 species and *Oecobius* (Oecobiidae) with 34 species (Wunderlich, 1992).

More widespread are the patterns of continental insularity, e.g. in the extensive cave systems along the Southern margin of the Alps (with many pseudoscorpions, millipedes and spiders), in the lava tubes of the Canary Islands (spiders: *Dysdera*), but also on mountain tops (e.g. arctic–alpine spiders). Extreme examples are the camaenid land snails in north-west Australia, with 27 species in three genera along the 52-km-long limestone Ningbing Range. The average range for one species is 1 km^2 and species ranges are mostly allopatric.

15.7 THE TEMPORAL DIMENSION

Many studies approaching the time dimension of species have been made in terrestrial snails. Sometimes (e.g. *Mandarina* from the Bonin Islands) there is evidence of polymorphism and hybridization patterns extending through time since the Late Pleistocene (Chiba, 1993).

For the groups we are discussing, evidence from fossils is restricted, in practice, to terrestrial snail shells, which are common at many Quaternary sites. These shells can be exploited for palaeoecological reconstructions, but the limited morphological evidence they provide has but scanty information content, in comparison with what can be obtained from many Quaternary insect remains (Elias, 1994). We cannot hope to get from these materials any sound evidence as for delimitation of species. However, in individual cases this fossil evidence may prove crucial in choosing among alternative hypotheses. Gould and Woodruff (1990) studied an example of an 'area effect' within the land snail *Cerion columna* Pilsbry and Vanatta on Great Inagua, Bahamas. A local population of this snail exhibits unusually squat and flat-topped shells, very different from those of the common kind, occurring both east and west of this area. Gould and Woodruff explained that area effect as the result of introgression from propagules of a related species, *C. dimidiatum* (Pfeiffer) from Cuba. This explanation, already suggested by multivariate morphometry, was most convincingly supported by the finding of fossils of *C. dimidiatum* cemented into soil crusts within the region inhabited by the unusual *C. columna*.

15.8 UNIPARENTALS

Animals with uniparental reproduction offer particularly intractable problems to the taxonomist. Parthenogenesis in earthworms is common, as is uniparental reproduction in land snails both as parthenogenesis and as autogamy. Further examples of parthenogenesis are known among isopods (e.g. *Trichoniscus pusillus* Brandt), centipedes (some *Lamyctes*), millipedes (e.g. *Nemasoma varicorne* C.L. Koch) and scattered groups of arachnids, from scorpions to mites.

In the opinion of Giusti and Manganelli (1992: 158) in genera such as *Vallonia, Lauria, Columella, Pagodulina, Chondrina, Abida,* etc., 'the distinction between an ecophenotype and a subspecies, or a subspecies and a species, is nearly always left to the personal philosophical or practical convictions of the individual researchers'. In *Vallonia,* for example, shell morphology allows the recognition of about seven European species. Of these some are well defined; others, however, live in mixed populations where intermediate phenotypes occasionally occur. To disentangle these messy groups, malacologists cannot resort to the usual strong morphological evidence of genitalia, because nearly all of these *Vallonia* are completely devoid of the penial complex and reproduce through parthenogenesis or self-fertilization. In other genera such as *Abida* and *Chondrina,* however, things are not so bad, in spite of the variability of some species (E. Gittenberger, personal communication).

Bisexual and parthenogenetic forms coexist in what are currently called *Nemasoma varicorne* and *Polyxenus lagurus,* two widespread European millipedes.

Variability in parthenogens has been extensively studied in terrestrial oligochaetes. In *Eiseniella tetraedra* and *Dendrobaena octaedra,* two obligately parthenogenetic earthworms, there is extensive morphological variation, not correlated with enzyme patterns (Terhivuo *et al.,* 1994; Terhivuo, 1988). In other earthworms, such as the obligate parthenogenetic *Octolasion tyrtaeum* (Örley) and *O. cyaneum* (Savigny) the number of clones occurring in populations at the northern edge of the range (Eastern Fennoscandia) is high (24 clones found in 238 individuals from eight localities) in the former, spontaneously dispersing species, but low (only two clones found in 134 individuals from four localities) in the other, anthropochorous species. But polymorphism has been documented in other obligate parthenogens, such as the enchytraeid worm *Fridericia striata* (Levinsen), whose sexual ancestor is unknown. In this animal, the diploid chromosome number is restored by terminal fusion, i.e. through the fusion of the products of the second meiotic division. Of 27 loci studied electrophoretically by Christensen *et al.* (1989), 13 were polymorphic, all but one in homozygotes. Two clones even differed in all polymorphic loci, probably due to the polyphyletic origin of this parthenogenetic agamospecies. But what is an agamospecies? According to Ghiselin (1984) agamospecies are 'heaps of leaves that have fallen off the tree that gave rise to them'. These clones, recognizable within a morphologically uniform parthenogen, may differ in subtle habitat requirements, for example in the woodlouse *Trichoniscus pusillus* and the oligochaetes *Dendrobaena octaedra* and *Fridericia galba* (Hoffmeister) (Christensen and Noer, 1986; Terhivuo and Saura, 1990; Christensen *et al.,* 1992).

In one group, the Desmonomata, a taxon of 'lower' oribatid mites, thelytokous parthenogenesis is perhaps the only manner of reproduction, thus presenting us with a taxonomic and evolutionary puzzle similar to that of the better-known bdelloid rotifers. According to conventional taxonomy, this group consists of seven families, 32 genera and about 400 nominal species (Palmer and Norton, 1992). Males may occur, in some of these mites, but they seem to be non-functional. Variability of these diploid parthenogens is low, but no population found comprised only one clone. No close sexual relative of these uniparental mites is known pointing to a remote origin of parthenogenesis. By implication, Palmer and Norton (1992) believe that genetic variability in these forms did arise after the loss of sex.

15.9 EPILOGUE

To sum up, difficulties with species limits are common among terrestrial non-insect invertebrates. However, these groups, also offer many opportunities for demonstrating the usefulness of biological species concepts.

We hope not to join the ranks of those authors, who, according to Brignoli (1988) are not too rare among the taxonomists of spiders and other groups, whose species identification criteria heavily rely on topographical data of museum labels, typological numerical values, and the good luck of the taxonomist!

Acknowledgements

We are grateful to Michael F. Claridge for inviting us to the Systematics Association Symposium on 'The Units of Biodiversity: Species in Practice'. We are also very indebted towards Henrik Enghoff (Copenhagen), Edmund Gittenberger (Leiden) and Konrad Thaler (Innsbruck) for useful comments on a previous draft. The work of one of the authors (A.M.) has been partly supported by grants of the Italian National Research Council (CNR) and the Italian Ministry of University and Scientific and Technological Research (MURST).

15.10 REFERENCES

Barnett, N. and Telford, S.R. (1996) Sperm competition and the evolution of millipede gentalia. *Mémoires du Muséum national d'Histoire naturelle, Paris*, **169**, 331–9.

Barros, F., Carracedo, A. and Diaz Cosin, D.J. (1992) Karyological and allozymic study of *Allolobophora molleri* complex and related taxa. Initial results. *Soil Biology and Biochemistry*, **24**, 1241–5.

Bernini, F., Avanzati, A.M. and Petrucci, R. (1988) Genetic variation in the soil phthiracarid mite *Steganacarus (S.) magnus* (Nicolet, 1855) (Acarida, Oribatida) (Notulae Oribatologicae XL). *Zeitschrift für zoologische Systematik und Evolutionsforschung*, **26**, 104–13.

Bieler, R. (1992) Gastropod phylogeny and systematics. *Annual Review of Ecology and Systematics*, **23**, 311–38.

Blandin, P. (1977) Le problème de l'espèce chez les araignées, in *Les problèmes de l'espèce dans le règne animal. Tome II (Mémoire n. 39)* (eds C. Bocquet, J. Génermont and M. Lamotte), Société Zoologique de France, Paris, pp. 13–56.

Blanke, R. (1986) Homologien in Fortpflanzungsverhalten von Krenzspinnen (Araneae, Araneidae) und deren Interpretation im Kontext von Systematik und der Existenz von Artbarrieren. *Actas do X Congreso Internacional de Aracnologia, Jaca/España*, **1**, 69–94.

Boato, A. (1988) Microevolution in *Solatopupa* landsnails (Pulmonata: Chondrinidae): genetic diversity and founder effects. *Biological Journal of the Linnean Society*, **34**, 327–48.

Brignoli, P.M. (1988) Ragni, specie e filosofia, in *Il problema biologico della specie*, Mucchi, Modena, pp. 293–307.

Carlquist, S. (1974) *Island Biology*, Columbia University Press, New York and London.

Chiba, S. (1993) Modern and historical evidence for natural hybridization between sympatric species in *Mandarina* (Pulmonata: Camaenidae). *Evolution*, **47**, 1539–56.

Christensen, B. and Noer, H. (1986) Spatial and temporal components of genetic variation in triploid parthenogenetic *Trichoniscus pusillus* (Isopoda: Crustacea). *Hereditas*, **105**, 277–85.

Christensen, B., Hvilson, M.M. and Pedersen, B.V. (1989) On the origin of clonal diversity in parthenogenetic *Fridericia striata* (Enchytraeidae, Oligochaeta). *Hereditas*, **110**, 89–91.

Christensen, B., Hvilson, M. and Pedersen, B.V. (1992) Genetic variation in coexisting sexual diploid and parthenogenetic triploid forms of *Fridericia alba* (Enchytraeidae, Oligochaeta) in a heterogeneous environment. *Hereditas*, **117**, 153–62.

Cobolli Sbordoni, M., De Matthaeis, E., Alonzi, A., Mattoccia, M., Omodeo, P. and Rota, E. (1992) Speciation, genetic divergence and palaeogeography in the Hormogastridae. *Soil Biology and Biochemistry*, **24** (12) (special issue), 1213–21.

Coddington, J.A. and Levi, H.W. (1991) Systematics and evolution of spiders (Araneae). *Annual Review of Ecology and Systematics*, **22**, 565–92.

Costa, F.G. and Francescoli, G. (1991) Analyse expérimentale de l'isolément reproductif entre deux espèces jumelles et sympatriques d'araignées: le *Lycosa thorelli* (Keyserling) et le *Lycosa carbonelli* Costa et Capocasale. *Canadian Journal of Zoology*, **69**, 1768–76.

Croll, R.P. (1983) Gastropod chemoreception. *Biological Reviews*, **58**, 293–319.

den Hollander, J. and Dijkstra, H. (1974) *Pardosa vlijmi* sp. nov., a new ethospecies sibling of *Pardosa proxima* (C.L. Koch, 1848) from France, with description of courtship display (Araneae: Lycosidae). *Beaufortia*, **22**, 57–65.

Dondale, C.D. (1964) Sexual behaviour and its application to a species problem in the spider genus *Philodromus* (Araneae: Thomisidae). *Canadian Journal of Zoology*, **42**, 817–27.

Eberhard, W.G. (1990) Function and phylogeny of spider webs. *Annual Review of Ecology and Systematics*, **21**, 341–72.

Elias, S.A. (1994) *Quaternary Insects and their Environment*, Smithsonian Institution Press, Washington and London.

Enghoff, H. (1982) The millipede genus *Cylindroiulus* on Madeira – an insular species swarm. *Entomologica Scandinavica, suppl.*, **18**, 1–142.

Enghoff, H. (1983a) *Acipes* – a Macaronesian genus of millipedes (Diplopoda, Julida: Blaniulidae). *Steenstrupia*, **9**, 137–79.

Enghoff, H. (1983b) Adaptive radiation of the millipede genus *Cylindroiulus* on Madeira: habitat, body size and morphology (Diplopoda, Julida: Julidae). *Revue d'Ecologie et Biologie du Sol*, **20**, 403–15.

Enghoff, H. (1992a) *Dolichoiulus* – a mostly Macaronesian multitude of millipedes with the description of a related new genus from Tenerife, Canary Islands (Diplopoda, Julida, Julidae). *Entomologica Scandinavica, suppl.*, **40**, 1–158.

Enghoff, H. (1992b) Macaronesian millipedes (Diplopoda) with emphasis on endemic species swarms on Madeira and the Canary Islands. *Biological Journal of the Linnean Society*, **46**, 153–61.

Enghoff, H. and Baez, M. (1993) Evolution of distribution and habitat patterns in endemic millipedes of the genus *Dolichoiulus* (Diplopoda: Julidae) on the Canary Islands, with notes on distribution patterns of other Canarian species swarms. *Biological Journal of the Linnean Society*, **49**, 277–301.

Enghoff, H., Dohle W. and Blower G. (1993) Anamorphosis in millipedes (Diplopoda) – the present state of knowledge with some developmental and phylogenetic considerations. *Zoological Journal of the Linnean Society*, **109**, 103–234.

Ghiselin, M.T. (1984) Narrow approaches to phylogeny: a review of nine books on cladism, in *Oxford Surveys in Evolutionary Biology* (eds R. Dawkins and M. Ridley), Oxford University Press, Oxford, pp. 209–22.

Gillespie, R.G. (1991) Hawaiian spiders of the genus *Tetragnatha*: 1. Spiny leg clade. *Journal of Arachnology*, **19**, 174–209.

Gillespie, R.G. (1992) Hawaiian spiders of the genus *Tetragnatha*: 2. Species from natural areas of windward East Maui. *Journal of Arachnology*, **20**, 1–17.

Gillespie, R.G. (1993) Biogeographic pattern of phylogeny in a clade of endemic Hawaiian spiders (Araneae: Tetragnathidae). *Memoirs of the Queensland Museum*, **33**, 519–26.

Gillespie, R.G. (1994) Hawaiian spiders of the genus *Tetragnatha*: 3. *Tetragnatha acuta* clade. *Journal of Arachnology*, **22**, 161–8.

Gittenberger, E. (1988) Sympatric speciation in snails: a largely neglected model. *Evolution*, **42**, 826–8.

Giusti, F. and Manganelli, G. (1992) The problem of the species in malacology after clear evidence of the limits of morphological systematics. *Proceedings of the ninth International Malacological Congress*, pp. 153–72.

Giusti, F., Grappelli, C., Manganelli, G., Fondi, R. and Bullini, L. (1986) An attempt of natural classification of the genus *Medora* in Italy and Jugoslavia on the basis of conchological, anatomical and allozymic characters (Pulmonata: Clausiliidae). *Lavori della Societtà Italiana di Malacologia*, **22**, 256–341.

Gould, S.J. and Woodruff, D.S. (1990) History as a cause of area effects: an illustration from *Cerion* on Great Inagua, Bahamas. *Biological Journal of the Linnean Society*, **40**, 67–98.

Grabner, R. and Thaler, K. (1986) Ein weiteres Stridulationsorgan bei manchen Arten der *mughi*-Gruppe der Gattung *Lepthyphanthes* (Araneae, Linyphiidae)? *Mitteilungen der Schweizer Entomologischen Gesellschaft*, **59**, 15–21.

Grasshoff, M. (1986) Die Radnetzspinnen-Gattung *Neoscona* in Afrika (Arachnida: Araneae). *Zoologische Wetenschaapen*, **250**, 1–123.

Hoagland, K.E. and Davis, G.M. (1987) The succineid snail fauna of Chittenango Falls, New York: taxonomic status with comparisons to other relevant taxa. *Proceedings of the Academy of Natural Sciences, Philadelphia*, **139**, 465–526.

Holman, E. (1987) Recognizability of sexual and asexual species in rotifers. *Systematic Zoology*, **36**, 381–6.

Johnson, M.S. (1987) Adaptation and rules of form: chirality and shape in *Partula suturalis*. *Evolution*, **41**, 672–5.

Johnson, M.S., Murray, J. and Clarke, B. (1986) High genetic similarities and low heterozygosities in land snails of the genus *Samoana* from the Society Islands. *Malacologia*, **27**, 97–106.

Johnson, M.S., Murray, J. and Clarke, B. (1987) Independence of genetic subdivision and variation for coil in *Partula suturalis*. *Heredity*, **58**, 307–13.

Johnson, M.S., Clarke, B. and Murray, J. (1990) The coil polymorphism in *Partula suturalis* does not favor sympatric speciation. *Evolution*, **44**, 459–64.

Kemperman, T.C.M. and Degenaars, G.H. (1992) Allozyme frequencies in *Albinaria* (Gastropoda: Pulmonata: Clausiliidae) from the Jonian Islands of Kephallinia and Ithaka. *Malacologia*, **34**, 33–61.

Krafft, B. and Roland, Ch. (1980) Quelques remarques au sujet de la communication chimique chez les araignées. *Comptes Rendues du V Colloque d'Arachnologie*, **9**, 129–34.

Levi, H.W. (1983) The orb-weaver genera *Argiope*, *Gea* and *Neogea* from the Western Pacific region (Araneae: Araneidae, Argiopinae). *Bulletin of the Museum of Comparative Zoology, Harvard*, **150**, 247–338.

Martens, J. (1969a) Die Sekretdarbietung während des Paarungsverhaltens von *Ischyropsalis* C.L. Koch (Opiliones). *Zeitschrift für Tierpsychologie*, **26**, 513–23.

Martens, J. (1969b) Die Abgrenzung von Biospezies auf biologisch-ethologischer und morphologischer Grundlage am Beispiel der Gattung *Ischyropsalis* C.L. Koch 1839 (Opiliones, Ischyropsalididae). *Zoologische Jahrbücher, Abteilung für Systematik*, **96**, 133–264.

Minelli, A. and Bortoletto, S. (1990) Segmentation in centipedes and its taxonomic implications, in *Proceedings of the 7th International Congress of Myriapodology* (ed. A. Minelli), E.J. Brill, Leiden, pp. 81–8.

Minelli, A., Negrisolo, E. and Fusco, G. (1996) Developmental trends in the post-embryonic development of lithobiomorph centipedes. *Mémoires du Muséum National d'Histoire Naturelle, Paris*, **169**, 351–8.

Mrsic, N. (1991) *Monograph on earthworms (Lumbricidae) of the Balkans*, Slovenska Akademija Znanosti in Umetnosti, Ljubljana.

Murray, J., Stine, O.C. and Johnson, M.S. (1991) The evolution of mitochondrial DNA in *Partula*. *Heredity*, **66**, 93–104.

Okuma, C. (1988) Redescriptions of the Hawaiian spiders of *Tetragnatha* described by Simon (Araneae, Tetragnathidae). *Journal of the Faculty of Agriculture, Kyushu University*, **33**, 77–86.

Palmer, S.C. and Norton, R.A. (1992) Genetic diversity in thelytokous oribatid mites (Acari; Acariformes: Desmonomata). *Biochemical Systematics and Ecology*, **20**, 219–31.

Pedroli-Christen, A. and Scholl, A. (1990) Ecological and genetic studies on parapatric *Rhymogona silvatica* (Roth.) and *R. cervina* Verh. (Diplopoda:

Craspedosomatidae) with special reference to hybrid populations in a zone of contact. *Revue Suisse de Zoologie*, **97**, 349–59.

Pomponi, G., Cicolani, B. and Bullini, L. (1988) Ricerche elettroforetiche su cinque specie del genere *Macrocheles* (Acarina: Mesostigmata). *Atti del XV Congresso nazionale italiano di Entomologia, L'Aquila 13–17 giugno 1988*, pp. 233–7.

Roewer, C.F. (1954) Araneae Lycosaeformia I. (Agelenidae, Hahniidae, Pisauridae). *Exploration du Parc National de l'Upemba, Bruxelles*, **30**, 1–420.

Schilthuizen, M. and Gittenberger, E. (1994) Parallel evolution of an *sAat*-'hybrizyme' in hybrid zones in *Albinaria hyppoliti* (Boettger). *Heredity*, **73**, 244–8.

Sierwald, P. (1987) Revision der Gattung *Thalassius* (Arachnida, Araneae, Pisauridae). *Verhandlungen des naturwissenschaftlichen Vereins Hamburg*, **29**, 51–142.

Smallridge, M.A. and Kirby, G.C. (1988) Competitive interactions between the land snails *Theba pisana* (Müller) and *Cernuella virgata* (Da Costa) from South Australia. *Journal of Molluscan Studies*, **54**, 251–8.

Sonenshine, D.E. (1984) Tick pheromones, in *Current Topics in Vector Research* (ed. K.F. Harris), Praeger, New York, pp. 225–63.

Takeda, N. and Tsuruoka, H. (1979) A sex pheromone secreting gland in the terrestrial snail *Euhadra peliomphala*. *Journal of Experimental Zoology*, **207**, 17–26.

Terhivuo, J. (1988) Morphological and morphometric variation of the parthenogenetic earthworm *Dendrobaena octaedra* (Sav.) (Oligochaeta, Lumbricidae) in Eastern Fennoscandia. *Annales Zoologici Fennici*, **25**, 303–20.

Terhivuo, J. and Saura, A. (1990) Allozyme variation in parthenogenetic *Dendrobaena octaedra* (Oligochaeta: Lumbricidae) populations of Eastern Fennoscandia. *Pedobiologia*, **34**, 113–39.

Terhivuo, J., Saura, A. and Hongell, K. (1994) Genetic and morphological variation in the parthenogenetic earthworm *Eiseniella tetraedra* (Sav.) (Oligochaeta: Lumbricidae) from South Finland and North Norway. *Pedobiologia*, **38**, 81–96.

Thorpe, J.P. (1983) Enzyme variation, genetic distance and evolutionary divergence in relation to levels of taxonomic separation, in *Protein Polymorphism: Adaptive and Taxonomic Significance* (eds G.S. Oxford and D. Rollinson), Academic Press, London, pp. 131–52.

Uetz, G.W. and Stratton, G.E. (1982) Acoustic communication and reproductive isolation in spiders, in *Spider Communication: Mechanisms and Ecological Significance* (eds P.N. Witt and J.S. Rogner), Princeton University Press, Princeton, pp. 123–59.

Woodruff, D.S. (1989) Genetic anomalies associated with *Cerion* hybrid zones: the origin and maintenance of new electromorphic variants called hybryzymes. *Biological Journal of the Linnean Society*, **36**, 281–94.

Woodruff, D.S. and Solem, A. (1990) Allozyme variation in the Australian camaenid land snail *Cristilabrum primum*: a prolegomenon for a molecular phylogeny of an extraordinary radiation in an isolated habitat. *The Veliger*, **33**, 129–39.

Wunderlich, J. (1992) Die Spinnen-Fauna der makaronesischen Inseln. Taxonomie, Ökologie, Biogeographie und Evolution. *Beiträge zur Araneologie*, **1**(1991).

Zanazzo, G., Filippucci, M.G., Boato A., and Minelli, A. (1994) Allozymic and karyological evidence concerning Italian populations of the *Eupolybothrus fasciatus*-group (Chilopoda: Lithobiomorpha). *Zoologischer Anzeiger*, **232**, 77–95.

16

Species concepts in systematics and conservation biology – an ornithological viewpoint

J. Cracraft

Contacting address: Department of Ornithology, American Museum of Natural History, Central Park West at 79th Street, New York 10024, USA

ABSTRACT

The biological species concept (BSC) has dominated within ornithology since the 1930s, but the past decade has seen increased application of the phylogenetic species concept (PSC). The central role of species concepts is to delineate the units of nature and thus provide the essential framework for understanding biological diversity. The PSC does this more objectively than the BSC. Many conservation biologists, particularly those who manage *in situ* and *ex situ* breeding programmes, have recognized that the BSC is inappropriate for this task. Their solution, 'evolutionary significant units (ESUs)', has gained wide support within the conservation community, yet it has significant problems. There is no general support on how to define ESUs nor apply the concept objectively. Perhaps more important, ESUs have no status within formal taxonomy, hence they have no standing within those legal instruments designed to conserve and use sustainably biological diversity. Phylogenetic species, as basal diagnosable units, are effective functional equivalents of ESUs, have standing in formal taxonomy, and have many advantages over biological species when applied to conservation and management problems. It is suggested that the concept of ESU be abandoned and that the PSC become the taxonomic currency of conservation biology.

16.1 INTRODUCTION

The discipline of ornithology has had a large influence on the debates over species concepts. Ornithologists such as O. Kleinschmidt and E. Hartert,

Species: The Units of Biodiversity. Edited by M.F. Claridge, H.A. Dawah and M.R. Wilson. Published in 1997 by Chapman & Hall. ISBN 0 412 63120 2

and later Erwin Stresemann and Bernard Rensch, were among the first to see the importance of species concepts and begin the shift toward the so-called polytypic, biological concept. In their midst was a young systematist, Ernst Mayr, who became the leading advocate for the biological species concept (BSC) and who remains an active partisan over 50 years later (Mayr, 1942, 1963, 1970, 1992, 1993). Since Mayr's influential work of 1942, many avian systematists whose primary interest was geographic variation and speciation analysis adopted the BSC (Short, 1969; Mayr and Short, 1970; Selander, 1971; Bock, 1987; Haffer, 1992).

In recent years support for the BSC within ornithology has waned as systematists have adopted a phylogenetic species concept (PSC; Cracraft, 1983, 1989; McKitrick and Zink, 1988). Although the latter is not strictly a cladistic concept – in fact, there has been debate within cladistics over species concepts – the notion of phylogenetic species has found strong support among those avian systematists who see hypotheses of the history of taxic differentiation as being at the conceptual centre of the analysis of geographic variation and speciation.

The purpose of this chapter is not to revisit the debates over species concepts within ornithology (see Haffer, 1992, for a review from the perspective of a supporter of the BSC). The reasons for adopting the phylogenetic species concept within ornithology are compelling and have been discussed elsewhere (Cracraft, 1983, 1989; McKitrick and Zink, 1988; see also many papers cited below). The first section, instead, touches on some conceptual and linguistic arguments that tend to obfuscate understanding of the differences and implications of adopting alternative species concepts, no matter which group of organisms might be considered. The second section attempts to dispel several myths about the phylogenetic species concept. Following this, a section is devoted to a brief comparison of the BSC and PSC to illustrate their different implications for systematic and evolutionary biology. Finally, the last section examines the role of species concepts within conservation biology. Biologists have begun to recognize the implications of alternative species concepts when considering the units underlying conservation action and sustainable development, yet systematists have played little role in shaping this discussion.

16.2 SPECIES IN THEORY AND PRACTICE

16.2.1 The language of discourse

One underlying reason for the ongoing debates over species concepts and definitions, and what they mean theoretically and empirically, is the structure of the discourse itself. How species might be recognized becomes confused with how species might be defined. The evidence used in recognizing species is often taken to be more or less equivalent to the definition of a species. Biologists speak of morphological species, genetic species, genotypic species, or behavioural species as if they are conceptually on the same level as biological species, phylogenetic species, or evolutionary species. The latter, more theoretical, definitions are often conflated with the evidence perceived to be at the heart of applying those definitions. Thus, biological species are taken to be genetic whereas phylogenetic species are characterized as being morphological. Inasmuch as genetic, morphological, behavioural, or other kinds of character data can be used to delimit species boundaries within the context of any view of species, such language serves only to confuse the dialogue over alternative species concepts. Notions of genetic species, genotypic species, or morphological species are inherently confusing and often non-sensical.

Equally unhelpful is the use of pejorative language in characterizing some species concepts as being typological, non-populational, or non-biological as is sometimes done by advocates of the BSC (Mayr, 1992; Haffer, 1992). Proponents of all species concepts use populational thinking inasmuch as they would not knowingly consider placing different sexes, morphs, or different stages of a life-cycle in separate species. Mistakes can happen under the guise of any species concept, but the biological species concept, simply because the word 'biological' appears in its name, is not inherently more biological, populational, or genetic than is the phylogenetic species concept. Likewise, the latter is not more inherently morphological. It is also not useful to characterize some concepts as being purely theoretical, in contrast to others that might be said to be practical or empirical. All definitions, no matter how empirical they may sound, are theory laden and rely on some conceptual understanding of other terms.

16.2.2 The structure of species definitions

Unless species concepts are used to individuate real, discrete entities in nature, they will have little or no relevance for advancing our understanding of the structure and function of biological phenomena involving those things we call species (Cracraft, 1987). Debates over species concepts rarely include much discussion about this. Perhaps systematists just take for granted that species are discrete entities, but this is not evident in some of the debates (nor has it been true in practice, especially within palaeontology where a nominalist-like view of species has long prevailed). If species are not considered to be discrete real entities – and holding this view does not mean that delimiting boundaries will always be straightforward or that the boundaries cannot be fuzzy depending on the spatiotemporal scale of the observer – then it implies that evolutionary and systematic biology would be based largely on units that are fictitious, whose boundaries, if drawn, are done so arbitrarily. It would also mean that most, if not all, of the processes that we ascribe to species are concoctions of the mind and have no objective reality. Entities of postulated processes must be real and discrete if those processes are to have much meaning. Having a theory about the behaviour of electrons, for example, makes no sense if electrons do not exist as discrete entities; the same is true for species. The notion of discreteness, clearly, must be contextualized with respect to a certain spatiotemporal frame of reference, and different perceptions of the latter can lead biologists to see species as being discrete, real things, on the one hand, or as arbitrary segments of evolutionary continua, on the other (Cracraft, 1987).

Unless a species concept can be used to individuate real-world entities, that concept will have limited utility for systematists getting on with their task of sorting out and understanding biological diversity. Both the BSC and PSC are meant to be guides to the practice of recognizing species, but they do not do so in the same way or equally effectively. Some biologists think the primary task of species concepts is to understand process-level phenomena. Important though that task may be, the view taken here is that the central role of species concepts is to delineate the units of nature and thus provide the essential framework for understanding life's diversity.

Species definitions should have three key elements if they are going to be useful for systematic practice:

1. They need to mention or imply reproductive cohesion (be populational) in order to provide a conceptual basis for including males and females of the same population in the same species.
2. They must have some notion of diagnosability so that populations or groups of populations can be distinguished one from another.
3. They must include some criterion for ranking these populations at the species-level as opposed to some other level of the Linnean hierarchy.

The definition of biological species – a group of interbreeding natural populations that is reproductively isolated from other such groups (Mayr, 1963) – has the first two of these key elements. 'Interbreeding natural populations' implies reproductive cohesion, and the phrase 'reproductively isolated' provides for some basis of diagnosability (if one assumes there are character-based differences that lead to reproductive incompatibility). Yet the definition is inherently flawed because it does not provide any specific framework for ranking. Proponents would like to use the conjunction of 'interbreeding' and 'reproductively isolated' to rank an entity at the species-level. The definition, however, sets no bound on where species limits might be drawn, only that it will stop at some point of 'reproductive isolation'. Yet, few supporters of the BSC would claim that all populations capable of interbreeding should be included in the same biological species, and considerable difference of opinion exists among supporters of the BSC over how much (or little) interbreeding is necessary and/or sufficient to justify uniting populations in the same species. In fact, it is the difficulty in reconciling these two components of the BSC that has led many biologists over the years to abandon it in practice.

In contrast, the definition of phylogenetic species – the smallest population or group of populations within which there is a parental pattern of ancestry and descent and which is diagnosable by unique combinations of character-states (Eldredge and Cracraft, 1980; Nelson and Platnick, 1981; Cracraft, 1983; Nixon and Wheeler, 1990) – has all three key elements. The phrase 'parental pattern of ancestry and descent' implies reproductive cohesion over time; the element of diagnosability is specifically mentioned; and the statement referring to the smallest population establishes the basis for ranking (that is, the boundary to species limits is the smallest population or group of populations that is diagnosably distinct).

16.3 THE PHYLOGENETIC SPECIES CONCEPT: SOME MYTHS AND MISREPRESENTATIONS

Growing support for the PSC within systematics has resulted in various reactions from those – primarily non-systematists – whose traditional allegiance has been to the BSC. This reaction has included a significant amount of misunderstanding about the PSC: what it is, how it might be applied, and what might be the consequences of using it broadly within systematics and evolutionary biology. Several of the most important misunderstandings deserve discussion.

As already noted, linguistic characterization can go far in casting aspersions on a particular species concept without actually facilitating a rational dialogue about it. Describing the PSC as being purely morphological or non-populational (Mayr, 1992), and therefore typological, for example, does little to further scientific discourse about species concepts because the description is patently false. The PSC is defined in terms of populations and their diagnosability, and nowhere in the definition does the word morphological appear. Indeed, systematists will want to use all relevant data when determining species limits under the PSC. The fact remains that most of the information available to systematists – no matter what definition they adhere to – is morphological, yet this does not make a definition, including even the BSC, non-biological or non-populational. A strength of the PSC is that it can be applied using only morphological data. The PSC is centred around the notion of diagnosability, which can be inferred more or less directly from morphological data; the BSC, in contrast, relies on an understanding of reproductive isolation, which can be inferred only very indirectly from morphological data.

The PSC is sometimes mistakenly interpreted as being a cladistic concept (Haffer, 1992; Mallet, 1995), and although cladists have been among the strongest proponents of the PSC within systematics, over the years a number of species concepts have been used by cladists. One perception that has arisen from the association of the PSC with cladism is that phylogenetic species are defined by apomorphies, or derived characters (Mallet, 1995), that they are (or must be) monophyletic (Haffer, 1992), or that they are somehow based on phylogeny (Mallet, 1995). Whereas it is true that some have supported defining phylogenetic species in terms of the smallest populations having apomorphic characters (Rosen, 1978, 1979) and that phylogenetic species should be monophyletic (McKitrick and Zink, 1988), others have argued that not all diagnosably distinct populations may have characters inferred to be derived yet those populations still deserve taxonomic recognition [Nelson and Platnick, 1980; Cracraft, 1983 (contra Mallet, 1995: 298), 1989]. Although the populations included within a phylogenetic species that is diagnosed only in terms of primitive characters may actually represent more than one phylogenetic species, such a mistake is simply a matter of available evidence. All designated species, whether delimited by apomorphic characters or not, are hypotheses subject to revision when new evidence arises. Inasmuch as phylogenetic species are basal, application of the concept of monophyly is superfluous and unnecessary. In addition, even though the PSC is the clear choice when the goal is to reconstruct phylogeny accurately, this does not imply that the definition of phylogenetic species is dependent upon any assumptions of cladistics or, contrary to Mallet (1995), is dependent on phylogeny. The PSC is a mechanism for sorting and interpreting character (broadly interpreted) variation within and among populations in order to recognize basal diagnosable taxa, and is largely independent of assumptions about process.

Critiques of the PSC have argued that a consequence of defining species as the smallest, diagnosably distinct population will be a proliferation of species taxa, since each small population can be found to be distinct for some character, particularly at the molecular level. This criticism can be carried to extremes when not thinking in terms of taxa: 'with detailed morphology or modern molecular techniques, one can find apomorphies for almost every individual' (Mallet, 1995: 298; see also Avise and Ball, 1990; Avise, 1994; among others).

Two misunderstandings are hidden in this criticism. First, it is argued that each individual organism is distinct, therefore the PSC is inapplicable. It may well be the case that each individual organism can be distinguished using one method or another –who would be surprised at this? – but such a finding is irrelevant because the PSC is not about the diagnosability of individuals but of populations. Such a criticism forgets that species concepts are populational concepts that are used to delimit basal taxa. The concept of apomorphy, moreover, has no meaning at the level of an individual but only at the level of taxa. Delimiting species taxa is a problem for systematics not population biology (Nixon and Wheeler, 1990; Wheeler and Nixon, 1990), although obviously population-level data are relevant.

Seeing species as taxa, and as being different from populations and individual organisms, is fundamental for describing and understanding biological diversity. Some critics of the PSC have seemingly not understood this distinction.

A second misunderstanding related to the first is the belief that the PSC will lead to an inordinate inflation of species names. The belief that having too many names would be inconvenient or pernicious has had a long history within systematics, and the consolidation of names was given as one reason why the BSC was a particularly useful innovation: with the introduction of the BSC the 'total number of [bird] species to be memorized by the taxonomist has thus been cut by two-thirds [from 27 000 to 8500; but see below]. The practical advantage of this simplification is so obvious that nothing more needs to be said' (Mayr, 1942: 127). One's memory capacity aside, such views have little scientific relevance for our attempts to describe taxic diversity accurately. It is essential – for many reasons – to individuate diagnosably distinct taxa in nature, including those that are basal (species) and if nature has a multitude of these, so be it.

There are several reasons for thinking that the PSC will not lead to an extraordinary proliferation of names. First, though they may not specifically acknowledge the use of a PSC, most systematists already apply the notion of a basal diagnosable taxon when delimiting species. This is true within most of entomology and those disciplines concerned with the highly speciose groups of non-vertebrates. As a consequence, extending the PSC to all groups of organisms, most of which are far less diverse, will not make a significant difference in global species numbers.

Second, even in those groups such as birds in which the BSC has had considerable influence, virtually all diagnosable taxa have already been described – as subspecies – and thus application of the PSC will not greatly affect the number of names, just potentially their ranking. Within the birds-of-paradise (Paradisaeidae), for example, applying the PSC to a group whose taxonomy has been dominated by the BSC increased the number of recognized species by only about two-fold (Cracraft, 1992). These results can possibly be generalized across all birds. When one randomly samples the world's biological species of birds, evaluates those species and their included subspecies as to whether they are basal diagnosable taxa, the estimated number of phylogenetic species in the world is again only about twice that of the current number (9000 or so) of putative biological species (G. F. Barrowclough, J. Cracraft and R. M. Zink, unpublished data).

There is no question that improved methods of resolving variation will increase our ability to recognize more diagnosably distinct taxa. Critics of the BSC seem to bemoan this fact; biologists interested in seeing nature described and interpreted with accuracy and precision will, in fact, applaud the use of the PSC.

16.4 TWO SPECIES CONCEPTS: A COMPARISON

The BSC and PSC provide very different lenses with which to see the world of systematic and evolutionary biology. How species are defined influences one's interpretations of patterns and processes (Table 16.1; see Cracraft, 1989, for additional discussion). Under the BSC, diagnosably distinct populations will sometimes be recognized as separate, monotypic species, but often those populations are united together under a single species name if the diagnosable differences are not judged to be significant. Within the context of the PSC, on the other hand, diagnosably distinct populations would always be accorded specific status. As a consequence, some biological species, those consisting of multiple diagnosable taxa of uncertain relationships to one another, will confound an accurate reconstruction of history if used as terminal taxa in a phylogenetic or biogeographic analysis (Rosen, 1978, 1979; Cracraft, 1983, 1989; Frost and Hillis, 1990; among many others). Phylogenetic species, in contrast, are appropriate terminal taxa for such analyses and cannot, in and of themselves, lead to a misrepresentation of history.

In principle, there is no gene flow among biological species because they are assumed to be reproductively isolated. In practice, however, proponents of the BSC will recognize two entities as biological species, even though there may be gene flow, as long as it is judged to be minor. Because phylogenetic species are defined in terms of diagnosability and not reproductive isolation, it is not uncommon, especially in plants, for there to be extensive gene flow across species borders. If the two entities are diagnosably distinct, they will be recognized as phylogenetic species even though there may be a hybrid zone within which gene flow might be extensive. In almost all cases the latter situation would result in the recognition of a single biological species. Under both concepts, however, populations that are reproductively isolated will also be distinct taxonomic entities.

Finally, the two concepts have markedly different implications for biogeography. If species are taken to be the units of analysis for recognizing areas of endemism, the use of biological species unless they are all monotypic and thus equivalent to phylogenetic species will always lead to a less precise classification of those areas than will be the case with the use of phylogenetic species. Furthermore, adoption of the BSC can sometimes confound the reconstruction of history and consequently lead to inaccurate depictions of the relationships among areas of endemism.

16.5 TAXONOMIC UNITS AND CONSERVATION BIOLOGY

16.5.1 The concept of evolutionary significant unit should be abandoned

As already noted, there is a growing awareness on the part of conservation biologists that species concepts have relevance for determining what to protect, how to protect it, and how to facilitate the sustainable use of

Table 16.1 Criteria for recognition and some comparisons between the biological and phylogenetic species concepts

Criterion	*Biological species concept*	*Phylogenetic species concept*
Diagnosably distinct populations recognized as separate species	Sometimes	Always
Species unit often includes diagnosable allopatric populations	Yes (subspecies concept widely applied)	Never (subspecies concept not relevant)
Species represent terminal taxa that can be used in phylogenetic and biogeographic analysis	Sometimes	Always
Inherently capable of misleading historical analysis	Yes	No
Gene flow among species	Rarely	Sometimes
Reproductively isolated populations recognized as separate species	Always	Always
Extensively hybridizing, diagnosably distinct populations recognized as separate species	Rarely	Almost always
'Potential' interbreeding of allopatric populations important for establishing species status	Yes	Never
Delimitation of areas of endemism	Coarse, less precise	Fine, more precise

biodiversity (Ryder, 1986; Avise, 1989, 1994; Rojas, 1992; Moritz, 1994a,b, 1995; Vogler and DeSalle, 1994; Grant, 1995; Barrowclough and Flesness, 1996). Many of these discussions, however, have not taken into account the full implications of the debates about species concepts within the systematic literature over the past 10 or 15 years. Instead, much of the dialogue has centred on the relevance of 'evolutionary significant units' (ESUs) and their use in conservation studies.

Discussion at the 1985 meeting of the American Association of Zoological Parks and Aquariums (AAZPA) sharpened the debate over the issue of the units of conservation (Ryder, 1986). With the introduction of the term ESU for those taxonomic entities having a distinct evolutionary history, the AAZPA aimed to identify groups that were most in need of conservation action, particularly those in captive breeding programs. Subsequent writers have given tacit support to the importance of ESUs in conservation (Dizon *et al.*, 1992; Moritz, 1994a,b, 1995; Vogler and DeSalle, 1994; Barrowclough and Flesness, 1996).

The theoretical and practical goals of the ESU concept are important: to provide an objective basis for the definition and recognition of management units in conservation activities. A primary difficulty is that no clear agreement has been reached on what constitutes an ESU. Thus, Vogler and DeSalle (1994: 356) see ESUs as 'clusters of organisms that are evolutionarily distinct and hence merit separate protection'. And, Moritz (1994b: 373; see also 1994a, 1995) argues that to be an evolutionarily 'significant' unit implies that 'the set of populations has been historically isolated and, accordingly, is likely to have a distinct potential'. Moritz goes on to note (1994b: 373) that '*ESUs should be reciprocally monophyletic for mtDNA alleles and show significant divergence of allele frequencies at nuclear loci*' [italics in original]. The confusion and ambiguity over how to define an ESU is so great – ranging from populations that are significantly differentiated, to biological species, to phylogenetic species – that its objective use is virtually precluded (see Grant, 1995, for a description of the confusion).

Both Vogler and DeSalle (1994) and Moritz (1994a,b, 1995) recognize earlier arguments [Cracraft, 1991; Barrowclough and Flesness, 1996 (in a paper circulated since 1993)] that ESUs are essentially equivalent to phylogenetic species. Despite this acceptance, Vogler and DeSalle (1994) and Moritz (1994a,b) would maintain the concept of ESU, even though the criteria used to recognize historically distinct and significant units are largely those that individuate phylogenetic species. Yet, the ambiguities implied by the various definitions of the ESU remain, thus leading one to conclude that it should be abandoned by conservation biologists.

Such a solution is also strongly supported by another consideration. The use of ESUs within conservation biology undercuts the scientific foundation and results of that discipline. ESUs have no scientific status within systematics, and it is systematics that provides the linguistic and historical framework for the study of biodiversity. The results of formal taxonomy – as reflected in species-level taxa – are now codified in an enormous series of national and international legal instruments. As Geist (1992: 274) remarks: 'courts and solicitors' offices are allowed to rule on taxonomy. Judges may now decide on matters such as the definition of species or subspecies, the criteria for establishing taxa, which taxa are valid, and which populations can be legally protected. The implications for conservation, but also for biology in general, are profound and worrying'.

Because ESUs have no formal systematic status, they will rarely have any legal status. The Convention on Biological Diversity will be the chief international legal instrument affecting the conservation and sustainable use of biodiversity for the foreseeable future. The use of ESUs in meeting the goals of the Convention would be difficult at best because of the lack of international standards of scholarship and formal nomenclatural rules over those units. Despite the fact that there may be arguments over species concepts, species-level taxonomy and its rules of nomenclature have broad acceptance within the systematic community and among those biodiversity sciences that use taxonomic information. No such framework exists for ESUs.

16.5.2 Phylogenetic species are the most relevant units for conservation biology

A comprehensive programme of conservation and sustainable use of biological diversity will depend upon having all taxonomically distinct, diagnosable populations identified and named. A comparison of the BSC with the PSC demonstrates that phylogenetic species, not polytypic biological species, are the most appropriate units for conservation (Table 16.2).

Table 16.2 Possible implications of different species concepts for conservation and sustainable development

Implication	*Biological species concept*	*Phylogenetic species concept*
Equivalent to 'evolutionary significant units' of conservation biology	Sometimes	Essentially always
Estimates of diversity	Underestimates numbers of differentiated taxa	Yields accurate estimates of basal taxa given available data
Delineation of areas of of endemism	Broadly defined; underestimates numbers of areas; confounds their historical relationships	Narrowly defined on basis basal taxa; more finely resolved;historical relationships not confounded
Demographic analysis	May tend to overestimate population sizes of endangered taxa	Accurately estimates population sizes of basal taxa
Apportionment of genetic variation	Confounded	Accurately apportioned
Captive breeding of endangered taxa (and reintroductions)	Units of programme ambiguous; increased danger of mixing distinct evolutionary units	Units of programme are basal taxa; less risk of mixing distinct evolutionary units
Trade in endangered taxa	Enforcement loosely interpreted; fewer taxa protected; higher risk due to trade	Enforcement stringently interpreted; more taxa protected; lower risk due to trade
Political impact for endangered taxa	Less	More

Phylogenetic species meet the objectives envisioned for evolutionary significant units. The desire of conservation biologists to have a unit that is relevant for management purposes is fulfilled by the PSC but not the BSC. As numerous authors have noted, many biological species are composites of diagnosably distinct as well as arbitrarily demarcated races or subspecies. As judged by the managers of captive breeding programmes, this inconsistency will lead to innumerable problems in assigning conservation priorities (Ryder, 1986).

Different species concepts can result in different estimates of species diversity. The use of the BSC will underestimate the numbers of diagnosably (evolutionarily) distinct units as compared with the PSC. Comparisons of diversity are important for countries in setting conservation priorities and are of interest to funding agencies that support activities to implement those priorities.

Many countries and conservation organizations are currently concerned with creating protected areas and ecosystem management zones using measures of endemism to set priorities. If species limits are drawn differently under different species concepts, then so too will their areas of endemism. The use of biological species results in an underestimate of the numbers of areas of endemism, tends to overestimate their size, and confounds the analysis of their historical interrelationships. All these problems are avoided by the use of the PSC.

Species concepts have important implications for demographic analyses. Inasmuch as different concepts can allocate populations to species taxa in opposing ways, numbers of individuals within populations of those species will be estimated differently. Many decisions about conservation priorities and actions depend on demographic information. Phylogenetic species allow for an accurate count of individuals within basal taxa, whereas biological species do not necessarily identify basal taxa and can lead to inappropriate or inaccurate estimates. If all the populations of a polytypic biological species are included in a demographic assessment of that species, the numbers of individuals within endangered basal taxa may be overestimated or ignored altogether.

For the same reasons, the use of phylogenetic species will allow genetic variation to be apportioned within and between taxa more accurately than will biological species. If diagnosably distinct basal taxa are united into a single biological species, there is a risk of ignoring the levels of genetic variation within those distinct taxa in favour of creating an estimate of variation within the biological species as a whole, a situation that is likely to compromise effective management of biodiversity.

The concept of the ESU grew out of attempts by conservation biologists to provide an effective framework for managing *in situ* and *ex situ* breeding programmes for endangered taxa. The ESU was born because the BSC failed to provide that framework. The PSC, in contrast, provides the formal taxonomic context for managing breeding programmes.

The BSC also does not provide an effective basis on which to manage trade in endangered taxa. To be effective, management programmes will need to monitor and regulate trade of all distinct taxa. If distinct and endangered basal taxa are lumped with more common forms under the BSC, the danger is that sufficient protection will not be extended to the former. This problem is avoided with the PSC because each distinct basal taxon would be ranked at the species-level and therefore assume a heightened legal status over unnamed populations or subspecies.

Finally, although some national and international legal instruments extend protection to taxonomic units below the level of species, many do not. Taxa of species rank are still the primary currency for conserving and managing biological diversity. Endangered, distinct populations ranked at subspecies, or not given formal rank at all, will generally carry less political and conservation significance than those ranked at the level of species. Because the PSC provides for specific rank for all diagnosable populations, the importance of the latter is magnified relative to lumping those populations into a larger biological species.

Acknowledgements

I should especially like to thank Professor Michael Claridge for inviting me to participate in this symposium. I am also grateful for the helpful comments of Drs Rob DeSalle and Robert Zink on the manuscript.

16.6 REFERENCES

Avise, J.C. (1989) A role for molecular genetics in the recognition and conservation of endangered species. *Trends in Ecology and Evolution*, **4**, 279–81.

Avise, J.C. (1994) *Molecular Markers, Natural History and Evolution*, Chapman & Hall, New York.

Avise, J.C. and Ball, R.M. (1990) Principles of genealogical concordance in species concepts and biological taxonomy. *Oxford Surveys Evolutionary Biology*, **7**, 45–67.

Barrowclough, G.F. and Flesness, N.A. (1996) Species, subspecies, and races: the problem of units of management in conservation, in *Wild Mammals in Captivity* (eds M. Allen and H. Harris), University Chicago Press, Chicago, pp. 247–54.

Bock, W.J. (1987) Species concepts, speciation, and macroevolution, in *Modern Aspects of Species* (eds D. Iwatsuki, P. Raven, and W.J. Bock), University Tokyo Press, Tokyo, pp. 31–57.

Cracraft, J. (1983) Species concepts and speciation analysis, in *Current Ornithology*, vol. 1, Plenum Press, New York, pp. 159–87.

Cracraft, J. (1987) Species concepts and the ontology of evolution. *Biology and Philosophy*, **2**, 329–46.

Cracraft, J. (1989) Speciation and its ontology: the empirical consequences of alternative species concepts for understanding patterns and processes of differentiation, in *Speciation and its Consequences* (eds D. Otte and J. A. Endler), Sinauer Associates, Sunderland, Massachusetts, pp. 28–59.

Cracraft, J. (1991) Systematics, species concepts, and conservation biology. Abstract, 5th Annual Meeting, Society of Conservation Biology, Madison, Wisconsin.

Cracraft, J. (1992) The species of the birds-of-paradise (Paradisaeidae): applying the phylogenetic species concept to complex patterns of diversification. *Cladistics*, **8**, 1–43.

Dizon, A.E., Lockyer, C., Perrin, W.F., Demaster, D.P. and Sisson, J. (1992) Rethinking the stock concept. *Conservation Biology*, **6**, 24–36.

Eldredge, N. and Cracraft, J. (1980) *Phylogenetic Analysis and the Evolutionary Process*, Columbia University Press, New York.

Frost, D.R. and Hillis, D.M. (1990) Species in concept and practice: herpetological applications. *Herpetologica*, **46**, 87–104.

Geist, V. (1992) Endangered species and the law. *Nature*, **357**, 274–6.

Grant, W.S. (1995) Multi-disciplinary approaches for defining evolutionarily significant units for conservation. *South African Journal Science*, **91**, 65–7.

Haffer, J. (1992) The history of species concepts and species limits in ornithology. *Bulletin British Ornithologists' Club Centenary Supplement*, **112A**, 107–58.

Mallet, J. (1995) A species definition for the modern synthesis. *Trends in Ecology and Evolution*, **10**, 294–9.

Mayr, E. (1942) *Systematics and the Origin of Species*, Columbia University Press, New York.

Mayr, E. (1963) *Animal Species and Evolution*, Harvard University Press, Cambridge, Massachusetts.

Mayr, E. (1970) *Populations, Species, and Evolution*, Harvard University Press, Cambridge, Massachusetts.

Mayr, E. (1992) A local flora and the biological species concept. *American Journal of Botany*, **79**, 222–38.

Mayr, E. (1993) Fifty years of progress in research on species and speciation. *Proceedings of the California Academy of Sciences*, **48**, 131–40.

Mayr, E. and Short, L.L. Jr (1970) Species taxa of North American birds, a contribution to comparative systematics. *Publications Nuttall Ornithological Club*, **9**, 1–127.

McKitrick, M.C. and Zink, R.M. (1988) Species concepts in ornithology. *Condor*, **90**, 1–14.

Moritz, C. (1994a) Applications of mitochondrial DNA analysis in conservation: a critical review. *Molecular Ecology*, **3**, 401–11.

Moritz, C. (1994b) Defining ' evolutionarily significant units' for conservation. *Trends in Ecology and Evolution*, **9**, 373–5.

Moritz, C. (1995) Uses of molecular phylogenies for conservation. *Philosophical Transactions Royal Society London*, **349B**, 113–18.

Nelson, G. and Platnick, N.I. (1981) *Systematics and Biogeography: Cladistics and Vicariance*, Columbia University Press, New York.

Nixon, K.C. and Wheeler Q.D. (1990) An amplification of the phylogenetic species concept. *Cladistics*, **6**, 211–23.

Rojas, M. (1992) The species problem and conservation: what are we protecting? *Conservation Biology*, **6**, 170–8.

Rosen, D.E. (1978) Vicariant patterns and historical explanation in biogeography. *Systematic Zoology*, **27**, 159–88.

Rosen, D.E. (1979) Fishes from the uplands and intermontane basins of Guatemala: revisionary studies and comparative geography. *Bulletin American Museum Natural History*, **162**, 267–376.

Ryder, O.A. (1986) Species conservation and systematics: the dilemma of subspecies. *Trends in Ecology and Evolution*, **1**, 9–10.

Selander, R.K. (1971) Systematics and speciation in birds, in *Avian Biology* (eds D. Farner and J. R. King), vol. 1, Academic Press, New York, pp. 57–147.

Short, L.L. (1969) Taxonomic aspects of avian hybridization. *Auk*, **86**, 84–105.

Vogler, A.P. and DeSalle, R. (1994) Diagnosing units of conservation management. *Conservation Biology*, **8**, 354–63.

Wheeler, Q.D. and Nixon, K.C. (1990) Another way of looking at the species problem: a reply to de Quieroz and Donoghue. *Cladistics*, **6**, 77–81.

17

The species in mammals

G. B. Corbet

Contacting address: Little Dumbarnie, Upper Largo, Leven, Fife KY8 6JQ, Scotland, UK

ABSTRACT

The biological species concept has been widely used in the discrimination of mammalian species and continues to provide a valuable theoretical framework. Sympatric and parapatric cryptic species should be recognized as such however subtle the differences, provided that these are likely to be stable rather than ephemeral; that there is good evidence of lack of hybridization; and that sampling has been adequate to extrapolate from specimens examined to the entire ranges of the species. Allopatric taxa should be considered conspecific if they differ only in ways that are analogous to those found within interbreeding populations and there is no other evidence of reproductive incompatibility. The use of subspecies is valuable when they can be diagnosed and shown to have objective geographical boundaries.

Phylogenetic species concepts are not more objective since they are equally subject to errors in extrapolation from sample to natural population, and depend upon interpretation of differences in terms of hypothetical history. Data from karyology and molecular techniques can play a valuable part in detecting species limits provided they are integrated with other data and that geographical sampling is adequate to extrapolate conclusions to the whole species.

17.1 INTRODUCTION

The delimitation of species in mammals shows many parallels to the situation in birds, and as in birds the problem is not complicated by asexual reproduction. The species-level taxonomy of mammals has been considerably influenced by that of birds in which much of the variation is more visible and sample size can more easily be enhanced by field observation. In these respects some groups of mammals closely resemble birds, e.g. among the diurnal and colourful primates and squirrels. However, most

Species: The Units of Biodiversity. Edited by M.F. Claridge, H.A. Dawah and M.R. Wilson. Published in 1997 by Chapman & Hall. ISBN 0 412 63120 2

mammals are small, communicate by scent, are nocturnal or otherwise elusive, and the collection of adequate samples has often been difficult.

An important factor in determining the nature of species is dispersive ability. In this respect mammals show a considerable range that can be reduced to the following main categories, although these are by no means discrete and each includes considerable variation:

1. Flying species (i.e. bats) capable of colonization across considerable stretches of water.
2. Large and highly mobile terrestrial species, especially those occupying the more continuous habitats, such as hares (*Lepus*) in the steppe and savannah zones.
3. Smaller, less mobile terrestrial species and those specialized for more discrete habitats (the majority).
4. Subterranean species with minimal dispersive ability, e.g. moles, gophers and mole-rats.
5. Oceanic species with good dispersive ability, e.g. most whales and dolphins.
6. Coastal marine species with more limited dispersive ability, e.g. most pinnipedes and some dolphins and porpoises.

The limitations of dispersive ability mean that most species of mammals have a moderately stable range, i.e. each species has a distinctive spatial pattern and shape.

17.2 HISTORICAL REVIEW

Linnaeus (1758), and most of his successors during the following hundred years, recognized species intuitively as groups of individuals within which clear-cut divisions could not easily be made but which appeared to be clearly definable in relation to each other. Species were recognized as being variable, but any variation tended to be interpreted as deviation from a fixed 'essence' or 'type'. Linnaeus himself in the 10th edition of his *Systema Naturae* (1758) described varieties within some species, but mainly domestic ones such as the dog, with eleven varieties, and the sheep, with seven.

From the total of 184 species of mammals described by Linnaeus in 1758 the number of known and described species rose rapidly, partly as a result of geographical exploration but helped enormously by the ease with which the Linnean system allowed additions to be named and added to the inventory in an orderly way. By the time Darwin's *Origin of Species* was published in 1859 the concept of the species as a discrete natural entity was already wavering because of the sheer number of new species being described that appeared to be progressively filling the gaps between hitherto readily distinguishable species. The burst of interest in evolution and the 'transmutation of species' further eroded the concept of species as discrete entities.

The main development in species-level taxonomy of mammals during the latter half of the 19th century came however not from the new evolutionary theories but from a growing interest in field natural history. Travelling became easier and this led to a shift from classifications that were concerned only with museum collections to classifications that were, it was hoped, applicable to the living populations from which the museum samples came. The increasing intensity of geographical sampling made it clear that in certain groups the 'gap-filling' between described 'species' demonstrated a continuum of variability, with a spatial component. Geographical subspecies, conceived as subdivisions of species and designated by trinomials, were first used for birds about 1845. In mammals J.A. Allen was one of the first to study geographical variation, for example of North American mammals (Allen, 1876). He argued, for instance, that the American grizzly and brown bears and the Eurasian brown bears were best considered as races of one species, *Ursus arctos*, a view now generally held although the intervening years have seen a great proliferation of species described within this group.

Towards the end of the 19th century a great boost was given to the collection and description of mammals by the work of C. Hart Merriam as part of the Biological Survey of North America, organized by the US Department of Agriculture. Merriam introduced the technique of preparing dry 'study skins' and skulls of small mammals in a standard way and building up series with precise geographical data in order to take intraspecific variation into account in comparing species. Merriam was emulated at the British Museum (Natural History) by Oldfield Thomas, who over the period from 1890 to the 1920s organized and encouraged similar collecting, especially of small mammals, throughout almost the entire world other than North America.

The only comprehensive enumeration of mammalian species around this period was by Trouessart (1897–1905) who listed 9381 species of which about 60% were fossil, leaving about 3750 living species.

A common perception of species at that time was expressed by Miller (1901), mammalogist at the US National Museum, who wrote: 'Species and subspecies, to use for the present these obsolete terms in the absence of the single word needed to replace them, are not invariably separated from their allies by characters easy to describe'. Between then and 1940 the description of species, and increasingly of subspecies, continued, but with little progress towards any kind of consensus as to how these groups were to be defined.

During the period 1920 to 1940 a fresh concept of the species as an objective entity began to emerge, especially in ornithology where field observation could more easily supplement museum investigation in detecting the overall patterns of variability in bird populations, but also helped by advances in the fields of genetics, ethology and ecology. The biological species concept that emerged was epitomized by Mayr's (1940) definition of a species as 'a group of populations which replace each other geographically or ecologically and of which the neighbouring ones intergrade or hybridize wherever they are in contact, or which are potentially capable of doing so in those cases where contact is prevented by geographical or ecological barriers'. In other words, stress was laid upon the gaps separating species and the limitations imposed upon variability within species by the retention of interfertility.

The relevance of this concept to mammals was expounded by Hall (1943) and put into practice in his revision of North American weasels (Hall, 1951), in which he recognized three 'polytypic' species in the subgenus *Mustela* in place of the 25 species listed by Miller (1924). This concept was also applied in *The mammals of North America* (Hall and Kelson, 1959) and in the checklists of Palaearctic and Indian mammals (Ellerman and Morrison-Scott, 1951) and of southern African mammals (Ellerman *et al.*, 1953). In general terms it has stood the test of time, although since the 1950s there has been a tendency towards the recognition of a larger number of more narrowly defined species, for several quite separate reasons.

These can be categorized as: (i) the refinement of morphological analysis; (ii) the development of karyology; (iii) the development of molecular biology; and (iv) improvement of geographical sampling. All have contributed to the discovery of sibling and cryptic species. When these have been sympatric there is little difficulty in applying the biological species concept and this has generally been done. Allopatric taxa show a wider range of cryptic differences that has led to many cases of differing opinion as to how many species should be recognized. Examples are discussed below.

17.3 SYMPATRIC CRYPTOSPECIES

17.3.1 Well-resolved examples

The first task of the taxonomist in delimiting sympatric species is to recognize those differences that are correlated with reproductive isolation amongst the noise created by other kinds of variability, e.g. that related to sex, age and season, and polymorphisms that do not interfere with reproduction. Examples from the European fauna of closely related pairs of species that were resolved on purely morphological grounds are the mice *Apodemus sylvaticus* and *A. flavicollis*, and the shrews *Crocidura russula* and *C. suaveolens*. However, although these are easily discriminated in the well-studied parts of their sympatric ranges in western Europe, they still give trouble in interpreting variability on the fringes of their ranges, beyond the sympatric zone, where character displacement tends to occur.

Satisfactory resolution of two pairs of sympatric species of European bats, *Plecotus auritus/P. austriacus*, and *Myotis mystacinus/M. brandti*, was first achieved by the discovery of differences in the male genitalia previously overlooked (Corbet, 1964; Gaukler and Kraus, 1970).

All the above examples are sufficiently differentiated to coexist over large parts of their ranges and presumably have different ecological requirements. Many other examples of sibling pairs have only been satisfactorily discriminated by karyological study, usually followed by biochemical work and the elucidation of subtle morphological differences. Most such cases that have been well studied are of basically parapatric pairs or series but with overlapping sympatric areas forming minor parts of the geographical ranges. Well-studied examples are mice of the genus *Mus*, the voles *Microtus arvalis/M. rossiaemeridionalis*, and shrews of the *Sorex araneus* complex in Europe. The last can be used as an example.

The common shrew, *Sorex araneus*, was first found to show Robertsonian chromosome polymorphism in Britain (Sharman, 1956). Subsequent karyological work has shown three sibling species in continental Europe, with substantial zones of sympatry: *S. araneus* (north and east), *S. coronatus* (France, etc.), and *S. granarius* (Iberia), reviewed by Hausser *et al.* (1985). These species differ by three pericentric inversions, three paracentric inversions and one reciprocal translocation, plus a few secondary translocations. Allozyme analysis has shown these three species to form a discrete group, with little differentiation between them – a mean genetic distance (Nei, 1972) of only 0.055. Attempts at morphological discrimination have failed to find differences allowing unequivocal identification of all individuals, but indices based upon cranial measurements allow about 96% to be identified (Hausser *et al.*, 1985; Handwerk, 1987).

Geographical sampling of this complex in western Europe has probably now been sufficient to demonstrate a pattern of ranges that is likely to remain reasonably stable and these three species have been accepted as such in the *Handbuch der Säugetiere Europas* (Niethammer and Krapp, 1990). However, *S. araneus* extends east to central Siberia. Of the 20 or so other currently defendable species of *Sorex* in Eurasia, 15 are clearly distinct from the *S. araneus* complex (Dannelid, 1991). Since sampling has been less intensive in eastern Europe and Asia it is quite probable that further cryptic species will be found associated with eastern *S. araneus.*

17.3.2 Partially resolved examples

Many cases where karyological work has indicated the presence of sibling species are still unresolved, some in spite of considerable study. A partially resolved example concerns the African mice of the genus *Mastomys*. Intensive work in southern Africa, summarized by Skinner and Smithers (1990), has revealed two widely sympatric species, *M. natalensis* and

M. coucha, differing in chromosome number (2n = 32 and 36 respectively) and in haemoglobin (by electrophoresis). Although further differences have been detected by multivariate analysis of cranial measurements, in small details of penis and spermatozoa, and in reproductive behaviour and ultrasounds, no easy means of identification by morphology have been found, making it difficult to use existing museum collections (which are very extensive) and to extend resolution of the species to the rest of Africa, although similar sibling species have been resolved in West Africa.

A further example, also of considerable economic and medical importance, involves the rats of the *Rattus rattus* complex in south-eastern Asia. Careful morphological work in recent decades has demonstrated that several forms previously included in *R. rattus* are in fact specifically distinct, e.g. *R. argentiventer* and *R. tiomanicus*, both important agricultural pests (reviewed by Corbet and Hill, 1992). This still leaves about a hundred named forms within *R. rattus s.l.*. Although karyological work has indicated the apparent coexistence of sibling chromosome species (Yosida, 1980), this has rarely been accompanied by any indication of the morphological characters nor by the preservation of material for morphological study. A recent study in southern India (F. Catzeflis *et al.*, unpublished results) has demonstrated a good correlation between chromosomal, genic and cranial characters, allowing the recognition of two sympatric species, provisionally called *R. rattus* (with 2n = 38) and *R. satarae* (2n = 42). However, both species are polychromatic in pelage characters, they cannot be confidently identified by external appearance, and geographical variation makes it difficult to apply the results far beyond the study area. Forms with 2n = 42 occur throughout south-east and east Asia and have been equated with *R. tanezumi* Temminck, 1844 from Japan (Musser and Carleton, 1993). However, so few populations have been sampled for chromosome data that there is a long way to go before these two species can be satisfactorily diagnosed and their ranges determined.

17.4 PARAPATRIC SERIES: SUBSPECIES OR SPECIES?

Several more or less well-studied examples of series of parapatric cryptic taxa are known, especially among subterranean mammals. The best documented is the group of mole-rats of the genus *Nannospalax* (or *Spalax*) in Israel and adjacent areas, reviewed by Nevo (1989). Four chromosomal forms of the *N. ehrenbergi* complex form a parapatric series with very limited hybridization in zones varying from 320 to 2800 m in width. They are not distinguishable by morphology in spite of intensive study and show very slight genetic differences (maximum 0.012). Although usually referred to as 'chromosome species' these have never been separately named. Since they can easily be referred to as, for example, '*Nannospalax ehrenbergi* 2n = 52', this is a practical solution.

It seems likely that similar situations are repeated elsewhere in the spalacine mole-rats (Savic and Nevo, 1990), the North American gophers of the genus *Thomomys* (Thaeler, 1985) and the South American tuco-tucos of the genus *Ctenomys* (Ortello, 1995).

Problems with the resolution of parapatric series are not confined to inconspicuous small mammals. There are many cases where parapatric pairs or series of taxa that have been recognized as distinct on the grounds of morphology or colour are variously treated as species or subspecies. The gibbons (*Hylobates*) include two series of parapatric taxa, with considerable instability in the number of species recognized (summarized by Corbet and Hill, 1992). In the subgenus *Hylobates* these are *H. pileatus* in Eastern Thailand and Cambodia; *H. lar* from Burma to peninsular Malaya and Northern Sumatra; and *H. agilis* in parts of peninsular Malaya and Southern Sumatra. These abut without intergradation and with only rare interbreeding, and there appears to be a consensus that they represent three distinct species. The situation on Borneo is less certain. The gibbons on most of Borneo are generally agreed to constitute a distinct species, *H. muelleri.* In the south is a parapatric form, *albibarbus*, which, on morphology, would appear to be *H. muelleri*, but is distinguished by a quite distinct territorial song that suggests that it represents *H. agilis* of Sumatra. The two forms are separated by major rivers such as the Barito but a small degree of hybridization takes place at the headwaters (Marshall and Sugardjito, 1986).

In the colourful leaf monkeys of the genus *Presbytis* recognition of three parapatric species on Sumatra has been based largely on vocalization although the situation is confused by considerable geographical variation in colour and pattern within species (Aimi and Bakar, 1992; Corbet and Hill, 1992).

On a larger scale the baboons, *Papio*, form a series of more or less parapatric forms spanning the length of Africa. Traditionally treated as about five distinct species, ranging from *P. hamadryas* in Ethiopia to *P. ursinus* in South Africa, the most recent review recognizes only one species (Groves, 1993). An equally controversial example concerns the sportive lemurs of Madagascar, *Lepilemur*, where the recognition of six parapatric species has been advocated on the basis of karyological data (Ishak *et al.*, 1992) although all have been treated as conspecific.

The African elephants, *Loxodonta*, have most often been treated as a single species, *L. africana.* However, the status of the smaller forms in the forests of west and central Africa has always been uncertain. Claims of two species have inevitably been hampered by small sample size and sparse geographical data (Eisentraut and Böhme, 1989; Böhme and Eisentraut, 1990). On the basis of multivariate analysis of cranial measurements Groves *et al.* (1993) supported the recognition of two species in spite of about six probable hybrids among 45 from the contact zone. More integrated presentation of the data and detailed mapping of the contact zone will be needed to clarify the situation.

In the case of pelagic cetaceans the problem of discriminating species is exacerbated by their mobility as well as the difficulty of collecting and preserving adequate samples. The genus *Delphinus*, for example, has been variously considered to include one species, the Common dolphin, *D. delphis*, or two: a short-beaked species, *D. delphis* and a longer-beaked one, *D. capensis* (or *D. tropicalis*) (van Bree and Gallagher, 1978). These have recently been shown to be discrete species in Californian waters (Heyning and Perrin, 1994) but there are difficulties in extrapolating the results to the rest of the world.

17.5 ALLOPATRY

Allopatric populations that have been shown to differ slightly, whether in morphology, colour, karyotype or in biochemical characters, present all the problems and uncertainties seen in parapatric forms, with the additional problem that there is no opportunity to determine the degree of hybridization, if any, under natural conditions. Many examples of insular populations that were originally distinguished from their continental relations as separate species have subsequently been treated as conspecific. This may be on the ground that the differences (often in size and colour) are little or no greater than those seen among clinal variation within continental populations, or on the basis of experimental attempts at interbreeding. Examples among small mammals of the British Isles are the Orkney voles, *Microtus arvalis orcadensis* and *M. a. westrae*, which are interfertile with continental *M. arvalis*; and the Skomer vole, *Clethrionomys glareolus skomerensis* which is easily distinguishable from but fully interfertile with *C. glareolus* of mainland Britain.

In many cases an abundance of data seems to confuse rather than clarify the question of where the species line should be drawn. Two species of chimpanzee, *Pan troglodytes* and *P. paniscus*, are generally recognized on the basis of a suite of morphological and behavioural characters. These are effectively allopatric, being separated by the Congo River. Studies of mitochondrial DNA gene sequences in all the great apes and humans has shown greater divergence within the gorilla, *Gorilla gorilla*, than between the two species of *Pan*, and even greater differences between the two island populations of orang utan, *Pongo pygmaeus*, on Borneo and Sumatra (Ruvolo *et al.*, 1994).

A more problematical example concerns the right whales of the genus *Eubalaena* (sometimes included in *Balaena*). Three allopatric populations are found in the temperate north Atlantic (*glacialis*), the temperate north Pacific (*sieboldi* or *japonica*) and the south temperate ocean (*australis*). In spite of earlier recognition as separate species more recent studies have failed to detect differences between the two northern populations (Omura, 1958; Omura *et al.*, 1969). Most recent authors continue to treat northern and southern populations as separate species (*E. glacialis* and *E. australis*) but even in the most comprehensive recent handbook (Cummings, 1985) no diagnostic differences are given. One cranial difference has been reported, concerning the shape and extent of the alisphenoid bone which was described as 'strikingly different' between the two forms by Muller (1954) but on the basis of only four specimens of each form. Using mitochondrial DNA Schaeff *et al.* (1991) reported a genetic distance of 1.82% 'suggesting that the two diverged between c. 0.9 to 1.8 million years ago'. Since the reproductive cycle of the two populations is 6 months out of phase these authors concluded that they must be reproductively isolated and therefore considered them separate species.

A similar uncertainty exists in the case of the minke whales *Balaenoptera acutirostrata s.l.* of the northern and southern oceans which have been considered separate species on the basis of allozyme analysis (Wada and Numachi, 1991).

17.6 DISCUSSION

17.6.1 The biological species concept

The biological species concept has, I believe, stood the test of time as the best theoretical framework for the description and diagnosis of species in living mammals (and indeed in most other sexually reproducing animals). We should accept that different species concepts are needed in organisms with very different breeding systems and therefore very different patterns of variation. It can easily be added to in order to take account of species in time (see section 17.6.5).

For living mammals a useful definition is: 'A species is a set of contemporary individuals that are capable of interbreeding to an extent that precludes the coexistence in the same place of discrete subsets'. This remains a useful concept, even when no direct data are available on interfertility or reproductive behaviour, since there are sufficient well-studied examples to enable us to use morphological, ethological, karyological, biochemical or other characters as indicators of the probable degree of reproductive isolation involved. Even in the best-studied examples we can only examine a tiny sample of individuals so that our conclusions must always carry, explicitly or not, a degree of probability that our sample is representative of the species.

It is probably no accident that many of the most intractable problems of defining mammalian species involve taxa of economic or medical importance – in other words the closer we look, the more complexity we find. Example are *Mus musculus*, etc., *Rattus rattus*, etc., and *Pan* spp.

Ideally, any judgement in borderline cases should be based on a suite of different kinds of data, e.g. on morphology, ethology, karyology, hybridization and genetics. Where results from a number of sources are congruent a judgement based on the biological species concept is relatively easy, especially in the case of sympatric and parapatric taxa. Where only morphological data are available we must make the best of it and assess the probable relationship by analogy with better-known taxa. Where data on karyotype and/or biochemistry are available these should be integrated with existing morphological or other data before decisions are made with regard to the rank of the taxa concerned. Consideration also needs to be given to the question whether the samples subjected to analysis are adequate to allow extrapolation of the results to the species as a whole, given its known geographical range.

Keeping these limitations in mind, recommendations can be made according to the following categories (Figure 17.1):

- Sympatric forms (Figure 17.1(a)): where two forms coexist with little or no hybridization they should be treated as species, however subtle the differences. However, formal naming is best delayed until it is reasonably certain that the differences are not ephemeral, and care needs to be taken to avoid extrapolation beyond the area from which samples have been examined. Examples of moderately well-resolved cryptic species in this category are the shrews *Sorex araneus*, *S. coronatus* and *S. granarius* in western Europe; and the multimammate mice *Mastomys natalensis* and *M. coucha* in southern Africa.

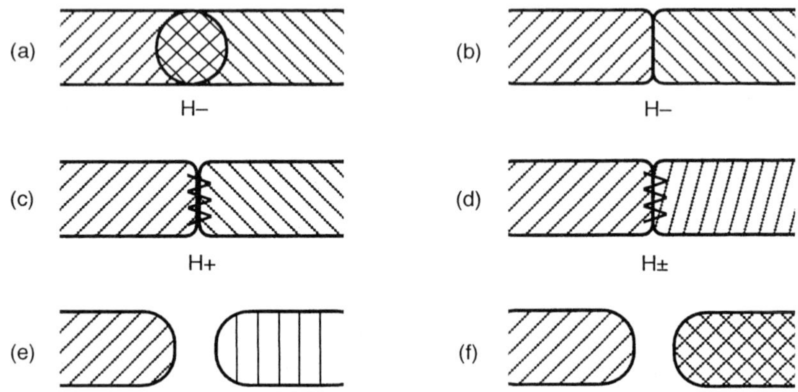

Figure 17.1 Spatial and phenetic relationships between sister taxa. H: ± hybridization. See text.

- Diagnosable parapatric forms with minimal hybridization (Figure 17.1(b)): these should be treated as separate species if they are diagnosable on the basis of a suite of differences, or even on a single character provided that sampling has been sufficiently intense for us to be reasonably certain that it is diagnostic. Example: the hedgehogs *Erinaceus europaeus* and *E. concolor* in Europe.
- Diagnosable parapatric forms with substantial hybridization or intergradation (Figure 17.1(c)): these are best treated as subspecies, e.g. the house mice *Mus musculus musculus* and *M. m. domesticus* in Europe and the African elephants *Loxodonta africana africana* and *L. a. cyclotis*.
- Parapatric forms with minimal differences and some hybridization, e.g. the mole-rat *Nannospalax ehrenbergi* in Israel (Figure 17.1(d)): these are better treated as subspecies, diagnosed by karyotype, since they are very unlikely to be able to coexist and such situations are difficult to detect without intensive study.
- Allopatric forms diagnosable on the basis of differences of a kind that are commonly found within interbreeding populations (Figure 17.1(e)): subspecies, e.g. many insular rodents such as the voles *Microtus arvalis orcadensis* and *Clethrionomys glareolus skomerensis*.
- Allopatric forms diagnosable on a suite of characters or on a single character difference that is unlikely to occur within an interbreeding population (Figure 17.1(f)): species, e.g. the chimpanzees *Pan troglodytes* and *Pan paniscus*, and possibly the right whales *Eubalaena glacialis* and *E. australis* (but variability of the sole diagnostic character in the latter case is unknown and may prove not to justify specific rank).

17.6.2 The phylogenetic species concept

The phylogenetic species concept defines a species as, for example, 'the smallest possible group of a sexually reproducing organism that possesses at least one diagnostic character that is present in all group members but is absent from all close relatives of the group' (Quicke, 1993). This has both theoretical and practical shortcomings. It assumes monophyly of the diagnostic characters, and assumes a degree of sampling intensity within the taxa concerned that can rarely be achieved other than in birds and a very few equally conspicuous taxa. It also depends upon the recognition of hypothetical past branching patterns that are notoriously difficult to establish with a high degree of confidence, especially at low taxonomic levels. The application of parsimony may enable the most probable cladogram to be found, but only among many other almost equally probable ones. Although parsimony is a valuable concept in phylogenetic reconstruction, evolution has had plenty of time to be messy.

In practice, application of the phylogenetic species concept would lead to many forms now recognized as insular subspecies being raised to species rank on the basis of characters that are unique in the samples available from the island but that are not yet known from the continental range of the species, even where these insular populations are known to be completely interfertile with the continental population. The consequence would be greater instability than results from application of the biological species concept, although both are vulnerable to the problems of inadequate sampling. There is no reason why 'phylogenetic' subspecies should not be treated as parts of 'biological' species.

To be practicable, the phylogenetic species concept needs the addition of a scale, other than 'smallest possible group'; the choice of an appropriate scale could make it virtually identical to the biological species concept in practice.

17.6.3 Domestic taxa

Domesticated forms of mammals do not fit comfortably into definitions of species or subspecies based upon wild forms, whichever concept is used, and their classification and nomenclature have been correspondingly unstable. Proposals to improve stability were made by Corbet and Clutton-Brock (1984) who advocated that domesticated forms that are readily recognizable as such, e.g. the domestic dog, should be treated as species separate from their wild ancestors (in this case the wolf, *Canis lupus*). Strict adherence to either biological or phylogenetic species concepts would be both theoretically irrelevant and grossly disruptive to longstanding nomenclature.

17.6.4 Subspecies

Subspecies have been widely used in mammals but also widely misused – the great majority of subspecific names used in the literature are virtually meaningless in that they cannot be related to discrete diagnosable taxa. The recorded ranges of contiguous continental subspecies can be quite spurious because of the chance siting of type localities (Corbet, 1966: 7–9; 1970).

Many versions of the biological species concept have also, unnecessarily, encouraged a spurious view of subspecific variation by describing species as, for example, 'groups of actually or potentially interbreeding natural populations' (Mayr, 1963). Most studies of variation in species occupying continuous habitat have shown a pattern of incongruent clines rather than discrete, definable 'populations' or subspecies, e.g. in the American marten, *Martes americana* (Hagmeier, 1958).

Nevertheless, I believe that the concept of subspecies is a valuable one when applied to allopatric taxa that differ only to a degree that is commonly found within interbreeding populations or that have been demonstrated experimentally to be highly interfertile; and to parapatric taxa with a considerable degree of hybridization.

17.6.5 The dimension of time

Many small mammals are effectively annuals. Major changes in gene frequency can therefore take place very quickly, e.g. through founder effects (Corbet, 1975). Consideration therefore needs to be given to whether observed patterns of variation are ephemeral or stable. This may be particularly applicable in the case of complexes of allopatric or parapatric chromosomal forms, as in *Mus musculus* in the Italian Alps (Corti *et al.*, 1986).

Species of course need to be definable in palaeontological time. The biological species concept has the advantage that when applied to living animals it reflects the contemporary pattern and is not dependent upon any hypothetical reconstruction of past events. Nevertheless, it can be extended to include those ancestors that are not separated from the living species by any discrete gaps or episodes of rapid change. A workable definition to accommodate ancestors is:

'A species is a set of contemporary individuals that are capable of interbreeding to an extent that precludes the coexistence in the same place of discrete subsets, along with their descendants and ancestors that are demonstrably part of the same lineage and are not separated from the set by a discrete episode of rapid change'.

17.7 CONCLUSION

Although recent techniques of karyology, molecular biology and cladistic analysis have significantly increased the objectivity of classification, most studies still suffer from fundamental problems of inadequate sampling, making it difficult to answer such questions as: how representative of the entire genome are the few genes sequenced?; or how representative of the entire species are the individuals examined? It will often be difficult to answer such questions but at least they should always be asked.

Whatever species concept is used, there is a need to bridge the gap between observed sample and the living population. We need to weigh up the probability that two morphologically distinct forms are interfertile; or that two diagnosable samples represent two equally diagnosable taxa in nature. Mathematical techniques may assist in the assessment of such probabilities, but will never be a complete substitute for human judgement based on a wide knowledge of the taxonomy, ecology and behaviour of the organisms concerned.

New data relative to the discrimination of mammalian species are appearing at an accelerating rate, much of it from sources unconnected with traditional taxonomy. However, unless these data are integrated with the existing body of taxonomic knowledge, following the disciplines learned over the past 250 years, much will not effectively enhance our understanding of biodiversity.

17.8 REFERENCES

Aimi, M. and Bakar, A. (1992) Taxonomy and distribution of *Presbytis melalophos* group in Sumatra, Indonesia. *Primates*, **33**(2), 191–206.

Allen, J.A. (1876) Geographical variation among North American mammals especially in respect to size. *Bulletin of the U.S. Geological and Geographical Survey of the Territories*, **2** (4).

Böhme, W. and Eisentraut, M. (1990) Zur weiteren Dokumentation des Zwergelefanten. *Zeitschrift Kölner Zoo*, **33**(4), 153–8.

Catzeflis, F., Corbet, G.B., Furano, A., Orth, A., Usdin, K. and Volobuev, V. Morphological, chromosomal and genic differences between *Rattus rattus* and a sibling, sympatric species of South India (unpublished).

Corbet, G.B. (1964) The grey long-eared bat *Plecotus austriacus* in England and the Channel Islands. *Proceedings of the Zoological Society of London*, **143**, 511–15.

Corbet, G.B. (1966) *The Terrestrial Mammals of Western Europe*, Foulis, London.

Corbet, G.B. (1970) Patterns of subspecific variation. *Symposia of the Zoological Society of London*, **26**, 105–16.

Corbet, G.B. (1975) Examples of short- and long-term changes of dental pattern in Scottish voles (Rodentia: Microtinae). *Mammal Review*, **5**, 17–21.

Corbet, G.B. and Clutton-Brock, J. (1984) Appendix: taxonomy and nomenclature, in *Evolution of Domesticated Animals* (ed. I.L. Mason), Longman, London, pp. 434–8.

Corbet, G.B. and Hill, J.E. (1992) *The Mammals of the Indomalayan Region: a Systematic Review*, Oxford University Press, Oxford.

Corti, M., Capanna, E. and Estabrook, G.F. (1986) Microevolutionary sequences in house mouse chromosomal speciation. *Systematic Zoology*, **35**, 163–75.

Cummings, W.C. (1985) Right whales *Eubalaena glacialis* (Müller, 1776) and *Eubalaena australis* (Desmoulins, 1822), in *Handbook of Marine Mammals* (eds S.H. Ridgeway and R. Harrison), vol. 3, Academic Press, London, pp. 275–304.

Dannelid, E. (1991) The genus *Sorex* (Mammalia, Soricidae) – distribution and evolutionary aspects of Eurasian species. *Mammal Review*, **21**(1), 1–20.

Eisentraut, M. and Böhme, W. (1989) Gibt es zwei Elefantenarten in Afrika? *Zeitschrift Kölner Zoo*, **32**(2), 61–8.

Ellerman, J.R. and Morrison-Scott, T.C.S. (1951) *Checklist of Palaearctic and Indian Mammals*, British Museum (Natural History), London.

Ellerman, J.R., Morrison-Scott, T.C.S. and Hayman, R. (1953) *Southern African Mammals 1758 to 1951: a Reclassification*, British Museum (Natural History), London.

Gaukler, A. and Kraus, M. (1970) Kennzeichnen und Verbreitung von *Myotis brandti* (Eversmann, 1845). *Zeitschrift für Säugetierkunde*, **35**, 113–24.

Groves, C.P. (1993) Order Primates, in *Mammalian Species of the World/ a Taxonomic Review*, 2nd edn (eds D.E. Wilson and D.M. Reeder), Smithsonian Institution Press, Washington, DC, pp. 243–77.

Groves, C.P., Grubb, P. and Bricknell, S. (1993) Taxonomy of African elephants, *Abstracts, 65th International Theriological Congress*, p.120.

Hagmeier, E.M. (1958) Inapplicability of the subspecies concept to North American marten. *Systematic Zoology*, **7**, 1–7.

Hall, E.R. (1943) Criteria for vertebrate subspecies, species and genera: the mammals. *Annals of the New York Academy of Sciences*, **44**, 141–4.

Hall, E.R. (1951) American weasels. *University of Kansas Publications, Museum of Natural History*, **4**, 446 pp.

Hall, E.R. and Kelson, K.R. (1959) *The Mammals of North America*, Ronald Press, New York.

Handwerk, J. (1987) Neue Daten zur Morphologie, Verbreitung und Ökologie der Spitzmäuse *Sorex araneus* und *S. coronatus* im Rheinland. *Bonner Zoologische Beiträge*, **38**, 273–97.

Hausser, J., Catzeflis, F., Meylan, A. and Vogel, P. (1985) Speciation in the *Sorex araneus* complex (Mammalia: Insectivora). *Acta Zoologica Fennica*, **170**, 125–30.

Heyning, J. and Perrin, W.F. (1994) Evidence for two species of common dolphin (genus *Delphinus*) from the eastern North Pacific. *Contributions in Science, Natural History Museum Los Angeles County*, **442**, 35 pp.

Ishak, B., Warter, S., Dutrillaux, B. and Rumpler, Y. (1992) Chromosomal rearrangements and speciation of sportive lemurs (*Lepilemur* species). *Folia Primatologia*, **58**, 121–30.

Linnaeus, C. (1758) *Systema Naturae*, (10th edn), Stockholm.

Marshall, J. and Sugardjito, J. (1986) Gibbon systematics, in *Comparative Primate Biology, Systematics, Evolution and Anatomy* (eds D.R. Swindler and J. Erwin), vol. 1, Alan R. Liss, New York, pp. 137–85.

Mayr, E. (1940) Speciation phenomena in birds. *American Naturalist*, **74**, 249–78.

Mayr, E. (1963) *Animal Species and Evolution*, Harvard University Press, Cambridge, Massachusetts.

Miller, G.S. (1901) [Review of *A Synopsis of the Mammals of North America and the Adjacent Seas*, by D.G. Elliot], *Science*, NS, **14**, 25–7.

Miller, G.S. (1924) List of North American Recent mammals. *Bulletin of the United States National Museum*, **128**, xvi + 673 pp.

Muller, J. (1954) Observations on the orbital region of the skull of the Mystacoceti. *Zoologische Mededeelingen*, **32**(23), 279–90.

Musser, G.G. and Carleton, M.D. (1993) Family Muridae, in *Mammal Species of the World: A Taxonomic Review* (eds D.E. Wilson and D.M. Reeder), Smithsonian Institution Press, Washington, DC, pp. 501–755.

Nei, M. (1972) Genetic distance between populations. *American Naturalist*, **106**, 283–92.

Nevo, E. (1989) Modes of speciation: the nature and role of peripheral isolates in the origin of species, in *Genetics, Speciation and the Founder Principle* (eds L.V. Giddings, K.Y. Kanshiro and W.W. Anderson), Oxford University Press, New York, pp. 205–36.

Niethammer, J. and Krapp, F. (ed.) (1990) *Handbuch der Säugetiere Europas*, band 3/1, *Insektenfresser-Insectivora/ Herrentiere – Primates*, Aula-Verlag, Wiesbaden.

Omura, H. (1958) North Pacific right whale. *Scientific Reports of the Whales Research Institute, Tokyo*, **13**, 1–52.

Omura, H., Ohsumi, S., Nemoto, T., Nasu, K. and Kasuya, T. (1969) Black right whale in the North Pacific. *Scientific Reports of the Whales Research Institute, Tokyo*, **21**, 1–78.

Ortello, M.O. (1995) Phylogenetic analysis of G-banded karyotypes among the South American subterranean rodents of the genus *Ctenomys*. *Biological Journal of the Linnean Society*, **54**, 43–70.

Quicke, D.L.J. (1993) *Principles and Techniques of Contemporary Taxonomy*, Blackie A & P, London.

Ruvolo, M., Pan, D., Zehr, S., Goldberg, T., Disotell, T.R. and Dornum, M. (1994) Gene trees and hominid phylogeny. *Proceedings of the National Academy of Science, USA*, **91**(19), 8900–4.

Savic, J.R. and Nevo, E. (1990) The Spalacidae: evolutionary history, speciation and population biology, in *Evolution of Subterranean Mammals at the Organismal and Molecular Levels. Progress in Chemical and Biological Research*, **335**, 129–53.

Schaeff, C., Kraus, S., Brown, M., Perkins, J., Payne, R., Gaskin, D., Boag, P. and White, B. (1991) Preliminary analysis of mitochondrial DNA variation within and between the right whale species *Eubalaena glacialis* and *Eubalaena australis*. *Reports of the International Whaling Commission, Special Issue*, **13**, 217–23.

Sharman, G.B. (1956) Chromosomes of the common shrew. *Nature*, **177**, 941–2.

Skinner, J.D. and Smithers, R.H.N. (1990) *The Mammals of the Southern African Subregion*, University of Pretoria, Pretoria.

Thaeler, C.S. (1985) Chromosome variation in the *Thomomys talpoides* complex. *Acta Zoologica Fennica*, **170**, 15–18.

Trouessart, E.L. (1897–1905) *Catalogus Mammalium tam Viventum quam Fossilium*, Friedländer, Berlin.

van Bree, P.J.H. and Gallagher, M.D. (1978) On the taxonomic status of *Delphinus tropicalis* van Bree, 1971. *Beaufortia*, **28**, 1–8.

Wada, S. and Numachi, K. (1991) Allozyme analysis of genetic differentiation among the populations and species of the *Balaenoptera*. *Reports of the International Whaling Commission, Special Issue*, **13**, 125–54.

Yosida, T.H. (1980) *Cytogenetics of the Black Rat*, University of Tokyo Press, Tokyo; University Park Press, Baltimore.

18

The ideal species concept – and why we can't get it

D. L. Hull

Contacting address: Department of Philosophy, Northwestern University, Evanston, IL 60208, USA

ABSTRACT

Ideally scientists would like their concepts to be as general, applicable and theoretically significant as possible. Unfortunately, these goals tend to conflict with each other, that is, one goal can be realized only at the expense of other, equally desirable goals. For example, theoretical significant species concepts tend not to be very operational. Attempts to make them more operational result in their being theoretically less significant. Recent suggestions for improving the species concept can be seen as attempting to realize one or more of these three goals. The test cases for each of the species concepts examined are asexual organisms, hybrid species and the inclusion of males and females with each other and their offspring in the same species.

18.1 INTRODUCTION

Cracraft (1997: Chapter 16) compares the search for the ideal species concept to the quest for the Holy Grail, and the comparison is not far off the mark. The temptation has always been to hope that, if we can only formulate just the right definition, all our problems will be solved. Enough time has passed and enough energy expended to convince quite a few of us that no magic bullet exists for the species concept (Mishler and Theriot, 1997). Any species concept, no matter which one we choose, will have some shortcoming or other. Either it is only narrowly applicable, or if applicable in theory, not in practice, and so on. One problem is that different systematists have different goals for their species concepts, but even those systematists who agree in principle on what a species concept

Species: The Units of Biodiversity. Edited by M.F. Claridge, H.A. Dawah and M.R. Wilson. Published in 1997 by Chapman & Hall. ISBN 0 412 63120 2

should do frequently prefer different species concepts. The trouble is that we have several criteria that we would like an ideal species concept to meet, and these criteria tend to conflict. Most importantly, if a species concept is theoretically significant, it is hard to apply, and if it is easily applicable, too often it is theoretically trivial (for additional general treatments of these issues, see Slobodchikoff, 1976; Otte and Endler, 1989; Ereshefsky, 1992; and Sterelny, 1994 critical notice).

In this chapter I set out three of the most common criteria that concepts are supposed to meet in science and see how well various species concepts meet these criteria. These criteria are universality (or generality), applicability, and theoretical significance. Scientists want their concepts to be as general as possible. For example, physicists intend their concept of physical element to encompass all physical substances, not just a subset. Biologists have had a more difficult time formulating a species concept that encompasses all organisms. The two phenomena that have proved to be the most intractable for species definitions are asexual reproduction and hybridism. One sign that scientists really do value generality is the increasing attention that these two problem cases are receiving (Cracraft, 1983: 171; Donoghue, 1985: 179; Templeton, 1989: 8, 10, 11; Echelle, 1990: 111; Nixon and Wheeler, 1990: 219; McDade, 1990, 1992; Vrana and Wheeler, 1992: 70; Mishler and Theriot, 1997). In this chapter I treat these two phenomena as test cases for proposed species concepts.

An issue closely connected to generality is monism. In its most extreme form monism is the view that a single way exists for dividing up the world into kinds and organizing these kinds into a single hierarchy of laws. Although no such monistic explanation of the empirical world currently exists, it is the goal toward which many scientists strive. Since Einstein physicists have attempted to produce a unified field theory. Right now monism is decidedly out of favour in certain philosophical circles. The sophisticated, not to mention politically correct, position is pluralism.

Pluralism comes in a spectrum of forms from promiscuous pluralism to more moderate positions. The major claim that pluralists make is that no unified picture can be presented of nature. We can view living things from a variety of perspectives, and each of these perspectives is legitimate. We can view living organisms from a genealogical perspective and classify them accordingly. Or we can produce ecological classifications that ignore genealogy. Or we can organize organisms into diagnosable units without any attention to genealogy, ecology, embryology, etc. Pluralists maintain that some of these ways may be preferable to others for certain purposes, others may be preferable for other purposes, but none of these perspectives is any more fundamental than any other (Kitcher, 1984, 1989; Dupré, 1993).

Applicability and theoretical significance tend to be in opposition to each other. The more theoretically significant a concept is, the more difficult it is to apply. A repeated theme in the taxonomic literature is that species concepts should be as operational as possible. Species should be defined with an eye to the sorts of data available to systematists. Since the most easily available and widespread data are patterns of morphological variation, species should be defined in terms of some sort of morphological similarity. Other systematists insist that such characters are not ends in themselves but are evidence to be used to infer something else – something a good deal more theoretically important than morphological similarity (Endler, 1989 uses theoretical versus operational as one of his four criteria for evaluating species definitions).

Temporal considerations tend to come into play at this juncture. Both observation and theory are important in science, but a belief common among scientists is that, to be genuinely scientific, scientists should begin with observations and only much later proceed to speculate about more theoretical issues. Thus, in evaluating alternative species concepts, theoretical significance poses two problems: theoretical input and theoretical output. Epistemically conservative systematists will allow that classifications can be legitimately used as a basis for theoretical speculations but that no theoretical considerations should be allowed into the formulation of these classifications. Other systematists insist that no such thing as a theory–neutral classification exists. Theoretical considerations should and do enter into classification right from the start.

The epistemically conservative position is reflected in the chapters by several authors in this volume. For example, Gornall (1997: Chapter 8) recommends using a working definition of species that is devoid of much, if any theoretical background. Hawkes (1997: Chapter 9) thinks that systematists tend to agree with each other, except when they are encumbered by theoretical baggage. Claridge *et al.* (1997: Chapter 1) concur, complaining that objectivity may often be clouded by adherence to particular theories of speciation. Most authors in this volume see no reason why our knowledge of the empirical world – all of it – cannot be used in constructing our classifications.

Classifications in terms of observable characters are not only more certain, so epistemically conservative authors claim, but in addition they are also more practical or useful. Systematists frequently justify their continued existence by reference to practical uses of classifications, e.g. identifying insect vectors. Such justifications have some point. For example, in the United States we have an Endangered Species Act. Lumpers recognize widespread species that are unlikely to be endangered, while splitters identify not only more species but more restricted species that are likely to be more vulnerable to extinction. Thus, as Mayden (1997: Chapter 19) has argued, which species concept we adopt has very direct and important effects on something as practical as the mass extinction currently under way. More importantly, if systematists give the impression that species recognition is largely an arbitrary affair, the justification for attempting to save endangered species is weakened. Thus, even the most philosophical exercises in science can have very practical effects (Eisner *et al.*, 1995).

But all the effort devoted to honing our species concepts also has effects on science in general. For example, the title of this volume is *Species: The Units of Biodiversity*, but as John and Maggs (1997: Chapter 5) point out, accurate estimates of biodiversity require sound species concepts. We often hear about the error of comparing apples and oranges, but if different workers use different species concepts in estimating biodiversity, then the results of all their efforts will not be comparable. They will have truly been comparing apples and oranges. Typically, higher taxa are not very comparable. Patterns of distribution of the same organisms differ for genera, families and orders (Signor, 1985). If species are no more comparable than higher taxa, then we are all in real trouble. To put the point more strongly, since species are not comparable, then we are in deep trouble if we do not take this fact into account (for a more general evaluation of systematics and the species category than provided here, see Frost and Kluge, 1994).

18.2 PRESENT-DAY SPECIES CONCEPTS

In this chapter I propose to evaluate the most salient species concepts in use today on the three criteria listed above to see how well each of these concepts does. I classify these species concepts into three basic kinds. First, one group of species concepts can be lumped together because they all require similarity of some sort. Similarity concepts include traditional morphological species concepts, the phenetic species concept, as well as certain molecular concepts. In each case, systematists look for some sort of overall similarity and/or gaps in character distributions. The chief difference between the phenetic species concept and earlier morphological species concepts is that early systematists tended to be typologists in the sense that each species (not to mention each higher taxon) must be defined in terms of universally covarying characteristics, while the pheneticists treat all Operational Taxonomic Units as being polythetic, that is, they can be defined only in terms of statistically covarying characteristics (see Van Regenmortel, 1997: Chapter 2).

When the notion of polythetic definitions of taxa was first introduced into systematics, it looked very promising (Hull, 1965). However, it has turned out to have some major drawbacks. Chief of these is that any group of organisms can be classified in indefinitely many ways using various clustering techniques, and no reasons internal to these methods exist for choosing among these classifications (see also de Queiroz and Good (1997) for a more extensive critique of phenetic clustering mechanisms).

The second group of species concepts that I evaluate includes Mayr's (1969) biological species concept, Simpson's (1961) evolutionary species concept, especially as it has been reworked by Wiley (1981), Paterson's (1981) species mate recognition concept, and Templeton's (1989) cohesion concept. What all these concepts have in common is theoretical commitment, especially a commitment to evolutionary theory. Because the evolutionary process operates in a particular way, species are to be individuated accordingly. Each of these theoretically committed species concepts is designed to highlight some significant level of organization in the evolutionary process.

I term my third group of species concepts phylogenetic to indicate their historical connection to Hennig's phylogenetic systematics (Mishler and Theriot, 1996). The two most prominent phylogenetic species concepts are the monophyletic (or autapomorphic) species concept and the diagnostic species concept. The monophyletic species concept as developed by Mishler and Donoghue (1982), Mishler and Brandon (1987), and de Queiroz and Donoghue (1988, 1990) among others, has its origins in Hennig as well as Rosen (1978, 1979). According to the monophyletic species concept, a species is the least inclusive monophyletic group definable by at least one autapomorphy. The reference to autapomorphies explains why this concept is also termed the autapomorphic species concept.

The diagnostic species concept also has its historical roots in the work of Hennig and of Rosen (1978, 1979), but it has taken its own direction; e.g. Platnick (1977), Eldredge and Cracraft (1980), and Nelson and Platnick (1981). The most influential formulation of the diagnostic species concept was first formulated by Cracraft (1983) and then further developed by McKitrick and Zink (1988), Nixon and Wheeler (1990), Wheeler and Nixon (1990), and Vrana and Wheeler (1992) among others. According to Cracraft (1983: 170), a species is the smallest diagnosable cluster of individual organisms within which there is a parental pattern of ancestry and descent. Cracraft includes reference to parental pattern of ancestry and descent (or cohesion) to avoid one of the prevailing problems of defining species purely in terms of similarity – the failure sometimes to include males and females in the same species with each other as well as with their progeny.

By now, the reader should have noticed that my classification of species concepts involves a degree of cross-classification. The monophyletic species concept is as theoretically committed as are the concepts of Mayr, Simpson, Wiley, Paterson and Templeton. However, I have separated it from these earlier concepts because of its connection to cladistics. Conversely, the diagnostic species concept could have easily been classified with the morphological and phenetic species concepts because of its concern to be as operational as possible. Once again, I have classified it along with the monophyletic species concept to highlight its connection to Hennig.

Although the phylogenetic species concepts were developed by Hennig's intellectual descendants, they form two quite distinct conceptual transformation series (Platnick, 1979; Nixon and Wheeler, 1990). They both start with concerns important to Hennig but then take paths of their own. For example, the diagnostic species concept eventually came to look very similar to pre-Darwinian morphological species concepts. In general, phylogenetic cladists such as Wiley and Donoghue remain truer to Hennig's original system than do such pattern cladists as Nelson and Platnick. In fact, pattern cladists have reworked Hennig's system so radically that it begins to look like the systems of ideal morphology that Hennig was concerned to refute. But in the identification of conceptual transformation series, what counts is descent, not similarity. As far as similarity is concerned, the diagnostic species concept may look suspiciously like the species concepts of Hennig's opponents, but as far as descent is concerned, it belongs in an Hennigian transformation series.

The most famous of the theoretically committed species concepts is Mayr's biological species concept. According to Mayr (1969: 26), species are groups of interbreeding natural populations that are reproductively isolated from other such species. Mayr's species concept is designed to distinguish between those groups of organisms in which gene flow is significant and those in which some isolating mechanism or other has interrupted gene flow.

One common criticism of the biological species concept is that it is not sufficiently operational and will be discussed in the next section. Critics have also alleged that each species according to Mayr must possess an isolating mechanism capable of insulating it against all possible species. To the contrary, each species needs to have a mechanism that isolates it only from those species with which it happens to come into contact. If it possesses such a mechanism, it remains distinct from this other species. If not, the two merge into a single species (Horvath, in press). The biological species concept is also non-dimensional. It applies only to organisms during a relatively short time span. A biological species is a time-slice of an evolving lineage. It also, quite obviously, applies only to organisms that reproduce sexually. (The notion of the temporal dimensionality of species has not been given the attention it properly deserves. For now, all it indicates is some unspecified but relatively short period of time.)

Simpson's evolutionary species concept extends the biological species concept through time. According to Simpson (1961: 153) evolutionary species evolve separately from other such lineages and possess their own unitary roles and tendencies. Wiley (1981: 25) reworked Simpson's evolutionary species concept, replacing unitary roles and tendencies with evolutionary tendencies and historical fate. Wiley's interpretation of species also differs from Simpson's interpretation in that Wiley is concerned to make his species concept compatible with the phylogenetic system of

Hennig. For example, Simpson was willing to subdivide a gradually evolving lineage into successive chronospecies while Wiley is not.

Paterson (1981) caused quite a stir with his specific mate recognition concept. According to Paterson, each species has its own fertilization system which enables organisms belonging to that species to mate. The chief virtue of this definition according to Paterson is that it defines each species in terms of what binds it together rather than in terms of indefinitely many isolating mechanisms that are supposed to keep it separate from indefinitely many other species (Horvath, in press). For this reason, he terms the biological species concept the isolation concept. One common response to Paterson's species concept is that the isolation and recognition concepts are just two sides of the same coin (Templeton, 1989: 7), a contention which obviously does not sit all that well with Paterson (1981: 26). One point to notice is that the species mate recognition concept, like the biological species concept, applies only to organisms that reproduce sexually at least sometimes.

Finally, Templeton formulated his cohesion species concept in order to synthesize the preceding theoretically committed species concepts. According to Templeton (1989: 12), a species is the most inclusive population of individuals having the potential for phenotypic cohesion through intrinsic cohesion mechanisms. Among these intrinsic cohesion mechanisms are gene flow and natural selection as well as ecological, developmental and historical constraints. The cohesion species concept is more general than either the biological species concept or the species mate recognition concept because it applies to asexual organisms via their adaptation to particular ecological niches.

According to Hennig (1966) his principles of phylogenetic systematics come into play only when reticulation gives way to splitting (phylogenetic relations). This is also the point, according to Hennig, at which species are to be recognized. According to Hennig, all higher taxa must be strictly monophyletic. Species to the contrary cannot be monophyletic because monophyly does not even apply to species. However, a group of Hennig's intellectual descendants have argued that monophyly can be extended far below the species level. For example, a series of populations can exhibit a tree-like structure. The result is the monophyletic species concept. Species are monophyletic groups of organisms characterized by at least one autapomorphy.

In addition, the monophyletic species concept is synchronous with respect to its grouping criterion because monophyly in the sense of sister-group relations has no temporal dimension. All taxa are arrayed along the top of a cladogram. Some may be older than others, some may even be an ancestor of one or more of the other taxa included in the cladogram, but the only relationship indicated in a cladistic classification is that of sister-group relations and they are synchronous. However, the monophyletic species concept assumes at least a minimal time-dimension because its ranking criteria do take at least some time to function.

The monophyletic species concept is, in principle, monistic with respect to its grouping criterion: monophyly and only monophyly matters. However, advocates of the monophyletic species concept also allow the naming of metaspecies, the smallest unresolved groups that have not yet been shown to be monophyletic, just as long as they are distinguished symbolically from true monophyletic species (Donoghue, 1985: 179). The monophyletic species concept is also pluralistic in a weak sense with respect to its ranking criteria because it makes reference to more than one ranking criterion. Pluralism in the strongest sense requires that quite different and incompatible species definitions be considered equally good in their own domain. Here, the authors are providing a single definition but include more than one way of fulfilling the requirements of this definition.

The diagnostic species concept differs from the monophyletic species concept in that the diagnostic characters that define species need not be autapomorphies. Instead, any unique combination of primitive and derived characters that have become fixed in a group will do (Cracraft, 1983: 170). In a later reformulation, Cracraft (1989: 34–35) modifies the diagnostic species concept to refer to a phylogenetic species as being an irreducible (basal) cluster of organisms which are diagnosably distinct from other such clusters (see also Nixon and Wheeler, 1990). Diagnostic characters need not be present in all the organisms that belong to a species throughout their development. Instead they can be present in just males, just females, or at some particular stage in their development. However, with these provisions in mind, diagnostic characters must be universally distributed, although not uniquely derived characters (autapomorphies).

As the preceding brief characterizations of theoretically committed species concepts have indicated, some are dimensional with respect to time; that is, species are explicitly treated as lineages extended through time (e.g. the evolutionary species concepts of Simpson and Wiley). Terming the concepts of Mayr, Paterson and Templeton non-dimensional is a bit too strong. The processes by which they define species take at least some time to operate. Gene flow, for instance, is not instantaneous. In addition, in various places, advocates of these species concepts, explicitly state that their non-dimensional species are intended to be time-slices of evolving lineages (Templeton, 1989: 20; Nixon and Wheeler, 1990: 218, 220; Mishler and Theriot, 1997).

The phenetic species concept, however, is radically non-dimensional. Phenetic species are not intended to be lineages or time-slices of lineages. Time simply does not enter into the equation. Phenetic species are defined by means of phenetic clusters of characters with no attention whatsoever to time. The importance of the distinction between dimensional and non-dimensional species concepts is attested to by Endler (1989) selecting it to be one of his four criteria of evaluation.

Because species are evolving lineages according to the evolutionary species definition, this definition has both retrospective and prospective dimensions. What counts as a species is determined in part by what has happened to it in the past, in part by what happens to it in the future. Reference to evolutionary fates seems to require that we can predict the future. Thousands of times a population begins to bud off from the main body of its species only to be reabsorbed later. However, every once in a while, one of these populations becomes established and is transformed into a new species. In retrospect, that founding population is considered part of this new species. If speciation had not been successful, it would simply be included as an ordinary population with the rest of its species (for details, see Kornet, 1993 and Graybeal, 1995).

But, critics note, one thing that we cannot do in biological evolution is predict the course of particular lineages. Of all the species of fruit flies, which will go extinct first, which will last the longest without going extinct, and which will speciate? I think I am safe in saying that no one knows. We can make some general observations, e.g. that small, genetically homogeneous species are more likely to go extinct than their larger, more heterogeneous relatives, but that is all. It is difficult enough to decide when a particular lineage has its own evolutionary role and tendencies. How can we possibly tell which lineages will retain their own fate? The answer is that we cannot. But this is the price that we pay for treating species as lineages. Lineages can be recognized only in retrospect (Sober, 1984: 339; de Queiroz and Donoghue, 1988: 330).

O'Hara (1993: 238–242) argues that the evolutionary species concept is clearly a lineage concept, but even such apparently non-dimensional species concepts as the biological, recognition, monophyletic and diagnostic species concepts include notions of fate, temporariness and permanence. As a result, they too cannot be applied with certainty in the present because they all depend on the future. However, he does not regard future dependence as a flaw of these species concepts; it is simply an uneliminable characteristic of them. I might add that the retrospective character of evolving species is only to be expected if species are individuals (Ghiselin, 1974; Hull, 1976).

18.3 UNIVERSALITY AND MONISM

Given the brief descriptions of current species concepts provided above, the question becomes how well each of these species concepts does according to the criteria commonly used to evaluate scientific terms – universality, applicability and theoretical significance. As I mentioned previously, scientists would like their concepts to be as general as possible.

For example, any definition of element in physics should apply to all matter, not just a restricted subset. Physicists would be less than pleased if their element concept applied only to metals or to non-radioactive substances. One virtue of atomic number is that all material substances are made up of atoms, and every atom has some atomic number or other. Atomic number applies across all substances and divides them into strictly comparable kinds. The question is whether a species definition can be found in biology that does the same for biological kinds. Does a single level of organization exist across all organisms that deserves to be recognized as the species level? Anyone answering the preceding question in the affirmative is a monist.

The choice between monism and pluralism is not easy, especially if attention is paid to self-reference. Scientists who are monists with respect to their own subject matter typically are also monists with respect to the philosophical issue of monism. For example, Mayr has argued for decades that the biological species concept and only the biological species concept is adequate for understanding the living world. And each subject matter must be explained in one and only one way. The evolutionary perspective is basic. Pluralists surprisingly also tend to be monists when it comes to pluralism. Pluralists argue that scientists do and should produce a variety of incommensurable classifications from a variety of perspectives. Any discomfort that scientists may feel about such conceptual riches simply stems from their inexcusable bias toward monism. However, as monistic as it may sound, pluralists insist that pluralism is the only acceptable view of science (Kitcher, 1984, 1989; Dupré, 1993). In general, investigators find pluralism attractive in someone else's discipline but tend to be monists with respect to their own (see also Mishler and Brandon, 1987; de Queiroz and Donoghue, 1988; and Sterelny, 1994 for criticisms of Kitcher-type pluralism).

With respect to the monism/pluralism issue, several of the species concepts discussed above appear to be strictly monistic. For example, the phenetic species concept refers to phenetic similarity and only phenetic similarity. It is also universal because it applies to all organisms regardless of any other considerations. It applies to asexual organisms as readily as to those that reproduce sexually. Because Sokal and Sneath (1963: 174) are not attempting to represent phylogeny in their phenetic classifications, the only problem that interspecific hybrids pose for them is a possible increase in borderline cases between parental clusters.

As long as the pheneticists assume that some one thing exists out there in nature properly termed overall similarity, then the phenetic species concept is monistic. Species are the smallest group exhibiting the appropriate degree of overall similarity. However, quite early on, pheneticists began to have their doubts about overall similarity and the general-purpose classifications that supposedly reflect overall similarity. For example, Ehrlich and

Ehrlich (1967) concluded that there is no such thing as overall similarity or general-purpose classifications.

Such critics as Johnson (1968) and Ghiselin (1969) added their voices to these suspicions. For a group like the pheneticists, who view themselves as being hard-headed and highly empirical, overall similarity eventually came to look very much like a metaphysical concept. Even Sneath (1995) seems to have given up the notion of overall similarity. In defence of phenetic taxonomy, he argues that the power of overall similarity measures to construct taxonomic groups, to determine evolutionary relationships, and for identification, has been amply borne out, but he adds, even if somewhat different forms of similarity may be needed for different purposes.

From the start, one of the most serious problems with respect to the phenetic species concept was that no reasons could be given to choose one level of phenetic similarity over any other to term the species level. Even so, all that mattered was phenetic similarity. In this respect the pheneticists were monists. But now that phenetic similarity comes in different forms, pheneticists have joined the pluralist camp in a weak sense of pluralism. However, the phenetic species concept remains universal. To the extent that it can be applied at all, it applies to all organisms.

In the past, certain limitations to prevailing species concepts, though acknowledged, tended to be relegated to the last paragraph or two in a chapter on the species concept. Such is the case with asexual reproduction and species of hybrid origin. Authors tended either to argue these problem cases away or to claim that they occur very rarely, if at all. For example, advocates of the biological species concept attempt to minimize this problem by observing that most organisms reproduce sexually, at least some times, but it should be kept in mind that very little in the way of gene exchange occurred during the first half of life on Earth, and meiosis evolved even later. According to the biological species concept, no species existed for at least the first half of life on Earth. Evolution occurred but in the absence of species. As strange as this observation may sound, it is no stranger than the recognition that multicellularity also did not evolve until quite late. We can no more insist that all organisms form biological species than we can insist that all organisms be multicellular.

Species of hybrid origin do not present as serious a problem to the biological species concept as do asexual organisms. According to the Mayr, reproductive isolation builds up very slowly. If the biological species concept is applied sequentially through time, a single unproblematic species will be found to give way to a series of problem cases until two distinct species emerge. However, if reproductive isolation can build up, it might also break down. The breakdown of reproductive isolation provides the same problem as does its original build up. In this case, two unproblematic species either gradually merge into one or produce hybrid individuals through mechanisms such as allopolyploidy.

When Mayr (1969: 32) addresses the issue of hybridism, he does his best to downgrade its pervasiveness and importance. He notes that the taxonomic literature records a number of instances in which two parental species have supposedly merged into a single new species but regards these instances as not being established unequivocally. Introgression is a populational affair, but hybrids can arise at the level of organisms as well (e.g. via allopolyploidy). Mayr is unable to dismiss these examples so readily. Life would be easier if reproductive isolation, once established, never broke down, but it does so, especially in plants.

Two points are at issue with respect to interspecific hybrids: one is empirical while the other is conceptual. The empirical issue is whether or not hybrid species actually occur in nature. As inconvenient as they may be, it seems as if they do. The conceptual issue is how to classify hybrids if they do occur. Right from the start, the formal similarity between evolution and classification seemed deceptively obvious. On the one hand, evolution is primarily a matter of successive splittings. On the other hand, the Linnaean hierarchy consists of a series of successive subdivisions. Hence, one should be easily mapped onto the other.

If Hennig and his descendants have done anything for the systematic community, it has been to show that this impression is seriously mistaken. It is easy enough to draw one line splitting in two; it is not so easy to represent such an occurrence in a Linnaean classification. Classifying common ancestors with their progeny in an unambiguous way has proven to be a major challenge to systematists. It is also easy to draw two lines merging into one, but such an occurrence poses exactly the same problem for Linnaean classifications as does the classification of common ancestors (Hull, 1979; Wiley, 1981). Although speciation is much more common than hybridism, the two processes pose the same problem for the comparative method. Their character distributions cannot be resolved unequivocally into a series of dichotomous divisions. Instead they produce unresolvable trichotomies.

Echelle (1990) and McDade (1990, 1992) show that the effects of introducing hybrid species into a classification are not quite as catastrophic as some early in-principle arguments implied. McDade (1992: 1329) concludes that hybrids are unlikely to cause breakdown of cladistic structure unless they are between distantly related parents. However, her results also indicate that cladistics may not be especially useful in distinguishing hybrids from normal taxa. In any case, advocates of more recent species concepts all take the ability of their favoured species concept to handle hybridism as a decided advantage (Cracraft, 1983: 171; Donoghue, 1985: 179; Templeton, 1989: 10; Nixon and Wheeler, 1990: 220; and Mishler and Theriot, 1997).

One reason for claiming that the biological species concept and the species mate recognition concept are two sides of the same coin is that asexual reproduction and species of hybrid origin pose the same problems for both. Asexual organisms possess no system for recognizing mates because asexual organisms do not have mates, and just as such recognition systems can build up, they can break down. Thus, like the biological species concept, the species mate recognition system is not universal. It is more clearly monistic in the sense that each species presumably has one and only one mate recognition system.

Templeton's cohesion species concept is monistic in the sense that all that matters is cohesion, but cohesion can be produced by several different mechanisms acting independently and in consort. One of these mechanisms (adaptation to a particular ecological niche) is designed primarily to accommodate asexual organisms. As Templeton (1989: 8) remarks, the asexual world is for the most part just as well (or even better) subdivided into easily defined biological taxa as is the sexual world. Species of hybrid origin pose the same problems for the cohesion species concept as they do for the other theoretical significant concepts.

Advocates of the monophyletic species concept take pains to point out that asexual organisms form monophyletic groups, whether at the level of individual organisms or at the level of more inclusive clones (Mishler and Donoghue, 1982: 491; Donoghue, 1985: 179; Mishler and Brandon, 1987: 406; Budd and Mishler, 1990; Mishler and Theriot, 1997). The monophyletic species concept is universal in that it applies to all organisms. It is also monistic with respect to its grouping criterion monophyly. Whether or not one wants to rank all least inclusive monophyletic groups as species is quite another matter. Mishler and Brandon (1987: 406) argue for a variety of criteria for ranking, depending on the causal agent judged to be most important in producing and maintaining distinct lineages in the groups in question. Hence, the monophyletic species concept is pluralistic with respect to ranking criteria. However, these authors make it clear that their sense of pluralism is quite distinct from the more radical views of Kitcher and Dupré.

Similarly, the diagnostic species concept is universal in that it applies to all organisms. One or more diagnostic characters and a parental pattern of ancestry and descent are all that is needed, and asexual organisms fulfil these requirements (Nixon and Wheeler, 1990: 219). This concept is also monistic in that no appeal to alternative mechanisms is made. Asexual organisms form a strictly branching pattern, while sexual organisms form a messy reticulation. Hybridism only adds to this reticulation at higher levels of organization. This is as pluralistic as the diagnostic species concept gets.

In sum, the authors discussed in this section acknowledge that generality is a virtue in a species concept. One of the most persuasive arguments for the phenetic species concept is that it applies equally to all organisms. No issues of splitting and merger arise because these notions play no role in phenetic taxonomy. Both Simpson and Wiley give as one reason for accepting an evolutionary species concept that it applies to sexual and asexual organisms alike. Cracraft (1997: Chapter 16) states explicitly that one reason why his diagnostic species concept is superior to other species concepts is that it is applicable to all groups of organisms.

Thus, we are left with the question with which we began. Is there a single level of organization across all organisms that deserves to be recognized as the species level? If the papers in this anthology do anything, they cast doubt on the monistic reply to this question. Some of the species concepts that I have discussed do apply to all organisms. For example, the monophyletic species concept is designed to mark the place where reticulation gives way to successive branchings. This transition is certainly important and is worth marking. The trouble is that this level varies from single organisms in asexual organisms, to single populations in some sexual species, to more traditional species, and finally to much larger groups where interspecific hybridization occurs. From the perspective of the replacement of reticulation with splitting, this transition forms a single level. However, from a more common-sense perspective, it wanders from level to level, depending on the circumstances; for a detailed discussion of reticulation in a particular group (corals), see Veron (1995).

The diagnostic species concept also applies to all organisms, but it subdivides organisms into smaller, more numerous species than systematists are used to recognizing. If a local population exhibits one or more diagnostic characters, it must be considered a species. However, several studies have shown that this in-principle problem is not as pervasive as one might think. For example, Cracraft estimates that the 8500 to 9000 species of birds recognized under the biological species concept would be expanded to about 18 000 under his diagnostic species concept (see also Cracraft, 1983: 173; Echelle, 1990: 110; Ereshefsky, 1989: 90; Nixon and Wheeler, 1990: 219)

A more fundamental problem is that the diagnostic species concept can divide up organisms into overlapping and incompatible species, depending on which characters are picked to be diagnostic. Cladistic characters must nest. Otherwise, they are not genuine characters. But the diagnostic characters of the diagnostic species concept need not be autapomorphies. As such, they need not nest.

Some species concepts (e.g. the biological and mate recognition concepts) apply to only a limited number of organisms. Where they apply, they do seem to mark important groups of organisms. They cut nature at her joints. But since these definitions do not apply to all organisms, there is some tendency not to use them even for the organisms to which they do apply. Pluralists would complain that this response is too monistic. Although I have a strong preference toward monism, I am forced to agree.

18.4 APPLICABILITY AND THEORETICAL SIGNIFICANCE

Scientists are strongly predisposed to make their concepts as operational as possible. Theoreticians are much less concerned. They formulate extremely abstract theories about space, time and energy or gene exchange and density-dependent selection and lesser mortals can worry about testing these theories. Philosophers tend to side with the theoreticians. The philosophy of science deals primarily with theories and their development. So does the history of science. The names of theoreticians that turned out to be right or at least wrong in big ways are chronicled at great length. We find it difficult to recall the names of more empirically minded scientists – or 'fact grubbers' as they are termed – with no fondness. We are all familiar with the distinction between batesian and müllerian mimicry, and most of us have heard about the paradigm example of batesian mimicry (monarch and viceroy butterflies) but few of us are likely to be able to recall the names of the authors who recently put this distinction to the test with disconcerting results (Ritland and Brower, 1991).

The philosophical arguments against operationism are decisive. Philosophers do not agree on much, but two propositions that have gained wide acceptance among philosophers of science are that no theoretically significant terms can be totally operationally defined, and that no terms are totally theory neutral. Although at one time I argued with all the enthusiasm of a graduate student that these propositions are well-taken (Hull, 1968), I now think that the attention that scientists pay to formulating operational criteria for theoretical concepts is appropriate. My early prejudice led me to be more critical of the operational tendency with respect to species concepts than I now think I should have been. Maybe philosophers don't have to apply scientific concepts, but scientists do.

In the previous section I discussed difficulties in applying concepts. In these discussions of applicability the issue was theoretical limitations, e.g. the biological and mate recognition concepts do not apply to asexual organisms. It is not that we lack techniques for applying these concepts but that they do not apply even in principle. The issues addressed in this section concern difficulties in applying concepts in practice to cases to which we know that they apply in principle. For example, as sure as we know anything, we know that some extinct organisms mated with each other and thus formed biological species, but such relations must now be inferred via various sorts of evidence, and sometimes these inferences are none too secure. In this section, it is this sort of applicability that I will be discussing.

The importance of applicability can be seen in the terms 'practical' or 'practice' appearing in the titles of five of the contributions to this volume. The term practice has a nice ring to it (Pickering, 1984). No fancy theories or idle speculations here. Just the facts and nothing but the facts. From our present-day perspective it is difficult to recall that the biological species concept was introduced to make species recognition more practical in the sense of objective. Early in the history of systematics, species recognition was largely intuitive. Experts in a particular group developed a knack for recognizing species in their group. I do not mean to denigrate such intuitive reasoning on the part of practising systematists. It still plays an important role in systematics. If I had limited money and time and needed a good rough-and-ready classification, I would surely turn to an experienced systematist, regardless of his or her methodological persuasion.

However, three problems plague intuitive species recognition and classification. One is that systematists who group their organisms according to implicit criteria on the basis of their individual expertise tend not to agree with each other, and there are no objective standards by which these disagreements can be arbitrated. Second, such intuitive methods are difficult to pass on to later generations. The only way to do so is the apprentice system. Extensive personal contact is necessary, and frequently even it is not enough. Disciples frequently decide that their master was all wrong and proceed to rework the classification. Finally, the most destructive feature of the intuitive species concept is that just when systematists get really good at what they are doing, they have the unfortunate habit of dying and with them go all their intuitive abilities. One reason why the biological species concept looked so attractive initially was that it provided a criterion for distinguishing between species – gene flow and its interruption.

Later workers take for granted the increased objectivity afforded by the biological species concept and proceed to raise their standards. They want concepts that are even more operational. In fact, a major motivation for nearly all later species concepts is the desire to make them more operational. The question then becomes how operational must a concept be in order to be scientifically acceptable? Certain species concepts are very theoretical; others are as operational as possible. But even those systematists who opt for theoretical species concepts acknowledge the need to provide operational criteria to facilitate their application. A second issue is whether or not these operational criteria must be included in the definition or whether they can just be appended to it.

Neither the philosophical nor the scientific literature provides a clear answer to the first question. Philosophers have argued conclusively that no theoretically significant term can be completely defined by means of the techniques used to apply it. Conversely, terms which lack any means of application are of doubtful use in science. As positivists used to put the point, such terms are metaphysical. But how about all the terms that fall somewhere in between these two extremes? Unfortunately, philosophers have not addressed this question in any detail, in large measure because the process of operationalizing concepts is highly particularistic. One can list example after example, but no general pattern emerges. The process of operationalizing concepts seems to be just one damned thing after another. Nor do scientists say much about this process. They just ***do*** it. For any student of science who is getting a bit bored with the Covering Law Model of Scientific Explanation or incommensurability, the topic of operationalizing concepts is wide open.

Answers to the second question seem more clear-cut. Scientists in general see no real need to include their criteria of application in their theoretical definitions. For example, physicists provide definitions of such theoretical terms as mass, force and length, but they do not include in these definitions all the different ways that they have developed to operationalize these concepts. Numerous ways exist for determining length. Including all these methods in the definition of length would be unduly cumbersome. Similarly, Mayr refers to isolation in his definition, but he does not include in his definition all possible isolating mechanisms or ways of determining their presence or influence. The same can be said for the evolutionary concepts of Simpson and Wiley. The appropriate criticism (if valid) is that they do not provide sufficiently detailed descriptions of how evolutionary roles and tendencies or evolutionary tendencies and historical fate are to be operationalized.

One virtue of Templeton's cohesion species concept is the lengths to which he goes to spell out all of the processes that can contribute to cohesiveness. Cracraft also refers to cohesion but does not state in his diagnostic species concept what he means by this term. He expands on this notion in the body of his text. Similarly, monophyly and autapomorphy are referred to in the monophyletic species concept but neither are defined or operationalized in that definition. One major feature of definitions is that they are relatively short. Putting too much terminological explanation or empirical data within the body of the definition would be counter-productive.

Certainly one of the major motivations for the phenetic species concept was to make it as operational as possible. All the taxa in phenetic taxonomy are to be operational. That is why Sokal and Sneath termed them Operational Taxonomic Units. In phenetic taxonomy, all elements used must be spelled out explicitly: the clustering technique employed, the list of characters used, and the level of similarity to be considered specific. As little as possible is left to personal judgement. However, pheneticists thought that this increased objectivity had to be purchased at the expense of theoretical significance. No general knowledge of the empirical world can be used in producing phenetic classifications. The fate of operational taxonomic units among systematists gives some indirect evidence for the value of such notions. Although quite a few systematists voice approval of making systematic method as operational as possible, they do not seem very enthusiastic about using the results of this process.

The authors of both the monophyletic and diagnostic species concepts prefer to call them phylogenetic, signalling their connection to Hennigian systematics. One would think, given the central role of autapomorphies in phylogenetic systematics, that no cladist would have any reservations about using autapomorphies to define species, but such pattern cladists as Nixon and Wheeler do. The reason revolves once again around epistemological issues. Nixon and Wheeler (1990: 217) argue that species concepts based on topological (cladogram) knowledge are problematic, because they cannot be implemented prior to cladistic analysis. The diagnostic species definition is precisely what is needed for cladistic analysis because diagnostic species can be recognized prior to a cladistic analysis, so that the analysis can be undertaken.

Previously, I mentioned the difference between phylogenetic and pattern cladists. The chief difference between the two groups is that pattern cladists want their conclusions to be more certain and their methods and definitions to be more operational than do the phylogenetic cladists. Thus, pattern cladists have been compared with the pheneticists in wanting classificatory process to be theory-free. Systematists must start with observations. This position is somewhat surprising given all the attention that pattern cladists have paid to Karl Popper, the author of the view that no terms in science, even the most observational, can be totally theory-free. Nor can scientists begin their investigations with observations and nothing but observations. In any case, the diagnostic species concept is more operational than the monophyletic species concept because its diagnostic characters need not be autapomorphies.

18.5 SUMMARY AND CONCLUSION

Several criteria are used commonly to evaluate scientific concepts. Others apply more narrowly to species concepts. How do the seven species concepts that I have discussed in this paper score on these criteria? As far as universality is concerned, the phenetic species concept is the most general concept because it does not purport to reflect very much about the natural world. Males and females need not be included with each other or with their progeny in the same lowest level operational taxonomic unit. At the other extreme, the biological and mate recognition concepts are the least general because they apply only to species that reproduce sexually. The other four species concepts are arrayed somewhere in between these two. The diagnostic species concept is closest to the phenetic species concept except that the phenetic species does require reproductive cohesion while the diagnostic species does not. The other three have more empirical content but not so much that any organisms are excluded.

Connected to universality is the distinction between monism and pluralism. Monism in the strong sense is the view that a particular concept is the correct concept. The supporters of all the concepts that have been discussed in this chapter are monists in this sense. They all insist that their concept is the correct concept. Monism and pluralism also come in weaker forms. A definition that includes alternative criteria or alternative ways of meeting a single criterion are pluralist in a weak sense. The biological, evolutionary, cohesion and monophyletic species concepts are pluralist in the sense that more than one force or mechanism can bring about the state specified in their respective definitions. For example, Templeton lists several forces that can increase cohesion. The monophyletic species concept is monistic with respect to its grouping criterion (monophyly) but pluralistic when it comes to its ranking criteria (not all minimally monophyletic taxa need to be ranked as species). The most monistic species concepts in this sense are the phenetic, diagnostic and mate recognition concepts.

Pluralism, in the sense of alternative forces and mechanisms, makes a concept look more universal than it actually is. For example, one might argue that all species are cohesive, and the only mechanism for cohesion is gene exchange. If this concept were actually applicable to all organisms, it would be universal in a very strong sense. However, one might argue that all species are cohesive and then go on to specify alternative mechanisms that promote cohesion. Instead of all species being X, all species are X or Y or Z. This concept is universal in the sense that it applies to all organisms but only by listing a series of alternative criteria. This is universality on the cheap.

Universality is generally considered a positive feature of a concept by both philosophers and scientists. As I mentioned earlier, some difference of opinion exists among philosophers about the virtues of monism and pluralism. However, among those systematists whose species concepts we have examined, monism is clearly valued quite highly. They push their preferred species concept for all its worth. Pluralism begins to look attractive only when it appears as if one's own preferred concept is losing in the battle over priority.

A special feature of species concepts is their dimensionality with respect to time. The evolutionary species concept is explicitly temporal. Species are lineages extended in time (space–time worms), while the phenetic species concept is as non-dimensional as you can get. Advocates of the other species concepts do not always explicitly state that their species concepts are designed to individuate time-slices of evolving lineages, but enough of them do to conclude that this is their intention. The pheneticists would certainly conclude that inclusion of temporal considerations in a species concept is a mistake and detracts from the value of this concept. However, advocates of other species concepts would disagree. Dimensionality is a virtue.

Applicability is a universally recognized virtue among philosophers and scientists alike, although scientists tend to have higher standards of applicability than do philosophers. Instead of the species concept discussed in this chapter falling into two or three groups, they are arrayed linearly with respect to how applicable (or operational) they happen to be. The phenetic species concept is clearly the most operational, since this after all was the main motivation for its formulation. The diagnostic species concept is the next most applicable concept. A systematist must know who tends to mate with whom, and what the results of these unions turn out to be, but little else save character covariation. The monophyletic species concept is as operational as the methods of cladistic analysis allow. The goal of cladistic analysis is the individuation of characters so that they nest perfectly. The literature on the difficulties involved in such an enterprise is huge. The mate recognition, biological, and cohesion concepts are even more difficult to apply because the forces and mechanisms that they specify are more difficult to discern. Finally, the evolutionary species concept is the most difficult to apply because it explicitly specifies that species are extended through time. At any one time it is impossible to say for sure that a particular group of organisms is a species, because species status can be determined only in retrospect, a characteristic of all historical entities.

With respect to theoretical significance, only the phenetic species concept is designed to be theory-neutral or theory-free. The diagnostic species concept assumes only some very low-level, unproblematic theories, while all the others are openly theoretical in their content. Whether or not one chooses to accept one or more of these theories turns on some very abstruse empirical issues, such as how effective gene flow is in producing cohesion. Once again, a difference of opinion exists among scientists, not to mention philosophers, on the virtues of theory-dependence. Most philosophers nowadays accept some version of theory-dependence. Theory neutral concepts are impossible, but even if they were possible, they would be undesirable. The pheneticists and pattern cladists have made ambiguous claims about the role of process theories in systematics. On the surface at least, they seem strongly opposed to anything that might be termed a theory to enter into the classificatory process, at least in the early stages. Advocates of the other species concepts discussed in this paper are more accepting of theories. They think that theories, especially process theories, are absolutely central to the science of systematics from beginning to end.

If the readers of this chapter so choose, they can make up a chart comparing the seven species concepts discussed on the criteria listed. One problem is that far from total agreement exists on which characteristics of species definitions are considered virtues and which vices. Scientists at least value universal, monistic, applicable concepts. The issues of temporal dimensionality, operationism, and theoretical significance are more controversial. However, I have made up charts on each of the possible permutations, and the depressing result is that all seven species concepts score about the same. This outcome may well explain why the species controversy continues unabated. Even if everyone agreed on what counts as a good species concept, no one concept is clearly superior to all the others – and not everyone agrees on what general criteria characterize a good species concept.

Acknowledgements

I would like to thank Michael Donoghue, Marc Ereshefsky, Chris Horvath, Rick Mayden, Brent Mishler and Ed Wiley for reading and commenting on early drafts of this paper.

18.6 REFERENCES

Budd, A.F. and Mishler, B.D. (1990) Species and evolution in clonal organisms – summary and discussion. *Systematic Botany*, **15**, 166–71.

Claridge, M.F., Dawah, H.A. and Wilson, M.R. (1997) Practical approaches to species concepts, in *Species: the Units of Biodiversity* (eds M.F. Claridge, H.A. Dawah and M.R. Wilson), Chapman & Hall, London, pp. 1–15.

Cracraft, J. (1983) Species concepts and speciation analysis, in *Current Ornithology*, vol. 1, Plenum Press, New York, pp. 159–87.

Cracraft, J. (1989) Speciation and its ontology: the empirical consequences of alternative species concepts for understanding patterns and processes of differentiation, in *Speciation and Its Consequences* (eds D. Otte and J.A. Endler), Sinauer Associates, Sunderland, Massachusetts, pp. 28–59.

Cracraft, J. (1997) Species concepts in systematics and conservation biology: an ornithological viewpoint, in *Species: the Units of Biodiversity* (eds M.F. Claridge, H.A. Dawah and M.R. Wilson), Chapman & Hall, London, pp. 325–39.

de Queiroz, K. (1990) Phylogenetic systematics and species revisited. *Cladistics*, **6**, 83–90.

de Queiroz, K. and Donoghue, M.J. (1988) Phylogenetic systematics and the species problem. *Cladistics*, **4**, 317–38.

de Queiroz, K. and Good, D.A. (1997) Phenetic clustering in biology: a critique, *Quarterly Review of Biology* (in press).

Donoghue, M.J. (1985) A critique of the BSC and recommendations for a phylogenetic alternative. *The Bryologist*, **83**, 172–81.

Dupré, J. (1993) *The Disorder of Things: Metaphysical Foundations of the Disunity of Science*, Harvard University Press, Cambridge, Massachusetts.

Echelle, A.A. (1990) In defense of the phylogenetic species concept and the ontological status of hybridogenetic taxa. *Herpetologica*, **46**, 109–17.

Ehrlich, P.R. and Ehrlich, A.H. (1967) The phenetic relationships of the butterflies. *Systematic Zoology*, **16**, 301–27.

Eisner, T., Lubchenco, J., Wilson, E.O., Wilcove, D.S. and Bean, M.J. (1995). Building a scientifically sound policy for protecting endangered species. *Science*, **268**, 1231–2.

Eldredge, N. and Cracraft, J. (1980) *Phylogenetic Patterns and the Evolutionary Process*, Columbia University Press, New York.

Endler, J.A. (1989) Conceptual and other problems in speciation, in *Speciation and Its Consequences* (eds D. Otte and A. Endler), Sinauer Associates, Sunderland, Massachusetts, pp. 625–48.

Ereshefsky, M. (1989) Where's the species? Comments on the phylogenetic species concept. *Biology and Philosophy*, **4**, 89–96.

Ereshefsky, M. (ed.) (1992) *The Units of Evolution: Essays on the Nature of Species*, MIT Press, New York.

Frost, D.R. and Kluge, A.G. (1994) A consideration of epistemology in systematic biology, with special reference to species. *Cladistics*, **10**, 259–94.

Ghiselin, M.T. (1969) The principles and concepts of systematic biology, in *Systematic Biology* (ed. C. G. Sibley), National Academy of Sciences, Washington, DC, pp. 45–55.

Ghiselin, M.T. (1974) A radical solution to the species problem. *Systematic Zoology*, **23**, 536–44.

Gornall, R.J. (1997) Practical aspects of the species concept in plants, in *Species: the Units of Biodiversity* (eds M.F. Claridge, H.A. Dawah and M.R. Wilson), Chapman & Hall, London, pp. 170–90.

Graybeal, A. (1995) Naming species. *Systematic Biology*, **44**, 237–50.

Hawkes, J. (1997) Cultivated plant diversity and taxonomy, in *Species; the Units of Biodiversity* (eds M.F. Claridge, H.A. Dawah and M.R. Wilson), Chapman & Hall, London, pp. 191–8.

Hennig, W. (1966) *Phylogenetic Systematics*, University of Illinois Press, Chicago.

Horvath, C. Phylogenetic species concepts: pluralism, monism, and history. *Biology and Philosophy* (in press).

Horvath, C. Some questions about identifying individuals: failed intuitions about organisms and species. *Philosophy of Science* (in press).

Hull, D.L. (1965) The effect of essentialism on taxonomy. *British Journal for the Philosophy of Science*, **15**, 314–326; **16**, 1–18.

Hull, D.L. (1968) The operational imperative – sense and nonsense in operationism. *Systematic Zoology*, **17**, 438–57.

Hull, D. (1976) Are species really individuals? *Systematic Zoology*, 25, 174–91.

Hull, D. (1979) The limits of cladism. *Systematic Zoology*, **28**, 416–40.

John, D.M. and Maggs, C.A. (1997) Species problems in Eukaryotic algae: a modern perspective, in *Species: the Units of Biodiversity* (eds M.F. Claridge, H.A. Dawah and M.R. Wilson), Chapman & Hall, London, pp. 83–107.

Johnson, L.A.S. (1968) Rainbow's end: the quest for optimal taxonomy, in *Proceedings of the Linnean Society of New South Wales*, **93**,1–45; reprinted 1970, *Systematic Zoology*, **19**, 203–39.

Kitcher, P. (1984) Against the monism of the moment: a reply to Elliott Sober. *Philosophy of Science*, **51**, 616–30.

Kitcher, P. (1989) Some puzzles about species, in *What Philosophy of Biology Is: Essays Dedicated to David Hull* (ed. M. Ruse), Kluwer, Dordrecht, pp. 183–208.

Kornet, D. (1993) *Reconstructing Species: Demarcations in Genealogical Networks*, Hortus Botanicus, Leiden.

Mayden, R.L. (1997) A hierarchy of species concepts: the denouement in the saga of the species problem, in *Species: the Units of Biodiversity* (eds M.F. Claridge, H.A. Dawah and M.R. Wilson), Chapman & Hall, London, pp. 381–424.

Mayr, E. (1969) *Principles of Systematic Zoology*, McGraw-Hill, New York.

McDade, L. (1990) Hybrids and phylogenetic systematics I. Patterns of character expression in hybrids and their implications for cladistic analysis. *Evolution*, **44**, 1685–700.

McDade, L. (1992) Hybrids and phylogenetic systematics II. The impact of hybrids on cladistic analysis, *Evolution,* **46,** 1329–346.

McKitrick, M.C. and Zink, R.M. (1988) Species concepts in ornithology. *Condor,* **90,** 1–14.

Mishler, B.D and Brandon, R.N. (1987) Individualism, pluralism, and the phylogenetic species concept. *Biology and Philosophy,* **2,** 397–414.

Mishler, B. and Donoghue, M.J. (1982) Species concepts: a case for Pluralism. *Systematic Zoology,* **31,** 491–503.

Mishler, B. and Theriot, E. (1997) Monophyly, apomorphy, and phylogenetic species concepts, in *Species Concepts and Phylogenetic Theory: A Debate* (eds Q.D. Wheeler and R. Meier), Columbia University Press, New York, (in press).

Nelson, G. and Platnick, N. (1981) *Systematics and Biogeography: Cladistics and Vicariance,* Columbia University Press, New York.

Nixon, K.C. and Wheeler, Q.D. (1990) An amplification of the phylogenetic species concept. *Cladistics,* **6,** 211–23.

O'Hara, R.J. (1993) Systematic generalizations, historical fate, and the species problem. *Systematic Biology,* **42,** 231–46.

Otte, D. and Endler, J.A. (eds) (1989) *Speciation and Its Consequences,* Sinauer Associates, Sunderland, Massachusetts.

Paterson, H.E.H. (1981) The continuing search for the unknown and unknowable: a critique of contemporary ideas on speciation. *South African Journal of Science,* **77,** 113–19.

Pickering, A. (1984) *Constructing Quarks,* University of Chicago Press, Chicago.

Platnick, N. (1977) Review of Concepts of Species (1976) (ed. C.N. Slobodchikoff). *Systematic Zoology,* **2,** 96–8.

Platnick, N. (1979) Philosophy and the transformation of cladistics. *Systematic Zoology,* **28,** 537–46.

Ritland, D.B. and Brower, L.P. (1991) The viceroy butterfly is not a batesian mimic. *Science,* **350,** 497–8.

Rosen, D. (1978) Vicariant patterns and historical explanation in biogeography. *Systematic Zoology,* **27,** 159–88.

Rosen, D. (1979) Fishes from the uplands and intermontane basins of Guatemala. *Bulletin of the American Museum of Natural History,* **162,** 267–376.

Signor, P.W. III (1985) Real and apparent trends in species richness through time, in *Phanerozoic Diversity Patterns: Profiles in Macroevolution* (ed. J. Valentine), Princeton University Press, Princeton, pp. 129–50.

Simpson, G.G. (1961) *Principles of Animal Taxonomy,* Columbia University Press, New York.

Slobodchikoff, C.N. (ed.) (1976) *Concepts of Species,* Hutchinson & Ross, Stroudsburg, Penn., Dowden.

Sneath, P.H.A. (1995) Thirty Years of Numerical Taxonomy. *Systematic Biology,* **44,** 281–98.

Sober, E. (1984) Discussion: Sets, Species, and Evolution: comments on Philip Kitcher's Species. *Philosophy of Science,* **51,** 334–41.

Sokal, R.R., and Sneath, P.H.A. (1963) *The Principles of Numerical Taxonomy,* W.H. Freeman, San Francisco.

Sterelny, K. (1994) The nature of species; review of Ereshefsky (1992). *Philosophical Books,* **35,** 9–20.

Templeton, A.R. (1989) The meaning of species and speciation, in *Speciation and Its Consequences* (eds D. Otte and J.A. Endler), Sinauer Associates, Sunderland, Massachusetts, pp. 3–27.

Van Regenmortel, M.H. (1997) Viral species, in *Species: the Units of Biodiversity* (eds M.F. Claridge, H.A. Dawah and M.R. Wilson), Chapman & Hall, London, pp. 17–24.

Veron, J.E.N. (1995) *Corals in Space and Time*, Comstock, Cornell University Press.

Vrana, P. and Wheeler, W. (1992) Individual organisms, as terminal entities: laying the species problem to rest. *Cladistics*, **8**, 67–72.

Wheeler, Q.D. and Nixon, K.C. (1990) Another way of looking at the species problem: a reply to de Queiroz and Donoghue. *Cladistics*, **6**, 7–81.

Wiley, E.O. (1981) *Phylogenetics: The Theory and Practice of Phylogenetic Systematics*, John Wiley and Sons, New York.

19

A hierarchy of species concepts: the denouement in the saga of the species problem

R. L. Mayden

Contacting address: Department of Biological Sciences, P.O. Box 0344, University of Alabama, Tuscaloosa, AL 35487, USA

ABSTRACT

At least 22 concepts of species are in use today and many of these are notably incompatible in their accounts of biological diversity. Much of the traditional turmoil embodied in the species problem ultimately derives from the packaging of inappropriate criteria for species into a single concept. This results from a traditional conflation of function of concepts with their applications, definitions with concepts, taxonomic categories with groups, and the ontological status of real species with teleological approaches to recover them. Analogous to classifications of supraspecific taxa, our forging inappropriate and ambiguous information relating to theoretical and operational discussions of species ultimately results in a trade-off between convenience, accuracy, precision, and the successful recovery of natural biological diversity. Hence, none of these expectations or intentions of species or classifications is attainable through composite, and possibly discordant, concepts of biological diversity or its descent.

Reviewing and evaluating the concepts of species for their theoretical and operational qualities illustrates that a monistic, primary concept of species, applicable to the various entities believed to be species, is essential. This evaluation reveals only one theoretical concept as appropriate for species, the Evolutionary Species Concept. This conceptualization functions as a primary concept and is essential in structuring our ideas and perceptions of real species in the natural world. The remaining concepts are secondary, forming a hierarchy of definitional guidelines subordinate to the primary concept, and are essential to the study of species in practice. Secondary concepts

Species: The Units of Biodiversity. Edited by M.F. Claridge, H.A. Dawah and M.R. Wilson. Published in 1997 by Chapman & Hall. ISBN 0 412 63120 2

should be used as operational tools, where appropriate, across the variance in natural diversity to discover entities in accord with the primary concept. Without this theoretical and empirical structuring of concepts of species our mission to achieve reconciliation and understanding of pattern and process of the natural world will fail.

19.1 INTRODUCTION

> 'I believe that the analysis of the species problem would be considerably advanced, if we could penetrate through such empirical terms as phenotypic, morphological, genetic, phylogenetic, or biological, to the underlying philosophical concepts. A deep, and perhaps widening gulf has existed in recent decades between philosophy and empirical biology. It seems that the species problem is a topic where productive collaboration between the two fields is possible'.
>
> (Mayr, 1957)

Little has changed with regard to the species problem since Mayr composed this piece. Some researchers argue for a particular concept of diversity known as species, while others prefer a pluralistic approach (Mishler and Donoghue, 1982). Today, the controversy continues over the conceptualization of species. This volume reflects some of this diversity of thought across multiple taxonomic groups. This seemingly timeless debate has generated a heterogeneous proliferation of concepts, most hoping to capture the operational and/or theoretical qualities of a good concept. The search has been for a concept–definition that is biologically relevant and meaningful, one that is easily applied, and one that encompasses natural biodiversity. That is, a concept of real species assisting in and ensuring their recognition and our understanding of them in nature. This goal has not been achieved for several reasons.

The 20th century history of biological classification illustrates why this so-called silver bullet species concept, one that will attend to all our perceived needs, has not yet been achieved. In phylogenetic systematics (or cladistics) the Linnaean classification scheme represents a hierarchical system of categories coordinate with a phylogenetic tree of named taxa. Represented in the classification is the idea of monophyly of taxa, or sister group (genealogical) relationship. Classifications are information retrieval systems about genealogical relationships. In evolutionary systematics the classification is purported to represent sister group relationship and evolutionary distinctiveness. Paradoxically, while this may be viewed as an expedient method to group information in a retrieval system, under this method one can never be sure which criteria are optimized at any part of a classification. Thus, confusion is inherent in an ambiguous information retrieval system. The ultimate trade-off of combining too many desired functions into a convenient method is that it is not always possible to isolate any one function (e.g. genealogy versus distinctiveness).

Much of the turmoil embodied in the species problem ultimately derives from our packaging inappropriate criteria for species into a single concept. This results from a traditional conflation of function of concepts with their applications, definitions with concepts, taxonomic categories with groups, and ontological status of real species with teleological approaches to recover them. Analogous to classifications of supraspecific taxa, our forging inappropriate and ambiguous information relating to theoretical and operational discussions of species ultimately results in a trade-off between convenience, accuracy, precision, and the successful recovery of natural biological diversity. None of these expectations or intentions of species or classifications is attainable through composite, and possibly discordant, concepts of biological diversity or its descent.

With this in mind can one tease apart the theoretical concepts and operational definitions of species and develop a primary concept applicable to the various entities believed to be species? I think this is possible through a hierarchical view of species concepts and their definitions. Below, I review the various species concepts and propose a hierarchical classification for them. Each of these concepts is briefly evaluated relative to their consequential qualities thought to be important in a concept (Hull, 1997: Chapter 18). This evaluation reveals only one appropriate primary and theoretical concept of species. The remaining definitions are secondary concepts, forming a hierarchy of definitional guidelines subordinate to this primary concept. The secondary concepts are engaged only as operational tools, where appropriate, across the variance in natural diversity to discover entities in accord with the primary concept.

19.2 METHODOLOGY

Probably more is written about species than any other topic in evolutionary biology. There are many opinions and studies addressing this question. Hence, an exhaustive survey of these is impossible. Concepts are ideas or intuitions uniquely developed in the minds of every person. Definitions of these concepts are the only form with which one can compare them. Sometimes, these definitions may be poorly developed or misinterpreted, ultimately leading to miscommunication of ideas. Regardless, I have endeavoured to understand the arguments on the various species concepts (Table 19.1), and compare and evaluate them. In section 19.7 I have also made an effort to identify synonyms of concepts; these are listed by assigned standard abbreviations or full titles. Where concepts were formerly identified as synonymous, credit is provided; in part refers to the observation that portions of concepts are equivalent.

Table 19.1 Species concepts and standardized abbreviations

1.	Agamospecies (ASC)	14.	Morphological (MSC)
2.	Biological (BSC)	15.	Non-dimensional (NDSC)
3.	Cohesion (CSC)	16.	Phenetic (PhSC)
4.	Cladistic (ClSC)	17.	Phylogenetic (PSC)
5.	Composite (CpSC)		1. Diagnosable Version (PSC_1)
6.	Ecological (EcSC)		2. Monophyly Version (PSC_2)
7.	Evolutionary Significant Unit (ESU)		3. Diagnosable and Monophyly Version (PSC_3)
8.	Evolutionary (ESC)		
9.	Genealogical Concordance (GCC)	18.	Polythetic (PtSC)
10.	Genetic (GSC)	19.	Recognition (RSC)
11.	Genotypic Cluster Definition (GCD)	20.	Reproductive Competition (RCC)
12.	Hennigian (HSC)	21.	Successional (SSC)
13.	Internodal (ISC)	22.	Taxonomic (TSC)

19.3 IMPORTANT QUALITIES TO CONSIDER

Few concepts can be viewed as more fundamental to the natural sciences than that of the species. Species as individuals (Hull, 1976) represent a unique level of organization of the natural world; they are self-organizing entities or particulars. This level of universality is the upper-most limit to involve tokogenetic relationships and the lower-most level participating in phylogenetic relationships. They are purported to be the highest level of integration to participate in natural processes while being spatiotemporally constrained. Thus, they are essentially fundamental units of evolution. In plain English, this means that for all disciplines using any species in pure or applied research, education, management, conservation, etc., success at accomplishing identified missions or deriving informative answers to particular questions is inextricably tied to a basic assumption that the species involved are **real** by-products of natural processes and not misguided fabrications of our own invention.

One method to compare and contrast the various concepts of species is through three criteria that have traditionally been employed to evaluate scientific concepts. Hull (1997: Chapter 18) provides such a review for seven of the most frequently used concepts for their theoretical significance, generality (or universality), and applicability (or operationality). Following a review of definitions, metaphysical topics, and species concepts, I extend a similar evaluation to all 22 of the various concepts of species. The relationships of the concepts within and among the three criteria, together with the question of monism versus pluralism, reveals a hierarchy of species concepts that should finally put the species problem to rest.

19.4 CONCEPTS, DEFINITIONS, GROUPS, CATEGORIES AND NAMES: THE UNFORTUNATE CONFLATION OF TERMINOLOGY

Important in the discussion and resolution of the species problem is the correct usage and understanding of ideas and terms related to the species issue. Historically, discussions of species have involved the use of four critical terms: concept, definition, group and category. These terms have been central to the fundamental melange that both scientists and philosophers have encountered with the species problem.

Communication of ideas or concepts in science is of utmost importance and hinges upon statements or definitions developed by persons formulating or discussing the concepts. Critically important to exact communication of ideas embodied in concepts demands that we do not obfuscate the terms or words used in definitions or statements of the concepts. Here, we confront difficulties both in the logical treatment and the evaluation of definitions.

Concepts of biological systems serve as fundamental links between pattern and process in nature, are employed in every discipline, and help guide our perception of natural systems. They are formulated by individual persons through observation, study and synthesis (impressions and imagination) of both theory and empirical data. Concepts may be real or abstract. Real concepts are those representing easily agreed discrete objects. Abstract concepts are those representing hypothetical and transient phenomena. A concept may be relayed from one person to one or more other persons by adapting it into a statement or definition, either verbally, in writing, or graphically. Such a definition may or may not induce the same concept in the mind of the other persons, depending upon the appropriateness, precision, and accuracy of the words used in the definition and the level of understanding of the other person. With real concepts (e.g. round versus square) one may compare statements developed by different observers with discrete objects to see if they agree. With abstract concepts (evolution, natural selection, species as taxa) it is difficult to know for sure if statements represent the same transient or hypothetical things, and the respective definitions can only be compared using previously agreed definitions of words used in the statement. One may also observe the effectiveness of such a concept through direct examination.

The term 'group' refers to a collection of objects or things. In the intersection of natural sciences, taxonomy, and systematics the term taxon is often used synonymously with the term group. A group can be real and have objective reality if it corresponds to qualities that are real and exclusive to it, and if it consists of things that have material existence. They may be arranged hierarchically, either as non-reticulate or reticulate groups. They may be represented at various levels of universality from groups of things to more inclusive groups of things, etc. They may be of any size and arise on the basis of intrinsic attributes and/or extrinsic decisions. Organisms can be members of any number of groups so long as they possess the attributes of the said groups. Groups, however, are not like concepts. Groups develop from sense impressions of concepts and can be agreed upon and definite if the statements about them are unambiguous and decisive.

Groups and categories are distinctly different and there is no real connection between them. The tradition in codes of nomenclature artificially forces the use of taxonomic categories in a hierarchy for groups. A biological classification is a contrived system of categories used for the storage and retrieval of information about biological diversity, taxa, or groups. The concept 'category' is a class and has no separate existence from its use in organizing objects or thoughts; categories have no reality. Unlike groups, categories have no attributes; things or objects are not members of categories, but are parts of groups; and organisms are not members of any taxonomic category. For example, Cyprinidae is a proper name given to a group of fishes possessing certain attributes. By taxonomic convention the -idae ending denotes a traditional level of universality in the zoological hierarchy. The group Cyprinidae can be a part of many other groups (Cypriniformes, Ostariophysi, Teleostei), but is only a member of one taxonomic category, Family. Because categories have no reality the Family Cyprinidae is not a member of any other more-inclusive categories, but Cyprinidae is.

Multiple classifications may exist for the same group of organisms, depending upon criteria being optimized in the classification. The Linnaean hierarchy imparts information regarding relationship, descent from hypothesized immediate common ancestors. Supraspecific categories are more inclusive than the species category. Groups are assigned to categories; assignment is based on the definition of the category. Historically, categories of this hierarchy were defined on the basis of distinctiveness; that is, distinctiveness of the group assigned to Genus was less than distinctiveness at the Family, and so on.

In the modern-day hierarchical system a distinction between the supraspecific categories and the species category is a dichotomy between phylogenetic and tokogenetic processes and relationships (Hennig, 1966; Frost and Kluge, 1994; Wiley and Mayden, 1997). Supraspecific categories are defined only as monophyletic groups having phylogenetic relationships and historical cohesion. The categorical rank assigned to a group satisfying this criterion is only a by-product of its level of inclusiveness. Assignment of a group to a supraspecific category is definite because of its historical existence, and our discovery and recognition of this previous existence requires demonstration of monophyly. Demonstration that a group shared an immediate common ancestor is by way of synapomorphies, one or more features inherited from and evolved in the immediate common ancestor to all known descendants. Only the concepts of monophyly and historical cohesion apply to all groups assigned to supraspecific categories.

The species category shares some, but not other qualities with supraspecific categories. The groups assigned to this category are different. Like other categories, the category species is an artificial construct used for organization of information. Unlike groups assigned to supra-specific categories those assigned to the species category have tokogenetic relationships (sexuals) or are tokogenetic vectors (clones), and may or may not be definite. Because species being classified today are potential future ancestors of groups to be placed in supraspecific categories of tomorrow, their existence does not necessitate demonstration via synapomorphies. These groups (species) are the types of entities once existing in historical communities (as ancestors) that modern-day systematists endeavour to document today by way of synapomorphies. Unlike supraspecific categories, there is a real connection between the species category and the groups assigned to this category. While the species category is a class construct like supraspecific categories, it is ontologically distinct from the class construct supraspecific categories. The species category is unique and contains only those groups of things conforming to the concept of this category. Herein lies the nucleus of the species problem. What is the appropriate concept for species and the species category?

Names applied to categories should not be confused with the names applied to the groups included in the categories. For example, there is a categorical level of the hierarchy to which is assigned the arbitrary name genus. A group (or group of groups) may be referred to this level of the hierarchy and referred to as a genus (see discussion below on twin meanings of species). The group of organisms is given a proper name distinct from the category name. For example, there is a group of fishes named *Cyprinella* and this group is assigned to a level of the hierarchy, namely genus. Problems arise when one defines the named category (genus) and confuses this definition with or extends this definition to the group of particulars (*Cyprinella*) being placed into the category.

19.5 THE TWIN MEANINGS OF SPECIES

The term species has two different meanings that, when not clearly differentiated, will result in confusion and misunderstanding relevant to real species, species concepts, species category and speciation. The term species is used to represent both a taxonomic category and those naturally occurring particulars that we discover, describe, and order into our classification system. Confusion of these terms is most detrimental to elucidating and understanding the importance of species concepts because they each have a very different ontological status (Hull, 1976). While this may seem obtuse, irrelevant, or strictly metaphysical to the working biologist, naturalist, or general scientist, it is not. Much of the confusion over species concepts relates directly to the conflation of these critically different meanings.

The two species terms are aligned in two different philosophical categories, Class and Individual (not to be confused with the formal taxonomic category or single organism, respectively). The taxonomic category species is a class. A category is spatiotemporally unbounded, lacks cohesion, is not self replicating, does not participate in any natural processes, has members, and can be defined. Members of a class may be classes, or not. A class can exist anywhere in the universe, so long as there is a definition for membership. Members of a class also can exist anywhere, if the definition applies (Hull, 1976).

Those tokogenetic and cohesive entities discovered and described in nature, referred to species, that we place in the category species, are individuals (or things, particulars). Individuals are spatiotemporally bounded, have intrinsic cohesion, are self-replicating, participate in natural processes, have part–whole relationships, but cannot be defined. Individuals change over time and can only be described. Individuals exist throughout the universe. Because individuals have no definitions, they do not have members; rather, they exist as parts of wholes. Parts of individuals may be other individuals (*Homo sapiens,* organisms, organs, tissues, cells, mitochondria, etc.), also resulting from various natural processes.

Thus, the twin meanings of species refer to two, radically different and basic metaphysical categories, classes and individuals, that when confused generates elementary problems for understanding. As a class, the category species is temporally unbounded, has a definition, and only those things fitting this definition can be included. Species as taxa change with time, have no definitions, they can only be described, identified, pointed to, etc. In our discussions of 'What is a species?', we reference Linnaean category species, the class concept, and what we decide should and should not be included in this category through a definition. In discussions of species as they exist in nature we reference the individual with a unique origin and no definition. These are particulars. If an organism is found on Mars that looks, acts, and speaks like *Homo sapiens* on Earth, it is not *H. sapiens* unless it descended from *H. sapiens* on Earth.

19.6 SPECIES CONCEPTS VERSUS EMPIRICAL DATA

It is not uncommon to find in discussions of species and species concepts researchers confusing empirical data used in the operation of recognizing a species with a conceptualization or definition of species. Empirical data can include such things as anatomy, morphology, genetics (DNA, proteins), behaviour, etc., all possibly evaluated and analysed in a variety of ways and with a variety of methods. Our abilities to gather these data are artificially constrained by technological advances; that is, we can only collect data that current technology permits. These artificial constraints on our ability to perceive variation in nature should not be confused with our desires, objectives, or attempts to illuminate natural variation. For example, it is often said that a particular group of organisms represents a distinct species based on a morphological or genetic species concept. If empirical methods are confused with concepts of species the logical outcome will be confusion. That is, species that fit the definition of a morphological species concept may not fit the definition of a genetic species concept, and *vice versa*. Assuming that we are mainly interested in identifying natural diversity resulting from historical processes encoded in genomes, empirically driven concepts are untenable. Given that any research study is grossly limited in the type and amount of empirical data available from all that is technologically possible, universally applicable concepts of species should not be bound by or confused with empirical evidence.

19.7 THE SPECIES CONCEPTS

The taxonomic, systematic, and evolutionary literature reveals that at least 22 concepts have been developed to characterize diversity (Table 19.1). Developed by researchers to suit individual needs, some are operationally or empirically motivated, some are galvanized by theoretical necessity, while some are motivated by peculiarities of organisms studied. Not all concepts have been equally well characterized or explicitly defined. Some have been essentially bequeathed to academic descendants of particular fields (e.g. population genetics, taxonomy, entomology, mammalogy, etc.), together with an awareness of the requisite qualities of species and necessary operations to be employed by researchers embracing them. Only abbreviated discussions of the various concepts are presented below alphabetically. Included are synonyms, definitions, discussion, and a synopsis as to the suitability of the concept.

19.7.1 Agamospecies concept (ASC)

Synonyms
Microspecies, Paraspecies, Pseudospecies, Semispecies.

Discussion
This concept refers specifically to taxa that do not fit the biparental, sexually reproducing mode. It serves as a general umbrella concept for all taxa that are uniparental and reproduce via asexual reproduction; often these species are the result of interspecific or intergeneric hybridization. These species may produce gametes but there is often no fertilization, except via hybridization. Ghiselin (1984a: 213) refers to these species as 'heaps of leaves that have fallen off the tree that gave rise to them'. Agamospecies may be part of a species complex wherein there also exist bisexually reproducing species. In these cases the agamospecies may be facultative or obligate apomicts. Obligate apomicts are sometimes referred to as microspecies.

In reality the composite of individual organisms of the species may often be polyphyletic, resulting from multiple crosses between parental, bisexual species. These taxa are most often diagnosed by features related to either morphology or chromosomes. Often, these species have very restricted ranges. Some authors only recognize them as species if their range includes at least 20 km diameter (Weber, 1981).

Synopsis

Because of the limited application of the ASC to asexually reproducing species the ASC should serve as a primary concept.

19.7.2 Biological Species Concept (BSC)

> 'A biological species is an inclusive Mendelian population; it is integrated by the bonds of sexual reproduction and parentage'. (Dobzhansky, 1970: 354)
>
> '...groups of actually or potentially interbreeding natural populations which are reproductively isolated from other such groups'. (Mayr, 1940)
>
> 'A species is a group of interbreeding natural populations that is reproductively isolated from other such groups'. (Mayr and Ashlock, 1991)

Synonyms

GSC, Isolation Species Concept (Paterson, 1993), Second Species Concept (Mayr, 1957), Speciationist Species Concept (Blackwelder, 1967),

Discussion

This concept has been reviewed by its strongest proponent, Mayr, in several publications and by several other authors (see Mayden and Wood, 1995). As recently espoused by Mayr and Ashlock (1991: 26–27) and Mayr (1997), species consist of reproductive communities wherein there is both an ecological and genetic unit. Individuals of a species seek and recognize one another for mating and thereby maintain an intercommunicating gene pool that, 'regardless of the individuals that constitute it, interacts as a unit with other species with which it shares its environment'. For Mayr (1997) 'each biological species is an assemblage of well balanced, harmonious genotypes and... indiscriminate interbreeding of individuals, no matter how different genetically, would lead to an immediate breakdown of these harmonious genotypes. As a result, there was a high selective premium for the acquisition of mechanisms, now called isolating mechanisms, that would favour breeding with conspecific individuals and inhibit mating with non-conspecifics. This consideration provides the true meaning of species. The species is a device for the protection of harmonious, well integrated genotypes. It is this insight on which the biological species concept is based'. Central to this concept, and the sole criterion for the reality of a species, is thus the idea of reproductive isolation of species from other such species. 'A species is a protected gene pool' that is 'shielded by its own devices (isolating mechanisms) against unsettling gene flow from other gene pools' (Mayr and Ashlock, 1991). The word interbreeding in the definition above 'indicates a propensity; a spatially or chronologically isolated population, of course, is not interbreeding with other populations but may have the propensity to do so when the extrinsic isolation is terminated' (Mayr, 1997). Accordingly, speciation **is** the process of achieving reproductive isolation (Mayr, 1963: 502; 1970: 288).

The BSC specifically excludes uniparental species even though they are known to exist, and some have relegated diversity of this type to pseudospecies (Dobzhansky, 1970). The concept also is viewed as being an operational definition in that 'taxa of the species category can be delimited against each other by operationally defined criteria, for example, interbreeding versus non-interbreeding of populations' (Mayr and Ashlock, 1991: 27). This concept is relational because 'A is a species in relation to B and C because it is reproductively isolated from them'. Finally, it is a non-dimensional concept that 'has its primary significance with respect to sympatric and synchronic populations..., and these are precisely the situations where the application of the concept poses the fewest difficulties. The more distant two populations are in space and time, the more difficult it becomes to test their species status in relation to each other but the more biologically irrelevant this status becomes'.

At least ten elements of this concept are viewed by Mayden and Wood (1995) as counter-productive toward discovering and understanding biodiversity. The BSC has received substantial criticism in recent years for issues dealing with: (1) the absence of a lineage perspective; (2) its non-dimensionality; (3) erroneous operational qualities as a definition; (4) its exclusion of non-sexually reproducing organisms; (5) indiscriminate use of a reproductive isolation criterion; (6) confusion of isolating mechanisms with isolating effects; (7) implicit reliance upon group selection; (8) its relational nature; (9) its teleological overtones; and (10) its employment as a typological concept, no different from the frequently criticized morphological species concept.

Synopsis

The nature of the unfavourable attributes inherent in the BSC preclude it from being considered a primary species concept.

19.7.3 Cladistic Species Concept (ClSC)

> '...that set of organisms between two speciation events, or between one speciation event and one extinction event, or that are descended from a speciation event'. (Ridley, 1989)

Synonyms
ISC (in part; Kornet, 1993), CSC (in part; Kornet, 1993).

Discussion
Ridley (1989) proposed this minimalistic lineage concept of species wherein species are treated as individuals, not classes. As subtheories, discussion of this concept incorporates the BSC and EcSC (within cladistic framework) to provide a more complete theory for understanding species. Ridley is one of the few authors discussing species that makes a clear distinction between theoretical and practical concepts. A species is a lineage and speciation produces two or more lineages via splitting. By definition, species cannot be paraphyletic, even if individual organisms of one or more of the descendant species are genealogically more closely related to individuals of one or more other descendant species. Rather, ancestral species necessarily become extinct following a speciation event. This concept is free from operational constraints of necessary defining attributes, typical of concepts treating species as Classes.

Synopsis
In some ways, this concept of species could serve as primary concept for biological diversity. It is a lineage concept, treats species as individuals, and places no constraints on necessary attributes that a species must possess in order to be validated. In this sense it is similar to the CpSC, ESC, ISC, and some versions of the PSC. However, there are important differences that preclude all of these concepts, except the ESC, from being considered a primary theoretical concept. With respect to the ClSC, ancestral species, by definition, become extinct following a speciation event and hence cannot be considered paraphyletic with respect to the organisms of ancestral and descendant species. Descendant species, by definition, are monophyletic; ancestral species, by definition, go extinct following speciation. This concept is criticized by Wilkinson (1990) for lack of specificity with regard to speciation, an issue related to the enforced monophyly of species. Kornet and McAllister (1993) compare the ClSC and CpSC and argue that discussions concerning the monophyly of species are inappropriate, but that organisms forming species involved in a speciation event will, in all probability, be paraphyletic relative to one another. Thus, the ClSC is inappropriate as a primary concept.

19.7.4 Cohesion Species Concept (CSC)

> '...the most inclusive population of individuals having the potential for phenotypic cohesion through intrinsic cohesion mechanisms'. (Templeton, 1989: 12)
> '...the most inclusive group of organisms having the potential for genetic and/or demographic exchangeability'. (Templeton, 1989: 25)

Discussion

Templeton (1989) developed this concept following a review and short critique of the ESC, BSC and RSC. Of these concepts only the ESC did not exclude known biological diversity. The BSC and RSC were rejected because of substantial inadequacies for all living organisms. The argument emphasized was that both concepts underscore a certain level of sexual reproduction and, as such, fail in recovering naturally occurring diversity. This was considered highly significant given that a whole spectrum of known and valid diversity never, or only rarely, employs sexual reproduction or many have too high of levels of sexuality to be validated using either the RSC or BSC.

Borrowing positive aspects of all three concepts reviewed, especially the ESC, Templeton (1989) provides specific guidelines representing mechanisms of cohesion (1989: 13, table 2) to be used in understanding species. Cohesion of a species includes various aspects classified as either genetic or demographic exchangeability. As demonstrated, there is no clear break between sexual and asexual reproduction in terms of mechanisms and its ultimate outcome to a population. As such, this concept accepts all reproductive modes, and species are evaluated and validated on the basis of cohesion, not isolation.

Synopsis

Templeton (1989: 5) noted that the ESC 'is not a mechanistic definition', and favoured the CSC because it was developed with operational mechanistic qualities in mind. While this criticism is valid when one seeks an operational concept of species, because the CSC provides extensive operational details and guidelines for recognizing species it must be specifically excluded as a primary concept of species. However, the comprehensive operational nature of the CSC makes it an important practical surrogate (secondary) for a primary concept.

19.7.5 Composite Species Concept (CpSC)

> '... all organisms belonging to an originator internodon, and all organisms belonging to any of its descendant internodons, excluding further originator internodons and their descendant internodons'. (Kornet and McAllister, 1993: 78)

Synonyms

PSC (in part).

Discussion

This concept has its origin in the ISC as formalized by Kornet (1993). Fundamental problems with the ISC, recognized by Kornet and McAllister (1993), lead to the formulation of this concept. An internodon is defined by Kornet and McAllister (1993: 78) as 'a set of organisms such that, if it contains some organism *x*, it contains all organisms which have the INT relation with *x*, and no other organism'. While any internodon could technically be referred to as a species under the ISC, these authors view species as historical conglomerates formed of internodons that are permanently isolated and morphologically divergent. A species is a 'set of organisms belonging to several consecutive internodons in the phylogenetic succession, identified and grouped together by some procedure' (page: 66). Composite species begin with the evolution of an 'originator internodon' possessing 'a morphological property shown by the internodons' member organisms' (page: 67). They close with the extinction of the latest internodon that is a descendant of the originator, that is not an originator itself, and where another originator between it and its originator has not evolved. These species may endure permanent splits in the network wherein no morphological property shown by member organisms may be detectable or permanent reproductive isolation has not evolved. In other words, composite species are diagnosable or reproductively isolated entities that, using the terminology traditionally applicable to phylogenetic trees, may be either monophyletic or paraphyletic groups of internodons. When permanent rifts in networks are not accompanied with anagenesis of morphology or changes in reproductive abilities, then these rifts will not be permanent or detectable. That species are largely paraphyletic groups of internodons is viewed as an essential element of this concept because species must be mutually exclusive entities. If species, as groups of internodons, were required to be monophyletic then species would not be mutually exclusive, but nested sets. This concept also sanctions the recognition of successional species (page: 84–85) and 'superposed' species (page: 85), two or more species evolving within a single internode of a composite species and defined by at least one morphological fixation each.

Synopsis

These authors provide a lucid comparison of the appropriateness of the terms monophyly, paraphyly, and polyphyly which refer directly to groupings of things (species, internodons, or organisms), but not to species as Individuals. In a practical sense, this concept is essentially inseparable from PSC_1. Theoretically the CpSC and PSCs are very different. Kornet and McAllister (1993) recognize that under the CpSC species are like higher taxa; that is, historical entities that cannot interbreed, lacking interspecies cohesion. Treatment of species as classes, the recognition of successional or superposed species, and the intolerance for unisexuals precludes the CpSC as a primary concept. Like the PSC, it may be a useful operational surrogate assisting with discovering some species diversity.

19.7.6 Ecological Species Concept (EcSC)

'... species is a lineage (or a closely related set of lineages) which occupies an adaptive zone minimally different from that of any other lineage in its range and which evolves separately from all lineages outside its range' (Van Valen 1976: 233).

Synonyms
ESC (*sensu* Simpson; Stuessy, 1990; Minelli, 1993).

Discussion
This concept views species as ecological units forming lineages through time in a competitive environment. It is an operational definition wherein differences in ecology constitute different, independently evolving species. It is tolerant of both bisexual and unisexual species, species that evolve via hybridization, and the species that exchange genes, so long as ecological distinction is maintained in the lineage. The equivalence of the Evolutionary Species Concept (ESC) and EcSC (Stuessy, 1990; Minelli, 1993) is inaccurate. These concepts are distinct, in that the ESC does not necessitate or outline any ecological divergence between sympatric species. Only in the original ESC of Simpson (1961) was species referred to in an evolutionary and ecological context.

Synopsis
There is no doubt that the possession of divergent ecologies among sympatric lineages warrants their recognition as distinct species. While a tolerant lineage concept, as an operational concept it cannot serve as a primary concept.

19.7.7 Evolutionary Species Concept (ESC)

'... a lineage (an ancestral–descendant sequence of populations) evolving separately from others and with its own unitary evolutionary role and tendencies'. (Simpson 1961: 153)
'... a single lineage of ancestor–descendant populations which maintains its identity from other such lineages and which has its own evolutionary tendencies and historical fate'. (Wiley, 1978)
'... an entity composed of organisms which maintains its identity from other such entities through time and over space, and which has its own independent evolutionary fate and historical tendencies'. (Wiley and Mayden, 1997)

Synonyms
ESU (in part; Mayden and Wood, 1995).

Discussion
This concept was championed originally by Simpson (1951, 1961) out of a general dissatisfaction with the non-dimensionality of the BSC. Wiley (1978, 1981) developed the concept further and argued for its general application to biological systems. Unlike other definitions reviewed herein,

the ESC largely was ignored, until recently. Frost and Hillis (1990), Frost and Kluge (1994), and Wiley and Mayden (1997) reviewed or further developed the concept. These authors argue that the ESC is the only available concept with the capacity to accommodate all known types of biologically equivalent diversity. Contrary to the perception of some (Minelli, 1993: 66–9) the ESC does not consider species as Classes or focus on species as ecological entities. The ESC is not equivalent to the EcSC. While Simpson (1961) advocated a lineage concept to species and ecological and evolutionary divergence, he also condoned the delineation of artifactual successional species. Thus, the logical corollaries of Simpson's ESC and Wiley's ESC are quite different.

The ESC is not an operational concept. However, it is a lineage concept that is non-relational. Thus, the attributes and patterns of species can be correctly interpreted with respect to their unique descent. The ESC accommodates uniparentals, species formed by hybridization, and ancestral species. It does not require knowledge of, nor specific changes in, a Specific Mate Recognition System (SMRS, see RSC, section 19.7.22). There is no threshold for particular attributes needed for the existence of a species. Finally, reproductive isolation, is considered a derived attribute from the plesiomorphic status of reproductive compatibility; reproductive success is thus largely uninformative.

Synopsis
The ESC is the most theoretically significant of the species concepts; it accommodates all 'types' of species known to date and thus has the greatest applicability. As such, the ESC can serve as a primary concept.

19.7.8 Evolutionary Significant Unit (ESU)

> '... a population (or group of populations) that 1) is substantially reproductively isolated from other conspecific population units, and 2) represents an important component in the evolutionary legacy of the species'. (Waples, 1991)

Synonyms
BSC (in part), ESC (in part; Mayden and Wood, 1995).

Discussion
The reliance upon criteria such as 'substantially reproductively isolated' and 'evolutionary legacy' incorporates attributes traditionally viewed as qualities of species from other concepts. It combines the isolation or mate recognition system of the non-dimensional BSC and RSC, and invokes the evolutionary lineage perspective of the ClSC, CSC, PSC, and ESC. These components are nothing more than the 'identities' of cohesive groups of

organisms through time and over space, possessing their own independent evolutionary fate and historical tendencies advocated in the ESC. While the ESU has been proposed as a concept targeted at revealing 'distinct' populations within species (Waples, 1991, 1996), the distinction between 'distinct' populations and species as natural, evolutionary entities is not made clear.

Synopsis
This concept excludes known biodiversity, thereby unduly biasing our perception of process. Incorrect assumptions about diversity targeted for protection, brought about by misconceived formulations, only obstructs efforts to understand and preserve it. While basically a lineage concept, its emphasis on genetics and isolation preclude its use as primary concept (see Mayden and Wood, 1995).

19.7.9 Genealogical Concordance Concept (GCC)

> '... population subdivisions concordantly identified by multiple independent genetic traits should constitute the population units worthy of recognition as phylogenetic taxa'. (Avise and Ball, 1990: 52)

Synonyms
BSC (in part), ClSC (in part), PSC (in part).

Discussion
Faced with the impending abandonment of the BSC for the PSC, the GCC is asserted by Avise and Ball (1990: 46) to be from the 'better elements of the PSC and BSC'. They proposed that the general principles of the new concept 'derive most easily from the theories and observations in molecular evolution, but can also be applied to hereditary, morphological, behavioural and other phenotypic attributes traditionally studied by systematists' (page: 46). They noted three problems with the exclusive use of the PSC. These include: (1) the number of species depends upon resolving power of analytical tools available, (2) unless persistent extrinsic (geographic) or intrinsic RBs [reproductive barriers] are present different gene genealogies will usually disagree in the boundaries of 'species' under the PSC, and (3) shared ancestry in sexually reproducing organisms implies historical membership in a reproductive community.

The arguments generated by Avise and Ball (1990: 45) between the PSC and the BSC are deceptive; they build a strawman argument of the PSC and portray it inaccurately. For example, Cracraft (1983) never required monophyly of species, only that species be diagnosable. Furthermore, in no discussion by proponents of the PSC has it been restricted to uniparental species, or the possibility that eventually individual organisms will qualify as species under the PSC. Monophyly as part of the PSC was a criterion developed after Cracraft's hypothesis (de Queiroz and Donoghue, 1988; McKitrick and Zink, 1988) and well before the GCC. The GCC specifies at least two or more apomorphies of a species, while the PSC does not. The specification of at least two apomorphies is no less arbitrary than is the specification of one, three, or more. The PSC does not advocate that one can find apomorphies for almost every individual, any more than one would by employing the BSC. Thus, there is essentially no difference between the GCC and monophyly formulations of the PSC, criticized by these authors. Other problems associated with the GCC are those identified with PSC_1 or PSC_2. While the GCC is defined and titled as a genealogical characterization of species, the criteria used by Avise and Ball (1990) for species recognition actually range from monophyly to geographic concordance to genetic differences without relevance to genealogy.

Synopsis

Avise and Ball (1990) emphasize that a problem with the PSC is resolution with available tools. There is no question that this is a limitation, but this limitation extends to all operational definitions of species, including the GCC. Emphasis on **differences** relegates the GCC to a concept that ignores differences between primitive and derived attributes and uses diagnosability as an operational guideline. Genetic differences can exist with respect to plesiomorphies that provide no relevant information on genealogy, making this essentially a typological concept. The general philosophy promulgated in the GCC is largely inseparable from that of the BSC. Thus, the GCC adopts with it all of the misgivings of the BSC, making it inappropriate as a primary concept of species.

19.7.10 Genetic Species Concept (GSC)

> '... group of organisms so constituted and so situated in nature that a heredity character of any one of these organisms may be transmitted to a descendant of any other.' (Simpson, 1943)
> '... the largest and most inclusive reproductive community of sexual and cross-fertilizing individuals which share in a common gene pool'. (Dobzhansky, 1950)
> '... members of a species form a reproductive community. The individuals of a species of animals recognize each other as potential mates and seek each other for the purpose of reproduction... The species, finally, is a genetic unit consisting of a large, intercommunicating gene pool [and these] properties raise the species above the typological interpretation of a 'class of objects'. (Mayr, 1969: 23)

Synonyms
BSC (Grant, 1981), PhSC, MSC.

Discussion
This concept is similar to the morphological species concept except that the method used to delineate species is a measure of genetic differences, presumed to reflect reproductive isolation and evolutionary independence. As a phenetic concept genetic distances and similarities are used to identify different species. Genetic independence is assessed using methods varying from chromatography, to protein electrophoresis, to sequencing.

While apparently operational, one of the basic problems with the GSC is that for the vast majority of diversity there exists no genetic information. Because divergence for any particular gene may not be at a uniform rate, in all likelihood there will never exist a standard distance for species. This concept rests on the assumption that for every speciation event there will be particular changes in all genes. If the researcher examines 200 genes and they are all identical between two species they would be considered the same species. However, the next gene may show tremendous revolution between sister species as a result of the speciation event. One divergent gene out of 200 monoallelic genes will result in a trivial genetic distance. On a linear scale, such a divergence will be trivial to a species comparison where five of 20 genes are divergent. Yet, in this heuristic example both species pairs are existing as evolutionarily independent and genetically independent species.

Synopsis
The GSC is essentially a surrogate, operational concept developed out of the BSC. A particular degree of genetic divergence is assumed to warrant species recognition. However, this operational definition, lacks any guidance for researchers as to how much difference is enough? This is largely because divergence for particular genes or across multiple genes is impossible to predict either within or between taxonomic groups. Using this non-evolutionary concept researchers are also misled to believe that the lack of divergence in genes that are merely available because of technology negates the reality of divergence that may be present for any other characters. As such, the reality of species with divergent and heritable morphologies may be naively questioned if divergence at readily accessible genes or proteins is wanting. While this concept has served as a traditional method for identifying species it is fatally flawed as a primary concept. The general paucity of data, combined with the enormous genetic variability observed between sister species, the questionable validity of relying exclusively upon genetic divergence for species validation, and the deficiency of a phylogenetic perspective in interpreting variation precludes the GSC from serving as a primary concept.

19.7.11 Genotypic Cluster Definition (GCD)

> '... clusters of monotypic or polytypic biological entities, identified using morphology or genetics, forming groups of individuals that have few or no intermediates when in contact'. (Mallet, 1995)

Synonyms
ASC, BSC, GSC, HSC, MSC, NDSC, PhSC, PtSC, PSC_1, SSC (in part), TSC.

Discussion
Mallet (1995) argues that a preferred alternative to the BSC is the GCD. While not stated directly, the GCD recognizes those clusters of monotypic or polytypic biological entities, identified using morphology or genetics, forming 'groups of individuals that have few or no intermediates when in contact'. This is a non-dimensional, polythetic, and phenetic concept of diversity serving largely as a surrogate of the BSC.

Synopsis
There are several evidential, philosophical, empirical and theoretical problems associated with this definition, precluding its use as a primary concept for species. Problems associated with the BSC, GSC, HSC, MSC, NDSC, PhSC, PtSC, SSC, and TSC hold true for this species concept.

19.7.12 Hennigian Species Concept (HSC)

> '... involving tokogenetic relationships"; a (potential) reproductive community'. (Hennig, 1950: 45–46)
> '... reproductively isolated natural populations or groups of natural populations [that] originate via the dissolution of the stem species in a speciation event and cease to exist either through extinction or speciation'. (Meier and Willmann, 1997)

Synonyms
BSC.

Discussion
This concept is a derivative of Hennig's (1950) earlier notion of species. It has been further developed by Willmann (1985a,b) and Meier and Willmann (1997). Importantly, however, the version advocated by these latter authors only incorporates some of Hennig's view of species. Their concept is an operational concept, and by their own admission, is 'identical to the biological species concept if absolute [reproductive] isolation is adopted as the criterion for contemporaneous populations, and the origin of the isolation of two sister species is used to delineate species boundaries

in time'. However, they do view this concept as different from the Mayr's BSC because 'he failed to provide a criterion that specifies how and when biospecies originate and cease to exist (if not by extinction)'. Intertwined in their discussion is the species concept issue and the significance of stem (ancestral) species. Logically following from this extreme version of the isolation concept (BSC) is that unisexuals are not species but are agamotaxa (*sensu* ASC), taxa not to be considered equivalent to bisexual species.

The HSC is rejected as an appropriate characterization of entities participating in speciation for many of the same reasons the BSC is rejected. The HSC should neither be employed for systematic questions nor issues of biodiversity. For some points, however, it is apparent that Meier and Willmann are more cognisant than Mayr of the fact that a concept of species is important to people other than just a 'cataloguer and curator of collections'. Thus, the HSC is characterized to be a dimensional concept to be used for allopatric or allochronic questions, and unlike the BSC, it acknowledges the importance of comparisons between sister taxa.

Synopsis

Regardless of any positive attributes over the BSC, the HSC is viewed as inappropriate for biological systems and developed out of a limited view of natural systems. Important problematic issues of this concept include the exclusion of some biological diversity, its relational nature, its heavy reliance upon operational criteria, its artificial advocation of isolation as a non-arbitrary demarcation of species, and its artificial contrivance of stem species.

19.7.13 Internodal Species Concept (ISC)

> '... individual organisms are conspecific in virtue of their common membership of a part of the genealogical network between two permanent splitting events or between a permanent split and an extinction event'. (Kornet, 1993: 28)

Synonyms

ClSC and HSC (in part; Kornet, 1993), PSC (in part).

Discussion

Formalized in philosophical and analytical detail by Kornet (1993), this concept identifies species solely on the basis of genealogical relationship. No criteria exist for conspecificity (e.g. morphological similarity, interbreeding) other than that species are mutually exclusive groups of organisms that derive from a permanent rift in genealogical connections. Permanence refers to separation of a lineage into two or more lineages that are never reunited by any level of interbreeding. This concept also precludes the origin of taxa via hybridization because such an event would terminate the independence of the lineages. While similar in some ways to the ClSC and HSC, Kornet and McAllister (1993: 64) advocate modified versions of each but admit that the concept has 'very limited practical value'.

Synopsis

The strict reliance upon permanent splits in genealogical networks, with no possibility for future exchange, and the non-acceptance of species of hybrid origin are unrealistic restrictions for a primary concept of species. Such a concept would eliminate many taxa that either maintain their independence through various mechanisms in spite of the fact that they freely interbreed with relatives or are divergent lineages of hybrid origin. This concept also confuses the phylogenetic lineages of species with the life spans of individual organisms in tokogenetic arrays, such that the death of one family unit would constitute a permanent split in the network and hence speciation. Thus, this 'concept does not approximate at all closely to our intuitions about the life span of species' (Kornet and McAllister, 1993: 64).

19.7.14 Morphological Species Concept (MSC)

> 'Species are the smallest groups that are consistently and persistently distinct, and distinguishable by ordinary means'. (Cronquist, 1978: 15)
> 'Species may be defined as the easily recognized kinds of organisms, and in the case of macroscopic plants and animals their recognition should rest on simple gross observation such as any intelligent person can make with the aid only, let us say, of a good hand-lens'. (Shull, 1923: 221)
> 'The smallest natural populations permanently separated from each other by a distinct discontinuity in the series of biotypes'. (Du Rietz, 1930: 357)
> 'A species is a community, or a number of related communities, whose distinctive morphological characters are, in the opinion of a competent systematist, sufficiently definite to entitle it, or them, to a specific name'. (Regan, 1926: 75)

Synonyms

Classical Species Concept, Linnaean Species Concept, Morphospecies Concept, PhSC, TSC. (Sokal, 1973; Grant, 1981; Stuessy, 1990).

Discussion

This is probably considered the most sensible and commonly used method of species definition by taxonomists, general biologists, and laypersons alike. Because in the vast majority of situations involving allopatric populations little or no information is available regarding reproductive independence, morphological distinctiveness serves only as a surrogate to lineage independence. This concept also bridges a decided gap inherent in some other concepts between sexual and asexual species, so long as morphological distinctiveness is heritable and is representative of lineage independence. Given that humans are a vision-oriented species, it is readily appealing as an operational concept. Kornet (1993) considers morphology in its widest sense wherein 'similarity between organisms may thus be perceived in macromorphology as well as in gene-structure, and may range from shared "sets of independent characters" for classical taxonomists to shared "unique combinations of character states" for pattern cladists'. In this case, some may consider the MSC to be synonymous with the PSC_1.

The only real problem with a morphological concept involves instances of sibling or cryptic species, or the retention of plesiomorphic morphologies. Here, little or no morphological divergence has accompanied the acquisition of lineage independence and two or more different species may appear similar. In such cases a morphological concept of species will underestimate biological diversity. Another potential problem with this concept is the inherent tendency to require an arbitrary level of morphological divergence. By employing such a criterion the researcher assumes that all morphological traits, especially those traditionally employed in a taxon, evolve at a constant rate of divergence. This is an unjustified assumption and is falsified by the observation that even within a taxonomic group morphological divergence is largely random.

Synopsis

This is a non-dimensional concept that treats species as classes, defining them on the basis of particular essential morphological attributes. Possession of these essential attributes provides for membership in the species. As such it does not allow the researcher to treat species as historical entities forming lineages. As individuals, the definition of every species will necessarily change as the essential attributes of a species at t_1 will be different from t_2 through descent. While this concept has served as a traditional method for identifying species it is fatally flawed as a primary concept.

19.7.15 Non-dimensional Species Concept (NDSC)

Synonyms

BSC, GSC, MSC, Palaeontological Species Concept, SSC, TSC.

Discussion

Several traditional concepts of species qualify as NDSCs, the most popular being the BSC. Concepts of this type have limited spatial and no temporal dimension of species in question. Thus, there is no evolutionary, phylogenetic, or lineage perspective with which one can view, perceive, or interpret descent of the taxa or their attributes (e.g. shared plesiomorphies or apomorphies, distances), including the ability or propensity to interbreed. Concepts of this nature may appear to be more operational than those incorporating temporal and geographic components. However, this convenience compromises both the accuracy and precision with which we are able to identify, quantify, and understand biodiversity. Finally, in this lack of accuracy we also lose our abilities to discover and understand the processes responsible for the evolution, functions, and maintenance of biodiversity.

Synopsis

Thus, while the non-dimensional species concept has been argued by some as a preferred operational concept of diversity, it has actually been a hindrance to the advancement of comparative and evolutionary biology. Concepts of this type should not be considered as primary concepts for species. Interestingly, in some areas of science (medicine) the non-dimensional concept has been perceived as grossly inferior to concepts incorporating spatial and temporal dimensions in discovering diversity (Paterson, 1993).

19.7.16 Phenetic Species Concept (PhSC)

> '... the species level is that at which distinct phenetic clusters can be observed'. (Sneath, 1976: 437)

Synonyms

BSC (in part), GCC (in part), GSC, GCD, MSC, NDSC, Palaeontological Species Concept, SSC, PtSC, TSC.

Discussion

This is a non-dimensional and strictly operational concept that may be likened to any concept where overall similarity is the primary criterion for the existence of species. Operationally, where variation in a set of characters is less within a group than between groups the entity is recognized as a distinct taxon. Species are treated as Classes under this concept; they do not exist as lineages and, if a species changes through descent, then the diagnosis will have to be revised.

Synopsis

While essentially the methodology employed by taxonomists, the barren theoretical nature of this concept precludes its use as a primary concept.

19.7.17 Phylogenetic Species Concept (PSC)

Currently at least three different concepts of species are identified as phylogenetic. These definitions represent an outgrowth of phylogenetic systematics and a general need among some researchers for an operational, lineage definition of species that is process-free. Some argue that with the growing popularity of phylogenetics it is critical to have a definition to identify the smallest units suitable for analysis (boundary between toko- and phylogenetic processes). For some, species is the smallest unit appropriate for analysis, and infraspecific units are inappropriate in this context (Nixon and Wheeler, 1990; Wheeler and Nixon, 1990). This same perspective holds that species diversity must be understood before a phylogenetic analysis is performed. Others defend the position that hierarchical patterns exist within species and phylogenetic methods are appropriate (de Queiroz and Donoghue, 1988, 1990; McKitrick and Zink, 1988).

Common to PSCs is an attempt to identify the smallest biological entities that are diagnosable and/or monophyletic. Species are thus the biological entities and unit product of natural selection and descent. Consequently, subspecies, fraught with ambiguities between convenience and naturalness, is not an appropriate evolutionary unit and has no ontological status (Cracraft, 1983; McKitrick and Zink, 1988; Warren, 1992). The different PSCs form three general Classes; one emphasizing monophyly, one emphasizing diagnosability, and one emphasizing both. Many similarities exist with the ISC, ClSC, CpSC and the PSC.

19.7.18 Diagnosable Version (PSC_1)

> '... a diagnosable cluster of individuals within which there is a parental pattern of ancestry and descent, beyond which there is not, and which exhibits a pattern of phylogenetic ancestry and descent among units of like kind'. (Eldredge and Cracraft, 1980: 92)
>
> '... the smallest diagnosable cluster of individual organisms within which there is a parental pattern of ancestry and descent'. (Cracraft, 1983: 170)
>
> '... simply the smallest detected samples of self perpetuating organisms that have unique sets of characters'. (Nelson and Platnick, 1981: 12)
>
> '... the smallest aggregation of populations (sexual) or lineages (asexual) diagnosable by a unique combination of character states in comparable individuals (semaphoronts)'. (Nixon and Wheeler, 1990)
>
> '... the smallest aggregation of (sexual) populations or (asexual) lineages diagnosable by a unique combination of character states'. (Wheeler and Platnick, 1997)

Synonyms
ClSC (in part), CpSC (in part; Kornet and McAllister, 1993), GSD, ISC (in part; Nixon and Wheeler, 1990; Kornet, 1993), PhSC, PtSC, TSC.

Discussion
This Class of definitions emphasizes the *a priori* diagnosability of species, irrespective of a criterion of monophyly. There are two purported benefits of this perspective. First, process is not invoked before pattern is observed. Second, phylogenetic methodologies are argued to be applicable only to genealogical relationships of species and supraspecific taxa, not below the level of integration of species wherein tokogenetic relationships of infraspecific entities are the norm (*sensu* Wheeler and Nixon, 1990; Nixon and Wheeler, 1990). To conduct a phylogenetic analysis below the level of species would confuse the reticulate tokogenetic relationships with the usual non-reticulate phylogenetic relationships.

For proponents of this concept, monophyly, paraphyly, and polyphyly apply only at a level of organization above species. Species are delimited by the distributions of fixed, diagnostic characters across populations. Where variability exists in an attribute within the taxon this attribute is considered inappropriate for that level of analysis where only tokogenetic, not phylogenetic, relationships exist. However, the operation(s) necessary for the practical delineation of tokogenetic and phylogenetic relationships is not developed explicitly by those favouring this concept. Without knowing if you are dealing with one or more species *a priori*, one is not likely to know if phylogenetic methods are appropriate. Likewise, the difference is unclear between the theoretical inapplicability of phylogenetic methods in tokogenetic systems versus using the same methods for resolving relationships of species derived via hybrid origin. Both contain reticulate patterns of history.

19.7.19 Monophyly Version (PSC_2)

> '... a geographically constrained group of individuals with some unique apomorphous character, is the unit of evolutionary significance'. (Rosen, 1978: 176)

Synonyms
Apomorphy Species Concept (Wheeler and Platnick, 1997), ClSC (in part), ISC (in part; Kornet, 1993).

Discussion
For Rosen (1978, 1979) and de Queiroz and Donoghue (1988, 1990) species have reality if they are monophyletic and supported by autapomorphies. Any biological entity possessing a uniquely derived character, of any type,

magnitude, or quantity, qualifies as a species. Those not possessing autapomorphic attributes do not constitute a species, as traditionally viewed, but are referred to as metaspecies by some. The application of this concept necessitates a phylogenetic analysis. A lucid discussion is offered in papers by de Queiroz and Donoghue.

19.7.20 Diagnosable and Monophyly Version (PSC_3)

Synonyms
ClSC (in part), CpSC (in part; Kornet and McAllister, 1993), ISC (Nixon and Wheeler, 1990; Kornet, 1993), SSC.

Discussion
The PSC of McKitrick and Zink (1988) is a modification of the PSC provided by Cracraft (1983) but incorporates the criterion of monophyly for species. While a definition was not provided by McKitrick and Zink (1988), they identified a species as the smallest diagnosable cluster of individual organisms forming a monophyletic group within which there is a parental pattern of ancestry and descent. Because in this conceptualization all recognized monophyletic taxa are diagnosable, this definition, the methods for the discovery of species, and any associated practical and theoretical limitations are equivalent to aspects of the PSC_1 and PSC_2.

Synopsis
Several positive aspects of the phylogenetic concepts make them particularly attractive as operations in discovering biodiversity, and resolving some of the perceived problems with other concepts (Mayden and Wood, 1995). In all versions the PSC is an operational definition, whether one uses diagnosability or monophyly. The set of operations necessary to discover diversity associated with species are clearly outlined. The concepts incorporate the notion of lineage(s), making them appropriate for reconstructing descent and interpreting evolution of attributes. The ability to interbreed is viewed as a shared–primitive attribute and not of consequence in the recognition of species as taxa. These concepts also have the ability to recognize both biparental and uniparental species, and possess no implied modes of selection nor speciation. Finally, in the execution of these concepts there is no inherently arbitrary divergence or distinction between species or subspecies in a polytypic species (Cracraft, 1983; Warren, 1992); subspecies have no ontological status.

There are some problems with the use of these concepts and these are reviewed by Mayden and Wood (1995). However, while there are problems with the exclusive use of any of the Classes of the PSC, there are also important positive operational aspects to these concepts over some others. I concur with the conclusions of Warren (1992: 34) in that the PSC serves as an excellent operational surrogate to a concept of species not implicated with as many variables limiting our potential to discover biodiversity. Yet, none of the versions of the PSC should serve as a primary concept.

19.7.21 Polythetic Species Concept (PtSC)

Synonyms
BSC (in part), GCD, MSC, NDSC, PSC (in part), PtSC, SSC, TSC.

Discussion
This concept derives essentially from what philosophers call cluster concepts. That is, species are defined by the statistical covariance of characters deemed important. A given individual belongs to a particular species if it possesses enough of the important characters for the species. This statistical and practical definition treats species as classes, not individuals. Often, species are delimited by their possession of a unique combination of characters, and these are usually phenotypic. Most individuals of a species may possess attribute A, while those not possessing A will still have attributes B, C or D, all features also viewed as characteristic of the species. Treated as natural kinds, species are not viewed as lineages.

Synopsis
While this concept may serve as a very useful operational recipe for the delineation of species, especially in situations with complex patterns of variability of characters, it has no theoretical basis for being considered a primary concept. Because species are both individuals and lineages, their diagnoses will necessarily have to be modified over time as their diagnostic attributes become modified through descent.

19.7.22 Recognition Species Concept (RSC)

> 'A species is that most inclusive population of individual, biparental organisms which share a common fertilization system'. (Paterson, 1993: 105)

Synonyms
BSC (Mayr, 1988).

Discussion
This concept was introduced Paterson (collective writings in Paterson, 1993). It was developed from a dissatisfaction with the BSC, a definition considered inadequate and inaccurate of natural patterns or processes, and inhibiting progress towards related goals. For Paterson the biological limits to the field for gene recombination are determined by the mate

recognition system, more precisely, a specific mate recognition system (SMRS), a series of coadapted signals and releasing properties exchanged between partners through complementary systems. The system is functional across a broad array of conceivable signal-reception methods from elaborate behaviours, including chemicals and pheromones, to cellular recognition by gametes. This coadapted complex is maintained by strong stabilizing selection as long as the species inhabits its natural habitat; this changes when the natural habitat for the species (perhaps ancestral) is changed through geographic or temporal disjunctions. At this point the coadapted complex of signals exchanged between partners may become altered *via* directional selection in the new habitats occupied by the descendant groups of daughter populations (or species). Paterson (1993: 33) argues that 'a new SMRS, derived in this way, determines a new gene pool and, hence, a new species. According to the recognition concept, species are populations of individual organisms which share a common specific-mate recognition system. Species are, thus, incidental effects of adaptive evolution'.

The RSC does not invoke a major role for selection in the evolution of positive assortative mating, the development of isolating mechanisms, and does not require sympatry and evolutionary reinforcement to complete speciation. The fallacy that selection is responsible for producing adaptations that, by design, are responsible for the isolation of gene pools is obvious from the observation that in large part the documented cases of speciation are the direct result of total allopatry, a speciation model that does not involve secondary contact and/or reinforcement of isolating mechanisms (see Mayden and Wood, 1995). Thus, if isolating mechanisms are products of descent they are the result of chance rather than design.

The general question for the RSC is not what are the characters and mechanisms that have evolved in the recognition or reproductive systems of a species that prevents successful matings and resulting ontogenetic development between sympatric species? Rather, what are the characters and mechanisms that have evolved in species that ensure effective syngamy, development, and future generations within a population occupying its preferred or natural habitat? (Paterson, 1993: 33).

Synopsis

While there are important positive theoretical and applied aspects to this concept permitting the identification of species in a largely process-free environment, there are important problems with a universal application of the RSC. These include: (i) strict reliance upon and knowledge of the SMRS; (ii) lack of a lineage perspective; and (iii) exclusion of uniparental species and species with retained–primitive SMRSs. These are reviewed by Mayden and Wood (1985). Thus, the RSC should not be viewed as a primary concept of species.

19.7.23 Reproductive Competition Concept (RCC)

> '... the most extensive units in the natural economy such that reproductive competition occurs among their parts'. (Ghiselin, 1974: 538)

Synonyms
BSC (in part; Ridley, 1989), Hypermodern species concept (Platnick, 1976).

Discussion
This is a non-dimensional and non-operational conceptualization of species. It is essentially limited to sexually reproducing species because of its focus on the intra- and interspecies competition for mates species. In its formalization, Ghiselin (1974, 1984a) likens species and evolutionary theory to firms, corporations, small businesses, craftsmen, etc. and economic theory.

Synopsis
The restriction of this concept to sexually reproducing organisms precludes its use as a primary concept of species. Should this restriction be eliminated, this concept could serve as a primary theoretical concept. However, competition for mates in reproduction is difficult to entertain for entities generally termed uniparentals.

19.7.24 Successional Species Concept (SSC)

Synonyms
Palaeospecies concept (Simpson, 1961), ESC (in part; Simpson, 1961), Chronospecies concept (George, 1956).

Discussion
This concept was devised as a surrogate for estimating divergence through time by researchers studying fossil taxa. Often these researchers have only fragmentary data both in specimens and through time to evaluate anagenesis and divergence.

In reality, the distinctions between successional species is an arbitrary delineation in time or strata based on divergent morphologies or gaps in morphologies or time. With anagenetic change within a lineage and only remnants surviving for study there is potentially an unlimited number of chronospecies throughout the history of what was once only a single self-integrating lineage behaving evolutionarily as a single species. The SSC is an operational concept, largely of convenience, to allow researchers of fossil taxa to communicate equivalent geological strata. Species identified using this concept should not be misconstrued as being biologically equivalent to species identified using most other concepts. This is not to say that there are not valid species that have been identified using this concept.

However, in general, palaeospecies are usually temporal forms of a single species' lineage. While Simpson's (1961) ESC did extend the non-dimensional BSC through time and provide much more of a lineage perspective to species, Simpson would argue for subdividing a single lineage into multiple chronospecies. The ESC of Wiley (1981) and Wiley and Mayden (1985, 1997), however, does not advocate chronospecies.

Synopsis

Because of the arbitrary and non-evolutionary nature of this concept it should not be considered a primary concept.

19.7.25 Taxonomic Species Concept (TSC)

> '... a species consists of all the specimens which are, or would be, considered by a particular taxonomist to be members of a single kind as shown by the evidence or the assumption that they are as alike as their offspring or their hereditary relatives within a few generations. When there is no evidence of the hereditary relationship, the taxonomist will rely on distinctions that have been found to be effective in segregating species among other [groups]'. (Blackwelder, 1967: 164)

Synonyms

ASC, GCD, MSC, PhSC (Sokal, 1973; Sneath, 1976), PSC (in part), PtSC.

Discussion

As described by Blackwelder (1967), 'these are the species of the taxonomist; they are not necessarily the species of the geneticist or the evolutionist'. This concept is probably used by most practising taxonomists as a working definition to segregate individual organisms in different taxa. It relies primarily on morphological attributes in the delineation of species because many other character bases have traditionally not been readily available to taxonomists. In practice, it is non-dimensional, treats species as classes, and lacks a lineage perspective.

Synopsis

The traditional character-based limitations for those in the field of taxonomy are less real in modern science. Many different types of characters are become increasingly more available and should be used in the delineation of taxa. However, given that humans are a vision-oriented species, the more convenient morphological attributes will probably remain the most used characters in deciphering taxonomic diversity. This truism, however, need not negate the existence of taxa identified using other types of characters (ecology, proteins, behaviour, sequences, etc.).

19.8 DISCUSSION

> 'An ideal species concept should meet the various intuitions that we have about species.... It is tempting to try to define a fully satisfying species concept which meets all the intuitions mentioned by somehow combining the definitions which address the different intuitive requirements. But part of the species problem originates in the fact that any attempt to combine [different] definitions into a more embracing concept, in which their criteria are given equal weight, is doomed to fail. This is because their criteria for conspecificity are incompatible; i.e., two organisms which are conspecific on the criterion of one concept are not necessarily so on that of another'. (Kornet, 1993: 29)
>
> 'To do justice to the intuition that species are historical entities, we required a species concept which defines species as entities with continuity in time between their origin and end'. (Kornet, 1993: 32)
>
> 'The species problem has often been approached with the presupposition that a single kind of entity exists in nature that corresponds to a species concept, just because the word 'species' exists in the language of biology. If this presupposition is dropped then the traditional species problem could be answered, at least in principle, by enumerating a heterogeneous list of the general characteristics that have been thought to bestow specific status to clusters of organisms'. (Wilkinson, 1990: 445)

In this discussion the following theories are taken as having reality in the natural world:

1. The notion of descent with modification is a unifying theory of natural sciences. Descent operates from kin-groups or populations to species as groups. Descent involves differential change in attributes or qualities originating through a variety of processes over time (generations) and space (geography).
2. Speciation results in the production of new species over time and space, a direct result of (1).
3. Classes have definitions, are spatiotemporally unrestricted, lack cohesion, and do not participate in natural processes.
4. Individuals lack definitions, are spatiotemporally restricted, have cohesion, and participate in processes.

There are at least five consequential factors that have fuelled the long-standing controversy over the species problem. These include: (i) a tradition of occupation; (ii) formalized rules of nomenclature; (iii) misunderstanding of terms; (iv) a persistent desire by humans for working definitions; and (v) the unique nature of those things that we hope to understand, i.e. species as taxa (or groups). Traditionally, the job of discovering and identifying diversity was left to the occupation of the taxonomist. For many of these researchers their responsibilities were viewed as finding different species and detailing attributes important for their identification (*sensu* TSC). In many ways this mode of operation relegates species as taxa to classes with essential features. While convenient for a user hoping to distinguish between the different things, this treatment ultimately leads to great difficulties with operationality and theory when species are known to participate in processes and evolve as either ancestors or descendants.

The formalized rules of nomenclature have reinforced the view of species as classes. The recognition of species requires not only their description and the designation of a type, but also its diagnostic features. For many, a diagnosis entails a listing of defining features, a prescription easily misunderstood as equivalent to essential traits. Thus, the operational necessity and emphasis on a diagnosis may be viewed as treating species taxa as classes. This, in concert with the traditional TSC, fosters great difficulties in the reconciliation of species as individuals.

This occupational and operational legacy has resulted in the confusion of the ontological categories classes and individuals (more recently Historical Groups). While some may view this aspect of the problem as purely metaphysical and without significant bearing on the issue, such a perspective is absolutely wrong and continually generates difficulties in resolving the controversy. As discussed herein, it is not merely an argument to distinguish between species as category and species as taxa. The delineation between classes and individuals is necessary, but not sufficient to resolve the problem. Species as taxa are individuals; species as category are classes. The former have no defining properties and can only be described; the latter can be defined through a series of desired properties for its members (species assigned to categories). Both diagnoses mandated by nomenclatural formalities and most species concepts treat species as taxa as if they are classes and immutable. In reality, they function only as operational guidelines or surrogate concepts for the discovery of those individual-like things that we think to be species. This exercise is of great necessity because the individual-type things are fuzzy and can only be diagnosed retrospectively. Most species concepts are functional constructs or definitions (class) employed link to our notion or concept of the species as taxon (individual).

19.8.1 Concepts and definitions of species and supraspecific categories

It is clear that the various categories used in biological classification are class concepts. In the way that we use these categories they are intimately linked to both theoretical and operational concepts. While there exists an infinite number of ways to organize groups of groups and assign them to supraspecific categories, the definition for Hennig (1950, 1966) and most others today is any historically formed group wherein the ancestor and all its descendants are included in a phylogenetic nexus, also known as monophyly. There is nothing operational about this definition; it is strictly a notion thought to be in harmony with the theory of descent with modification and our empirical observations of the end products of this process. Yet, most agree that while non-operational it is the appropriate theoretical concept for groups assigned to supraspecific categories. It is impossible to observe monophyly either historically or in real time because we do not witness the evolution of ancestors and their descendants, and the theoretical definition of monophyly provides no operational guidelines. Without direct observation, we are incompetent in our abilities to locate monophyly, unless we retain bridging principles through some type of operational concept. The concept that we now use as a suitable surrogate to monophyly is that of synapomorphy. That is, operationally we recognize supraspecific taxa or monophyletic groups through character analysis and the discovery of shared–derived characters inferred to have evolved in a common ancestor and retained or existing as homologues in immediate descendants. There are, however, other possible concepts of monophyly (e.g. percent similarity, ability to hybridize, etc.) that could be employed. These are rejected because of known inconsistencies between alternative concepts and empirical and theoretical observations of the real world. Families are no longer defined by a prescribed level of similarity greater than that expected of orders, but less than that of genera, because similarity does not always denote close relationship. Likewise, the ability to reproduce is an ancestral feature of lineages and is retained unless there is anagenesis of some attribute(s) closely linked to reproductive success.

Interestingly, the species problem can be seen to parallel the theoretical and operational issues traditionally associated with supraspecific taxa and categories. These separate problems are very similar if one is conscious of the ontological differences between species as taxa and species as category. Everyone has their own notion of what species as taxa really are. In many instances we use this concept to guide our perceptions, observations, and understanding about theoretical and empirical aspects of the natural world (diversity, character evolution, speciation, ecology, physiology, etc.). The class concept species as category used in the assignment of groups from nature thought to fit its definition. Is there, or should there be, a definition of species that is both theoretical and operational analogous to the relationship that exists for supraspecific taxa and categories? Should the concept of the species category be operational or should it be theoretical like the concept of monophyly for supraspecific groups? Either way, which concepts should be used? If operational, which of the available operations should be employed, if any? If theoretical, which theory should be used? Unlike the abundant discussions following from the longstanding issue of concepts and criteria for recognizing supraspecific taxa, there is limited discussion clearly focusing on these issues for species. To date, no progress has issued from the multitude of pages debating the species problem. It is my opinion that the difficulties associated with species in theory and species in practice derives directly from the conflation of these notions. Much to the disappointment of many, a coveted denouement in this long standing controversy will never follow until these questions are addressed with a sound metaphysical framework.

19.8.2 Evaluating concepts for important qualities

Hull (1997: Chapter 18) advocates a comparison of species concepts on three traditional criteria: theoretical significance, generality, and applicability (or operationality). I agree that this is an appropriately unbiased method of comparison. Herein, I regard applicability and operationality as separate criteria because a concept can be fully operational but not at all applicable to a problem, and *vice versa*. How do the various concepts fair in evaluations of these criteria?

(a) Theoretical significance

While difficult to measure across the various concepts, there are important theoretical differences between them. The most significant ingredient is the treatment of species as individuals rather than classes. Excluding the ESC, all concepts treat species as classes. Such concepts preclude sound interpretations of speciation, character evolution, etc. because species cannot be perceived as lineages. The deficiency of a lineage perspective eventually leads researchers to view all attributes and geographic locations of species as proximal and causal explanations. For example, if one of three taxa has an array of traits appearing intermediate between two other adjacent taxa, what explanation other than hybridization or introgression, could account for such a pattern? Phylogenetic intermediacy represents only one of several when species are viewed as lineages (Mayden and Wood, 1995).

(b) Generality

Several theoretical and empirical elements of species concepts, relative to species as taxa, may be considered under this criterion, including their tolerances of divergent lifestyles, modes of reproduction, modes of speciation, genetic exchange, distributional and character information, and finally, diagnoses. Not all concepts view evidentiary information pertinent to these elements equally. Informative comparisons of generality require some estimate of baseline diversity to be recovered, or things that we currently envisage as behaving like species. Tolerance limits for each concept must be compared with this baseline of diversity.

What is our working baseline of diversity? First, we know that species exist that encompass the entire gamut between sexual and asexual reproduction, with numerous intermediate conditions (Templeton, 1989). Numerous speciation modes have been hypothesized for organismic diversity, ranging from complete allopatry to complete sympatry (Wiley, 1981). Numerous examples exist wherein species exchange genetic information either in current communities or historical communities without condemnation of identities. In fact, some hypothesized historical genetic exchange between groups may be responsible for the evolutionary success of the involved groups, each going on to produce diverse clades (Mayden and Wood, 1995). Finally, the types of character information traditionally used to discover species is heterogeneously distributed across taxonomic groups. When viewed across all taxonomic groups all types of data from DNA and RNA sequences and similarity, to behaviour and ecology, protein variability, morphology, and other traits, are standard markers used to reveal species diversity.

Some concepts are basically intolerant of gene exchange between species and require sympatry before species can be validated (HSC, BSC, ISC, CpSC). Under the BSC, taxa in allopatry are sometimes considered semispecies. Because gene exchange is not tolerated, speciation via hybridization is also not a valid form of speciation under some concepts. Some of these concepts are also intolerant of uniparental reproduction. Some are intolerant of groups of individual organisms that may be paraphyletically related to one another relative to one or more descendants; that is, all surviving ancestral species (PSC, GCC, ClSC). Some usually only recognize species wherein there has been divergence at the morphological level (MSC, TSC, PhSC). Likewise, some demand divergence at the ecological (EcSC) or recognition system (RSC) level. One concept, the ESC demands only that speciation and evolution are natural processes involving lineages that maintain cohesion and have unique identities. The ESC has thus the greatest generality. All other concepts are less general and exclude real diversity.

(c) Operationality

This is one quality consistently argued in discussions of species, either implicitly or explicitly. That is, anyone should be able to follow a prescribed set of identifiable and repeatable operations and at the end of these operations be able to tell (with a certain level of confidence) if they have a species. The requirement of such an execution places limits on what is recognizable, defined by criteria of the operational concept. While this may be more convenient, convenience is not a criterion that should be optimized when attempting to discover and understand pattern and process in the natural world. Operationalism is a fundamental fault of any species concept adopting it. What is operational is determined strictly by the perceived reality of the viewer. If the viewer's senses perceive only a portion of reality and these are expressed in an operational definition of what reality consists of, then we will never know otherwise. If, however, the viewer is capable of perceiving or conceptualizing all of reality, then all of diversity can be discovered without placing limits on what can be recognized with an operational concept. For instance, it is a mistake for someone who is red–green colour blind to mandate a concept of species based on the operational criterion of colour. Anyone discussing species diversity of hummingbirds, flowering plants, or darters, with this person would continually be frustrated with what is reality.

Excluding the ESC, all of the other concepts are operational at some level. That is, with all of them one can conduct certain experiments and extract pertinent information about the criterion emphasized. Some are more operational than others but with this increasing operationality one necessarily sacrifices an ability to account for diversity. For example, the ASC, MSC, PhSC, SSC, or TSC are probably the most operational concepts guiding the discovery of species. These concepts, however, will necessarily exclude equally valid species that can and will be recognized using other concepts. The next most operational concepts would include ClSC, CpSC, EcSC, GCC, GCD, NDSC, versions of the PSC, and RSC. The BSC, HSC, and RCC are all minimally operational. The ESC is unique in not being an operational concept, a consequential quality for a primary concept. Nothing in the ESC, other than evolution produces species as lineages with identities and cohesion, is operational; this, however, is extremely difficult to apply without bridging principles.

(d) Applicability

Given that the various concepts were all formulated from research on patterns of diversity across a diversity of temporal and geographical situations using varied technologies, each attempting to unveil processes associated with descent, they are all applicable as concepts of species as taxa. However, applicability extends from those having lesser applicability and embracing only a subset of natural diversity, to those with greater applicability wherein the concept embraces most or all of diversity. The ESC has the greatest applicability because it is consistent with and embraces all known species diversity that has evolved through currently understood processes descent. All other concepts have lesser applicability because as class constructs they are capable only of embracing a lesser portion of natural diversity by excluding some forms of species (e.g. asexuals, ancestors, etc.).

19.9 HIERARCHY OF CONCEPTS: SPECIES IN THEORY AND PRACTICE

When Mayr (1957) discussed species concepts and definitions, he mentioned – but did not dwell on – a need for two different levels of concepts for species, these being primary and secondary concepts. 'All our reasoning in discussions of "the species" can be traced back to the stated three primary concepts. As concepts, of course, they cannot be observed directly, and we refer to certain observed phenomena in nature as "species", because they conform in their attributes to one of these concepts or to a mixture of several concepts. From these primary concepts, just discussed, we come thus to secondary concepts, based on particular aspects of species' (Mayr, 1957: 16).

As primary concepts, Mayr is referring to the typological, second and third species concepts discussed in that paper. From the discussion it is clear that he recognized that all species as taxa would fit one or a combination of these concepts, but that secondary concepts are those used to identify species and employ differences in morphology, genetics, behaviour, etc. to infer diversity consonant with primary concepts. It is unfortunate that more scientists from all disciplines had not read this passage in 1957 and bequeathed this philosophy to their academic descendants.

As fundamental links bridging observable patterns and inferred processes, concepts are employed in every discipline, assisting to guide our understanding, perception, and disclosure of natural systems. Given that descent with modification and speciation are undeniable processes of diversification and that individual species are the highest level of organization capable of participating in these processes, a monistic notion of species is not only natural but is logical. Descent and speciation are processes occurring in lineages. Individual organisms to populations, each with spatiotemporal cohesion, and only lineages with this type of integrity uniquely participate in speciation. A primary concept of species is fundamental to the whole of biological sciences, particularly for understanding species as taxa. Currently, the multiple concepts in operation are decidedly inconsistent with one another as to what constitutes diversity (also see Hull, 1997: Chapter 18) and most are inconsistent with the range of diversity acknowledged as species in different disciplines. Without a primary concept as a working hypothesis and to serve as a bridge between pattern and process, it is untenable that we can advance on many fronts. Heretofore, much of our effort has been expended struggling with the conceptualization of species. Many adopt only operational concepts that will produce contrived species diversity; unfortunately those searching for pattern and processes associated with this diversity may be deceived.

What then are the criteria we should be looking for in a primary concept? It should be consistent with current theoretical and empirical knowledge of diversification. It should be consistent with the ontological status of those entities participating in descent and other natural processes; that is, species as taxa must be referenced as individuals. Finally, it should be general enough to encapsulate all types of biological entities considered species as taxa by researchers working with supraspecific taxa. Only the ESC is suitable as primary concept, guiding our quest for species as taxa and our search for natural order. This concept is robust theoretically and is unique in its global generality. One drawback is that it is not operational. While this may be viewed as a possible shortcoming, it is not so for a primary concept. The ESC is maximally applicable because everything we currently understand about descent, speciation, and species are compatible with the intent of the ESC.

While the ESC is the most appropriate primary concept, it requires bridging concepts permitting us to recognize entities compatible with its intentions. To implement fully the ESC we must supplement it with more operational, accessory notions of biological diversity – secondary concepts. Secondary concepts include most of the other species concepts. While these concepts are varied in their operational nature, they are demonstrably less applicable than the ESC because of their dictatorial restrictions on the types of diversity that can be recognized, or even evolve. However, they serve as surrogates for the ESC and, together, further our understanding of descent, both anagenesis and cladogenesis, by recognizing any entity consistent with the primary concept. They represent the practical or applied definitions, guidelines, or tools, used by investigators to discover hypothesized real particulars, entities, individuals, or things that we accept as species. These secondary concepts can account for nearly all species diversity, with the possible exception of ancestral species, either surviving or extinct. Because the ISC and SSC are both capable of delineating diversity beyond real species, these definitions must be used with caution.

Together, the primary and secondary concepts form a hierarchical system displaying both their operational and theoretical inter-relationships (Figure 19.1). The primary concept is the ESC. Relationships among the secondary concepts may be envisioned in multiple forms; I have illustrated only one such system. The one criterion emphasized most throughout discussions of secondary concepts is sex. I have chosen to use this as a first level criterion among secondary concepts. Reproduction, similarity–dissimilarity, monophyly, diagnosability, apomorphy, and tolerances for gene exchange are other criteria used to further reveal relationships among secondary concepts (Figure 19.1). Some concepts (BSC, GSC, RCC) terminate at multiple locations in the hierarchy either because of different uses or ambiguities in the concepts. However, this is acceptable and one may view other concepts as having similar results.

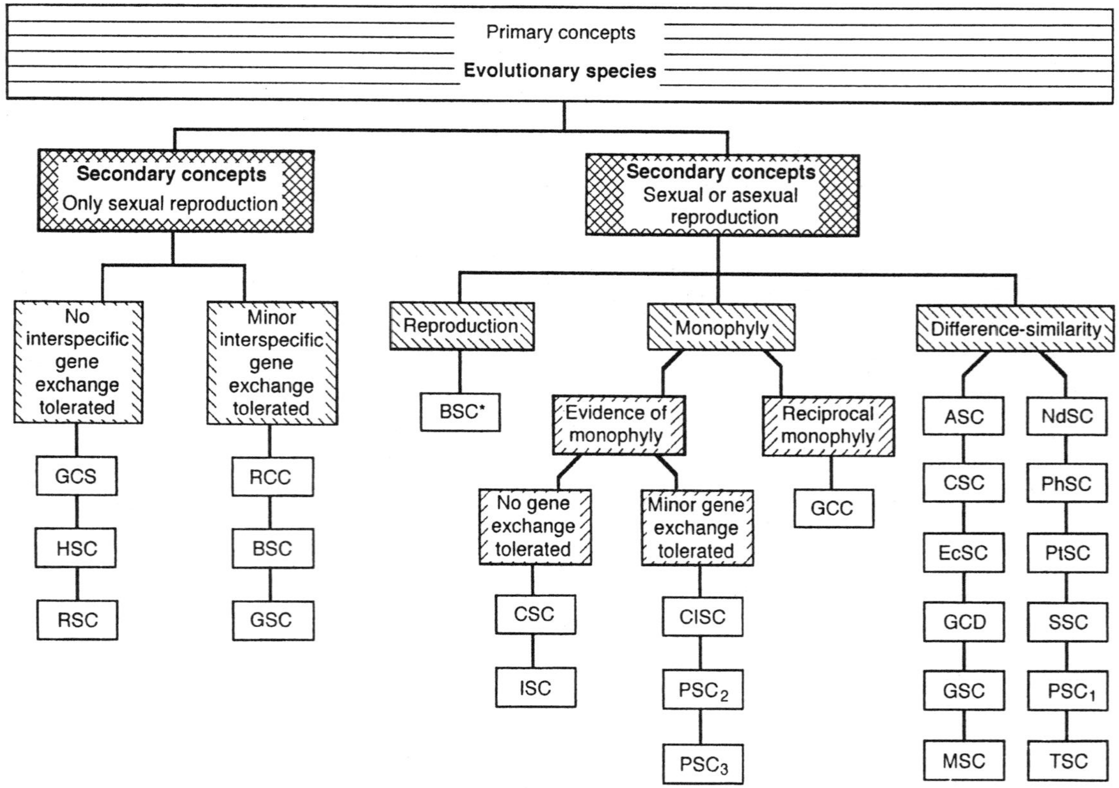

Figure 19.1 A hierarchy of primary and secondary species concepts. The non-operational Evolutionary Species Concept serves as the primary concept of species. The operational secondary concepts form a hierarchy below this primary concept based on their tolerances or requirements for modes of reproduction, gene exchange, monophyly, and diagnosability. Because some concepts represent hybrid versions of other concepts (mixed criteria) they may be depicted more than once in the hierarchy. Species concepts are listed alphabetically within any grouping. Asterisk denotes a version of BSC modified for asexual species. See Table 19.1 for concept abbreviations.

There are extraordinary advantages to accepting the premise of monistic primary and pluralistic secondary concepts of species. First, a primary concept of species that can be continually evaluated in light of new information ensures that all things behaving as species are potentially recoverable, given unavoidable constraints associated with available technology and extinction of taxa never observed. Second, with the possible exception of the ISC and SSC, all currently employed secondary concepts theoretically compatible with the primary concept can be mutually applicable in the discovery of species and the elucidation of pattern and process. While some concepts conflict in their intentions, they are all equally valid. When viewed together as guidelines in the detection of species they ensure that natural species diversity is neither unrecognized nor misunderstood. Thus, patterns observed in the natural world can be used by all disciplines to reveal natural processes in an uninhibited manner.

Our classification system for supraspecific taxa is analogous to the outlined system of primary and secondary concepts of species. Classifications are theories about the organization of biological diversity. What groups should be placed in the various supraspecific categories, and how and why should just these groups be recognized over other possible groups? One may choose to optimize various information in a classification, from overall similarity, ecological guilds, or modes of reproduction, to genealogical relationships, just to mention a few. In the current system the concept adopted for supraspecific categories is a particular genealogical relationship, specifically monophyly. Other criteria have been rejected as primary concepts because of ambiguity, inconsistencies or artificiality. Thus, we employ monophyly as a primary concept to bridge to natural groups in the classification of supraspecific taxa. Because we are unable to observe descent we adopt secondary concepts or definitions, particular homologies compatible with the intentions of the primary concept. Through character evaluation secondary concepts permit the organization of diversity into such groups. Inferences derived from phylogenetic systematics can either corroborate or falsify hypotheses of groups suspected to meet criteria outlined in our primary concept monophyly. That is, a secondary, operational concept, the discovery of synapomorphy, permits continual re-evaluation of monophyly of groups and our theory of descent represented through our classification. This is all done within the context of a theory that there is a history of descent, that characters are modified and inherited through this descent, and that pattern and process is recoverable. Here we have a primary conceptual basis for the type of supraspecific taxa that we wish to recognize in classifications. This concept is necessary and sufficient in our search for them. The concept of monophyly, like the ESC, is applicable but is in no way operational. Secondary concepts for both species and supraspecific categories are requisite in our discovery of species and supraspecific groupings, respectively.

19.10 REFERENCES

Avise, J.C. and Ball, R.M. Jr (1990) Principles of genealogical concordance in species concepts and biological taxonomy, in *Oxford Surveys in Evolutionary Biology* (eds D. Futuyma and J. Antonovics), Oxford University Press, Oxford, pp. 45–67.

Blackwelder, R.E. (1967) *Taxonomy: A Text and Reference Book*, John Wiley & Sons, New York.

Cracraft, J. (1983) Species concepts and speciation analysis, in *Current Ornithology*, vol. 1, Plenum Press, New York, pp. 159–87.

Cronquist, A. (1978) Once again, what is a species? in *BioSystematics in Agriculture* (ed. L. V. Knutson), Allenheld Osmun, Montclair, New Jersey, pp. 3–20.

de Queiroz, K. and Donoghue, M.J. (1988) Phylogenetic systematics and the species problem. *Cladistics*, **4**, 317–38.

de Queiroz, K. and Donoghue, M.J. (1990) Phylogenetic systematics and species revisited. *Cladistics*, **6**, 83–90.

Dobzhansky, T. (1950) Mendelian populations and their evolution. *The American Naturalist*, **74**, 312–21.

Dobzhansky, T. (1970) *Genetics of the Evolutionary Process*, Columbia University Press, New York.

Du Rietz, G.E. (1930) The fundamental units of biological taxonomy. *Svensk Botanisk Tidskrift*, **24**, 333–428.

Eldredge, N. and Cracraft, J. (1980) *Phylogenetic analysis and the Evolutionary Process*, Columbia University Press, New York.

Frost, D.R. and Hillis, D.M. (1990) Species concepts and practice: herpetological applications. *Herpetologica*, **46**, 87–104.

Frost, D.R. and Kluge, A.G. (1994) A consideration of epistemology in systematic biology, with special reference to species. *Cladistics*, **10**, 259–94.

George, T.N. (1956) Biospecies, chronospecies and morphospecies, in *The Species Concept in Paleontology* (ed. P.C. Sylvester-Bradley), Systematics Association, London, pp. 123–37.

Ghiselin, M.T. (1974) A radical solution to the species problem. *Systematic Zoology*, **23**, 536–44.

Ghiselin, M.T. (1984a) Narrow approaches to phylogeny: a review of nine books on cladism, in *Oxford Surveys in Evolutionary Biology* (eds R. Dawkins and M. Ridley), Oxford University Press, Oxford, pp. 209–22.

Ghiselin, M.T. (1984b) 'Definition,' 'Character,' and other equivocal terms. *Systematic Zoology*, **33**, 104–10.

Grant, V. (1981) *Plant Speciation*, 2nd edn, Columbia University Press, New York.

Hennig, W. (1950) *Grundzuge einer Theorie der Phylogentischen Systematik*, Aufbau Verlag, Berlin.

Hennig, W. (1966) *Phylogenetic Systematics*, University Illinois Press, Urbana.

Hull, D.L. (1976) Are species really individuals? *Systematic Zoology*, **25**, 174–91.

Hull, D.L. (1997) The ideal species concept – and why we can't get it, in *Species: the Units of Biodiversity* (eds M.F. Claridge, H.A. Dawah and M.R. Wilson), Chapman & Hall, London, pp. 357–80.

Kornet, D. (1993) Permanent splits as speciation events: a formal reconstruction of the internodal species concept. *Journal of Theoretical Biology*, **164**, 407–35.

Kornet, D.J. and McAllister, J.W. (1993) The composite species concept, in *Reconstructing Species: Demarcations in Genealogical Networks*. Unpublished PhD dissertation, Institute for Theoretical Biology, Rijksherbarium, Leiden, pp. 61–89.

Mallet, J. (1995) A species definition for the modern synthesis. *Trends in Ecology and Evolution*, **10**, 294–9.

Mayden, R.L., and Wood, R.M. (1995) Systematics, species concepts, and the evolutionarily significant unit in biodiversity and conservation biology. *American Fisheries Society Symposium*, **17**, 58–113.

Mayr, E. (1940) Speciation phenomena in birds. *The American Naturalist*, **74**, 249–78.

Mayr, E. (1957) Species concepts and definitions, in *The Species Problem* (ed. E. Mayr), American Association for Advancement of Science, Washington, DC, pp. 1–22.

Mayr, E. (1963) *Animal Species and Evolution*, Harvard University Press, Cambridge, Massachusetts.

Mayr, E. (1969) *Principles of Systematic Zoology*, McGraw-Hill, New York.

Mayr, E. (1970) *Populations, Species and Evolution*, Belknap Press of Harvard University, Cambridge, Massachusetts.

Mayr, E. (1988) The why and how of species. *Biology and Philosophy*, **3**, 431–41.

Mayr, E. (1997) The biological species concept, in *Species Concepts and Phylogenetic Theory: A Debate* (eds Q.D. Wheeler and R. Meier), Columbia University Press, New York (in press).

Mayr, E. and Ashlock, P.D. (1991) *Principles of Systematic Zoology*, McGraw-Hill, New York.

McKitrick, M.C. and Zink R.M. (1988) Species concepts in ornithology. *The Condor*, **90**, 1–14.

Meier, R. and Willmann, R. (1997) The Hennigian species concept, in *Species Concepts and Phylogenetic Theory: A Debate* (eds Q.D. Wheeler and R. Meier), Columbia University Press, New York (in press).

Minelli, A. (1993) *Biological Systematics: The State of the Art*, Chapman & Hall, New York.

Mishler, B.D. and Donoghue M.J. (1982) Species concepts: a case for pluralism. *Systematic Zoology*, **31**, 491–503.

Nelson, G. and Platnick N.I. (1981) *Systematics and Biogeography: Cladistics and Vicariance*, Columbia University Press, New York.

Nixon, K.C. and Wheeler Q.D. (1990) An amplification of the phylogenetic species concept. *Cladistics*, **6**, 211–23.

Paterson, H.E.H. (1993) *Evolution and The Recognition Concept of Species, Collected Writings of H. E. H. Paterson* (ed. S. F. McEvey), Johns Hopkins University Press, Baltimore, Maryland.

Platnick, N.I. (1976) Are monotypic genera possible? *Systematic Zoology*, **26**, 198–9.

Regan, C.T. (1926) *Organic Evolution*. Report British Association for Advancement of Science, 1925, pp. 75–86.

Ridley, M. (1989) The cladistic solution to the species problem. *Biology and Philosophy*, **4**, 1–16.

Rosen, D.E. (1978) Vicariant patterns and historical explanation in biogeography. *Systematic Zoology*, **27**, 159–88.

Rosen, D.E. (1979) Fishes from the uplands and intermontane basins of Guatemala: Revisionary studies and comparative biogeography. *Bulletin American Museum of Natural History*, **162**, 267–376.

Shull, G.H. (1923) The species concept from the point of view from the geneticist. *American Journal of Botany*, **10**, 221–8.

Simpson, G.G. (1943) Criteria for genera, species, and subspecies in zoology and paleontology. *Annals New York Academy of Science*, **44**, 145–78.

Simpson, G.G. (1951) The species concept. *Evolution*, **5**, 285–98.

Simpson, G.G. (1961) *Principles of Animal Taxonomy*, Columbia University Press, New York.

Sneath, P.H.A. (1976) Phenetic taxonomy at the species level and above. *Taxon*, **25**, 437–50.

Sokal, R.R. (1973) The species problem reconsidered. *Systematic Zoology*, **22**, 360–74.

Stuessy, T.F. (1990) *Plant Taxonomy. The Systematic Evaluation of Comparative Data*, Columbia University Press, New York.

Templeton, A.T. (1989) The meaning of species and speciation: a genetic perspective, in *Speciation and Its Consequences* (eds D. Otte and J. A. Endler), Sinauer Associates, Sunderland, Massachusetts, pp. 3–27.

Van Valen, L. (1976) Ecological species, multispecies, and oaks. *Taxon*, **25**, 233–9.

Waples, R.S. (1991). Pacific salmon, *Oncorhynchus* spp., and the definition of 'species' under the Endangered Species Act. *Marine Fisheries Review*, **53**, 11–22.

Waples, R.S. (1996) Evolutionary significant units and the conservation of biological diversity under the Endangered Species Act. *American Fisheries Society Symposium*, **17** (in press).

Warren, M.L. (1992) Variation of the spotted sunfish, *Lepomis punctatus* complex (Centrarchidae): meristics, morphometrics, pigmentation and species limits. *Bulletin Alabama Museum of Natural History*, **12**, 1–47.

Weber, H.E. (1981) *Sonderb.* Naturwiss. Ver., Hamburg, **4**, 1–229.

Wheeler, Q.D. and Nixon, K.C. (1990) Another way of *looking at* the species problem: a reply to de Queirox and Donoghue. *Cladistics*, **6**, 77–81.

Wheeler, Q.D. and Platnick N.I. (1997) The argument for phylogenetic species, in *Species Concepts and Phylogenetic Theory: A Debate* (eds Q.D. Wheeler and R. Meier), Columbia University Press, New York (in press).

Wiley, E.O. (1978) The evolutionary species concept reconsidered. *Systematic Zoology*, **27**, 17–26.

Wiley, E.O. (1981) *Phylogenetics: The Theory and Practice of Phylogenetic Systematics*, John Wiley & Sons, New York.

Wiley, E.O. and Mayden R.L. (1985) Species and speciation in phylogenetic Systematics, with examples from the North American fish fauna. *Annals Missouri Botanical Garden*, 72, 596–635.

Wiley, E.O. and Mayden R.L. (1997) The evolutionary species concept, in *Species Concepts and Phylogenetic Theory: A Debate* (eds Q.D. Wheeler and R. Meier), Columbia University Press, New York (in press).

Wilkinson, M. (1990) A commentary on Ridley's cladistic solution to the species problem. *Biology and Philosophy*, 5, 433–46.

Willmann, R. (1985a) *Die Art in Raum und Zeit*, Paul Parey Verlag, Berlin.

Willmann, R. (1985b) Reproductive isolation and the limits of the species in time. *Cladistics*, **2**, 336–8.

Index

Page numbers appearing in **bold** refer to figures; page numbers appearing in *italic* refer to tables.

Systematics Association Publications

1. Bibliography of key works for the identification of the British fauna and flora, 3rd edition (1967)†
   Edited by G.J. Kerrich, R.D. Meikle and N. Tebble
2. Function and taxonomic importance (1959)†
   Edited by A.J. Cain
3. The species concept in palaeontology (1956)†
   Edited by P.C. Sylvester-Bradley
4. Taxonomy and geography (1962)†
   Edited by D. Nichols
5. Speciation in the sea (1963)†
   Edited by J.P. Harding and N. Tebble
6. Phenetic and phylogenetic classification (1964)†
   Edited by V.H. Heywood and J. McNeill
7. Aspects of Tethyan biogeography (1967)†
   Edited by C.G. Adams and D.V. Ager
8. The soil ecosystem (1969)†
   Edited by H. Sheals
9. Organisms and continents through time (1973)†
   Edited by N.F. Hughes
10. Cladistics: a practical course in systematics (1992)
    P.L. Forey, C.J. Humphries, I.J. Kitching, R.W. Scotland, D.J. Siebert and D.M. Williams

†Published by the Association (out of print)

Systematics Association Special Volumes

1. The new systematics (1940)
   Edited by J.S. Huxley (reprinted 1971)
2. Chemotaxonomy and serotaxonomy (1968)*
   Edited by J.G. Hawkes
3. Data processing biology and geology (1971)*
   Edited by J.L. Cutbill
4. Scanning electron microscopy (1971)*
   Edited by V.H. Heywood
   Out of print
5. Taxonomy and ecology (1973)*
   Edited by V.H. Heywood
6. The changing flora and fauna of Britain (1974)*
   Edited by D.L. Hawksworth
   Out of print

7. Biological identification with computers (1975)*
   Edited by R.J. Pankhurst
8. Lichenology: progress and problems (1976)*
   Edited by D.H. Brown, D.L. Hawksworth and R.H. Bailey
9. Key works to the fauna and flora of the British Isles and north-western Europe, 4th edition (1978)*
   Edited by G.J. Kerrich, D.L. Hawksworth and R.W. Sims
10. Modern approaches to the taxonomy of red and brown algae (1978)
    Edited by D.E.G. Irvine and J.H. Price
11. Biology and systematics of colonial organisms (1979)*
    Edited by G. Larwood and B.R. Rosen
12. The origin of major invertebrate groups (1979)*
    Edited by M.R. House
13. Advances in bryozoology (1979)*
    Edited by G.P. Larwood and M.B. Abbott
14. Bryophyte systematics (1979)*
    Edited by G.C.S. Clarke and J.G. Duckett
15. The terrestrial environment and the origin of land vertebrates (1980)
    Edited by A.L. Pachen
16. Chemosystematics: principles and practice (1980)*
    Edited by F.A. Bisby, J.G. Vaughan and C.A. Wright
17. The shore environment: methods and ecosystems (2 volumes) (1980)*
    Edited by J.H. Price, D.E.G. Irvine and W.F. Farnham
18. The Ammonoidea (1981)*
    Edited by M.R. House and J.R. Senior
19. Biosystematics of social insects (1981)*
    Edited by P.E. House and J.-L. Clement
20. Genome evolution (1982)*
    Edited by G.A. Dover and R.B. Flavell
21. Problems of phylogenetic reconstruction (1982)*
    Edited by K.A. Joysey and A.E. Friday
22. Concepts in nematode systematics (1983)*
    Edited by A.R. Stone, H.M. Platt and L.F. Khalil
23. Evolution, time and space: the emergence of the biosphere (1983)*
    Edited by R.W. Sims, J.H. Price and P.E.S. Whalley
24. Protein polymorphism: adaptive and taxonomic significance (1983)*
    Edited by G.S. Oxford and D. Rollinson
25. Current concepts in plant taxonomy (1983)*
    Edited by V.H. Heywood and D.M. Moore
26. Databases in systematics (1984)*
    Edited by R. Allkin and F.A. Bisby
27. Systematics of the green algae (1984)*
    Edited by D.E.G. Irvine and D.M. John
28. The origins and relationships of lower invertebrates (1985)‡
    Edited by S. Conway Morris, J.D. George, R. Gibson and H.M. Platt
29. Infraspecific classification of wild and cultivated plants (1986)‡
    Edited by B.T. Styles

30. Biomineralization in lower plants and animals (1986)‡
Edited by B.S.C. Leadbeater and R. Riding
31. Systematic and taxonomic approaches in palaeobotany (1986)‡
Edited by R.A. Spicer and B.A. Thomas
32. Coevolution and systematics (1986)‡
Edited by A.R. Stone and D.L. Hawksworth
33. Key works to the fauna and flora of the British Isles and north-western Europe, 5th edition (1988)‡
Edited by R.W. Sims, P. Freeman and D.L. Hawksworth
34. Extinction and survival in the fossil record (1988)‡
Edited by G.P. Larwood
35. The phylogeny and classification of the tetrapods (2 volumes) (1988)‡
Edited by M.J. Benton
36. Prospects in systematics (1988)‡
Edited by D.L. Hawksworth
37. Biosystematics of haematophagous insects (1988)‡
Edited by M.W. Service
38. The chromophyte algae: problems and perspective (1989)‡
Edited J.C. Green, B.S.C. Leadbeater and W.L. Diver
39. Electrophoretic studies on agricultural pests (1989)‡
Edited by Hugh D. Loxdale and J. den Hollander
40. Evolution, systematics, and fossil history of the Hamamelidae (2 volumes) (1989)‡
Edited by Peter R. Crane and Stephen Blackmore
41. Scanning electron microscopy in taxonomy and functional morphology (1990)‡
Edited by D. Claugher
42. Major evolutionary radiations (1990)‡
Edited by P.D. Taylor and G.P. Larwood
43. Tropical lichens: their systematics, conservation and ecology (1991)‡
Edited by G.J. Galloway
44. Pollen and spores: patterns of diversification (1991)‡
Edited by S. Blackmore and S.H. Barnes
45. The biology of free-living heterotrophic flagellates (1991)‡
Edited by D.J. Patterson and J. Larsen
46. Plant-animal interactions in the marine benthos (1992)‡
Edited by D.M. Johns, S.J. Hawkins and J.H. Price
47. The Ammonoidea: environment, ecology and evolutionary change (1993)‡
Edited by M.R. House
48. Designs for a global plant species information system‡
Edited by F.A. Bisby, G.F. Russell and R.J. Pankhurst
49. Plant galls: organisms, interactions, populations‡
Edited by MichŠle A.J. Williams
50. Systematics and conservation evaluation‡
Edited by P.L. Forey, C.J. Humphries and R.I. Vane-Wright
51. The Haptophyte algae‡
Edited by J.C. Green and B.S.C. Leadbeater

52. Models in phylogeny reconstruction‡
    Edited by R. Scotland, D.J. Siebert and D.M. Williams
53. The Ecology of Agricultural Pests: biochemical approaches**
    Edited by W.O.C. Symondson and J.E. Liddell
54. Species: the units of diversity**
    Edited by M.F. Claridge, H.A. Dawah and M.R. Wilson

*Published by Academic Press for the Systematics Association
†Published by the Palaeontological Association in conjunction with Systematics Association
‡Published by the Oxford University Press for the Systematics Association
**Published by Chapman & Hall for the Systematics Association